GAS PHASE REACTIONS IN
ORGANIC SYNTHESIS

GAS PHASE REACTIONS IN ORGANIC SYNTHESIS

Edited by

Yannick Vallée

Professor of Chemistry, University Joseph Fourier, Grenoble, France,
Member of the 'Institut Universitaire de France'

Gordon and Breach Science Publishers

Australia • Canada • China • France • Germany • India • Japan • Luxembourg • Malaysia
The Netherlands • Russia • Singapore • Switzerland • Thailand • United Kingdom

CHEM
6 9894452

Copyright © 1997 OPA (Overseas Publishers Association) Amsterdam B.V. Published in The Netherlands under license by Gordon and Breach Science Publishers.

Amsteldijk 166
1st Floor
1079 LH Amsterdam
The Netherlands

British Library Cataloguing in Publication Data

Gas Phase reactions in organic synthesis
   1. Organic compounds – Synthesis
   I. Vallée, Yannick
   541.3'9 acv

ISBN 90-5699-081-0

MPC
4-24-01
NC

CONTENTS

CONTRIBUTORS

J-M. Denis
Synthèse et Electrosynthèse Organiques
Université de Rennes 1
France

F.W. Eastwood
Department of Chemistry
Monash University
Australia

A-C. Gaumont
Synthèse et Electrosynthèse Organiques
Université de Rennes 1
France

L.W. Jenneskens
Department of Physical
   Organic Chemistry
Utrecht University
The Netherlands

A.J.H. Klunder
Department of Organic Chemistry
University of Nijmegen
The Netherlands

N. Pelloux-Léon
LEDSS
Université Joseph Fourier
France

J-L. Ripoll
Laboratoire de Chimie Moléculaire
   et Thio-Organique
Université de Caen
France

Y. Vallée
LEDSS
Université Joseph Fourier
France

U.E. Wiersum
Akzo Nobel Central Research
The Netherlands

B. Zwanenburg
Department of Organic Chemistry
University of Nijmegen
The Netherlands

1. GAS PHASE PYROLYTIC METHODS FOR THE PREPARATION OF CARBON–HYDROGEN AND CARBON–HYDROGEN–OXYGEN COMPOUNDS

FRANK W. EASTWOOD

Department of Chemistry, Monash University, Clayton, Victoria, 3168 Australia

This chapter is concerned with the chemistry of hydrocarbon species and of oxygen containing compounds formed in the gas phase generally at high temperatures either using flow conditions in a carrier gas at pressures ranging from atmospheric (760 Torr) down to moderate vacuum (10^{-1} Torr), flash vacuum pyrolysis (FVP) at lower pressures (10^{-1}–10^{-4} Torr) or more specialised techniques. The arrangement of the material is based arbitrarily on the decreasing ratio of the number of carbon atoms to the number of hydrogen atoms and on the increasing proportion of oxygen but this is modified by grouping some material under particular reaction sequences.

In the hydrocarbon series it is in the high C/H ratio where many recent developments have occurred. At the highest ratios, precursors are required which fragment on pyrolysis to form small very stable molecules and concomitant highly reactive hydrocarbon species which then may undergo profound change to form stable products.

As the ratio of carbon to hydrogen decreases, stable precursors may automerize or isomerize in many ways. With aliphatic compounds a profusion of radical fragmentation processes may occur.[1b] Such chemistry only gets passing mention in this chapter. Books and reviews covering developments in the fields of flow pyrolysis and flash vacuum pyrolysis which include discussions of hydrocarbons are listed.[1] Polycyclic aromatic hydrocarbons (PAHs) which are formed together with soot in pyrolytic and combustion processes are the subject of a great deal of research because of economic considerations and because of their deleterious effects on the biosphere. An understanding of the mechanisms of the complicated reactions leading to PAHs is at an early stage of development but has recently been reviewed.[2c] Chapter 3 in this book also describes their chemistry.

1. DIDEHYDROBENZENES

1.1 The *o*-Benzyne (1,2-Didehydrobenzene) to Cyclopentadienylidenecarbene Rearrangement

o-Benzyne (**8**) can be formed in the gas phase by flash vacuum pyrolysis of phthalic anhydride (isobenzofuran-1,3-dione) (**1**), benzocyclobuten-1,2-dione (**2**), 2-(3',3'-dimethyltriazenyl)benzoic acid (**3**) (a safe modification of the benzyne precursor 2-diazoniabenzoate (**4**)), phthaloyl peroxide (**5**) and a variety of compounds which fragment to yield cyclopentadienylideneethenone (**6**) which decarbonylates to cyclopentadienylidenecarbene (**7**) and rearranges to benzyne (**8**).[3,4]

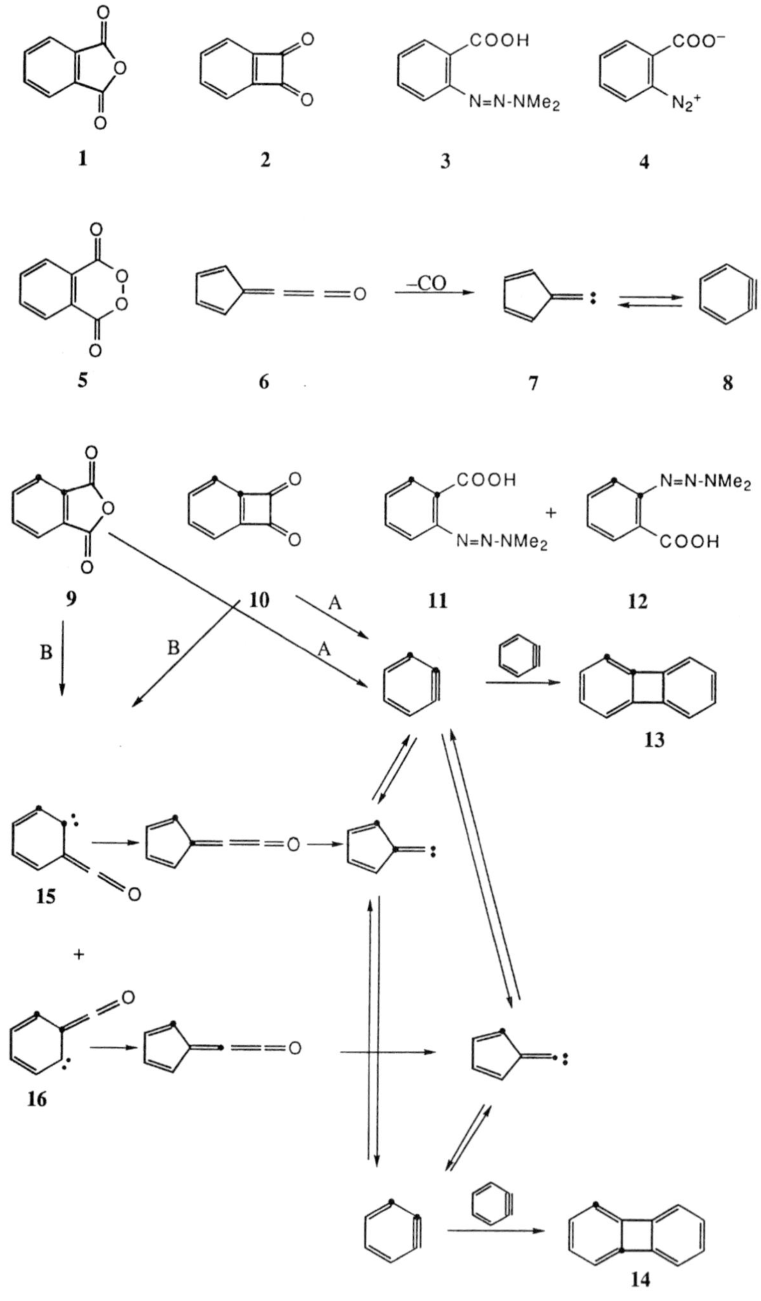

Scheme 1

The rearrangement of benzyne (**8**) to cyclopentadienylidenecarbene (**7**) was first established by FVP of $[1,6^{-13}C_2]$-phthalic anhydride (**9**) and $[2a,3^{-13}C_2]$-benzocyclobuten-1,2-dione (**10**), each of which gave an approximately equimolecular mixture of $[1,8b^{-13}C_2]$-biphenylene (**13**) and $[1,4b^{-13}C_2]$-biphenylene (**14**).[5] The mechanism proposed involved direct formation of benzyne followed by rearrangement and dimerization (Pathway A, Scheme 1). An alternative mechanism based on Wolff rearrangement of the ketene carbenes (**15**) and (**16**) was suggested by Wentrup and coworkers[6] (Pathway B, Scheme 1) which, because of the symmetry of the structures, required equimolecular mixtures of (**13**) and (**14**) to be formed at all temperatures at which decomposition took place.

That the reaction was a thermal rearrangement of benzyne was established by pyrolysis of a mixture of $[1,6^{-13}C_2]$- and $[2,3^{-13}C_2]$-2-(3',3'-dimethyltriazenyl)-benzoic acid, (**11**) and (**12**) respectively, which on FVP at 400° gave (**13**) but with increased temperatures gave an increasing proportion of (**14**).

Detailed arguments concerning the reactions in Scheme 1 have been set out in a recent paper which describes the detection of a nitrogen analogue of the ketenecarbene (unlabelled **15,16**). The ketenimine carbene (**18**) was formed on FVP of N-phenylphthalisoimide (**17**) and detected by means of its rearrangement products fluorenimine (**19**), phenanthridine (**20**) and 2-phenylbenzonitrile (**21**). The hypothesis was advanced that in Scheme 2, route A is the major pathway and route B the minor one.[8]

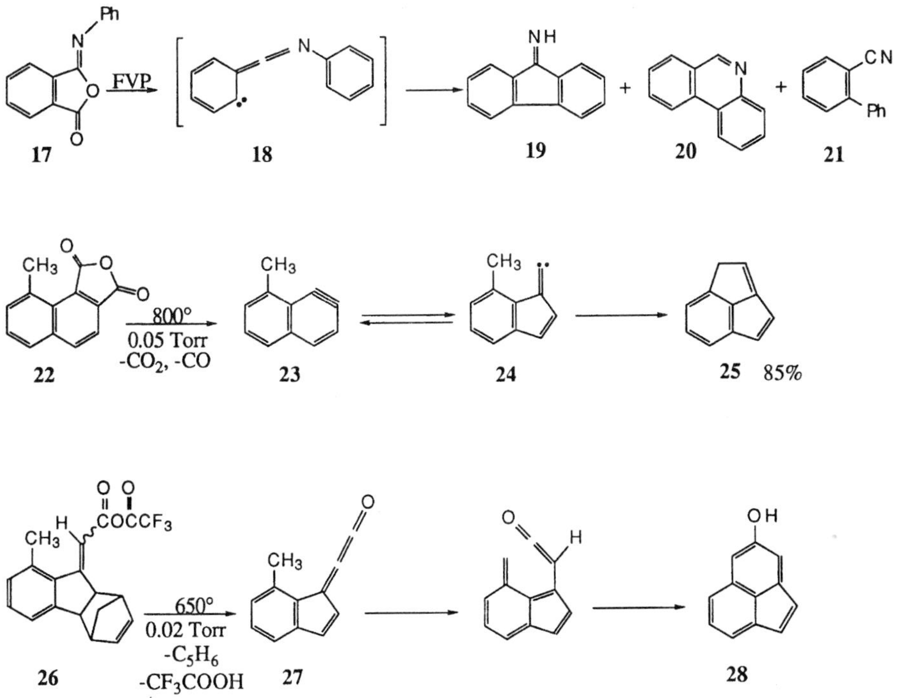

Scheme 2

29   **30**

Although the mechanism of formation of benzyne from phthalic anhydride and other precursors will remain an area for study, attention has shifted to the chemistry of the exocyclic carbene formed by ring contraction. Aromatic cyclic anhydrides, despite requiring high temperatures for fragmentation, are the initial choice as precursors of arynes because of their synthetic availability. A number of studies on pyrolyses of such compounds has been reported.

The first demonstration of trapping of a carbene formed in this way involved pyrolysis of 8-methylnaphthalene-1,2-dicarboxylic anhydride (**22**) to yield 1*H*-cyclopent[*c,d*]indene (**25**). In this reaction 1,2-didehydro-8-methylnaphthalene (**23**) rearranges to 7-methyl-1*H*-indenylidenecarbene (**24**).[9] In an examination of the mechanism of this reaction the indenylideneketene (**27**) was formed by pyrolysis of the mixed anhydride (**26**). This ketene did not decarbonylate to yield the carbene (**24**) and cyclopentindene (**25**), but cyclised to acenapthylen-4-ol (**28**) (Scheme 2).[10]

Ring contraction of a substituted benzyne was demonstrated when pyrolysis of tetraphenylphthalic anhydride (**29**) yielded 1,2,3-triphenylbenzopentalene (**30**).[11] The exocyclic carbene could be trapped by substituents in monosubstituted benzynes. Pyrolysis of 3-phenylphthalic anhydride (**31**) gave acenaphthylene (**32**) and benzopentalene (**33**) which was stable at low temperature but on warming to room temperatures gave a dimer, possibly (**34**).[12,1k] Three diphenylphthalic anhydrides undergo similar ring contraction but surprisingly yield as the principal product unsubstituted benzopentalene (**33**) and its dimer (**34**). This reaction involves replacement of a phenyl group by hydrogen. On pyrolysis of 3-*o*-tolylphthalic anhydride (**35**) the carbene was trapped by the methyl group to give a 1:1 mixture of benzindenes (**36**) together with fluorene (**37**) which was the unexpected major product wherein the aryne ring had not contracted.[13] 3-(1-Naphthyl)phthalic anhydride (**38**) also showed no evidence of ring contraction. On pyrolysis it was converted into fluoranthene (**39**) possibly by a radical process.[12] On the other hand, 3-vinylphthalic anhydride (**40**) was converted on pyrolysis into pentalene (**41**) which was detected as the known dimer (**42**).[12]

Returning to the naphthalene series, pyrolysis of 3-phenylnaphthalene-1,2-dicarboxylic anhydride (**43**) gave dibenzopentalene (indeno[2,1-*a*]indene) (**44**), and 8-phenylnaphthalene-1,2-dicarboxylic anhydride (**45**) gave aceanthrylene (**46**) and fluoranthene (**39**)[14] (see Section 4.10). In the more rigid benzo derivative, phenanthrene-3,4-dicarboxylic anhydride (**47**), pyrolysis gave a highly strained intermediate (**48**) which underwent ring cleavage to give 1-ethynyl (**49**) and 3-ethynylacenaphthylene (**50**) in low yield.[15,11]

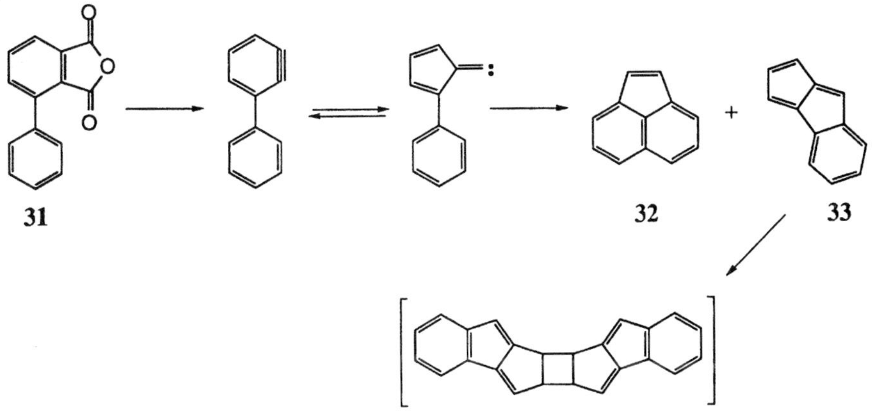

31   **32**   **33**

34

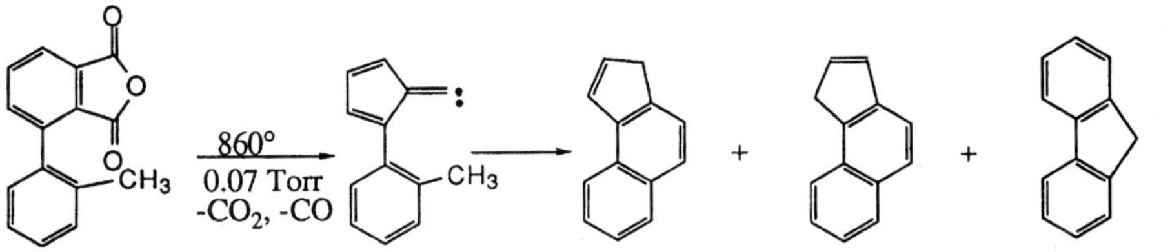

35   **36**   **37**

1 : 1, 22%   49%

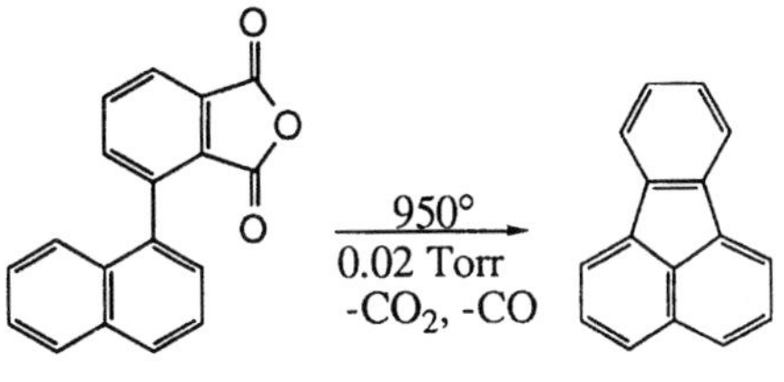

38   **39**

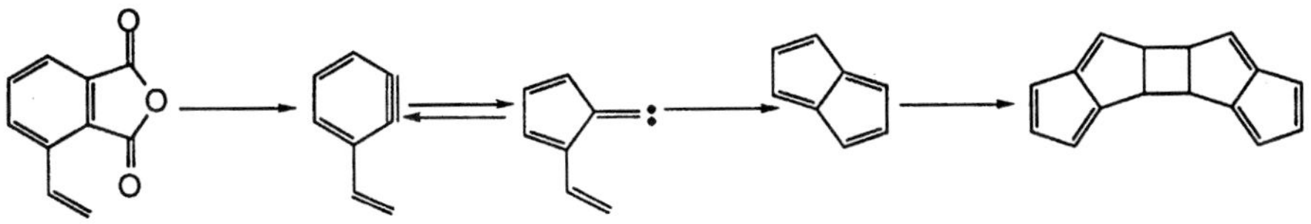

40   **41**   **42**

When substituted napthalene-2,3-dicarboxylic anhydrides were pyrolysed, ring contraction did not take place. 1-Phenylnaphthalene-2,3-dicarboxylic anhydride (**51**) gave fluoranthene (**39**) by a radical process and the benzo analogue 4-phenylphenanthrene-2,3-dicarboxylic anhydride (**52**) gave benzo[e]pyrene (**53**).[12]

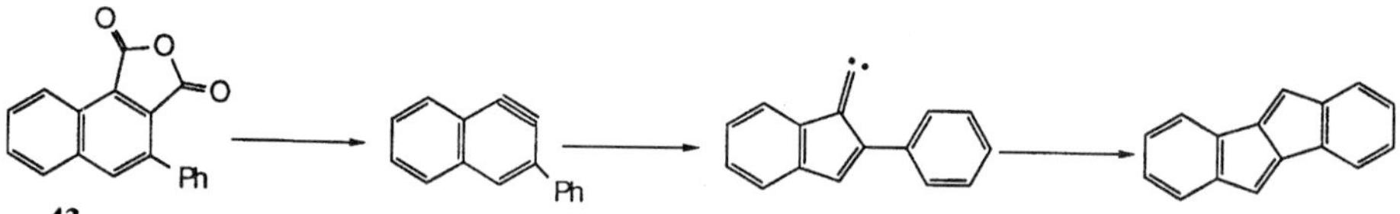

43 → → → 44

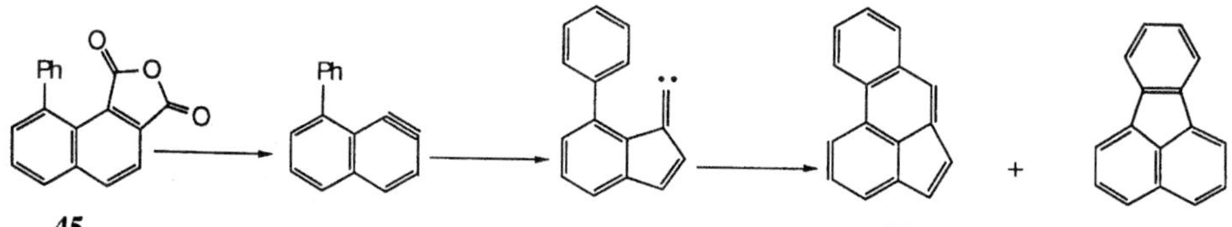

45 → → → 46 + 39

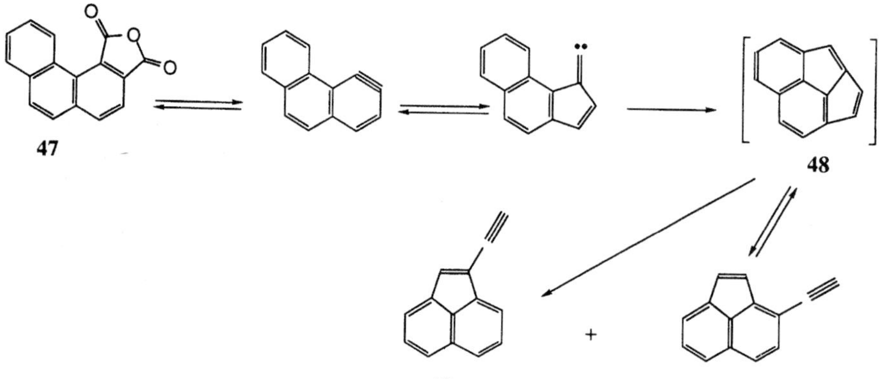

47 ⇌ ⇌ → 48

49 + 50

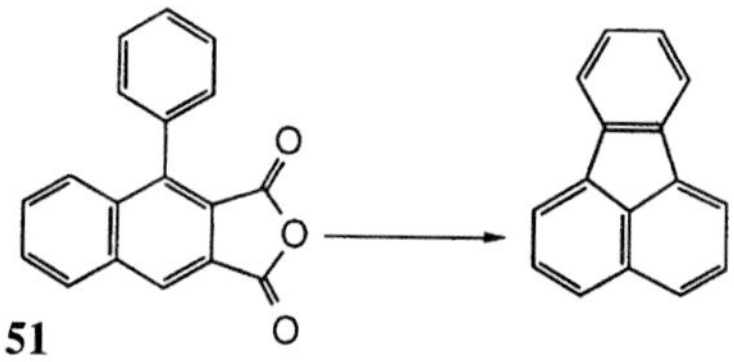

51 → 39

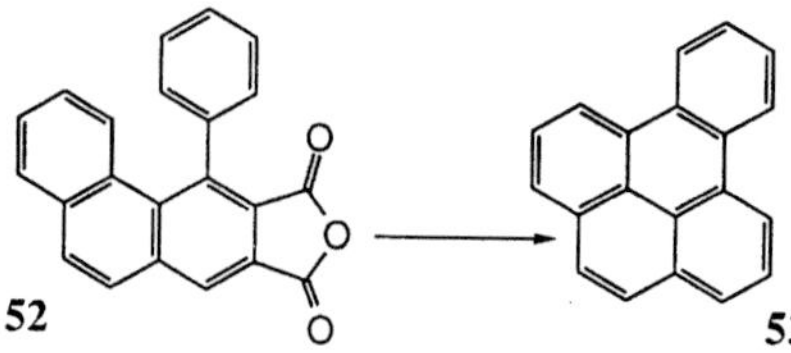

52 → 53

These reactions can be summarised on the basis of current knowledge by the statement that, where alternative routes are unavailable, benzyne (**8**) rearranges to cyclopentadienylidenecarbene (**7**), 1,2-didehydronaphthalene (**54**) rearranges to 1*H*-indenylidenecarbene (**55**) but 2,3-didehydronaphthalene (**56**) does not undergo ring contraction to (**57**) (Scheme 3).

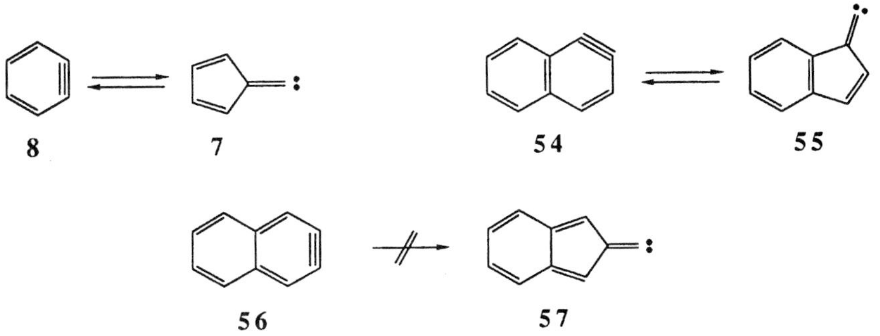

Scheme 3

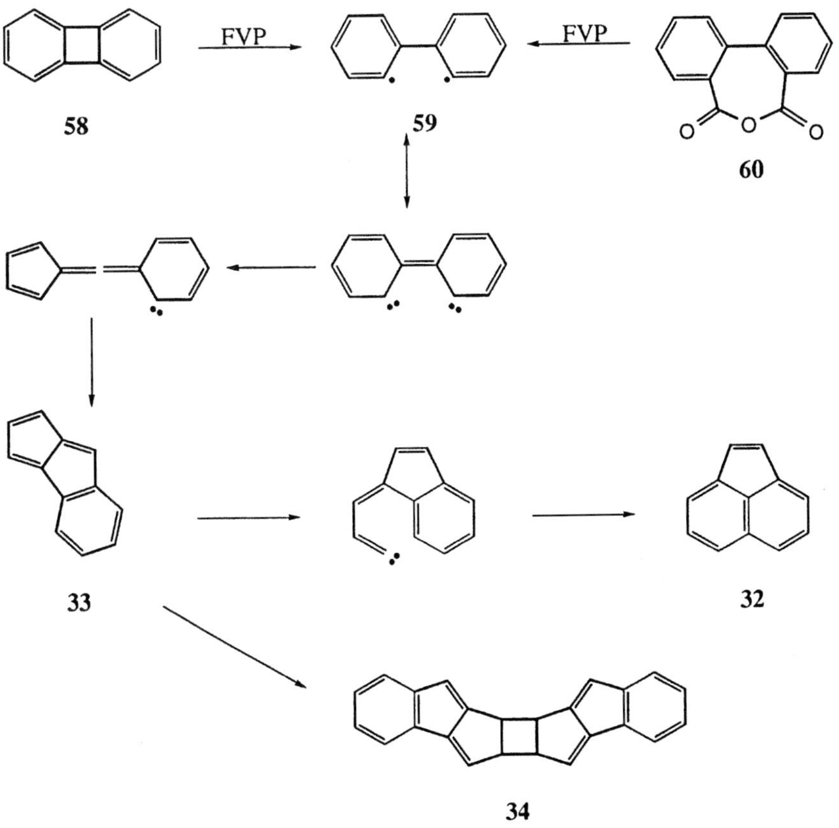

Scheme 4

1.2   2,3- and 2,2'-Didehydrobiphenyl

Pyrolysis of 3-phenylpthalic anhydride (**31**) at 900°/0.02 Torr leads to loss of CO_2
and CO and formation of 2,3-didehydrodiphenyl, $C_{12}H_8$, which rearranges to
acenaphthylene (**32**) and benzopentalene (**33**). Wiersum and Jenneskens[16a] have
shown that FVP of biphenylene (**58**) or diphenic anhydride (**60**) at 900° yields an
alternative $C_{12}H_8$ species, 2,2'-didehydrodiphenyl (**59**) which also yields acenaphthylene (**32**), benzopentalene (**33**) and the benzopentalene dimer (**34**) (Scheme 4).
The relationship between these $C_{12}H_8$ species has not yet been elucidated.[16b]

1.3 p-Benzyne (1,4-Didehydrobenzene) in Thermal Rearrangements

The original finding by Jones and Bergman[17] that pyrolysis of Z-[1,6-d$_2$]-hex-3-ene-
1,5-diyne (**61**) at 300° in a flow of nitrogen gave Z-[3,4-d$_2$]-hex-3-ene-1,5-diyne
(**63**) by interconversion through p-benzyne (1,4-didehydrobenzene) (**62**) (Scheme 5)
has been explored further to determine the radical nature of the intermediate.
Pyrolysis of Z-4,5-diethynyl-4-octene (**64**) at 370° in a flow of nitrogen gave
Z-dodec-6-ene-4,8-diyne (**66**), benzocyclooctene (**67**) and 2-allyl-1-propylbenzene
(**68**) (Scheme 5).[18,19] Pyrolysis of the enediyne (**66**) at 400° in a flow of nitrogen
gave a high yield of (**67**) and (**68**) and the formation of these compounds was
attributed to intramolecular radical transfer reactions. Thermolysis of the di-
ethynyloctene (**64**) in solvents inert to radical abstraction of hydrogen gave good
yields of the products (**66**), (**67**) and (**68**). In the presence of hydrogen donor
solvents, 1,2-dipropylbenzene (**69**) was formed and this evidence confirmed the
diradical nature of the 1,4-didehydro-2,3-dipropylbenzene (**65**). Further examination of this reaction using deuterium donor solvents showed that abstraction of
hydrogen from the solvent could take place at different stages in the overall
conversion. Supporting evidence for these proposals was obtained from studies
involving 2,3-diethyl-1,2-didehydrobenzene and 1,2-didehydro-2,3-dimethylben-
zene and the p-benzyne was shown to be a singlet diradical species (Scheme 5).[20]

1.4   Pyrolytic Dehydrogenation of Triphenylene

Triphenylene (**70**) is one of the stable end products of pyrolyses yielding
o-benzyne in the gas phase.[21] However, it has been reported by Wiersum[2c] that
FVP of triphenylene at 1000° yields cyclopent[h,i]acephenanthrylene (**77**)
(Scheme 6). It is argued that loss of a hydrogen atom from one of six positions
leads to 1-triphenylyl (**71**) which loses another hydrogen atom to form 1,2-
didehydrotriphenylene (**72**). Ring contraction to (**76**) and insertion leads to the
highly strained hydrocarbon (**75**) which undergoes ring cleavage (compare
intermediate (**48**)) to yield the acetylene $\rightleftarrows$ carbene intermediate (**74**). A further
insertion yields the product (**77**). FVP of triphenylene-1,2-dicarboxylic anhydride
(**73**) also yields the cyclopentacephenanthrylene (**77**), confirming the expected
behaviour of the aryne (**72**). An alternative mechanism based on 1,4-didehydro-
triphenylene (**79**) has been explored. FVP of the diallyl ester (**80**), which was
expected to give the diradical (**79**) and to open to the diacetylene (**78**), yielded

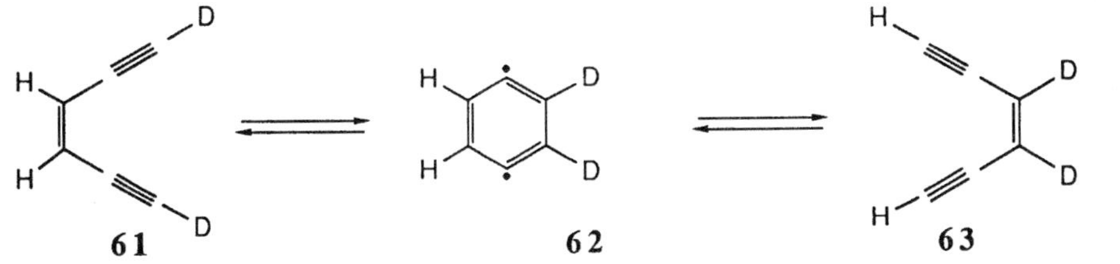

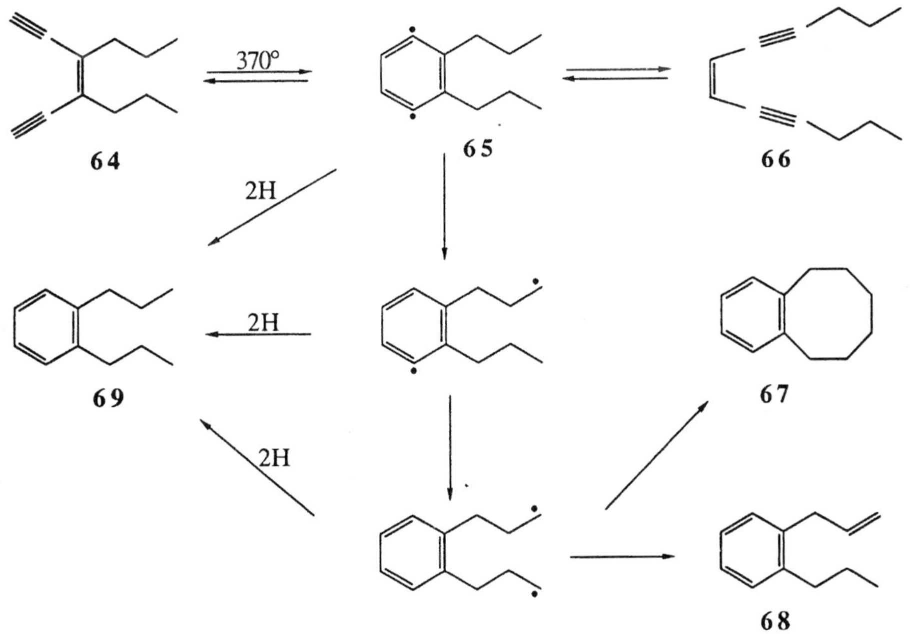

Scheme 5

compound (**77**).[22] A separate pyrolysis of the diacetylene (**78**) also gave (**77**) (Scheme 6).

2. THE CHEMISTRY OF CARBENES FORMED REVERSIBLY FROM ACETYLENES

2.1 Rearrangement of Acetylene

The unsaturated carbene $H_2C=C:$, variously termed methylenecarbene, ethenylidene or vinylidene, was distinguished as a product of photolysis of ethylene[23] and of abstraction by singlet C_2 of two hydrogen atoms from acetone.[24] The rearrangement of $HC\equiv^{13}CD$ to $DC\equiv^{13}CH$ under FVP conditions

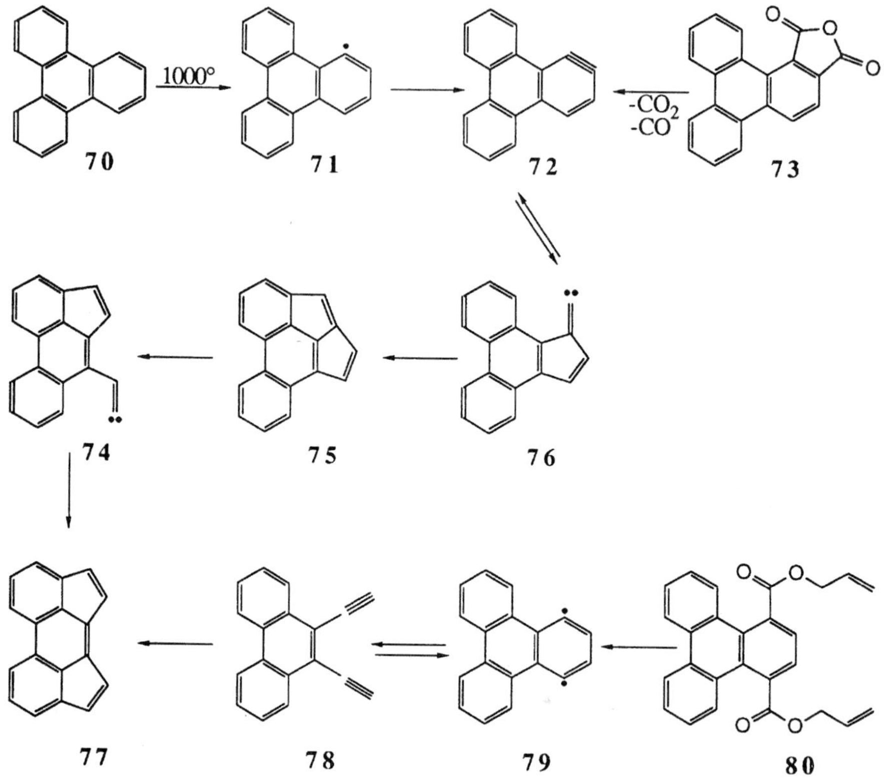

Scheme 6

(10% conversion at 715° rising to 40% at 850°) provided evidence for the formation of this species at high temperatures.[25]

Calculations suggested that vinylidene existed on the C_2H_2 energy surface as a shallow minimum (∼ 4 kcal/mol) about 40 kcal/mol above the ground state of the acetylene.[26] The high bond dissociation energy (homolysis) of the C–H bond in acetylene estimated to be 125 kcal/mol[27] further suggested that vinylidene rather than radical intermediates may be involved in high temperature reactions of acetylene. Colussi and his coworkers[28] expanded acetylene alone or gaseous mixtures of acetylene and a reactant into a fused silica reactor at ∼ 550–700° and determined the nature of the products and their rates of formation by on line modulated beam mass spectrometry. From rate measurements of the reactions of acetylene with benzene and with toluene, an average activation energy of 37 ± 2 kcal/mol was determined. Acetylene (**81**) was converted into vinylacetylene (**83**) by a mechanism involving vinylidene (**82**) and (**83**) was then converted into benzene. A mixture of acetylene and benzene at 550–700° gave styrene (**84**) and polyvinylbenzenes (**85**). A mixture of acetylene and toluene gave isomeric methylstyrenes (**86**). Use of ethylene as a source of vinyl radical in a pyrolysis with toluene provided evidence that radicals were not involved in the acetylene reaction.[28]

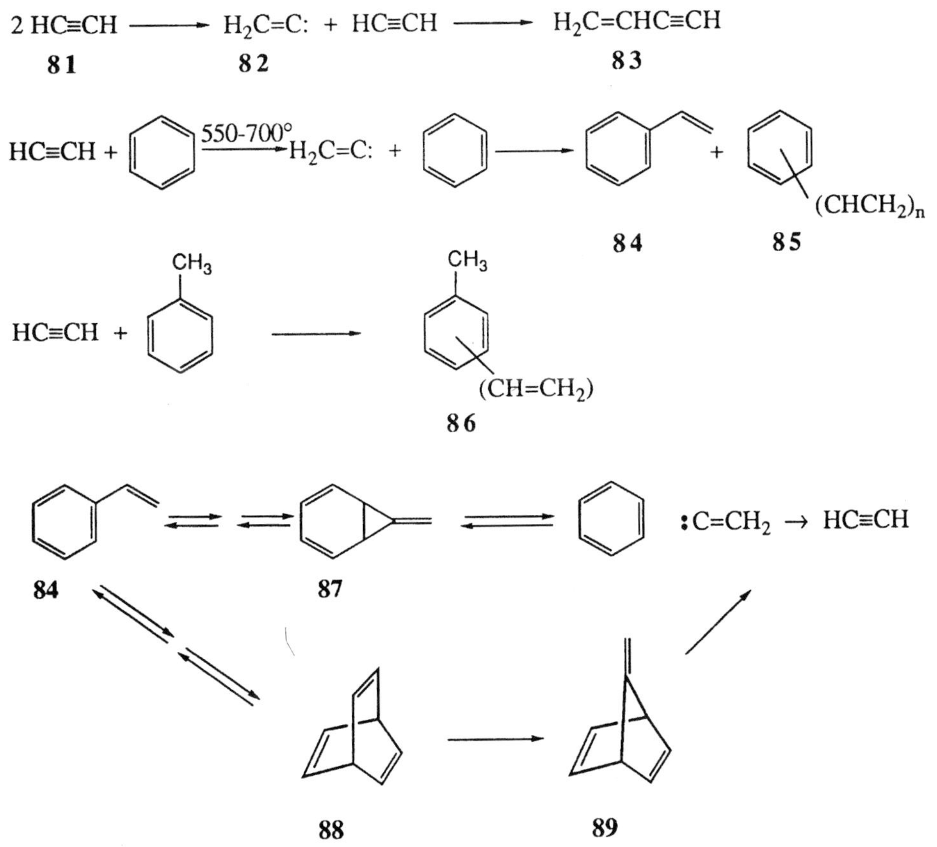

$$2\ HC{\equiv}CH \longrightarrow H_2C{=}C\text{:} + HC{\equiv}CH \longrightarrow H_2C{=}CHC{\equiv}CH$$

81   **82**   **83**

$HC{\equiv}CH + \,\bigcirc\,\xrightarrow{550\text{-}700°} H_2C{=}C\text{:} + \,\bigcirc\, \longrightarrow$ ... $+$ ...

$(CHCH_2)_n$

84   **85**

$HC{\equiv}CH +$ (toluene, CH_3) $\longrightarrow$ (CH_3) ($CH{=}CH_2$)

86

84 ... **87** ... $\text{:}C{=}CH_2 \rightarrow HC{\equiv}CH$

88   **89**

Scheme 7

Thermal fragmentation of styrene into benzene and acetylene was also shown by Colussi and coworkers[29] to proceed by a non-radical process in which the activation energy was compatible with the formation of benzene and of vinylidene as the immediate fragmentation products. The mechanism of the equilibrium reactions, styrene $\rightleftarrows$ benzene and acetylene, occurring at high temperatures was consequently interpreted as involving reversible addition of vinylidene to a double bond to yield (**87**). Also, since the activation energy for elimination of acetylene from barrelene (**88**) was compatible with its extrusion as vinylidene,[30] this reaction was proposed as an alternative pathway for the equilibrium reaction and triene (**89**) was considered as a possible intermediate (Scheme 7).

This detailed examination of styrene represents an extension of the mechanism of pyrolytic conversion of C_8H_8 compounds such as cyclooctatetraene, dihydropentalene, benzocyclobutene etc., to styrene and to benzene and acetylene. The formation of styrene from barrelene had previously been proposed (Section 6). In these reactions the final conversion into styrene has generally been shown as irreversible but automerization of styrene suggests that with sufficient activation energy styrene molecules can attain the energy surface. Styrene is a minimum on both the methylcycloheptatetraene and cyclooctatetraene energy surfaces and it is not yet clear which is involved in the automerization. Remmler et al.[31] have shown that FVP of [2'-^{13}C]-styrene (61%^{13}C) (90) at 930°/0.1 Torr gave 4% rearrangement to [1'-^{13}C]-styrene (91). In a gas flow reactor at 900°/0.6 Torr with nitrogen carrier gas, [2'-^{13}C]-styrene (90) gave benzene (1.5%), a mixture of benzocyclobutene (92) and phenylacetylene (93) (4.5%), and unknown C_{14} compounds (9%). The ^{13}C labelling in the recovered styrene was 7% in the 1'-position and 4% in the aromatic ring. The ^{13}C shift from the 2' to the 1' position was discussed in terms of carbene intermediates and from the 2' position into the aromatic ring involved benzocyclobutene but alternative processes involving barrelene, 1-vinylfulvene or methylcycloheptatetraenes may be involved.

2.2 Rearrangement of Monosubstituted Acetylenes

The reversible rearrangement of monosubstituted acetylenes was first demonstrated with [2'-^{13}C]-phenylacetylene[32] which equilibrated at 700°/0.05 Torr, and with [2'-^{14}C]-1-adamantylacetylene[33] which rearranged to the extent of 25% at 780°/0.02 Torr. These reactions were discussed in terms of the substituted vinylidene carbenes. Unsaturated carbenes formed from arylacetylenes could be trapped by ortho-substituents. o-Tolylacetylene (94) on FVP at 720°/0.02 Torr gave indene (95) (75%), and 1-ethynyl-8-methylnaphthalene (96) at 750°/0.5 Torr gave a mixture of phenalene (97) and the starting material in a ratio of 5:4.[32]

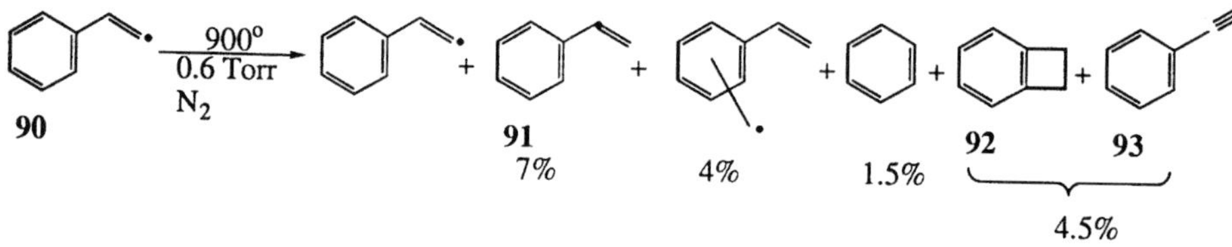

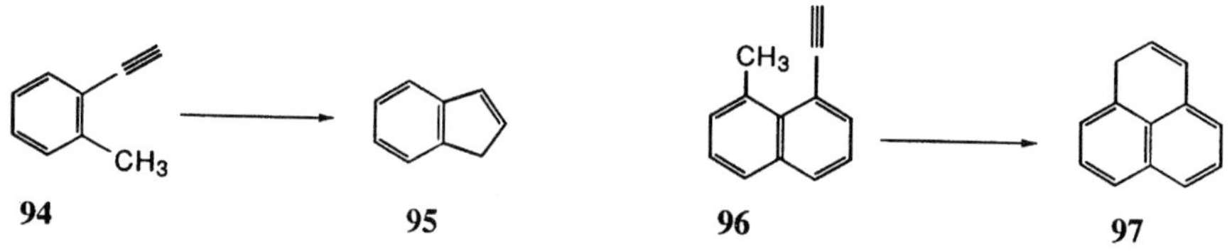

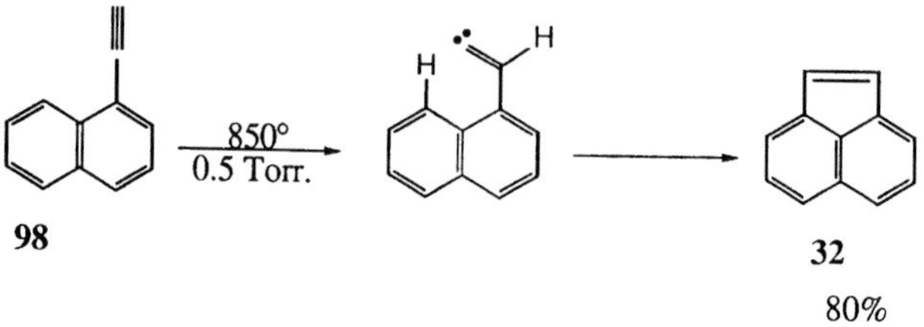

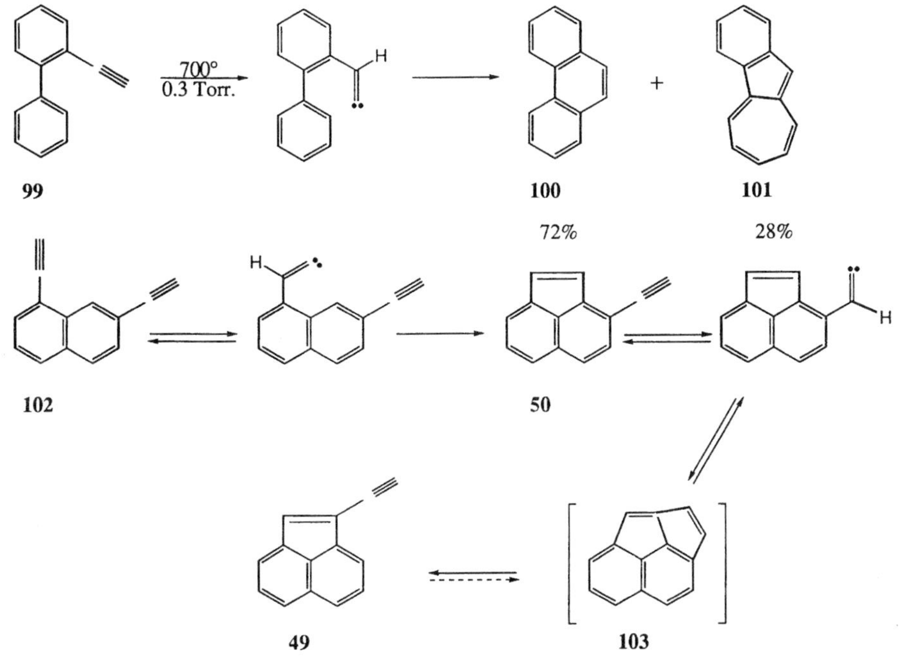

99   100   101

                                                72%   28%

102   50

49   103

Scheme 8

The intermediate carbenes can be trapped with an adjacent aromatic C–H bond and this procedure has proved useful in synthesis. 1-Ethynylnaphthalene (**98**) on FVP at 850°/0.5 Torr gave acenaphthylene (**32**) (80%), and biphenyl-2-ylacetylene (**99**) at 700°/0.3 Torr gave a mixture of phenanthrene (**100**) and 1,2-benzazulene (**101**) (72:28) (99%).[32] This insertion reaction has been used for the synthesis of 3-ethynylacenaphthylene (**50**) from 1,7-diethynylnaphthalene (**102**). Compound (**50**) was isomerised through the highly strained, probably non-planar, intermediate (**103**) to 1-ethynylacenaphthylene (**49**) at 1100° (Scheme 8).[34] Compounds (**49**) and (**50**) had been obtained previously from FVP of phenanthrene-3,4-dicarboxylic anhydride (**47**) (Section 1.1).[35]

2.3 Synthesis of Corannulene

Trapping of vinylidene carbenes was used by Scott and his coworkers in an elegant synthesis of corannulene (**105**).[36] FVP of 7,10-diethynylfluoranthene (**104**) at 1000° yielded corannulene by a process that required the intermediate carbenes to insert into the aromatic C–H bond while the molecule was temporarily in a non-planar geometry, resulting in the formation of the bowl-shaped corannulene (**105**). However, volatilisation of high molecular weight polyethynyl aromatic compounds leads to considerable loss through polymerization and corannulene was obtained in only 10% yield from this reaction.

The use of the more robust 2,2-dibromoethyl group, $Br_2C=CH—$, in place of acetylene proved profitable and pyrolysis of 7,10-bis-(2',2'-dibromoethenyl)-

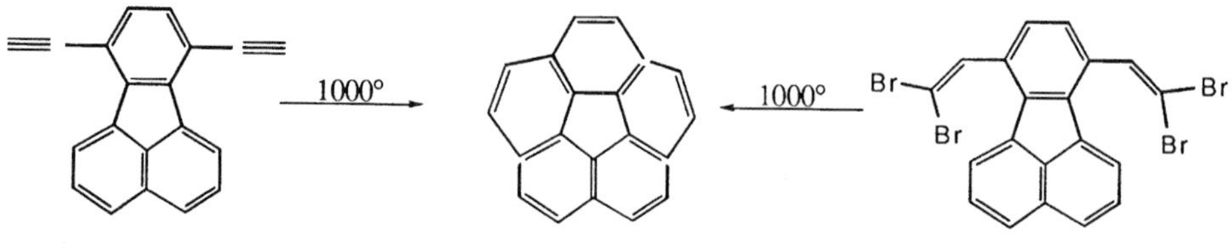

104 **105** **106**

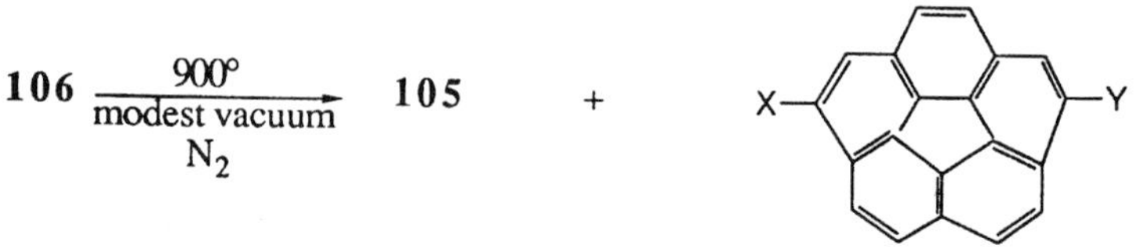

106 $\xrightarrow[\substack{\text{modest vacuum} \\ N_2}]{900°}$ **105** +

107 X = H, Y = Br
108 X = Y = Br
109 X = H, Y = C(CH$_3$)$_2$OH

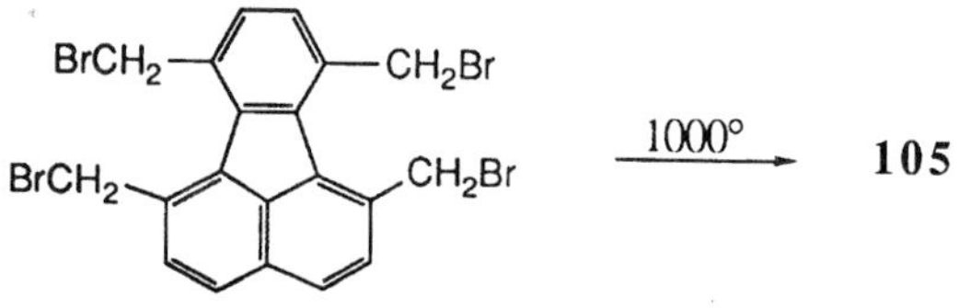

110 $\xrightarrow{1000°}$ **105**

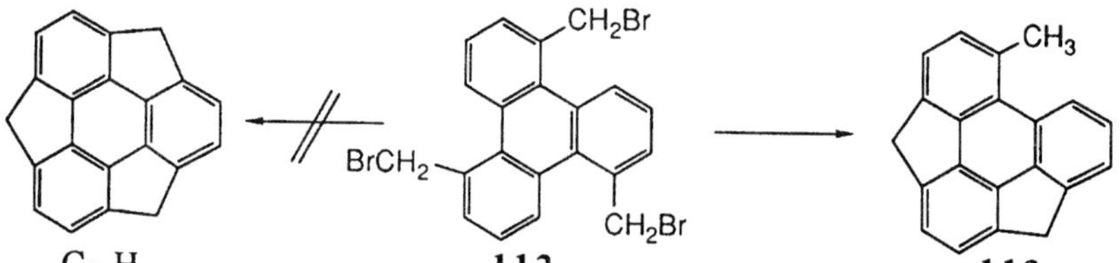

C$_{21}$H$_{12}$
111

112

113

fluoranthene (**106**) gave corannulene (**105**) in 40% yield through loss of bromine atoms. Pyrolysis of the tetra-bromo compound (**106**) at 900° under a modest vacuum with a slow bleed of nitrogen gave a mixture of corannulene (**105**), 1-bromocorannulene (**107**) and 1,6-dibromocorannulene (**108**).[37] Pathways for cyclisation and loss of bromine were discussed but no direct evidence concerning the mechanism of this reaction has been presented. The availability of 1-bromo-corannulene allowed synthesis of corannulenyldimethylcarbinol (**109**) and the determination of the inversion frequency of corannulene (2 × 10^6 times/sec).[37] Another pyrolytic approach to corannulene has been described by Siegel and coworkers[38] who synthesised 1,6,7,10-tetra(bromomethyl)fluoranthene (**110**) as a key intermediate allowing alternative methods of ring closure. FVP of (**110**) at 1000° yielded corannulene (**105**) in 18% yield.

The corannulene system, $C_{20}H_{10}$, can be considered as a segment of buckminsterfullerene, C_{60}. Metha and coworkers[39] attempted a pyrolytic approach to an alternative segment (**111**), $C_{21}H_{12}$, using 1,5,9-tris(bromomethyl)triphenylene (**112**). FVP of (**112**) resulted in closure of two rings and the loss of bromine to give the hydrocarbon (**113**).

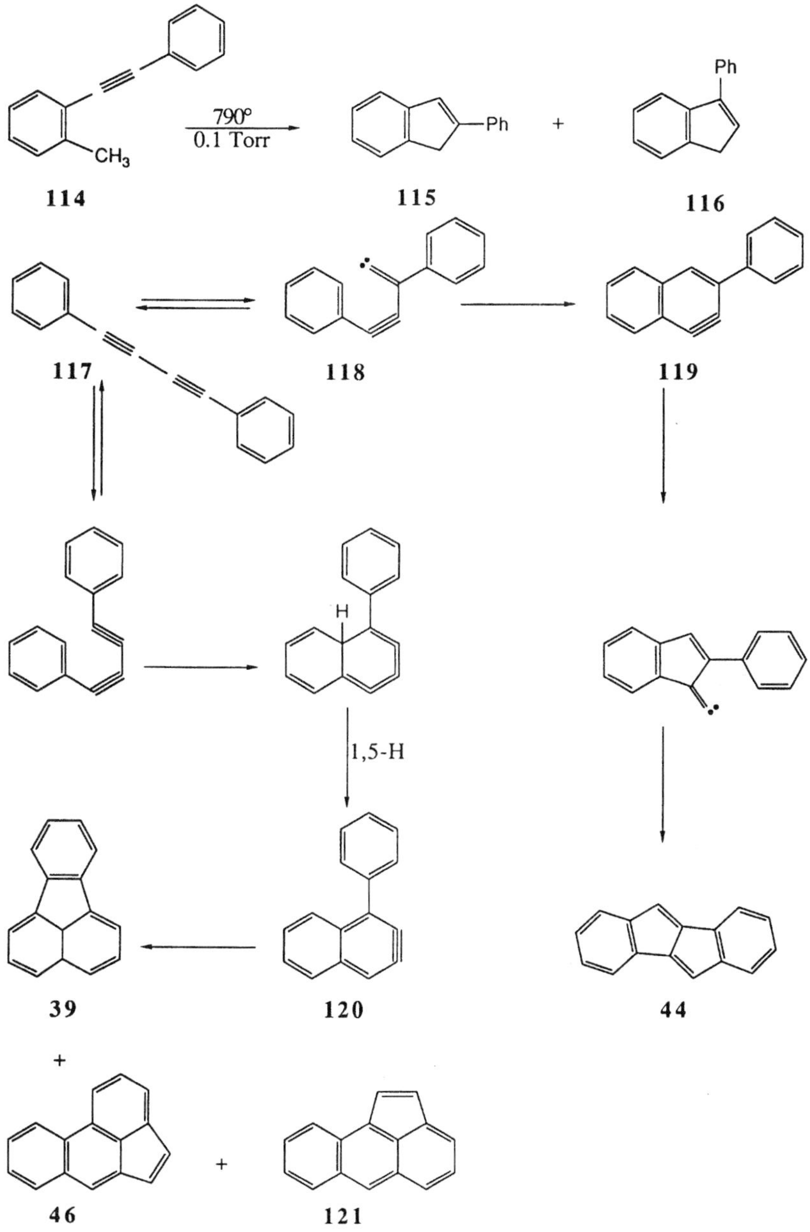

Scheme 9

2.4   Carbenes Formed from Disubstituted Acetylenes

Pyrolysis of (2-methylphenyl)phenylethyne (**114**) at 790°/0.1 Torr gave 2-phenylin-
dene (**115**) (38%) and 3-phenylindene (**116**) in a ratio of 12:7.[33] However, there
are several mechanistic possibilities for this reaction and carbene intermediates
may not be involved. On the other hand, FVP of 1,4-diphenylbutadiyne (**117**) at
1120°/0.03 Torr gave a pyrolysate (22% mass recovery) consisting of indeno[2,
1-*a*]indene (dibenzopentalene) (**44**) (19%), fluoranthene (**39**) (59%), acephenanthrylene (**46**) (13%), and aceanthrylene (**121**) (2%).[40] It was proposed that rearrangement of 1,4-diphenylbutadiyne (**117**) gave carbene (**118**) which by insertion gave
3-phenyl-1,2-didehydronaphthalene (**119**) and yielded the indeno[2,1-*a*]indene
(**44**). Fluoranthene (**39**) and its rearrangement products (**46**) and (**121**) were considered to arise from 1-phenyl-2,3-didehydronaphthalene (**120**) (Scheme 9).[14,41]

3.   REARRANGEMENT OF ARYLMETHYLENES

Formation of phenylmethylene (**122**), C_7H_6, by FVP of phenyldiazomethane and
other precursors leads to the 1,2,4,6-cycloheptatetraene (**125**)/cycloheptatrienyli-
dene (**128**) species. At 450–600° heptafulvene (**129**) is the principal product and
at 700–900° fulveneallene (**127**) is obtained through the bicyclic intermediate
(**126**) (Scheme 10).[1a] The intermediates, bicyclo[4.1.0]hepta-2,4,6-triene (**124**)
and cycloheptatrienylidene (**128**) have not been detected spectroscopically[42] but
the corresponding species in the naphthalene series have been observed (see
below). Ideas concerning these reaction pathways have been developed over a
considerable time but there are further observations which raise question concerning the mechanisms that have been proposed.

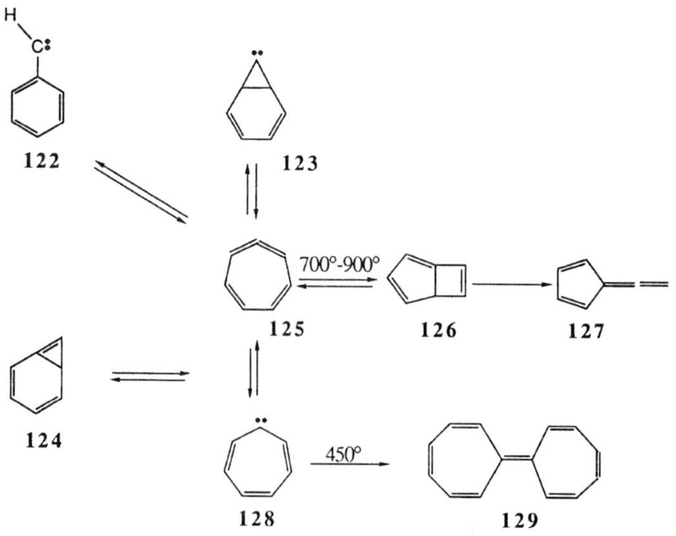

Scheme 10

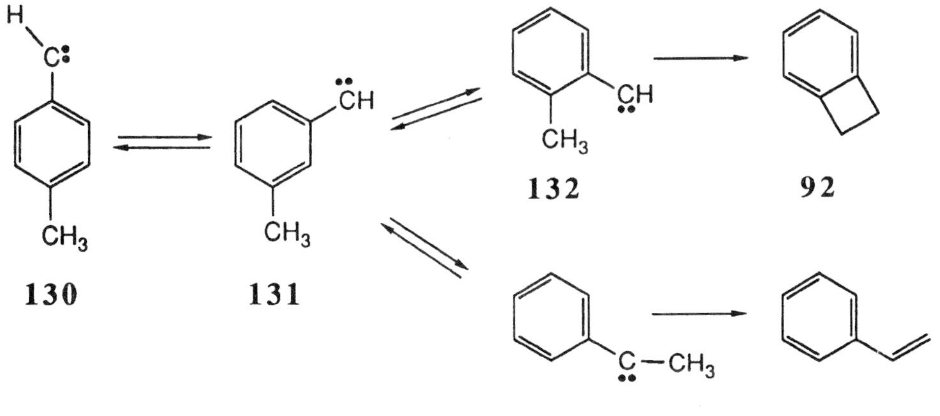

Scheme 11

It is known that under FVP conditions the stable end products arising from *o*-(**132**), *m*- (**131**) and *p*-tolylmethylene (**130**), C$_8$H$_8$, formed from the corresponding tolyldiazomethanes are benzocyclobutene (**92**) and styrene (**84**).[43] These products are formed in the ratio 2.8:1 from *o*-tolyldiazomethane and 0.8:1 from *m*- and *p*-tolyldiazomethanes. The development of the mechanistic ideas concerning these C$_7$H$_6$ and C$_8$H$_8$ species has been reviewed and new mechanisms which provide an explanation for the different ratios in the tolylmethylene chemistry have been proposed.[44]

To account for these results, Gaspar *et al.*[44] suggested that the mechanism must involve two pathways, one leading to *o*-tolylmethylene (**132**) and benzocyclobutene (**92**) and a second leading directly to α-methylphenylmethylene (**133**) and styrene (**84**) (Scheme 11). If this is correct, and provided the —CH group moves around the ring (Baron mechanism)[43] and not directly across the ring, then any α-methylphenylmethylene (**133**) formed leads quantitatively to styrene (**84**) and the —CH group is not able to move past a methyl group. This was tested using 3,5-dimethylphenyldiazomethane (**134**) which on FVP gave 4-methylbenzocyclobutene (**137**) and 3-methylstyrene (**140**) but no 3-methylbenzocyclobutene (**141**) (Scheme 12). It was concluded that (**138**) was not formed.[44]

If *m*-tolylmethylene (**131**) is converted directly to α-methylphenylmethylene (**133**), there is the possibility of detecting this change in competition with the formation of both the benzocyclobutene and the styrene. FVP of 3,4,5-trimethylphenyldiazomethane (**142**) gave the carbene (**143**) which can lead directly to (**144**) and 2,3-dimethylstyrene (**145**) and through 2,3,4-trimethylphenylmethylene (**148**) to 3,4-dimethylbenzocyclobutene (**149**) (Scheme 13). If before the carbene in (**148**) is trapped by the *ortho*-methyl group it can rearrange to the α-methyl-2,6-dimethylphenylmethylene (**147**) then 2,6-dimethylstyrene (**146**) will be formed. A small amount of this styrene was obtained.[44] To encompass these facts a norcaradienylidene intermediate (**123**) (Scheme 10) has been proposed as an alternative pathway for migration of the carbon atom around the ring.

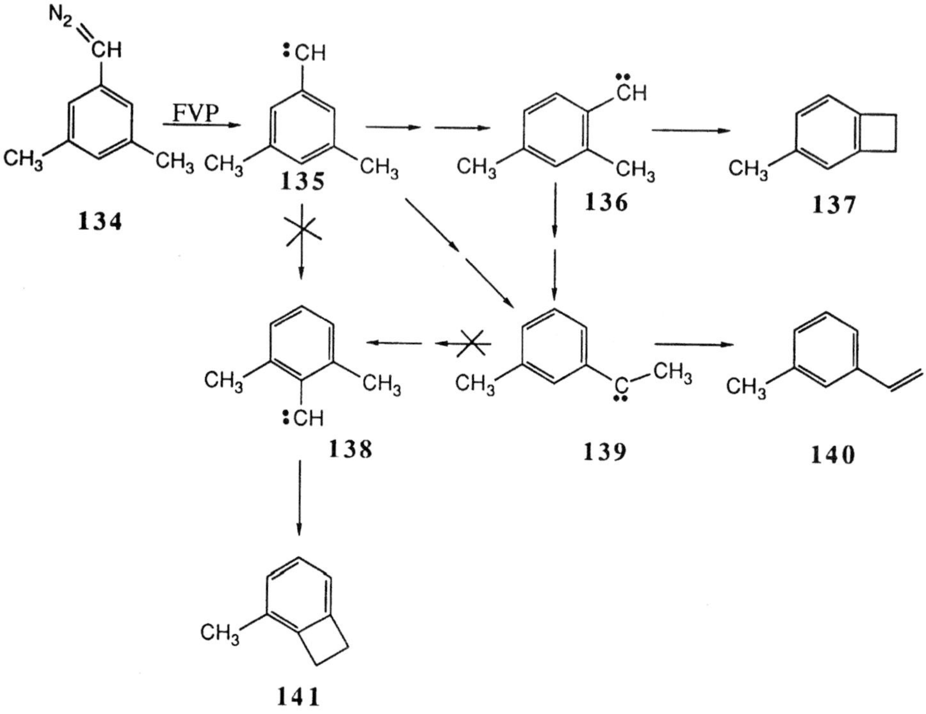

Scheme 12

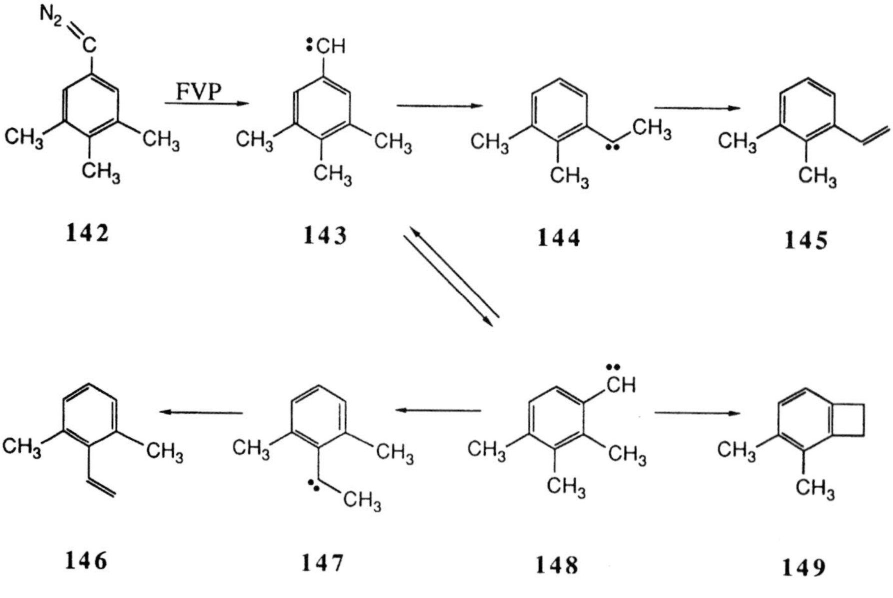

Scheme 13

More recently Jones, Shevlin and coworkers[45] used this type of rearrangement as a source of carbenes to show that thermal decompositions of carbene precursors give some products which are formed directly from the precursor and not from the carbene. The proportions of Z-1-phenylpropene (**151**) to E-1-phenylpropene (**152**) formed by FVP at 250–500° of the precursors (**150**), (**154**) and (**157**) were found to be in the range 28–31:72–79. On the other hand, FVP at 320–500° of the precursors (**153**), (**156**) and (**159**), which could only give rise to the products (**151**) and (**152**) through the p-ethylphenylmethylene (**158**) to α-ethylphenylmethylene (**155**) rearrangement, gave a different ratio 37–34:63–66. It was argued that the second ratio gave the proportions arising from hydrogen transfer in the true carbene and that the precursors (**150**), (**154**) and (**157**) underwent direct elimination. A further experiment involving deoxygenation of propiophenone with atomic carbon at 77 K, a process which should yield a true carbene, gave a ratio of (**151**) to (**152**) of 36:64 which fell within the expected range.

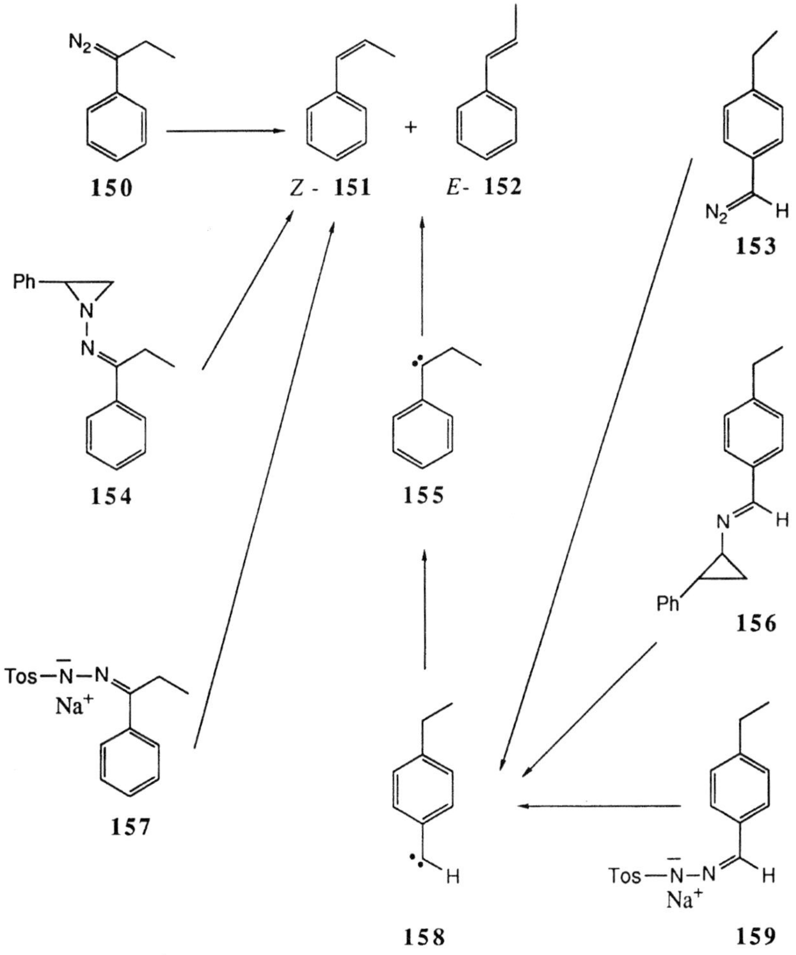

Scheme 14

Hoffmann and coworkers[46] have described a new entry to the C_7H_6 system involving pyrolytic elimination of acetic acid from the acetoxy compounds (**160**), (**161**) and (**162**). FVP of the acetoxytetracycloheptane (**160**) at 300° yields 7-acetoxynorbornadiene (**161**) which on further pyrolysis at $450°/10^{-3}$ Torr gave heptafulvene (**129**) (58%), acetic acid (100%) and benzene (10%). Pyrolysis of ester (**161**) at 700° gave fulveneallene (**127**) (20%), acetic acid (100%) and benzene (42%). These results strongly indicated the formation of the C_7H_6 intermediates (**125**)/(**128**) and the finding that FVP of 7-acetoxycycloheptatriene (**162**) at 450° also gave heptafulvene (**129**) (67%), acetic acid (100%) and benzene (10%) confirmed this possibility and provided an alternative mechanism (Scheme 15). Pyrolysis of the 7-deuterated precursor (**163**) gave heptafulvene (**129**) containing two deuterium atoms and acetic acid estimated to contain about 5% AcOD. The mechanism proposed involved thermal rearrangement of (**161**) to (**162**) followed by elimination of acetic acid, possibly from an ion pair as shown in Scheme 16, to give the intermediates (**125**) and (**128**) which lead to the observed products.

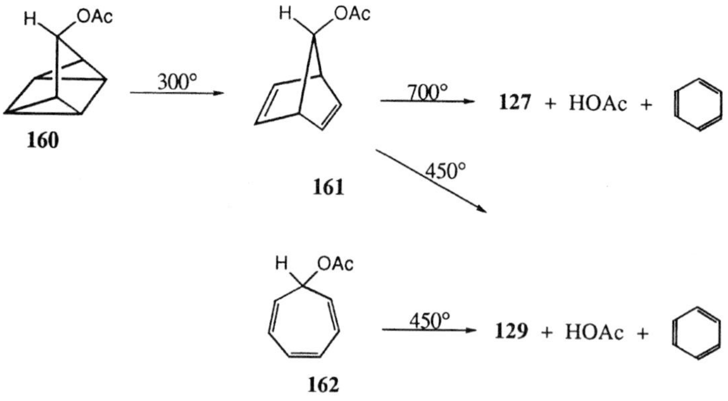

Scheme 15

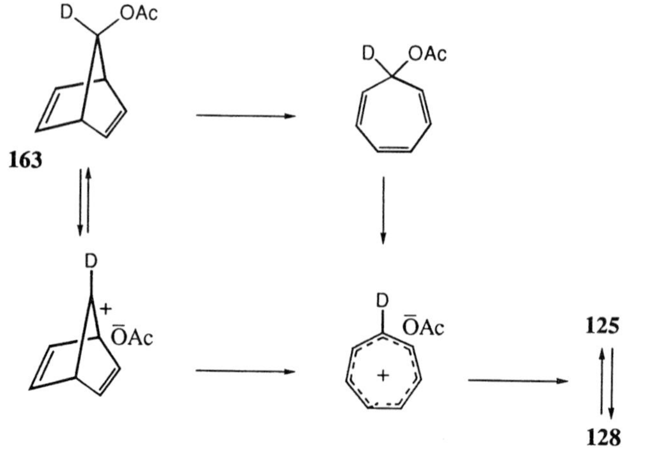

Scheme 16

The formation of 7-norbornadienylidene (**165**) by FVP of *N*-nitroso-*N*-(7-norbornadienyl)urea (**164**) resulted in the isolation of benzene by loss of a carbon atom and products arising from C_7H_6 intermediates. At 200° dimers (**166**) of 1,3,5-bicyclo[3.2.0]heptatriene (**126**) were isolated. Heptafulvene (**129**) was obtained at 350° and fulveneallene (**127**) at 400° (Scheme 17).[47]

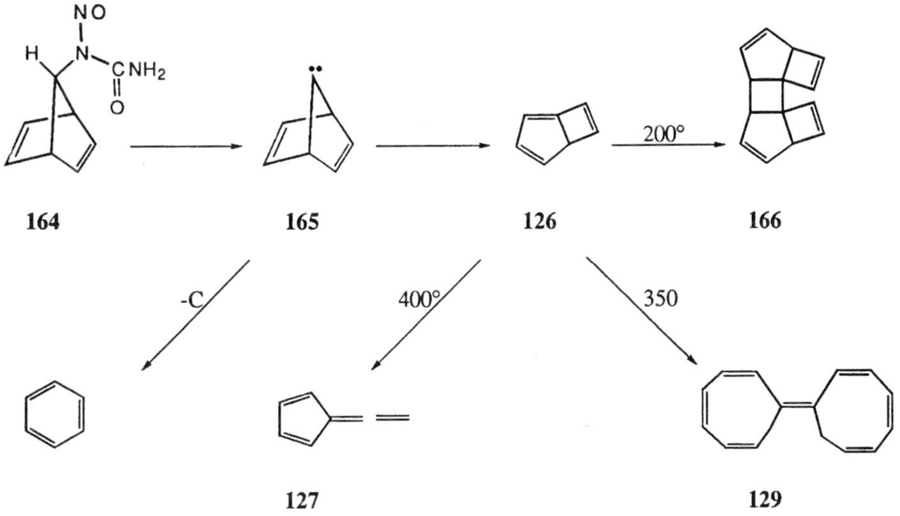

Scheme 17

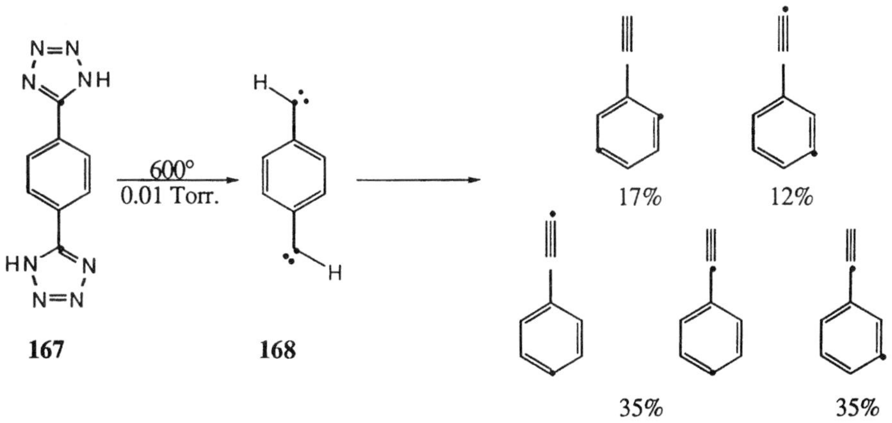

Scheme 18

Baum, Jones and coworkers[48] have investigated the C_8H_6 energy surface by pyrolysing doubly labelled 1,4-bis(5-tetrazolyl)benzene (**167**) (20% $^{13}C_2$) to form the dicarbene (**168**). Pyrolysis at 600°/0.01 Torr gave much char and nitriles and a 3% yield (25% relative yield) of phenylacetylene which constituted about 90% of the hydrocarbon product. Mass spectrometry showed the phenylacetylene to be doubly labelled and consequently the rearrangement to be intramolecular. ^{13}C NMR spectroscopy showed that all carbons were labelled except the ipso carbon C1. The pattern of labelling that was found is shown in Scheme 18. The rearrangement was discussed in terms of the movement of the phenylcarbenes around the ring. The possibilities of single ring expansion or sequential ring expansion leading to cyclooctapentaene were considered.

Becker and Wentrup[49] showed that FVP of 1- (169) or 2-naphthyldiazomethane (172) yielded 1H-cyclobuta[d,e]naphthalene (174) demonstrating that the carbene migrates from the 2- (171) to the 1-position (170) on the naphthalene and is trapped by insertion in the aromatic C–H bond. Engler and Schechter[50] used FVP of the methoxy trimethylsilane derivative (173) as a source of carbene (170) and isolated cyclobuta[d,e]naphthalene (174). The latter authors also pyrolysed the 7-methylnaphthalene derivative (175) (510°/0.1–0.2 Torr) and isolated an 11:1 mixture (25% yield) of 3-methyl-1H-cyclobuta[d,e]naphthalene (177) and 2-methyl-1H-cyclobuta[d,e]naphthalene (179). This demonstrated that migration of the carbene could take place from the 1- (176) to the 4- position (178) of naphthalene. They also prepared a number of peri-methanoarenes (anthracene, phenanthrene, pyrene) from arylmethylenes. One reaction involved rearrangement of 9-phenanthrylmethylene to 4-phenanthrylmethylene and ring closure to 4H-cyclopenta[def]phenanthrene.

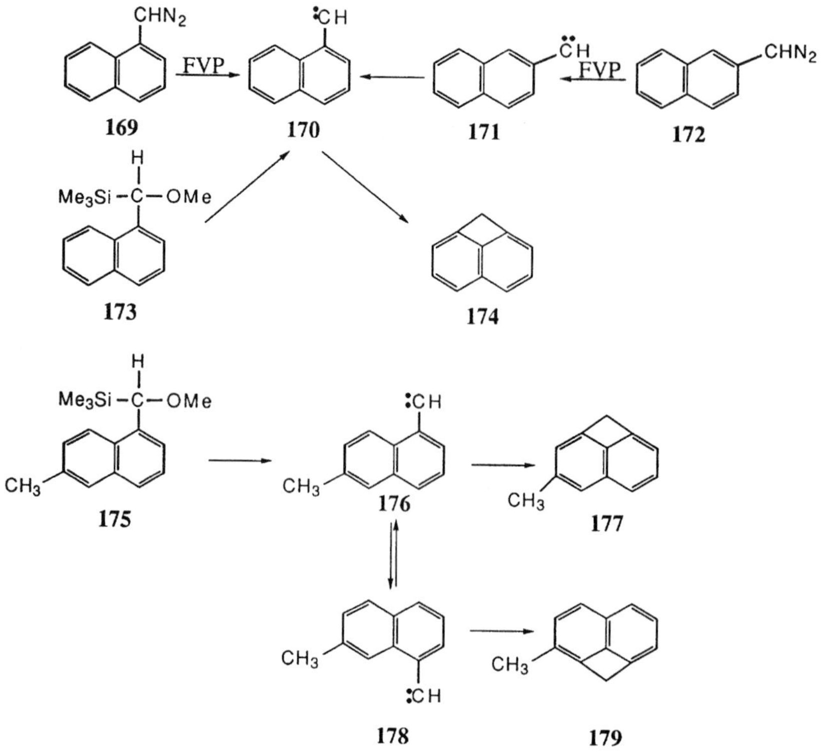

Scheme 19

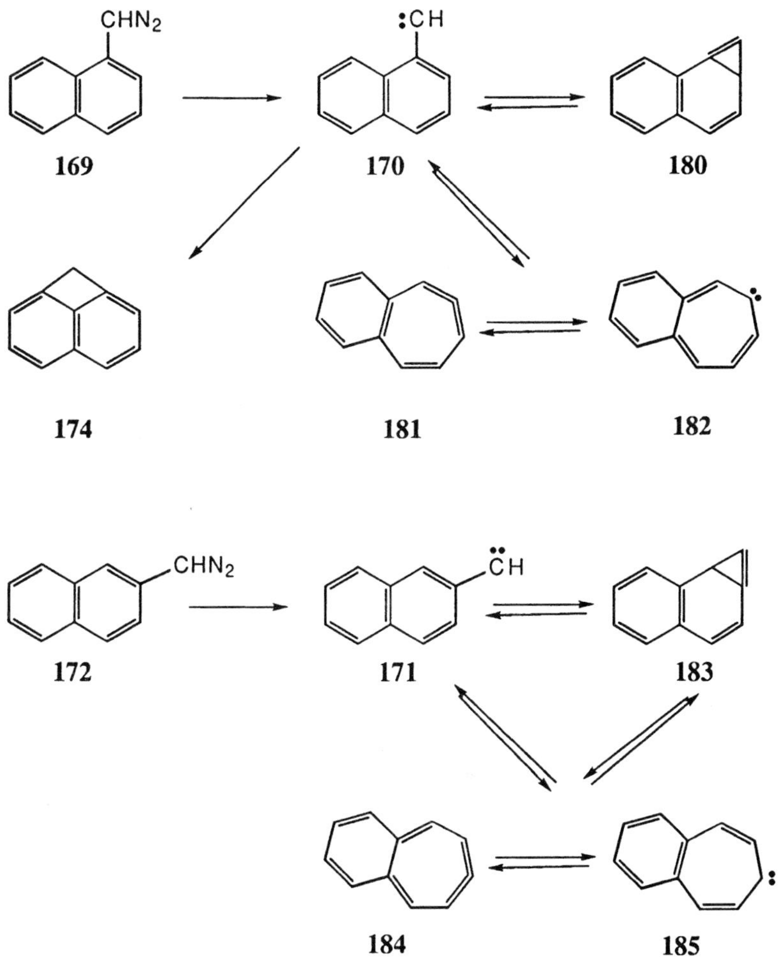

Scheme 20

Ideas concerning the intermediates formed from naphthylmethylenes continue to evolve.[51–53] Assuming stepwise transfer of the CH group around the ring, the species involved must lead to a symmetrical carbene (**185**) or allene (**184**) to permit transfer of the CH group from the 1- to the 4- position. Some intermediates formed photochemically and detected spectroscopically may be involved in the pyrolytic process but are not detectable under these conditions.

In argon matrix isolation experiments Chapman *et al.*[51] showed that irradiation (>510 nm) of 1-naphthyldiazomethane (**169**) slowly produced 4,5-benzobicyclo[1.1.0]hepta-2,4,6-triene (**180**) and a small amount of 1*H*-cyclobuta[*d,e*]naphthalene (**174**) whereas FVP of (**169**) and deposition of the pyrolysate in argon gave (**174**) but no (**180**). Similarly, 2-naphthyldiazomethane (**172**) on irradiation (>364 nm) gave 2,3-benzobicyclo[4.1.0]hepta-2,4,6-triene (**183**) and on FVP gave (**174**) by a route passing through 1-naphthylmethylene, but no (**183**) was observed (Scheme 20). Horn and Chateauneuf[53] have used laser flash photolysis to characterize 2-naphthylmethylene (171) in their study of the rearrangement of (171) to (185).

Wentrup and Benedikt[54] have described the rearrangement of another $C_{11}H_8$ species. FVP of 1-(2-diazoethylidene)indene (186) at 400° gave the carbene (187) which formed spiro[cyclopropene-1,1'-indene] (188). Pyrolysis of either (186) or (188) at 600° provided a route to the stereoisomeric carbene (189) which cyclised to 1H-cyclopent[c,d]indene (25). Pyrolysis of (186) at 800° gave 1-vinylidene-indene (190) which was identified by an infrared band at 1935 cm^{-1}, measured on a film of pyrolysate at $-196°$, and a ^{1}H NMR resonance at δ 5.4. Pyrolysis of 1-naphthyldiazomethane (169) at 800° also gave 1-vinylideneindene (190) in low yield.

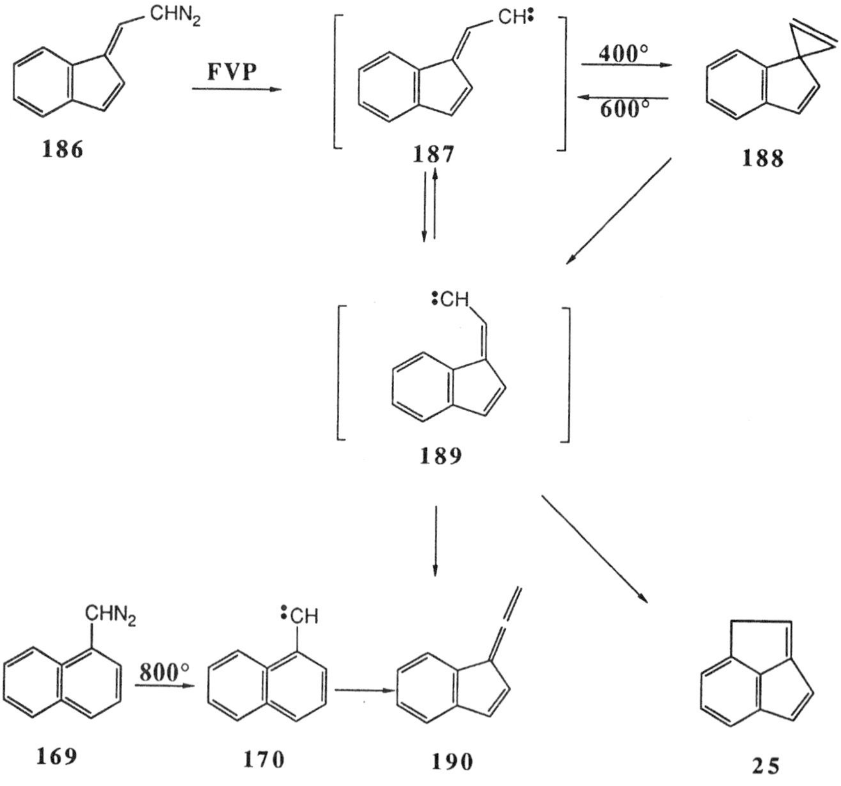

Scheme 21

Intramolecular trapping of carbenes in arylmethylenes has been demonstrated but retention of the 7-membered ring expanded structure has been less common. Wentrup and Becker[55] have shown that FVP (420°/10^{-1}–10^{-3} Torr) of 5-(2-fluorenyl)tetrazole (192) gave 2-cyanofluorene (191) (33%), by a commonly observed reversion of the tetrazole, benz[a]azulene (195) (44%) and 4,8-dihydrocyclopenta[d,e,f]fluorene (194) (4%). The reaction was shown to proceed through 2-fluorenyldiazomethane (**193**) and must involve migration of the carbene and trapping to give (**194**) and ring expansion and a hydrogen shift leading to (**195**) (Scheme 22).

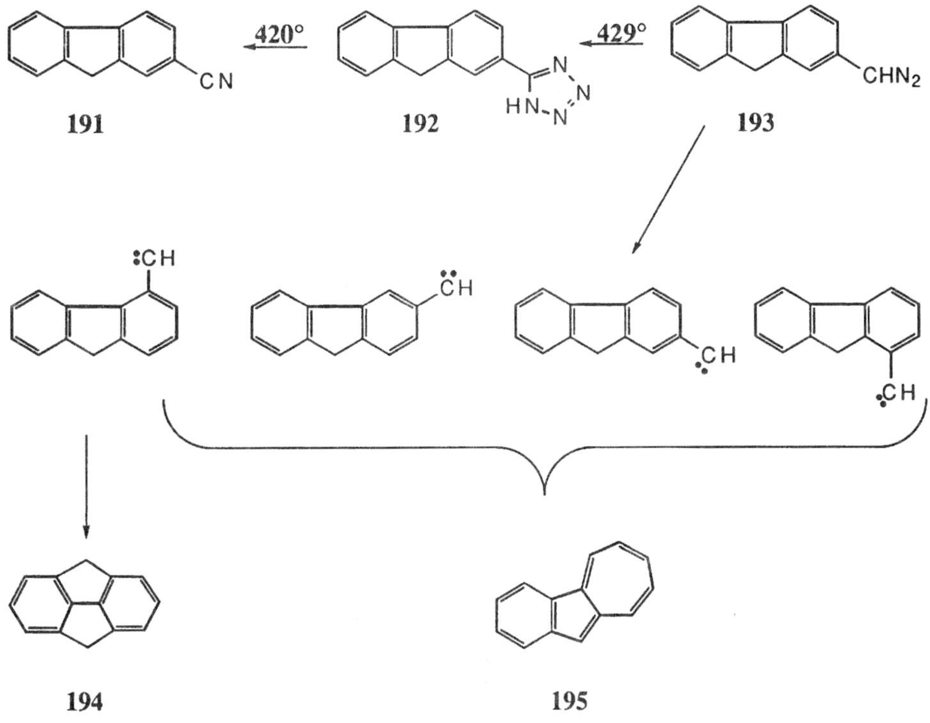

Scheme 22

4. AUTOMERIZATION AND ISOMERIZATION OF AROMATIC HYDROCARBONS

The thermal stability of aromatic hydrocarbons is part of the lore of chemistry. Isomerizations leading from less to more thermodynamically stable aromatic systems, such as the azulene to naphthalene rearrangement,[56,57] have been familiar to chemists for many years but the mechanisms of such rearrangements have remained unresolved problems. Considerable progress has been made in this field, as much by a proliferation of information as by increased insight into the processes involved. The interest engendered by the discovery of the buckminsterfullerenes has brought increased attention to the mechanisms of accretion of carbon fragments into fullerenes and of hydrocarbon fragments into polycyclic aromatic hydrocarbons. The PAHs are widely distributed in the environment and the subject of much discussion. At the present time, studies of thermally induced automerizations (degenerate skeletal rearrangements) and isomerizations are leading to an appreciation of the processes involved. The accretion process requires continual dehydrogenation and new findings involving dehydrogenation of aromatic compounds leading to more condensed though not necessarily fully aromatic systems is shedding new light on this area.[1e,2c,16,22]

In the following discussion, the experiments on automerization are first outlined and these are followed by isomerizations. Many of these experiments were designed to test particular mechanistic theories but the discussion of mechanism has been confined to a single section.

Automerization of benzene

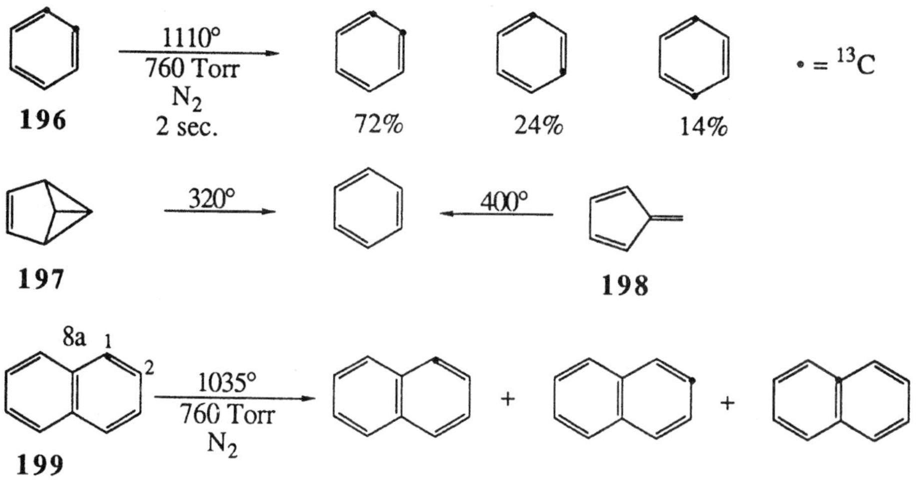

Scheme 23

	Carbon atom position		
Contact time (sec)	1	2	8a (4a)
1.0	78	18	4
2.0	71	26	3
5.0	58	40	2
8.0	55	43	2
11.0	52	46	2

4.1 C_6H_6, Automerization of Benzene

Benzene is extremely stable and requires high temperatures before rearrangement occurs. Scott and coworkers in 1987[58] described the pyrolysis of [1,2-$^{13}C_2$]-benzene (81% $^{13}C_2$) (196) through a packed quartz tube at 1110°/760 Torr using nitrogen carrier gas and with a contact time of 2 sec. The distribution of ^{13}C atoms shown was determined by nitration of the recovered benzene and examination of the isolated *m*-dinitrobenzene by ¹H and ¹³C NMR spectroscopy. The reaction was shown to be intramolecular and the ratios of products clearly illustrated that switching adjacent carbon atoms was the primary reaction pathway.

Hypotheses concerning the mechanism of this reaction have been examined using *ab initio*.calculations[59] and the findings will be considered below (Section 4.12).

4.2 C_6H_6, Isomerization of Benzvalene and Fulvene to Benzene

Benzvalene[60] (**197**) and fulvene[61,62] (**198**) are known to isomerise to benzene at 313–330° and 400° respectively.

4.3 $C_{10}H_8$, Automerization of Naphthalene

The study by Scott and Agopian in 1977[63] was the first demonstration of the occurrence of automerization as a phenomenon. [1-¹³C]-Naphthalene (90% ¹³C) (**199**) was sublimed in nitrogen carrier gas through a packed quartz tube at 1035°/760 Torr and the distribution of the ¹³C atoms in the recovered naphthalene was determined by mass spectral analysis of derived 1,4-naphthoquinone (experimental error ± 4%) and by ¹H and ¹³C NMR spectroscopy. The ratios of the labelled isotopomers obtained at different contact times are tabulated in Scheme 23. With increasing time there was increasing exchange between positions 1 and 2 but there was no exchange from the 1 or 2 carbons to the 8a (4a) position which remained unlabelled within experimental error during several half lives of 1,2 scrambling.

4.4 $C_{10}H_8$ Automerization and Isomerization of Azulene

The thermal conversion of azulene to naphthalene described in 1947[56,57] was the first example of isomerization of an aromatic hydrocarbon. The rearrangement was investigated by Alder *et al.* in the 1970s using substituents and labelling experiments.[65–67] Under static conditions (440°, 2–5.5 h) the rearrangement was shown to be catalysed by radicals. The reaction was examined by Wentrup, Zeller and coworkers in 1980[67] and automerization in azulene was established using ¹³C-labelling experiments. [4-¹³C]-Azulene (92% ¹³C) (**200**) on flash vacuum pyrolysis through a quartz tube at 1000°/10⁻³ Torr, gave only poor conversion and the ratio of azulene to naphthalene in the pyrolysate was 93.5:6.5. However, at 1180°, conversion was good and the ratio of azulene to naphthalene was 16.5:83.5. Examination of the recovered azulene by means of ¹³C NMR spectroscopy showed that automerization had occurred (Scheme 24). The label was distributed to positions 3 and 5 but not to 3a. The distribution of the label in the naphthalene obtained was determined by ¹³C NMR spectroscopy and is discussed below in relation to the mechanism of rearrangement.

Scott and Kirms[68] pyrolysed [1-¹³C]-azulene (90% ¹³C) (**201**) through an empty quartz tube in nitrogen at 760 Torr in the temperature range 700–900° with contact times of 0.3–3.1 min. At the higher temperatures or with longer contact times

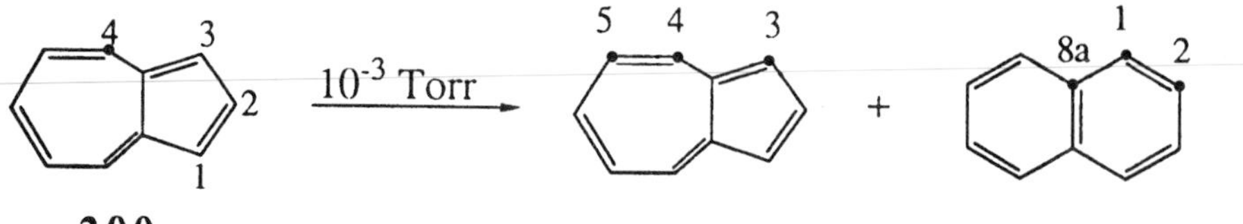

Temperature	C3(1), C4, C5	C1, C2, C8a (4a)
1000°	2.7 : 92 : 5.4	57 : 10 : 32
1150°		54 : 14 : 32
1180°		53 : 14.5 : 32

Scheme 24

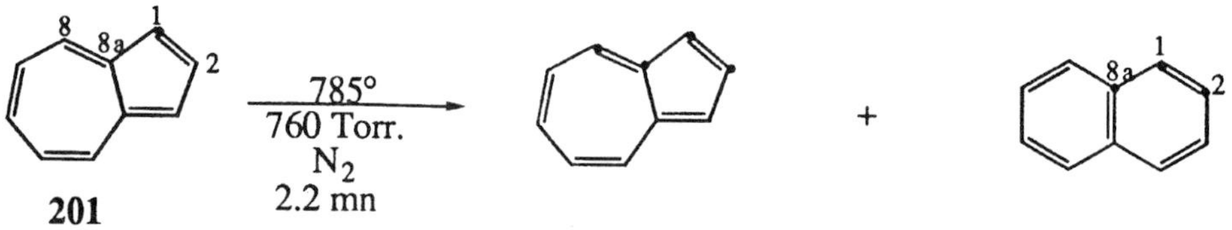

Position of ^{13}C	1, 2, 8a (3a), 8 (4)	1, 2, 8a (4a)
Relative conc.	95 : 2 : 2 : 1	53 : 39 : 6

Scheme 25

there was complete conversion into naphthalene while at lower temperatures some azulene was recovered. The distribution of labelling in azulene and naphthalene was determined with errors of ± 3% using ^{13}C NMR spectroscopy (Scheme 25).

At 785° and a contact time of 2.2 min the ratio of recovered azulene to naphthalene was 64:36 and there was about 5% automerization of the azulene distributed as shown in Scheme 25. When the [1-^{13}C]-azulene (18% ^{13}C) (201) was pyrolysed at the lower temperature of 725° with contact time 2.2 min, the naphthalene was shown to have the same distribution of label as shown in Scheme 25. At 900° with contact time of 0.3 min in unpacked and packed quartz tubes the distribution in naphthalene remained for positions 1, 2 and 8a as 53:39:6. Finally, with alternative labelling, [3a-^{13}C]-azulene (19% ^{13}C) (202) was pyrolysed at 835° with a contact time of 1.7 min to give naphthalene labelled as shown in Scheme 26.

Zeller[68a] has shown that gas phase thermolysis of 1-phenyl[1-^{13}C]azulene (203) at 400° gives 2-phenyl[2-^{13}C]azulene (204). In unsubstituted azulene there is very little automerization from the 1- to the 2- position (Scheme 25) but in the 1-phenyl substituted compound this is a major pathway. A bicyclo[1.1.0]butane intermediate was proposed.

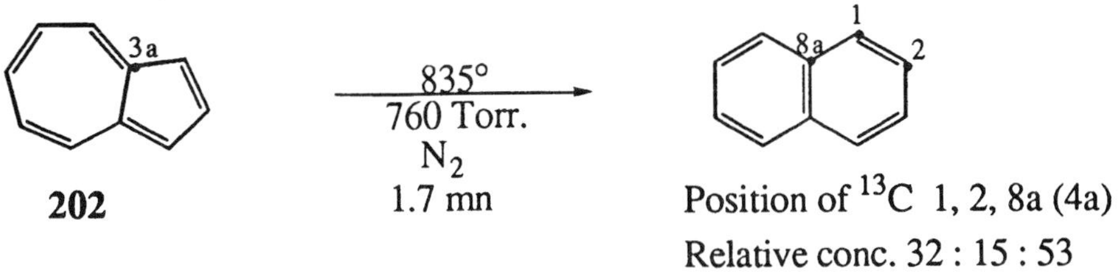

$$\xrightarrow[\substack{760 \text{ Torr.} \\ N_2 \\ 1.7 \text{ mn}}]{835°}$$

202

Position of ^{13}C 1, 2, 8a (4a)

Relative conc. 32 : 15 : 53

Scheme 26

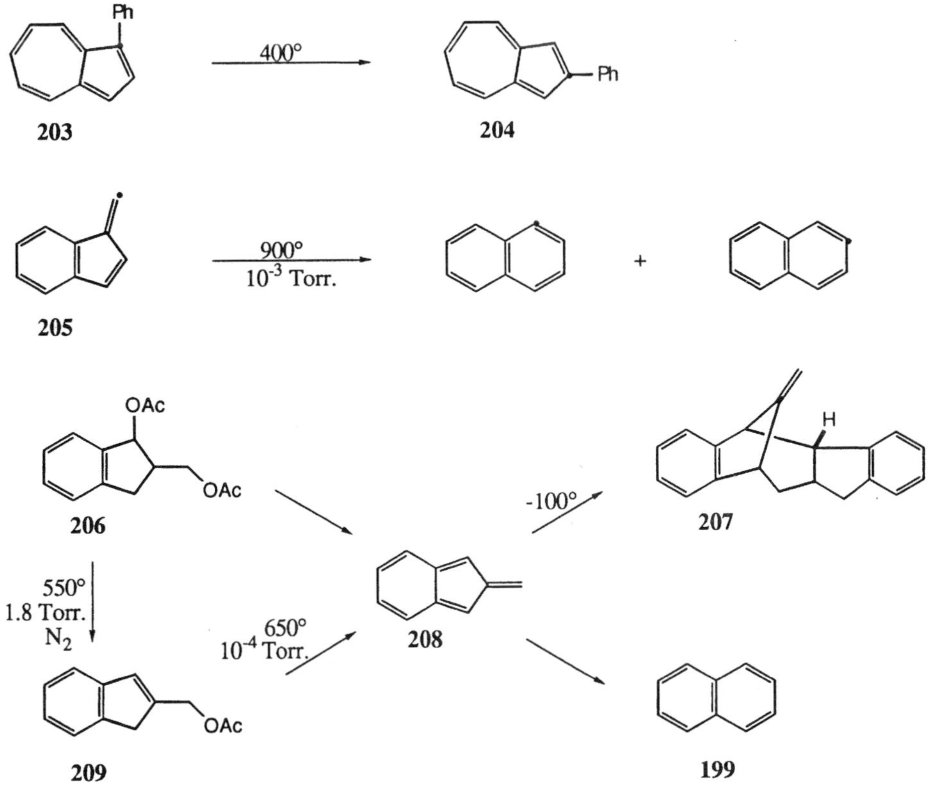

203 $\xrightarrow{400°}$ **204**

205 $\xrightarrow[10^{-3} \text{ Torr.}]{900°}$ +

206   -100°   **207**

$\substack{550° \\ 1.8 \text{ Torr.} \\ N_2}$ $\xrightarrow[10^{-4} \text{ Torr.}]{650°}$ **208**

209   **199**

Scheme 27

4.5 $C_{10}H_8$, Isomerization of Benzofulvene and Isobenzofulvene

Benzofulvene (unlabelled **205**) on FVP at $800°/10^{-2}$ Torr gave naphthalene as the exclusive product.[69] This reaction has been repeated and confirmed using 1-(methylene ^{13}C)-1H-indene (benzofulvene) (99% ^{13}C) (**205**).[70] Pyrolysis of compound (**205**) at $900°/10^{-3}$ Torr gave [1-^{13}C]-naphthalene and [2-^{13}C]-naphthalene in a ratio of 21:79. The pyrolysis of benzobenzvalene in the gas phase gives naphthalene but in solution at 150° it forms benzofulvene.[70a] The diacetate (206) on pyrolysis in nitrogen carrier gas at 550°/1.8 Torr gave the monoacetate (209) which on FVP gave isobenzofulvene (208).[71] Pyrolysis of (206) or (209) at 650°/10⁻⁴ Torr gave isobenzofulvene (208) as a blue hydrocarbon collected at −196°. It was identified by variable temperature photoelectron spectroscopy. When warmed to −100° the blue isobenzofulvene formed the colourless dimer (207). Pyrolysis at 700° of compounds (206) or (209) gave naphthalene (199), isobenzofulvene (208) and dimers. At 800° the isobenzofulvene (206) was converted completely to naphthalene (199) (Scheme 27).

4.6 $C_{12}H_8$, Automerization of Acenaphthylene

Based on the hypothesis that in condensed aromatic hydrocarbons there would be simultaneous expansion of a five-membered ring and contraction of a six-membered ring, it was predicted by Scott and Roelofs[72] that a dramatic shift of the labelled atom would occur on automerization of acenaphthylene.

[1-^{13}C]-Acenaphthylene (99% ^{13}C) (210) was pyrolysed through a base-washed quartz tube in nitrogen gas at 990°/760 Torr with a contact time of 2 sec. The recovered acenaphthylene was analysed by ^{13}C NMR spectroscopy (Scheme 28). The results showed that the [1-^{13}C]-acenaphthylene was converted by contraction of one 6-membered ring into [4-^{13}C]-acenaphthylene and of the other into [5-^{13}C]-acenaphthylene. At the higher temperature of 1070°, as predicted, [3-^{13}C]-acenaphthylene was formed by further rearrangement of the [5-^{13}C]-acenaphthylene. The rearrangements involve the sequence of steps 1→4, 5; 4→1, 4; 5→1, 3; 3→5, 3a; 3a→3, 5a, where the numbers define the position of labelling. On repetitive rearrangement this leads to distribution of the ^{13}C label around the periphery of the molecule.

4.7 $C_{14}H_8$ Automerization of Paracyclene (Cyclopent[gf]acenaphthylene)

Proposals concerning automerization of buckminsterfullerene, C_{60}, have included the interconversion on the truncated icosahedral surface of the paracyclene unit as shown for the paracyclene molecule (211). This was first proposed by Stone and Wales[73] and the idea was elaborated in considering the number of possible isomers for the fullerene, C_{84}.[74] Most recently Scott[75,1e] reported that ^{13}C labelled paracyclene does not automerize even at very high temperatures and this resistance to change has led to consideration of the relationship between bond order and the ability of that bond to participate in rearrangement.

4.8 $C_{16}H_{10}$, Automerization of Pyrene

Pyrene was investigated by Scott and coworkers[76] as an adjunct to the study of naphthalene automerization. [1-^{13}C]-Pyrene (90% ^{13}C) (212) was pyrolysed through quartz at 1100° in nitrogen carrier gas with a contact time of 2 sec. The

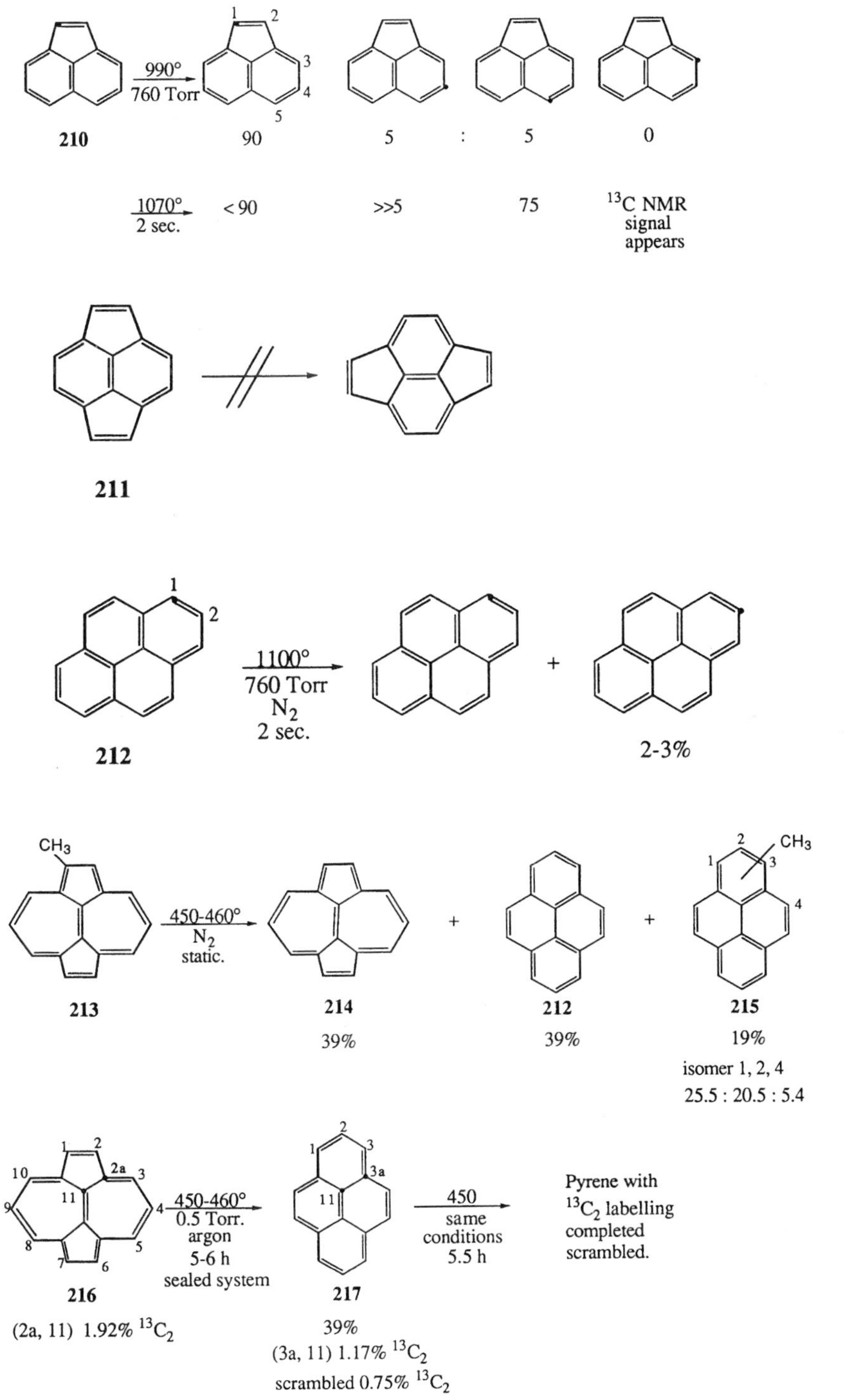
210
990°
760 Torr
1
2
3
4
5
90
5 : 5
0
1070°
2 sec.
< 90
>>5
75
^{13}C NMR signal appears
211
212
1
2
1100°
760 Torr
N_2
2 sec.
+
2-3%
CH_3
450-460°
N_2
static.
+
+
213
214
39%
212
39%
215
19%
isomer 1, 2, 4
25.5 : 20.5 : 5.4
CH_3
1
2
3
4
1
2
2a
3
4
5
6
7
8
9
10
11
450-460°
0.5 Torr.
argon
5-6 h
sealed system
216
(2a, 11) 1.92% $^{13}C_2$
1
2
3
3a
11
217
39%
(3a, 11) 1.17% $^{13}C_2$
scrambled 0.75% $^{13}C_2$
450
same conditions
5.5 h
Pyrene with $^{13}C_2$ labelling completed scrambled.

Scheme 28

recovered pyrene was purified by sublimation and investigated by ^{13}C NMR spectroscopy. The presence of [2-^{13}C]-pyrene was confirmed, but the conversion observed under these conditions was only 2–3%.

4.9 $C_{16}H_{10}$, Isomerization of Dicyclopenta[ef, ke]heptalene, Azupyrene, to Pyrene

This isomerization was studied using a sealed system but the findings are given here for completeness. Rearrangement of the azupyrene nucleus was first studied by Anderson et al.[77] using 1-methylazupyrene (213). Thermal isomerization (450–460°/N$_2$, static system) gave azupyrene (214) (39%) pyrene (212) (39%), a mixture of 1-, 2- and 4-methylpyrene (215) (19%) in a ratio of 25.5:20.5:5.4 and unchanged (213) (3%) (Scheme 28).

   No single mechanism could account for the formation of the three methyl-pyrene isomers and a more detailed study was undertaken by Anderson and Haddock[78] using [2a, 11-$^{13}C_2$]dicyclopenta[ef,kl]heptalene (216). The doubly labelled azupyrene was diluted to 1.915% $^{13}C_2$ and heated in a sealed quartz tube under 0.5 Torr pressure of argon at 450–460° for 5–6 h. Examination of the resultant pyrene (217) (45% yield) by ^{13}C NMR spectroscopy showed that of the 1.92% double labelling present, 1.17% was in the adjacent 3a and 11 positions of the pyrene and the remaining 0.75% was distributed equally around the peri-phery of the pyrene molecule (Scheme 28). The results show that in the major pathway of isomerization, the bond between the ring-junction and central carbon atom was not broken. The recovered pyrene (217) underwent automerization under the same conditions (450°/5.5 h) to give pyrene in which complete scrambling had occurred.

4.10 $C_{16}H_{10}$ Thermal Interconversion of Aceanthrylene, Acephenanthrylene and Fluoranthene

The isomerization of these three hydrocarbons by successive rearrangement of fused six- and five-membered rings to form five- and six-membered rings was established by Scott and Roelofs in 1987.[41] Pyrolysis of aceanthrylene (218), through a packed quartz tube at 1100° with nitrogen carrier gas at atmospheric pressure and a contact time of 2 sec gave a mixture of aceanthrylene (218) acephenanthrylene (217) and fluoranthene (39) in the ratio 40:14:46. By an iterative procedure a mixture of these three compounds in the ratio 17:9:74 was established as being unchanging on pyrolysis. Pyrolysis of any of the three isomers generated mixtures which approached this composition but at different rates. From these data a qualitative energy relationship could be established. In terms of thermodynamic stability, the order was fluoranthene > aceanthrylene > acephenanthrylene. From the ease of approach to the equilibrium mixture, the transition state energy for the acenanthrylene $\rightleftarrows$ acephenanthrylene transfor-mation was judged to be greater than that for the acephenanthrylene $\rightleftarrows$ fluoranthene interconversion.

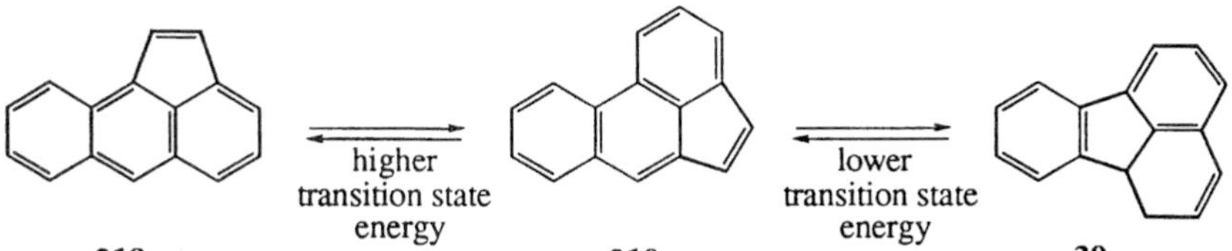
higher
transition state
energy
lower
transition state
energy

218 **219** **39**

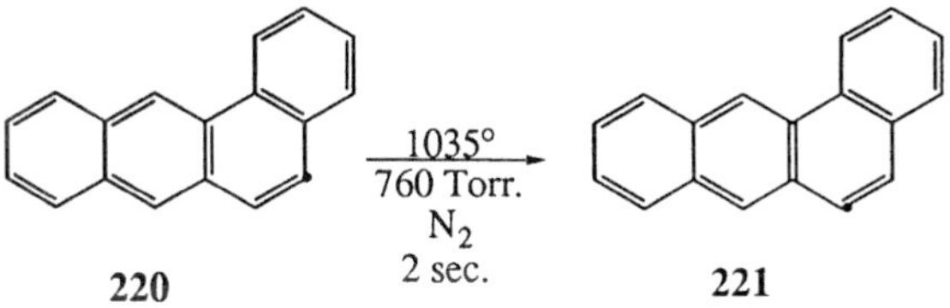
1035°
760 Torr.
N_2
2 sec.

220 **221**

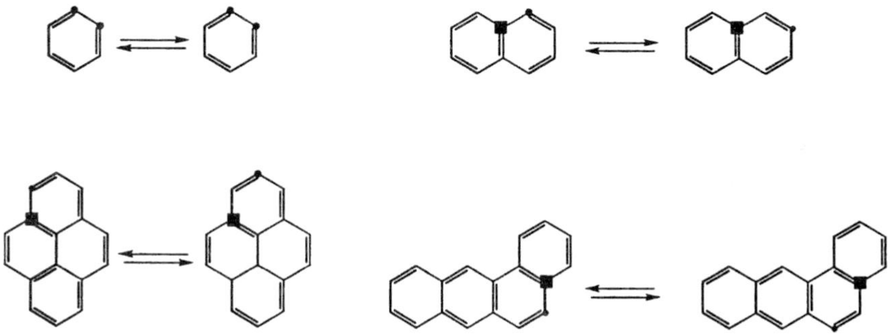

Scheme 29

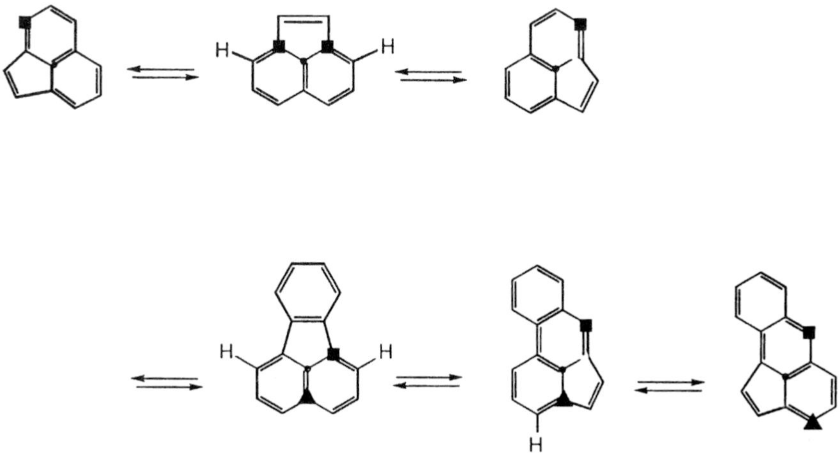
H
H
H
H
H

Scheme 30

4.11 $C_{18}H_{12}$ Automerization of Benzo[a]anthracene

In further studies related to naphthalene automerization, Scott and coworkers[79] pyrolysed [5-^{13}C]-benzo[a]anthracene (90% ^{13}C) (**220**) through quartz at 1035° in nitrogen carrier gas at atmospheric pressure with a contact time of 2 sec. The compound did not undergo skeletal rearrangement to yield isomeric compounds and the benzo[a]anthracene was recovered and examined by ^{13}C NMR spectroscopy. The compound automerized exclusively into [6-^{13}C]-benzo[a]anthracene (**221**), the 5- and 6-isotopomers being obtained in a ratio of 60:40.

4.12 The Mechanism of Automerization and Isomerization of Aromatic Hydrocarbons

It is apparent from the preceding examples (summarised in Scheme 29) that in benzene adjacent atoms change places with one another. In polycyclic aromatics the ring junction carbon atom changes its partner but does not leave its quaternary position. Where there is a cyclopenteno ring fused to the junction of two six-membered rings (acenaphthylene, fluoranthene etc.), the carbon atom at the junction of the two six-membered rings bonds to the adjacent carbon atom within a six membered ring and by so doing reverses the sizes of the adjoining rings. In all cases the carbon atom newly bonded to the ring junction was originally bonded to a hydrogen atom (Scheme 30). The lability of this CH bond must be a contributing factor in these rearrangements. The difficulty in determining the mechanism of these reactions is to define the skeletal changes and the role of the hydrogen atoms during those changes. In the conversion of azulene to naphthalene in static systems the reaction is known to be catalysed by radicals and mechanisms involving addition and loss of a hydrogen atom have been proposed. Such possibilities cannot be ruled out completely from flow and flash pyrolytic reactions but in general mechanisms based on molecular rearrangements have been sought.

Mechanistic ideas in pyrolysis are firmly based on transition state theory. The problem is to determine which mechanism provides the lowest energy pathway. If more than one mechanism is involved, labelling experiments may produce complex labelling patterns. This is the case with the azulene to naphthalene rearrangement but in other cases the findings suggest that specific mechanisms are the rule rather than the exception.

Seven mechanisms which do not involve any alteration in the atomic composition have been proposed for the changes in connectivity observed in hydrocarbon rearrangements. Two mechanisms have been proposed involving the addition and loss of hydrogen atoms to account for radical catalysis. Generally, intermediates in gas phase reactions have been considered to be neutral but the possibility of protonation, deprotonation or other forms of catalysis on the surface of the apparatus cannot be ignored.

The nine mechanisms that have been considered are illustrated in Scheme 31 using naphthalene as the basic molecule. These are all applicable to the rearrangement of polycyclic aromatics but not necessarily to benzene.[80] The first three mechanisms, the bicyclobutane, norcaradiene–vinylidene and benzvalene involve valence isomerism as the first step. The dyotropic mechanism involves simultaneous reversal of the bonding of two atoms attached at either end of a bond. Two further mechanisms, the hydrogen shift and the 1,2-carbon shift, lead to carbenes, which may undergo Wolff rearrangement or a hydrogen shift, respectively, to form a stable intermediate. A diradical intermediate formed by homolysis of the central bond (4a–8a) of naphthalene, which can be considered as an allenic annulene, has been proposed.[68] Finally, the spiran and the methylene walk mechanisms require addition of a hydrogen atom to the ring junction or the α-carbon respectively. These last two were proposed to account for the radical catalysis displayed in the azulene to naphthalene rearrangement.[64–66]

Merz and Scott have examined four mechanisms applicable to benzene, the 1,2-carbon shift, dyotropic, benzvalene and 1,2-hydrogen shift mechanisms (Scheme 32).[59] Results were reported of *ab-initio* calculations based on 6-31G* and 6-31G** basis sets with electron correlation, optimized geometries and thermodynamic corrections (at 1383 K, the pyrolysis temperature). For the 1,2-carbon shift mechanism no transition state was found between benzene and the carbene intermediate (**222**) which could itself collapse to fulvene (**223**) or benzvalene (**224**). In the dyotropic mechanism the free energy of activation ($\Delta_a G$) to form the transition state leading to fulvene was calculated as 132.7 kcal/mol and for the conversion of benzene to benzvalene it was 103.8 kcal/mol. In the 1,2-hydrogen shift mechanism the free energy of activation was estimated to require 85.8 kcal/mol to form the intermediate carbene (**225**).

The rate controlling step in this mechanism was the ring contraction of the carbene to fulvene with a free energy of activation of 100.5 kcal/mol. It was concluded that the activation energy for the 1,2-hydrogen shift mechanism was slightly less than that for benzvalene and both of these were significantly lower than that for the dyotropic shift. A similar conclusion had been arrived at for naphthalene, that is the hydrogen shift mechanism ((5) Scheme 31) was concluded to be the lowest energy process for automerization.[70]

The azulene (**226**) to naphthalene (**228**) rearrangement is an example of a reaction which involves more than one pathway. If the reaction is written as involving a norcaradiene intermediate (**227**) then this can be converted by a hydrogen shift to naphthalene (**222**) or it can be disconnected to the vinylidene (**231**). This was the basis of an elegant experiment by Wentrup, Zeller and coworkers[67] who formed the vinylidene intermediates (**230**) and (**231**) by two methods. FVP of the cinnamylideneisoxazolone (**229**) at 680–750°/10⁻¹–10⁻³ Torr to form the vinylidene (**230**) gave a mixture of *E*- (**232**) and *Z*-1-phenylbuten-3-ynes (**233**), naphthalene (**228**) and azulene (**226**) (27:15:36:1). The mixture of *E*- and *Z*-acetylenes was isolated and on FVP at 950–1000°/10⁻² Torr naphthalene and azulene were formed. In addition, FVP of azulene at 1100°/10⁻² Torr gave naphthalene as a major product together with small amounts of (**232**) and (**233**) (3% yield) in a ratio of 70:30. The relationship between these intermediates is shown in Scheme 33. Thus there is clear evidence for the norcaradiene–vinylidene rearrangement.

A radical mechanism is also known to be involved. However, there is at least one further major pathway and this may involve a 1,2-hydrogen shift mechanism and the intermediacy of benzofulvene but this has not been established with certainty.

A. Mechanism involving formation of valence isomers

1)   *bicyclobutane mechanism*

2)

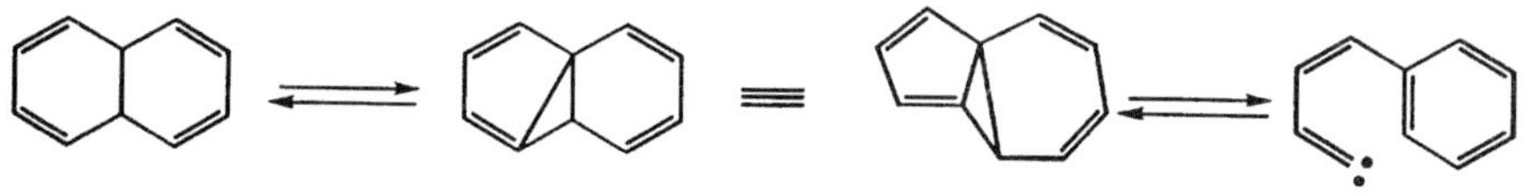

norcaradiene - vinylidene mechanism

3)   *benzvalene mechanism*

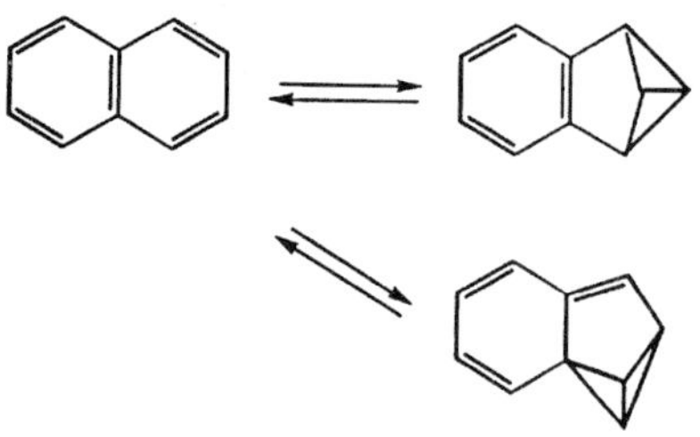

4)   *dyotropic mechanism*

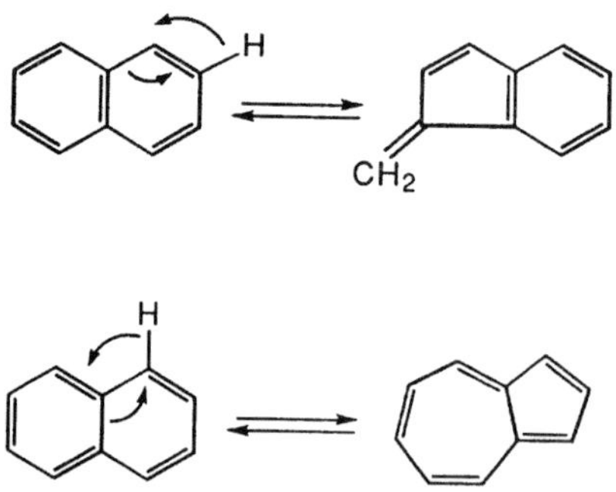

CH₂

B. Mechanisms involving carbene formation

5)

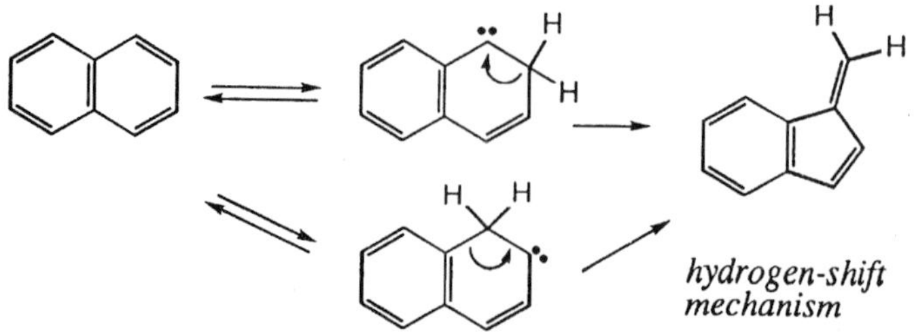

hydrogen-shift mechanism

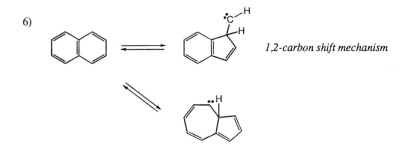

6) *1,2-carbon shift mechanism*

C. Mechanism involving diradicals

7)

D. Mechanism involving addition of a hydrogen atom

8) *spiran mechanism*

D. Mechanism involving addition of a hydrogen atom

9) *methylene walk mechanism*

Scheme 31

1,2-Carbon-shift mechanism

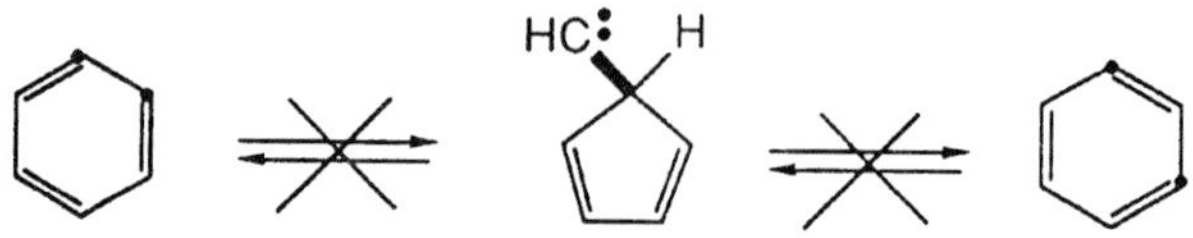

222

Dyotropic mechanism

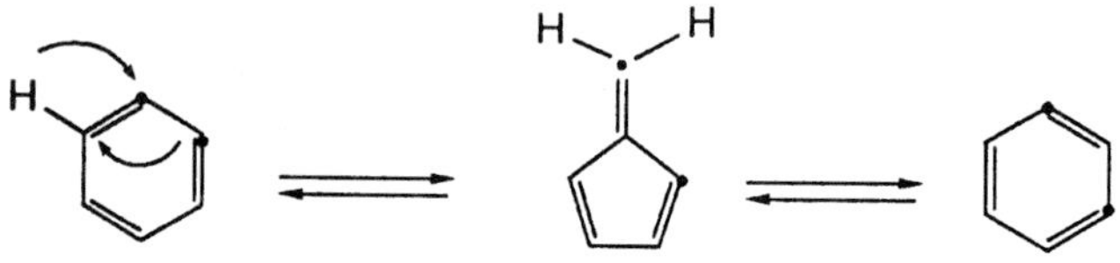

223

Benzvalene mechanism

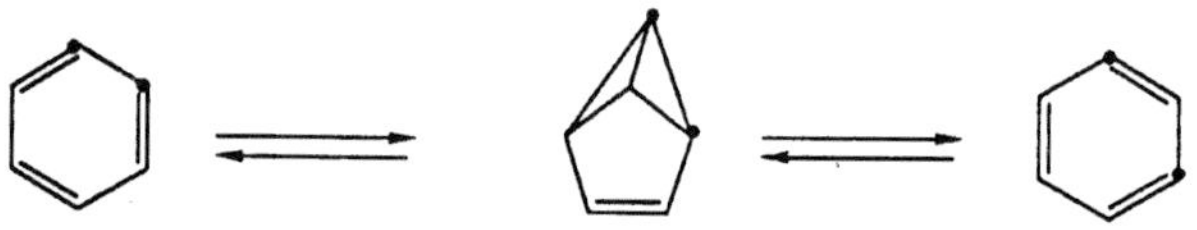

224

1,2-Hydrogen-Shift mechanism

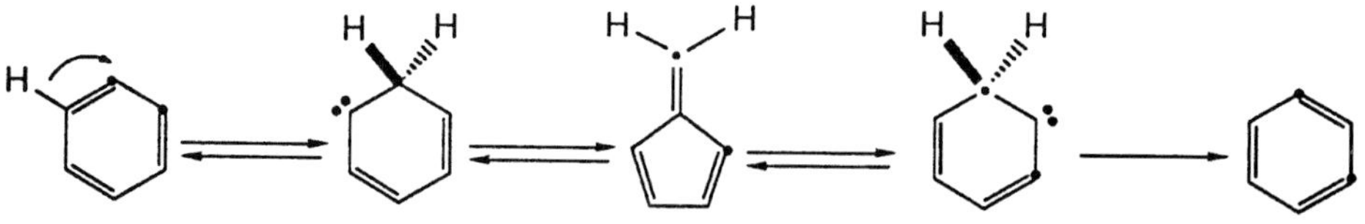

225

Scheme 32

5.   THERMAL REARRANGEMENT OF C_6H_6 HYDROCARBONS

Zimmermann, Hopf and coworkers[82] reported the gas-phase pyrolysis of 2-ethynyl-1,3-butadiene (**235**) which was known to be formed pyrolytically from hexa-1,2-dien-5-yne (**234**).[84,1c] They showed that (**235**) was also formed by loss of ethylene on pyrolysis of 1-ethynylcyclohexene (**236**) (625–700°, N_2, 760 Torr). The principal product from pyrolysis of (**235**) (500–600°, N_2, 760 Torr) was benzene, and fulvene (**238**) was the only other isomeric product. Numerous other

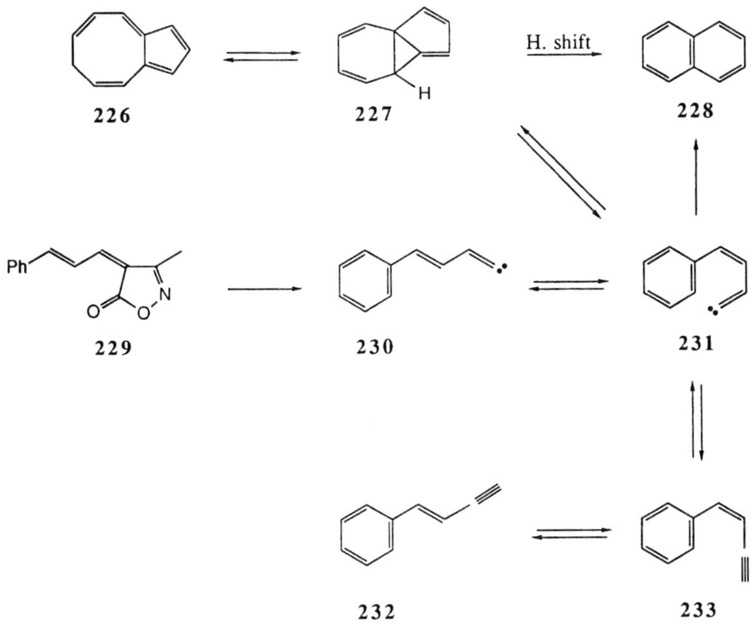

Scheme 33

compounds were formed. Deuterium labelling of the ethynyl group gave fulvene
in which the label was scrambled through the intermediacy of benzvalene (**237**)
or vinylcyclobutadiene (**239**).

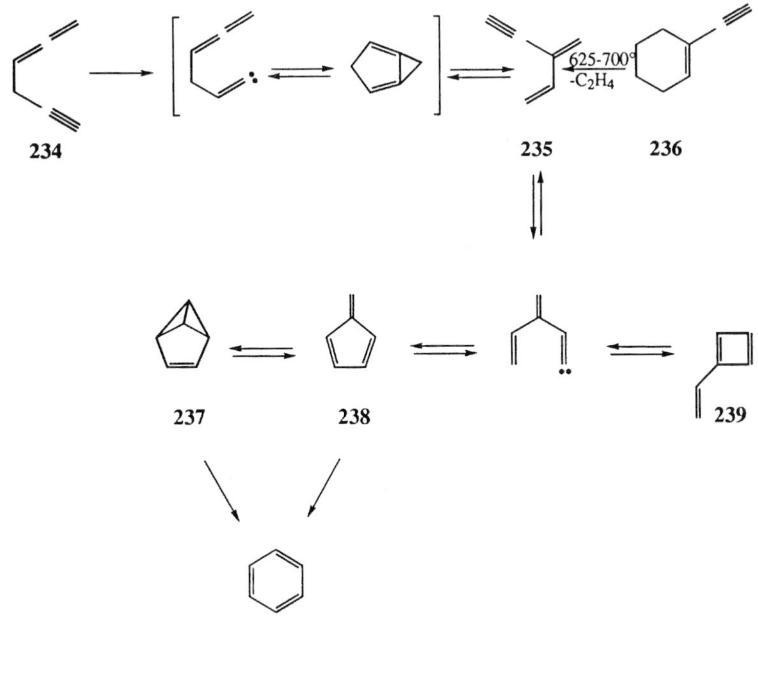

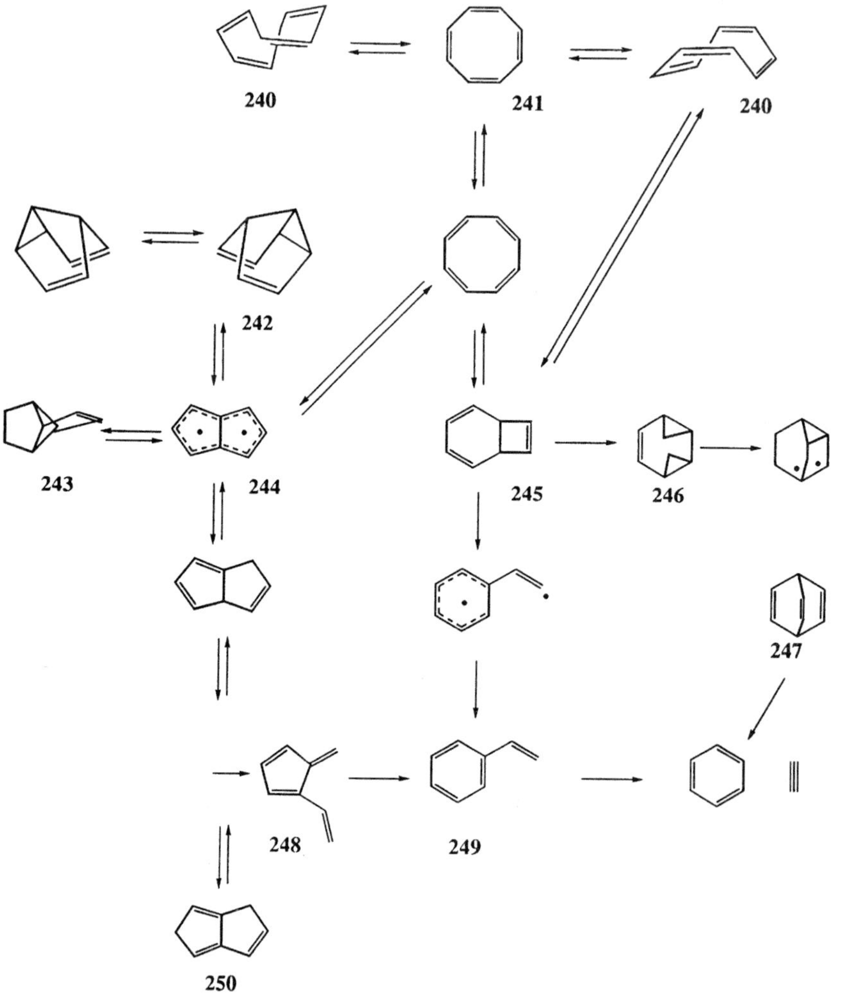

Scheme 34

6. THERMAL REARRANGEMENT OF C_8H_8 HYDROCARBONS

A detailed discussion of the thermal rearrangement of $(CH)_8$ hydrocarbons was published by Hassunrück, Martin and Walsh in 1989,[85] building on the earlier compilation of material by Gajewski,[1b] but generally omitting other C_8H_8 isomers. Only the principal interconversions of the $(CH)_8$ series are summarised here together with more recent material. Of the 17 possible constitutional isomers and 4-stereoisomers having the composition $(CH)_8$, 13 have been synthesised and another, (**246**), is strongly implicated as a reactive intermediate.[85] The core of this chemistry is the thermal rearrangement of 1,3,5,7-cyclooctatetraene (**240**) summarised in Scheme 34.

At temperatures below 100° cyclooctatetraene (240) is in rapid equilibrium with bicyclo[4,2,0]-octa-2,4,7-triene (245). It also undergoes ring inversion through a planar form (241) estimated to be 12–15 kcal/mol greater in energy than the puckered system (240). Bond switching through a planar intermediate also takes place. Between 100° and 400° vibrationally excited cyclooctatetraene undergoes valence isomerization most probably through the diradical (244) to semibulvalene (242). A degenerate route to a carbon scrambling process involving the intermediate (243) also comes into play and these new species lead to further isomeric compounds. At temperatures between 450° and 750° fragmentation occurs principally to give acetylene, benzene, 1,5-dihydropentalene (250) and styrene (249). Two routes are proposed for formation of styrene and an alternative route to benzene and acetylene involves barrelene (247).

This abbreviated interrelationship is shown in Scheme 34. The possibility that acetylene is lost in the form of vinylidene from styrene (249) and barrelene (247) is discussed in Section 2.1.[85a]

Griesbeck[86] has described a simple synthesis of cyclopentadiene-protected 6-vinylfulvene (251). On FVP at 450–520°/0.02 Torr this gave 6-vinylfulvene (propenylidenecyclopentadiene) (252) which cyclised to 1,2-dihydropentalene (253) and rearranged to 1,5-dihydropentalene (250). This intramolecular reaction had previously been established by Gajewski and Cavender.[87]

A more detailed study of the dihydropentalenes has been described by Meier and coworkers.[88,89] The enthalpies of formation of the six isomeric dihydropentalenes (250, 253–257) have been calculated using the MNDO method and the four most stable isomers (250) (253) (254) and (255) have been prepared by FVP of cyclooctatetraene under different conditions[88] (Scheme 35). The 1,5- (250) and 1,2-dihydropentalenes (253) have been shown to be interconvertible by the catalytic action of alumina. FVP of 1,2-dihydrofulvene (253) 700–740°/0.7 × 10⁻² Torr gave 1-vinylfulvene (248) which is irreversibly converted to styrene at higher temperatures.[89]

Four cyclooctadienynes, C_8H_8, have been investigated by Meier and coworkers. These are cycloocta-1,5-dien-3-yne (259),[90] -1,3-dien-5-yne (260),[91] -1,6-dien-3-yne (261)[92] and -1,3-dien-6-yne (262).[93] The first of these (259) was isolable but unstable and at room temperature was transformed quantitatively into cyclooctatetraene (240). Compound (262) was a colourless oil with a half life of 5 h at room temperature. Compound (260) could only be obtained in solution.

251   252   253   250

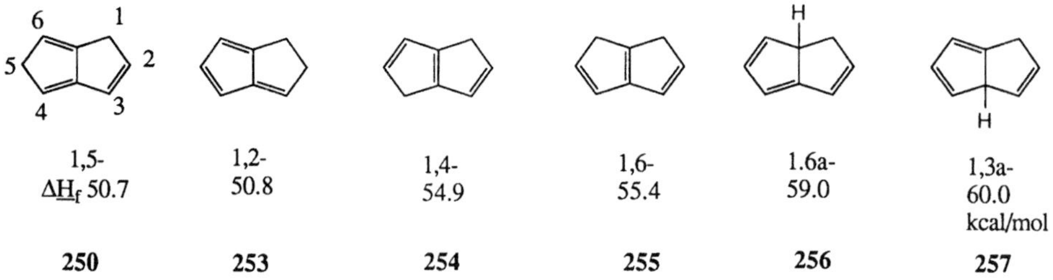
6
1
5
2
4
3
H
H
1,5-
ΔHf 50.7
1,2-
50.8
1,4-
54.9
1,6-
55.4
1.6a-
59.0
1,3a-
60.0
kcal/mol
250
253
254
255
256
257

Scheme 35

Attempted condensation of (**260**) in a cold trap led to recovery almost exclusively of benzocyclobutene (**258**). Compound (**261**) also could only be obtained in solution and at room temperature it isomerized rapidly to benzocyclobutene (**258**) and styrene (**249**) in a ratio of 22:78.

FVP of compounds (**260**), (**261**) and (**262**) at 450–640°/0.3 Torr gave 1,2-dihydropentalene (**253**), benzocyclobutene (**258**), and styrene (**249**) in the ratios (0:9:91) (60:3:37) and (0:0:100). Thus (**260**) and (**262**) gave principally styrene (**249**) and (**261**) gave 1,2-dihydropentalene as the major product (Scheme 36).

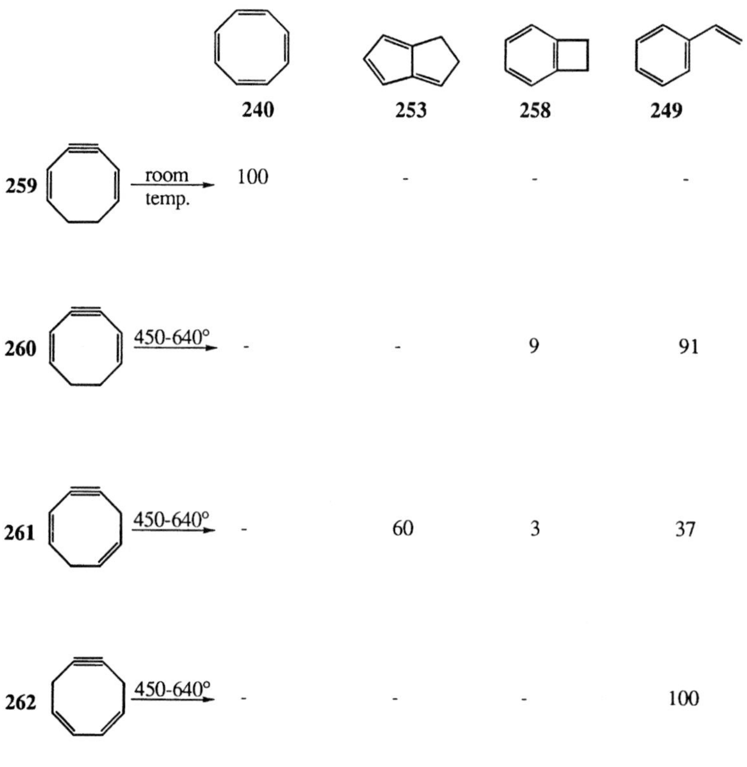
240
253
258
249
259
room
temp.
100
-
-
-
260
450-640°
-
-
9
91
261
450-640°
-
60
3
37
262
450-640°
-
-
-
100

Scheme 36

7. ISOMERIZATION OF BENZOBICYCLO[3.1.0]HEX-2-ENE, $C_{10}H_{10}$, AND PYROLYSIS OF DERIVATES OF NENITZESCUS HYDROCARBON (CH)$_{10}$

While not directed towards synthesis, the work of Lamberts and Laarhoven[94] showed that the direction of radical reactions could to some degree be controlled by adding phenyl substituents to allow formation of benzylic radical intermediates. They studied the pyrolytic behaviour of benzobicyclo[3.1.0]hex-2-ene (263), the 1-phenyl (264), 4-phenyl (267) and 5-phenyl (265) derivatives and the 6-phenyl as a mixture of isomers (266). In Scheme 37 the pyrolyses of two of these substrates are illustrated with the products shown in order of decreasing rates of formation starting at the temperature shown. It is evident from the products formed from (263) and (267) that the lowest energy pathways involved formation of o-divinyl compounds. At higher temperatures, 600–750°, dihydronaphthalenes became the principal products and at higher temperatures still these were dehydrogenated to naphthalenes. The products from pyrolysis of 4-phenylbenzobicyclo[3.1.0]hex-2-ene (267) had undergone a skeletal rearrangement leading to separation of the aromatic rings by two carbon atoms. The homolytic cleavage of the C1–C5 bond originally proposed did not provide a clear mechanism for formation of the divinyl or rearranged compounds. McNab[95] proposed that if this diradical underwent a neophyl-like rearrangement, it would lead to the divinyl compounds early in the rearrangement process and the phenyl shifts would be a direct consequence of this mechanism (Scheme 38). The initial product can cyclise to dihydronaphthalene derivatives under the reaction conditions.

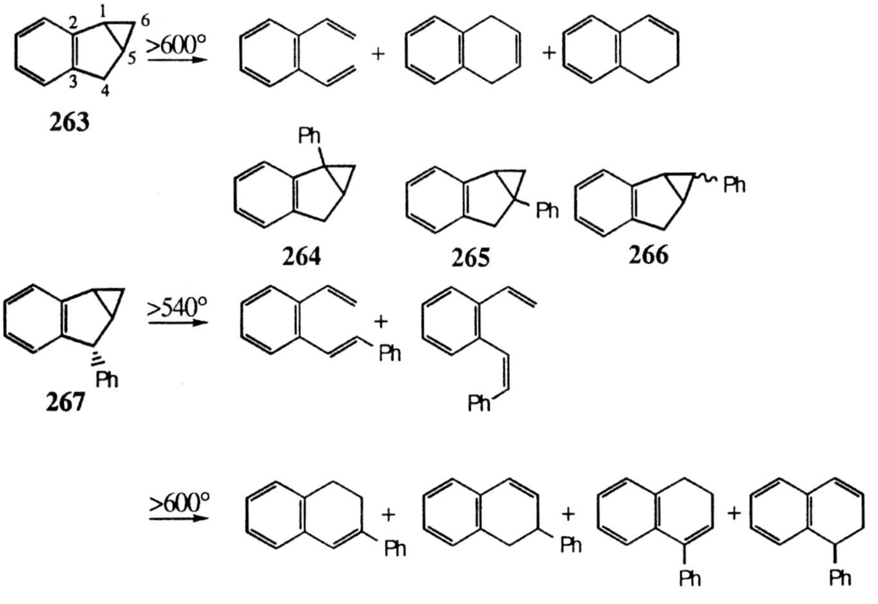

Scheme 37

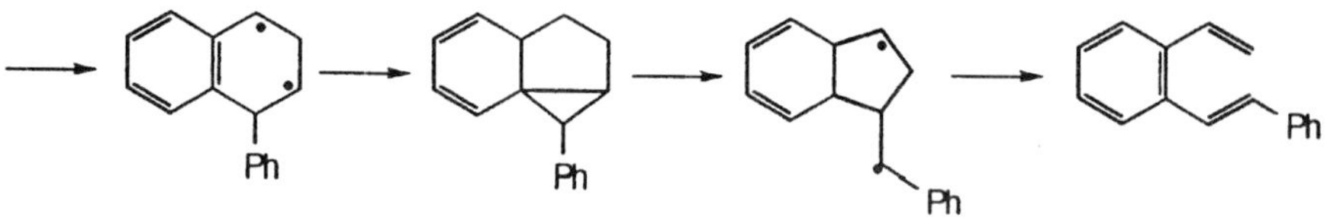

Scheme 38

A Bucharest group has described the pyrolysis of a dibenzo and a tribenzo derivative of Nenitzescu's hydrocarbon, tricyclo[4,2,2,0^{2,5}]deca-3,7,9-triene (**268**), (CH)$_{10}$.[96] The earlier studies have been reviewed.[97] Flow vacuum pyrolysis of the dibenzo[c,q] derivative (**269**) (argon, 1 Torr, 350–650°) gave naphthalene (**270**), the angular dimer of benzocyclobutene, Cava's dimer (**271**), benzo[a]biphenylene (**272**), dibenzo[a,e]cyclooctatetraene (**273**) and small amounts of benzo[a]anthracene and chrysene. The principal mode of cleavage involved cycloreversion to naphthalene and benzocyclobutene. Loss of acetylene and valence isomerism led to (**273**) and the remaining products probably arose from radical cleavage. Pyrolysis of the tribenzo[c,g,j] compound (**274**) gave anthracene (**275**) and Cava's dimer (**271**) by cycloreversion.

8.   ISOMERIZATION OF PAGODENE TO DODECAHEDRANE C$_{20}$H$_{20}$

Prinzbach and coworkers[98a] using MM2 calculations showed that conversion of pagodene (**276**), C$_{20}$H$_{20}$, to dodecahedrane (**279**) would involve a considerable reduction in enthalpy and strain energy. Thermolysis of pagodene (**276**) in a vertical pyrolysis tube at 700–750° gave naphthalene (**270**) as the only isolable product. This reaction was considered to take place by a radical process leading to the diene (**277**) which cleaved to the known triene (**278**) and this underwent loss of hydrogen to be transformed, possibly via azulene, into naphthalene (**270**). More significantly, pyrolysis in H$_2$/He carrier gas over a Pt/Al$_2$O$_3$ catalyst at 310° yielded dodecahedrane in 8% yield as one of six products (Scheme 39).[98b]

9.   ISOMERIZATIONS INVOLVING DIRADICAL INTERMEDIATES IN SOME COMPOUNDS C$_n$H$_{n+(1-6)}$.

9.1   Intramolecular Radical Coupling

The work of Meier and Schmidt[99] on cyclooct-1-en-5-yne (**280**), C$_8$H$_{10}$, contrasts with the Meier group's findings on C$_8$H$_8$ compounds (Section 6). They reported the ring contraction of the enyne (**280**) at 560°/0.04 Torr to yield 4-ethenylidene-cyclohex-1-ene (**282**) in 57% yield. The mechanism proposed involved homolytic cleavage of the C3–C4 bond in (**280**) to form an intermediate (**281**) containing two stabilised radicals which recombined to form the exocyclic allene (**282**).

Introduction of an oxiran in a 1,3-relationship to the triple bond allowed similar radical cleavage. FVP of the epoxide (**283**) at 540°/0.04 Torr gave a

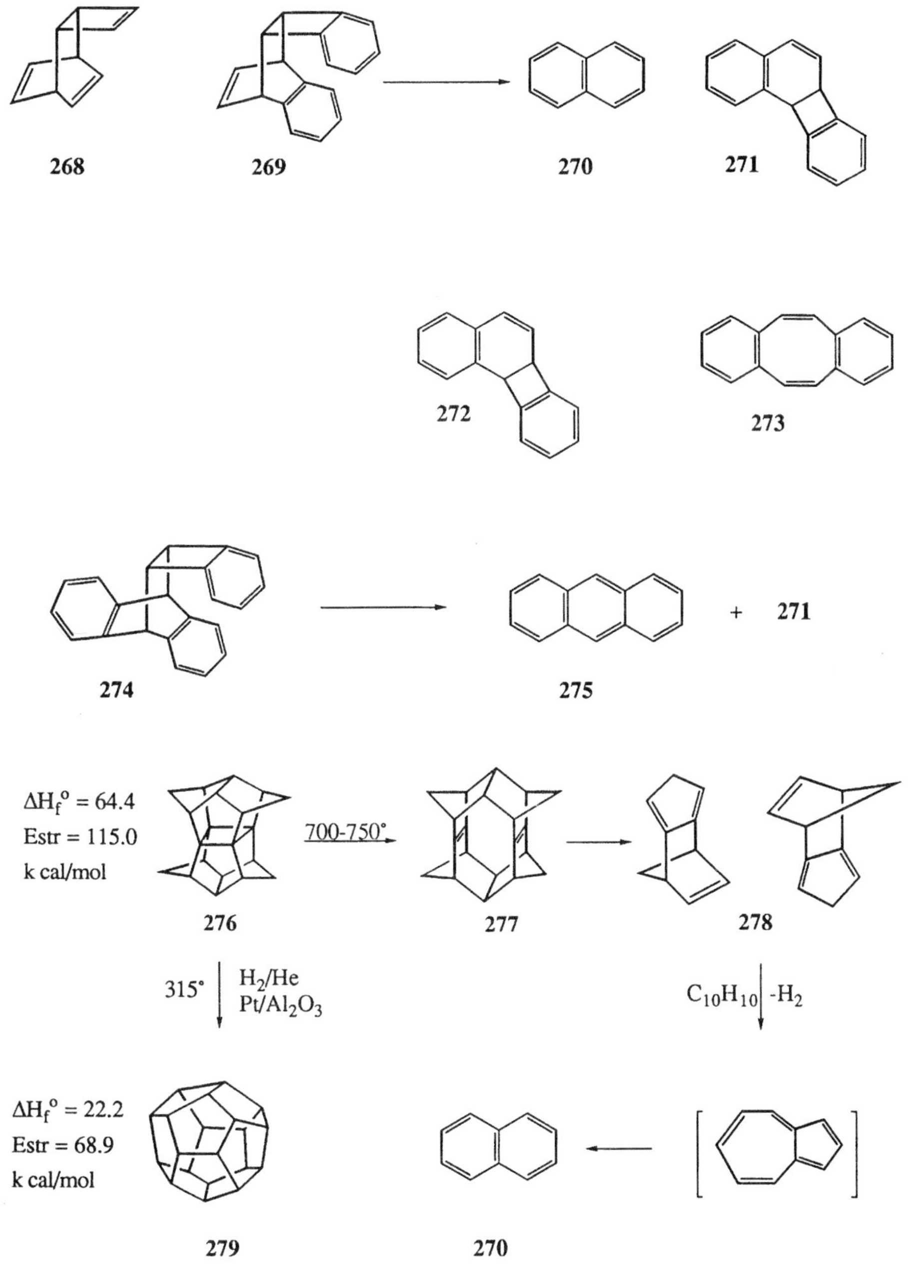

268   269   270   271

272   273

274   275   +   271

$\Delta H_f^\circ = 64.4$
Estr = 115.0
k cal/mol

700-750°

276   277   278

315° | H_2/He
         Pt/Al$_2$O$_3$

$C_{10}H_{10}$ | -H_2

$\Delta H_f^\circ = 22.2$
Estr = 68.9
k cal/mol

279   270

Scheme 39

diradical (**284**) in which the oxiranyl radical underwent a hydrogen shift and decarbonylation to give a terminal alkyl radical (**285**) which on recombination formed ethenylidenecyclopentane (**286**) and, by a [1,3-H] shift, 1-vinylcyclopentene (**287**) in a ratio of 38:62 and an overall yield of 72%.

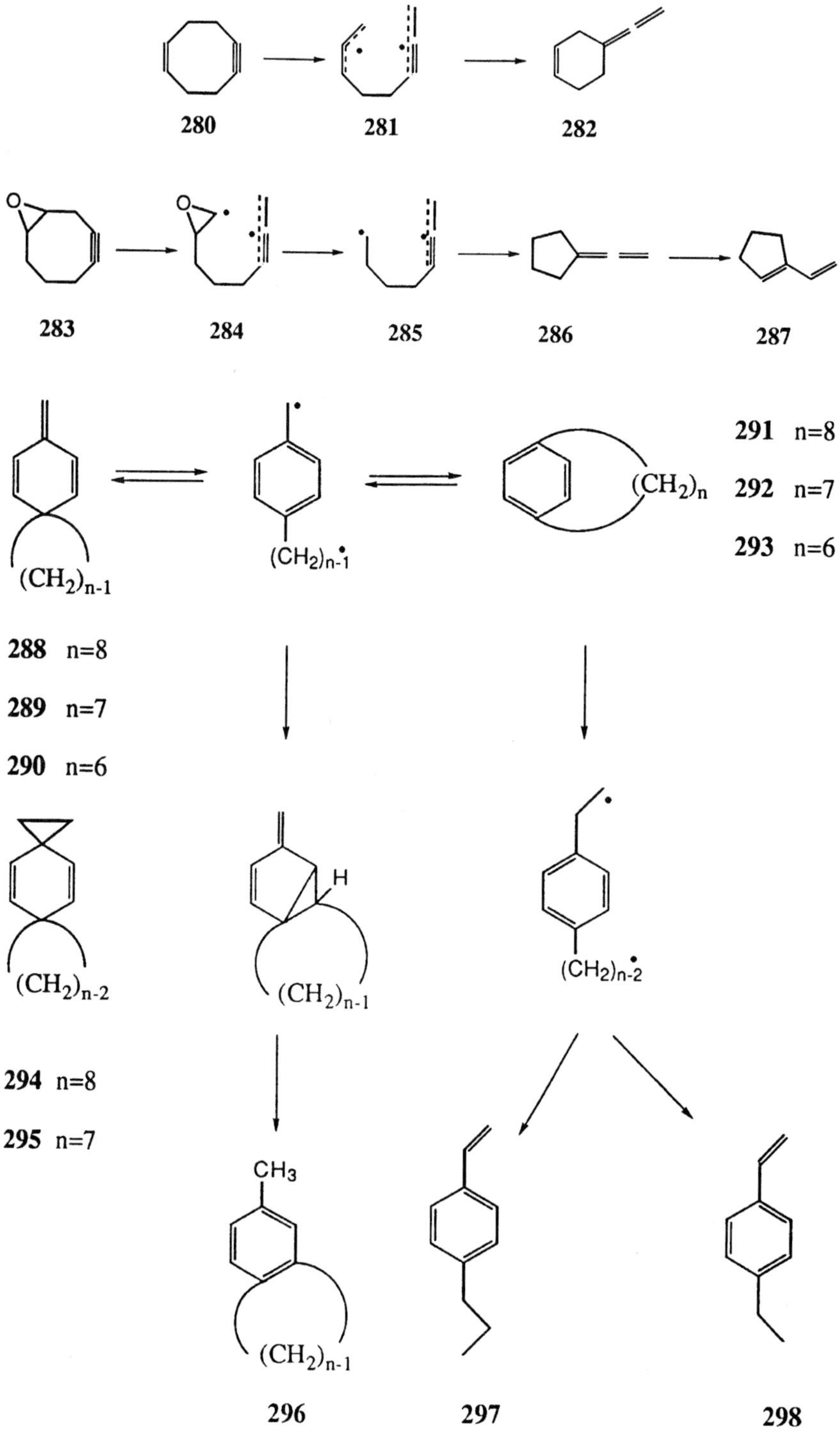

280   **281**   **282**

283   **284**   **285**   **286**   **287**

291 n=8
292 n=7
293 n=6

$(CH_2)_{n-1}$

288 n=8

289 n=7

290 n=6

$(CH_2)_{n-2}$

294 n=8

295 n=7

$(CH_2)_{n-1}$

CH_3

$(CH_2)_{n-2}$

$(CH_2)_{n-1}$

296   **297**   **298**

Scheme 40

9.2   Radical Coupling in Paracyclophane Synthesis

Bickelhaupt and coworkers[100] showed in 1977 that flow pyrolysis (packed Pyrex tube, 460–500°/10^{-1}Torr, N$_2$, residence time 0.5 sec) of 3-methylenespiro[5,7]-trideca-1,4-diene (**288**) yielded [8]-paracyclophane (**291**) in 20% yield together with 70% of polymer. 3-Methylene-spiro[5,6]dodeca-1,4-diene (**289**) gave [7]-para-cyclophane (**292**) (15%), 4-propylstyrene (**297**) (<5%) and polymer (∼70%) under the same conditions.

This pyrolytic approach to paracyclophanes was explored further and the precursors (**288**), (**289**) and (**290**) were subjected to FVP (aluminium oxide tube, 0.03 Torr, 100 mg/h of precursor) at different temperatures.[101] The recovery of monomeric products was significantly increased (60–85%) but was pressure dependent. At 550–600° precursor (**288**) was converted into [8] paracyclophane (**291**) in 80% yield, a marked improvement. However, compound (**289**) at 550° gave only 19% yield (by VPC) of [7]-paracyclophane (**292**) and this product was inseparable from the mixture. A useful yield of 7% for (**292**) was obtained under other conditions.

Pyrolysis of (**290**) gave no trace of [6]-paracyclophane (**293**). At higher temperatures other products were formed in all cases and examination of the processes involved in their formation has led to an understanding of the mechanisms. Compound (**288**) at 750° yields 4-propylstyrene (**297**) (27%), 4-ethylstyrene (**298**) (14%) and 9-methylbenzocyclononene (**296** n=8) (6%); compound (**289**) at 600° yields (**297**) 40%, and (**296**, n=7) (13%); and compound (**290**) at 600° yields (**297**) (47%) and (**296**, n=6) (2%). The formation of the styrene derivatives, in which there has been an apparent transfer of a carbon atom from one side of the ring to the other, was explained by formation and fragmentation of the paracyclophanes.

The key experiments were the pyrolysis of the available paracyclophanes. FVP of (**291**) at 700° and at 750° gave recovered (**291**) (88% and 65%), 4-propylstyrene (**297**) (0.7%, 12%), 4-ethylstyrene (**298**) (1%, 9%) and 9-methylbenzocyclononene (**296**, n=8) (6%, 11%). With increasing temperature there was an increasing proportion of styrenes and *o*-cyclised product. FVP of (**292**) at 700° gave 4-propylstyrene (**297**) (73%) and 8-methylbenzocycloctene (**296**, n=7) (21%). The difference in reactivity between (**291**) and (**292**) was accounted for by the greater strain in (**292**). For the formation of (**296**, n=8) and (**296**, n=7), cleavage to yield the benzyl radical was proposed and, for the formation of the styrene derivatives, cleavage of the paracyclophane ring at the β-carbon was postulated (Scheme 40).

In an attempt to form the intermediate radicals proposed, the spirocyclo-propanes (**294**) and (**295**) were pyrolysed but the products were different from those previously described.[102] A synthesis of [3,2] orthoparacyclophane (**299**) was attempted by pyrolysis of the spiro compound (**302**).[103] FVP of (**302**) (C$_{17}$H$_{18}$) at 650°/0.03 Torr gave the isomeric compounds (**306**) (26%), (**307**) (14%) and (**308**) (3%), and a compound, C$_{16}$H$_{18}$, identified as (**309**). None of the expected cyclophane (**299**) was obtained but it was considered to be an intermediate. Of the four products obtained, two, (**306**) and (**307**), could be considered to arise directly from the initially formed diradical (**303**). The other two were considered to be formed from the cyclophane (**299**). Cleavage of bond a in (**299**) would lead back to (**303**), of bond b would give rise to a diradical (**304**) which on hydrogen atom transfer would give (**308**). Cleavage of bond c followed by hydrogen transfer and cleavage of the molecule would lead to *o*-methylbenzyl which on coupling would yield (**309**) (Scheme 41).

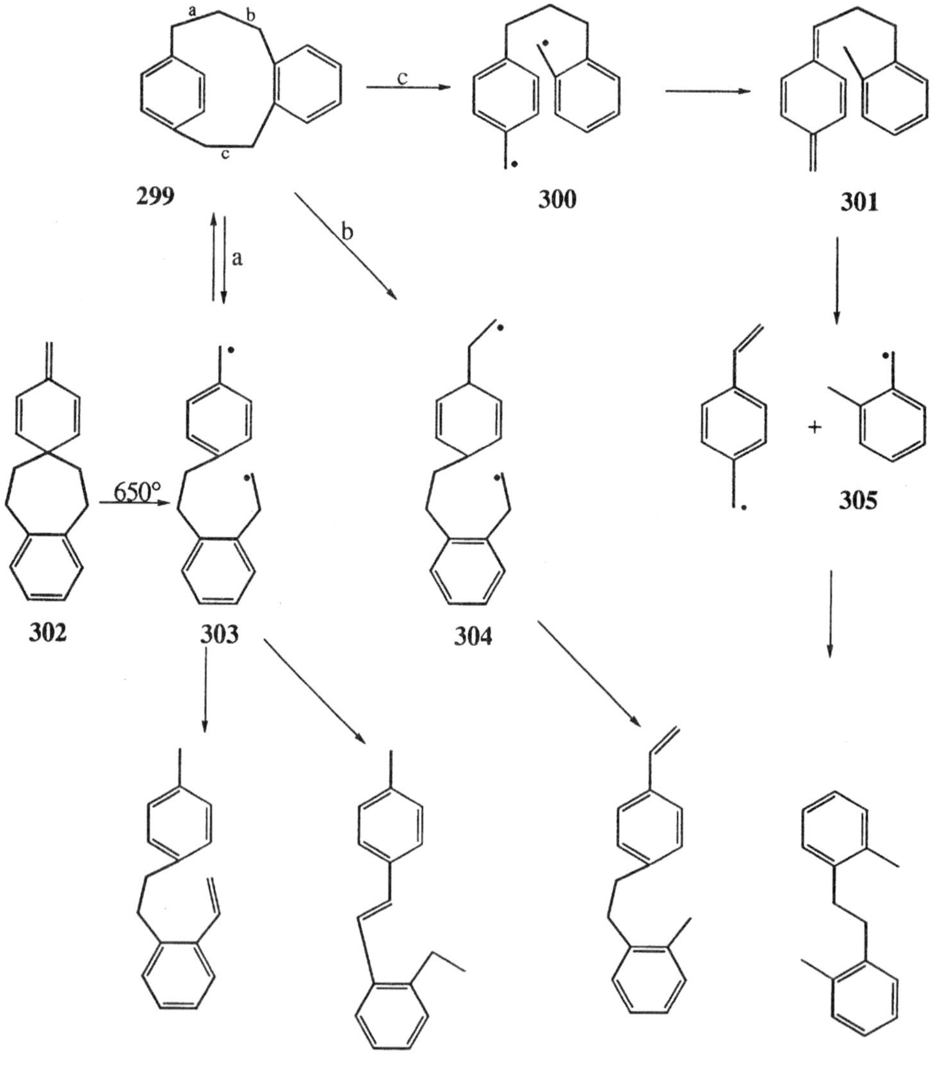

Scheme 41

Pyrolysis of 3-methylenespiro[6,5]dodeca-1,4,8-triene (**310**), the olefinic analogue of the benzo compound (**302**) gave entirely different results.[104] FVP at 440°/0.05–0.07 Torr gave some recovered (**310**) together with the tetraene (**313**). It was argued that bond cleavage gave rise to the diradical (**311**) and that attack by the terminal radical took place on the *ortho* position of the benzylic radical. This resulted in the formation of the *meta* bridged compound (**312**) which underwent a Cope rearrangement to give the product (**313**).

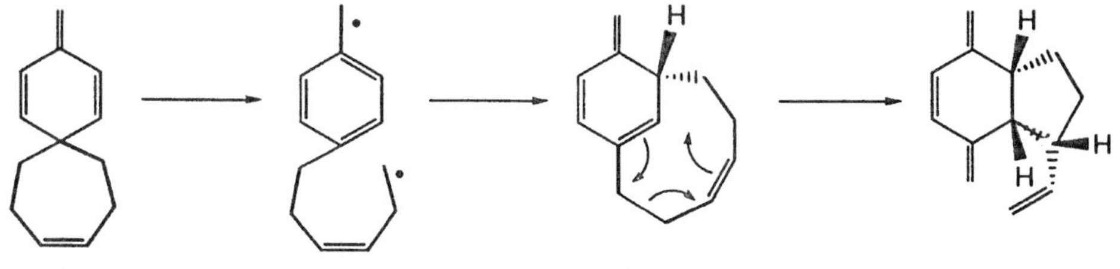

310   **311**   **312**   **313**

9.3   Pyrolytic Formation of Cyclopentane-1,3-diyl

Peterson and Carpenter[105] have used gas phase pyrolysis to demonstrate that an unstabilized cyclopentane-1,3-diyl can be intercepted by hydrogen intramolecular atom transfer. Pyrolysis of an equimolecular mixture of stereoisomers (**314a**) and (**314b**) at temperatures in the range 280–300° gave the diene (**315**) and bicyclooctanes (**316a**) and (**316b**). Side reactions could only be suppressed in Pyrex tubes by conducting the pyrolysis in the presence of potassium t-butoxide but the use of lead glass tubes prevented most side reactions. The three products were obtained in ratios between 3.0:3.4:1 and 4.2:2.9:1 respectively. It was argued on the basis of these and other experimental results that the hydrogen atom transfer to give (**318**) was evidence that the diyl intermediate (**317**) represented a true intermediate and not a transition state.

10.   GAS-PHASE SYNTHESIS OF STRAINED
CYCLIC CUMULENES AND ACETYLENES

The chemistry of strained cyclic cumulenes has been reviewed by Johnson.[106] Gas phase pyrolytic methods have been used to prepare 1,2-cyclononadiene but principally to form strained cyclic allenes as reactive intermediates. Adam *et al.*[107] showed that FVP of cyclopropane-1,2-dicarboxylic cyclic peroxide at 450° yielded allene by rearrangement of cyclopropylidene. The reaction was applied to the bicyclo[6,1,0]nonane derivative (**319**) and on FVP at 450°/0.16 Torr this was converted into the stable 1,2-cyclononadiene (**320**). The latter compound has been pyrolysed by Price and Johnson[108] to form non-1-en-8-yne.

The reverse reaction involving thermally induced coupling between terminal acetylene and alkene has been applied to oct-1-en-8-yne. The terminally deuterated compound (**321**) rearranged at 490°/760 Torr to [6-*d*]-oct-1-en-7-yne (**323**) *via* 1,2-cyclooctadiene (**322**).[109]

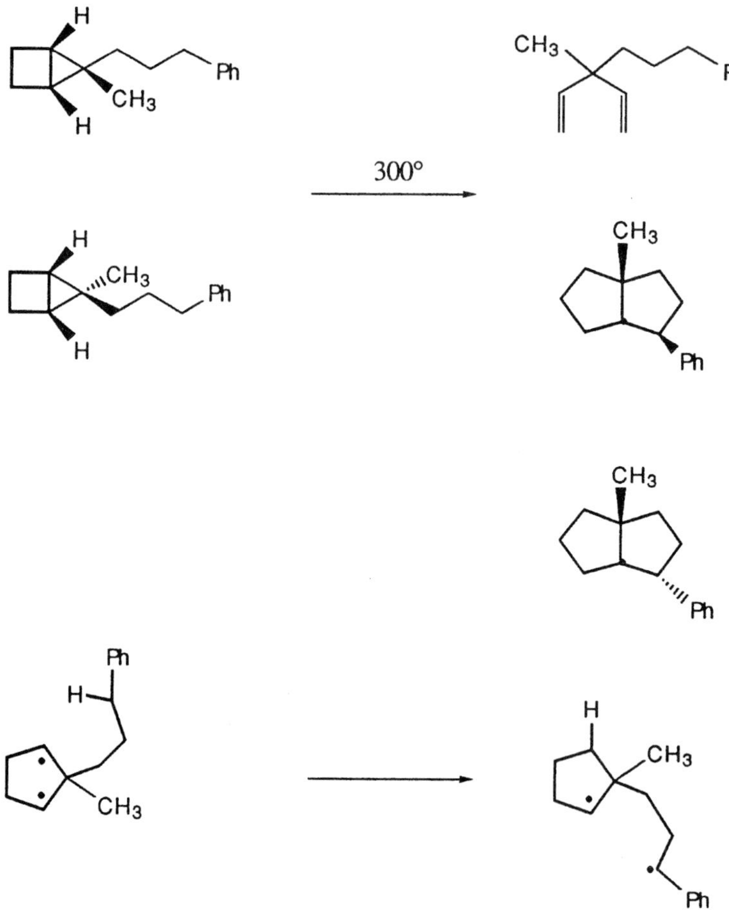

314a

314b

315

316a

316b

300°

317   318

Two pyrolytic approaches to the most strained allene, 1,2-cyclohexadiene (328) through bicyclo[3,1,0]hexan-6-ylidene (326) have been described. The chemistry of the reactive intermediate was essentially the same but there was a divergence in the spectroscopic evidence. Wentrup *et al.*[110] pyrolysed bicyclo-[3.1.0]hexane-6-carbonyl chloride (324) at 800°/10^{-4} Torr to form ketene (325) which decarbonylated to the carbene (326) and formed the allene (328). Trapped in an argon matrix at 11 K, the pyrolysate had ν_{max} 1886 cm^{-1}. At temperatures above $-170°$ the allene (328) dimerised to (329) (Scheme 42).

In an alternative approach Runge and Sander[111] pyrolysed the bromo stannane (327) in the temperature range 400–700° to form the carbene (326) and allene (328). The pyrolysate in argon at 10 K had ν_{max} 1829 cm^{-1} and on warming the same dimer (329) was formed. FVP of (327) at 550° gave 50% conversion to products and at 700° complete conversion principally to but-1-en-3-yne and ethylene together with 1,3-cyclohexadiene and its pyrolysis products, benzene and 1,3,5-hexatriene (Scheme 42).

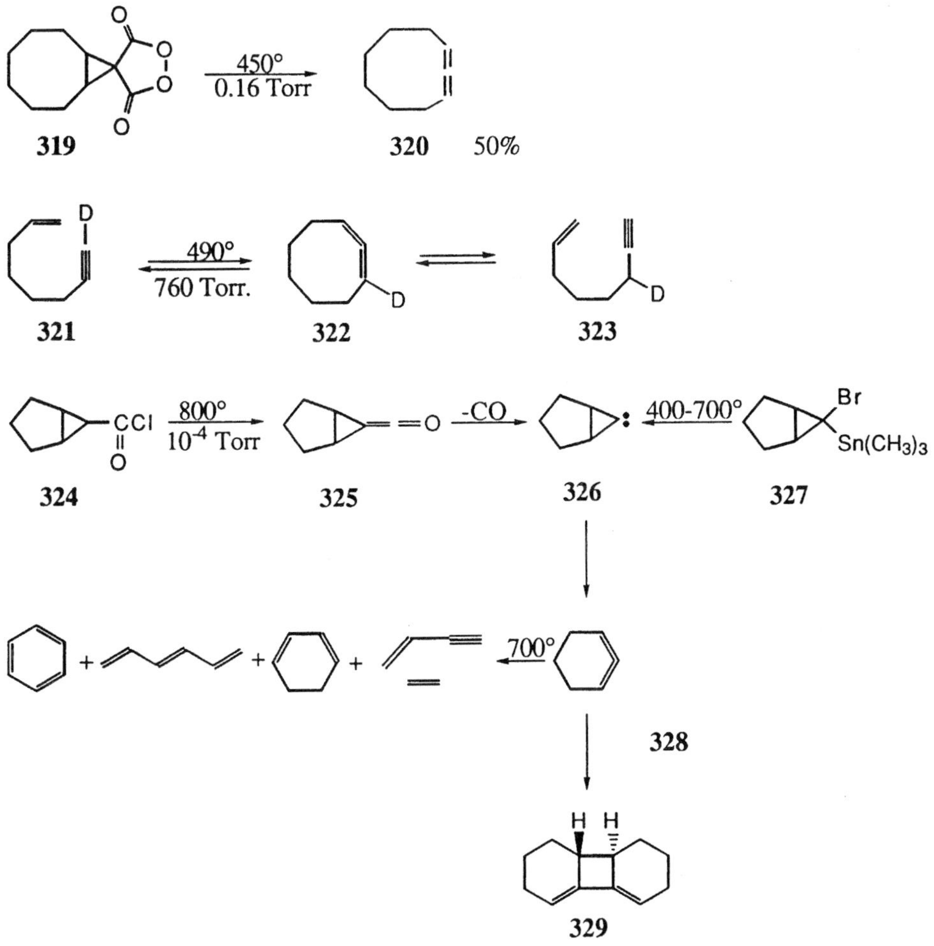

Scheme 42

Sander and Chapman[112] prepared tetramethylbutatriene (**332**) and the strained 3,3,6,6-tetramethylcyclohexyne (**331**) by pyrolytic methods. FVP of 2,2,5,5-tetramethylbicyclo[4.1.0]hept-1(6)-7-one (**330**) at 600° gave carbon monoxide, ethylene and tetramethylbutatriene (**332**). Pyrolysis of (**330**) at $370°/10^{-6}$ Torr with argon trapping at 12 K gave the cyclohexyne (**331**) which on warming to 45 K was converted into the dimer (**333**). Fragmentation of the cyclohexyne (**331**) was evident at pyrolysis temperatures above 450°.

11.   PYROLYSIS OF CYCLOPROPANE DERIVATIVES

Grimme, Warner and coworkers have investigated the kinetics of extrusion of benzene from the norcaradiene adducts (**335**) and (**337**).[113] These thermally

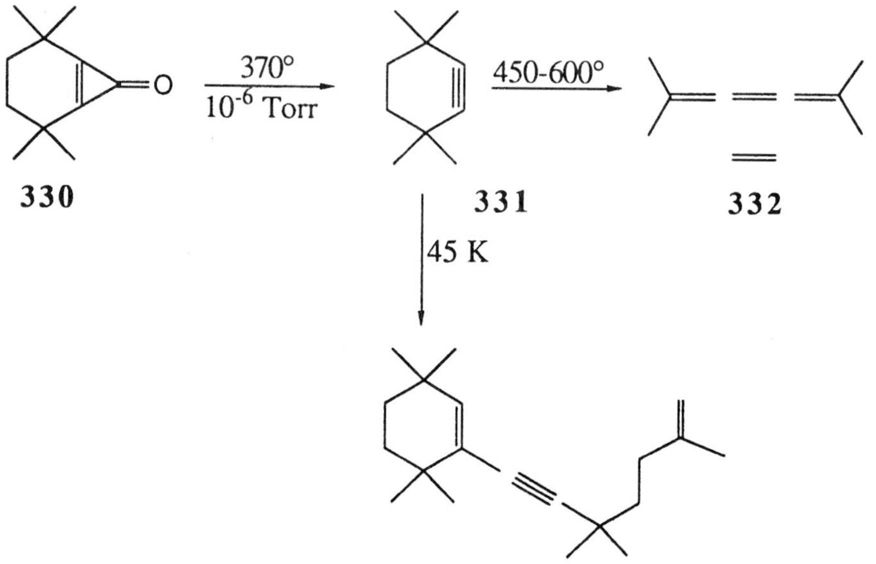

330   **331**   **332**

333

labile compounds were prepared by decarbonylation of the precursors (**334**) and (**336**) using FVP. Pyrolysis of (**334**) at 240°/0.9 Torr in a flow of argon gave (**335**), and of (**336**) at 270°/0.9 Torr/argon gave (**337**).

Wiberg and Walker[114] described rearrangement of [1.1.1]-propellane (**338**) in the gas phase at 114° to give methylenecyclobutene. Belzner and Szeimies have pyrolysed [1.1.1]-propellane (**338**) and the bridged derivatives (**340**) and (**343**).[115] At 430°/ Torr the [1.1.1] propellane gave dimethylenecyclopropane (**339**) as the only product. FVP of (**340**) at 370°/0.1 Torr gave a mixture of unreacted (**340**), the bridged dimethylenecyclopropane bicyclo[5.1.0]octan-1,6-diene (**341**) and the allene 6-ethenylidenebicyclo[3.1.0]hexane (**342**) in a ratio of 1:6:4. When the mixture was repyrolysed, the allene (**342**) was formed at the expense of the other products. FVP of (**340**) at 420°/0.1 Torr gave the allene (**342**) as the sole product. It was argued that the reaction pathway involved the sequence (**340**) → (**341**) → (**342**) but a direct conversion could not be excluded.

The bridged propellane (**343**) on FVP at 430°/0.1 Torr gave 1,2-divinylcyclopropene (**345**) presumably by Cope rearrangement of bicyclo[4.1.0]hepta-1,5-diene (**344**) although this was not detected.

12. VINYLCYCLOPROPANES, DIVINYLCYCLOPROPANES, VINYLOXIRANES AND DIVINYL OXIRANES

Hudlicky and coworkers have studied the thermal rearrangement of vinylcyclopropanes to cyclopentenes and of divinylcyclopropanes to cycloheptadienes. The

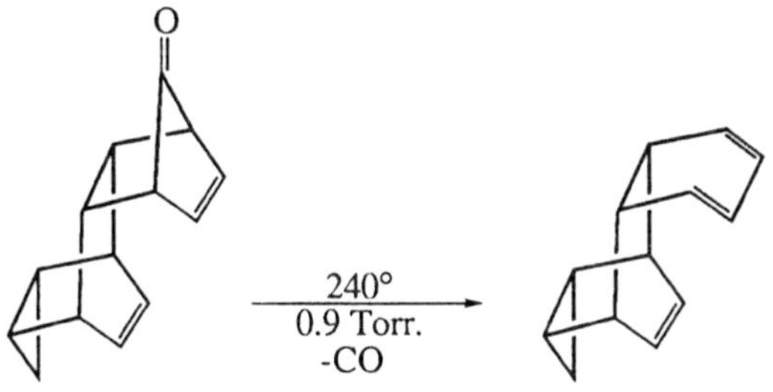

334 → 335

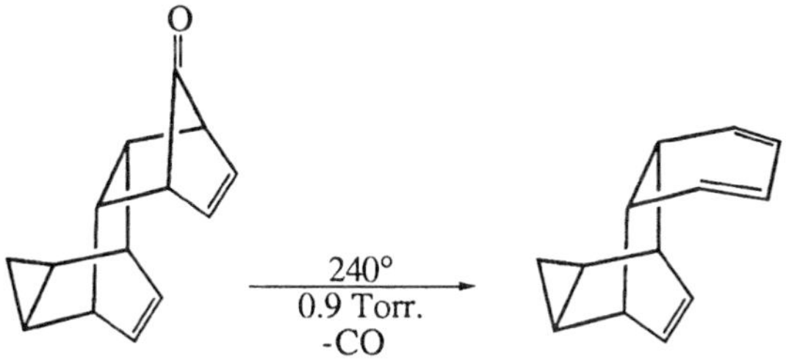

240°
0.9 Torr.
-CO

336 → 337

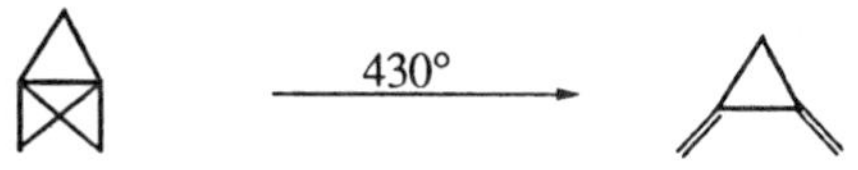

240°
0.9 Torr.
-CO

338 → 339

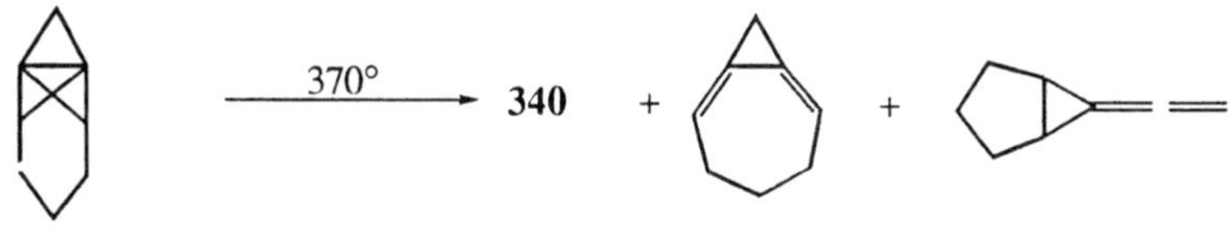

430°

340 → 340 + 341 + 342

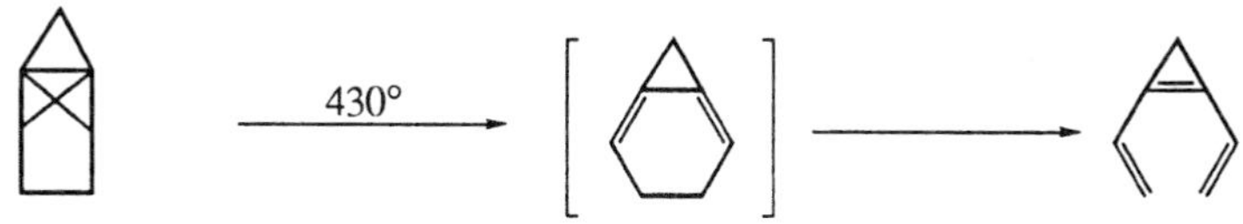

370°

343 → [344] → 345

430°

heterocyclic analogue of the latter reaction, the rearrangement of divinyloxiranes to dihydroxepines has also been developed. The first of these reactions was comprehensively reviewed in 1985.[116] The majority of the thermal rearrangements described were performed at moderate temperatures in solution or under static conditions but there are many examples reported of gas-phase isomerizations of vinylcyclopropanes done under conditions of flow pyrolysis, flash pyrolysis in an inert atmosphere and flash vacuum pyrolysis. Some examples of such reactions described since that review was published are given below.

The divinylcyclopropane–cycloheptadiene rearrangement was reviewed in 1992.[117] Of the thermally induced reactions, most proceeded at moderate temperatures and only a few were done under gas-phase conditions. Some recent examples are described. A general review on cyclopropanes in synthesis includes examples of these reactions.[118]

Hudlicky and his group have refined the pyrolytic rearrangement of vinylcyclo-propanes into a procedure for cyclopentane annulation and have applied this reaction to the synthesis of angular triquinanes such as (347 a, b and c). The opening of the cyclopropane ring at bonds 1 and 2 (structure 346) is generally considered to involve homolysis and the annulation reaction is competing with cleavage of bond 2. Good results were achieved by stabilising the cyclopentyl radical (as in 349) as part of an allylic system developed with the aid of an exocyclic methylene group. However, success in these reactions is dependent on the suppression of acid catalysed reactions and pyrolysis through base-washed Vycor tubing rinsed with a slurry of lead carbonate has been found to prevent undesired side reactions.[119] FVP of compound (346 a) gave the triquinane (347a) and the triene (348a) in a ratio of 30:70, and of the less sterically demanding (346b) gave the triquinane (347b) and triene (348b) in a ratio of 43:57.[120] Where the methoxy-methylene substituent was introduced, even better control was achieved and the rearrangement of the ester (346c) to the triquinane (347c) was the major pathway. The reactions are considered to proceed through the diradicals (349) and (350).

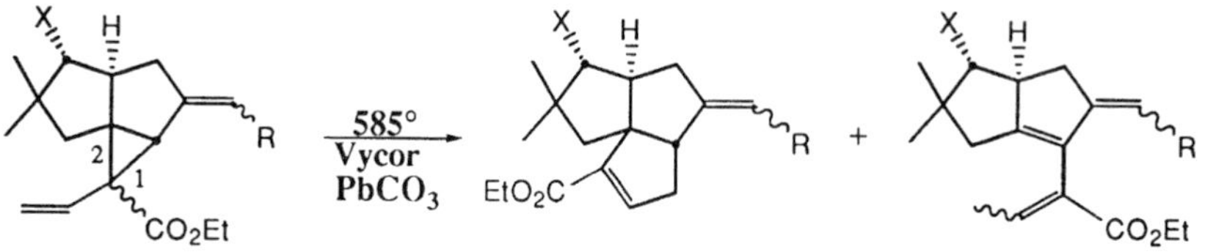

	346		347	348
a	X=OTBDMS	R=H		30 : 70
b	X=H	R=H		43 : 57
c	X=H	R=OMe		66 : 34

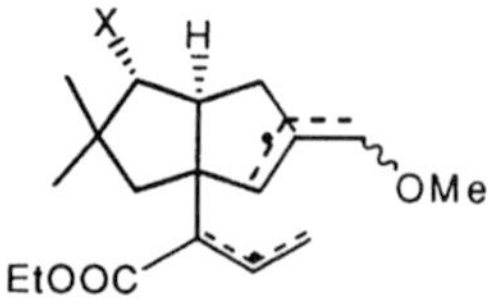

349

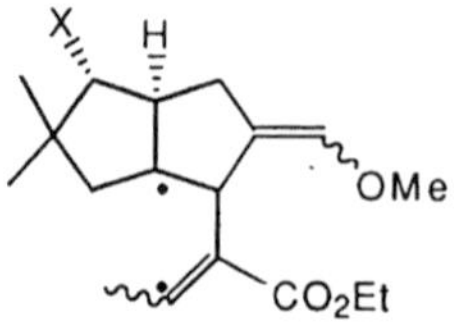

350

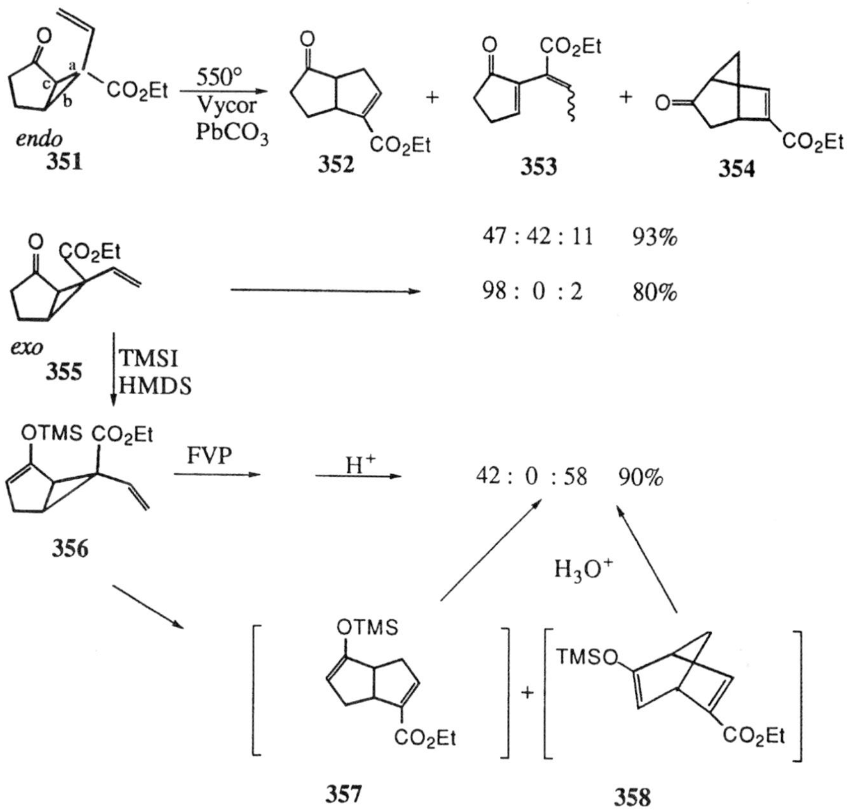

47 : 42 : 11   93%

98 : 0 : 2   80%

42 : 0 : 58   90%

In the vinylketoester (**351**) there is the further possibility of enolisation which permits Cope rearrangement through the resulting divinylcyclopropane system. FVP of the *endo* vinyl compound (**351**) at 550° through Vycor/PbCO₃ gave the products (**352**), (**353**) and (**354**) arising from cleavage of bonds, a, b and c respectively of (**351**).[121] Pyrolysis under the same conditions of the *exo*-isomer (**355**) gave predominantly the desired annulated product (**352**). The *exo*-isomer (**355**) was converted with trimethylsilyliodide (TMSI) and hexamethyldisilazan (HMDS) to form the trimethylsilyl enol ether (**356**). Pyrolysis of (**356**) and desilylation of the pyrolysate (**357**) and (**358**) gave (**352**) by interaction of one vinyl group with the cyclopropane together with a much increased proportion of (**354**) through interaction of the two double bonds with the cyclopropane.

The same research group has developed a synthesis of dihydrofurans based on the rearrangement of vinyloxiranes under FVP conditions.[122] The intermediates were prepared by addition of the lithium dienolate of ethyl 2-bromobut-2-enoate (**359**) to aldehydes at −105 to −60°. FVP of the vinyloxiranes (**360**) at 500–550°/ 0.02 Torr resulted in [2+3] annulation giving a series of dihydrofuran esters (**361a–e**). Application of the vinyloxiration reaction to acrolein gave a mixture containing the divinyloxiran (**362**). This underwent Cope rearrangements to yield oxepin (**363**) and some of the aldehyde (**364**). The 3-furyl derivative (**365**) also gave the Cope product (**366**).

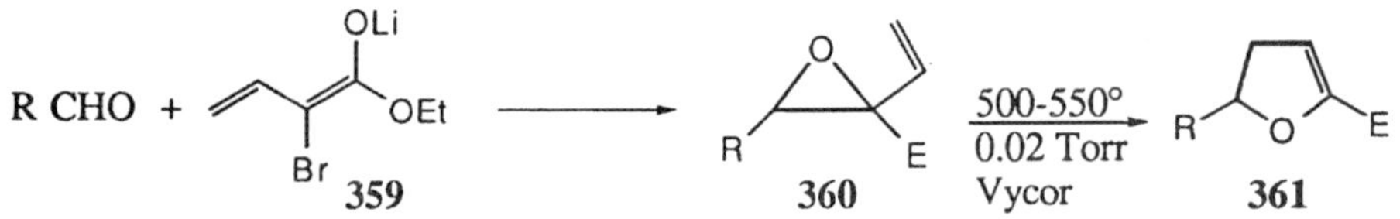

R = a) Pr, b) iPr, c) Ph, d) 2-furyl, e) 3-furyl

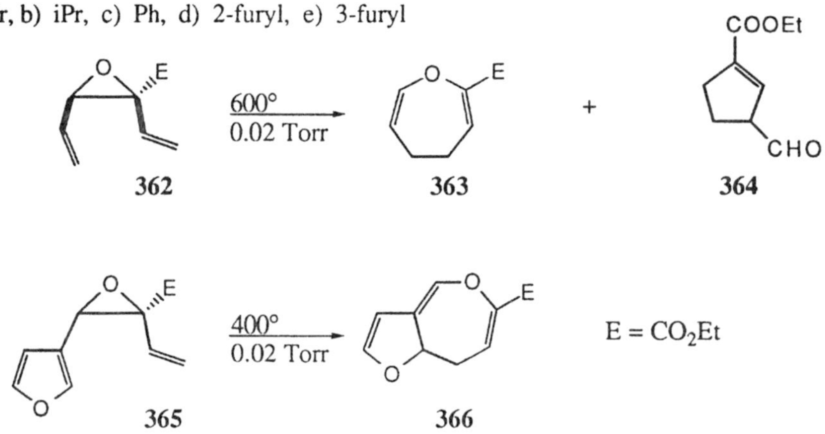

E = CO$_2$Et

13. ALKYNONE REARRANGEMENTS

Gas flow pyrolysis of conjugated alkynones leads *via* a 1,2-shift to acylvinylidene intermediates which may undergo carbene insertion into an adjacent site. This α-alkynone cyclisation was developed by Karpf and Dreiding[123] and it has provided a useful synthetic method for cyclopentenone annulation. Examples of the use of the reaction up to 1985 have been compiled by Karpf.[1g] These two authors and their coworkers have provided a further example of annulation and have explored the rearrangement in other precursors.

In a synthesis of (±) isocomene, an angular triquinane, the cyclopentenone (**368**) was prepared by flow pyrolysis of the alkynone (**367**). Pyrolysis at 540°/14 Torr in a stream of nitrogen gave the required ketone (**368**) in 77% yield together with a small yield (7%) of the elimination product (**369**).[124]

Gas flow pyrolysis of 1-isobutenyl alkynyl ketones (**370**) (**372**) and 2-methyl-phenyl alkynyl ketones (**371**) (**373**) has been investigated and four principal

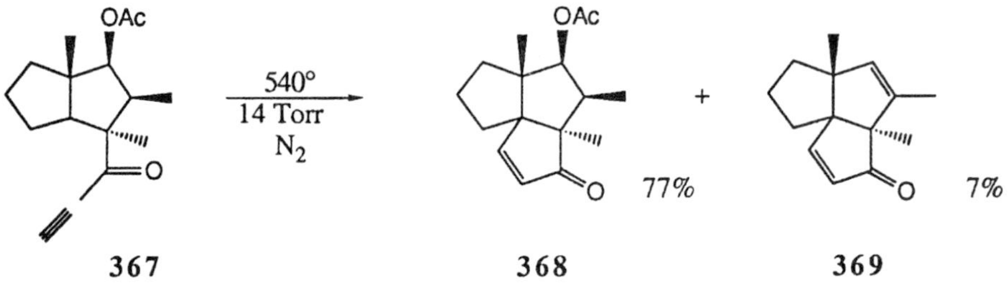

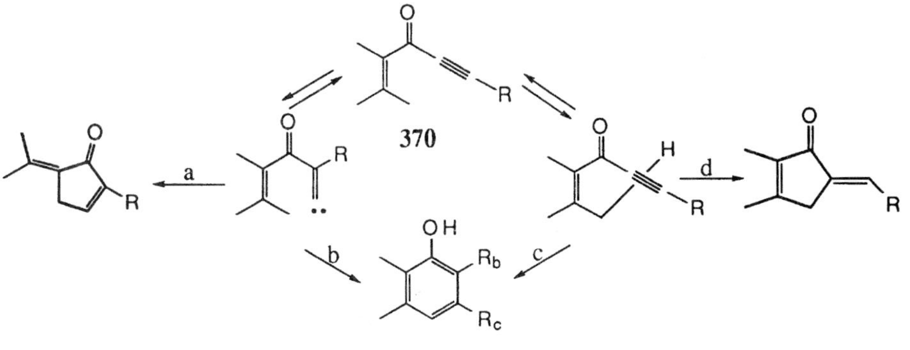

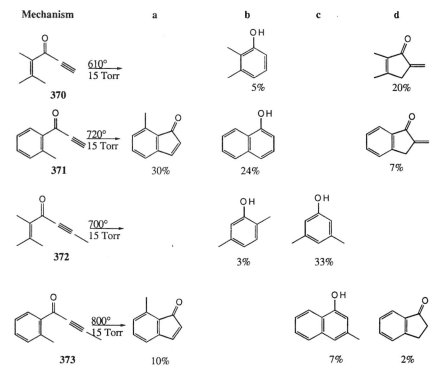

Scheme 43

reaction pathways have been demonstrated which involve either 1,2-rearrange-ment of the acetylene to the vinylidene or direct addition to the alkyne. This mechanistic relationship is shown in Scheme 43 and the four routes are desig-nated a, b, c and d. The principal products formed by pyrolysis of two olefinic and two aromatic species are tabulated as examples in Scheme 44. They can be seen to arise from the mechanistic pathways proposed.

Mechanism	a	b	c	d
370 610° 15 Torr		OH 5%		O 20%
371 720° 15 Torr	30%	OH 24%		O 7%
372 700° 15 Torr		OH 3%	OH 33%	
373 800° 15 Torr	10%		OH 7%	O 2%

Scheme 44

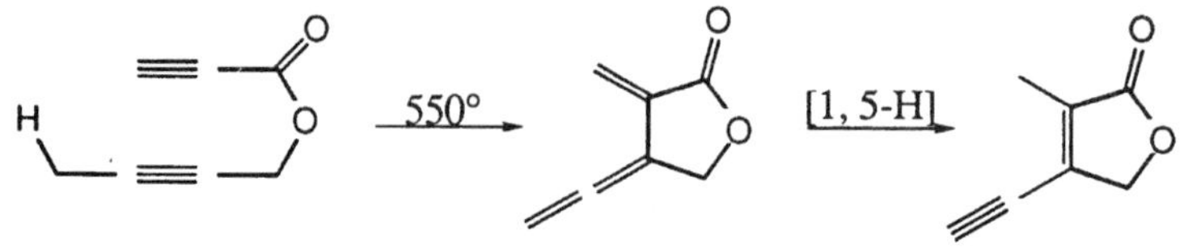

374 **375** **376**

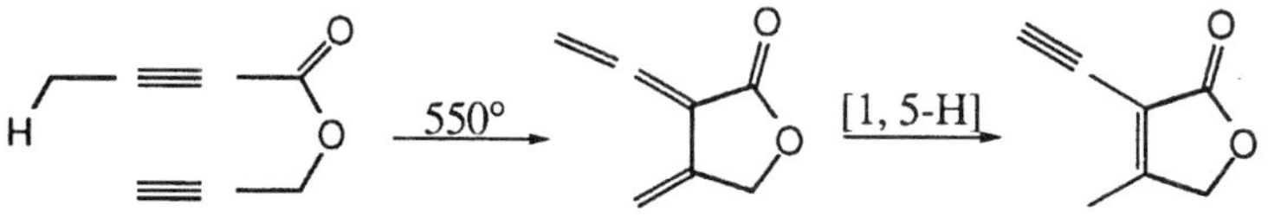

377 **378** **379**

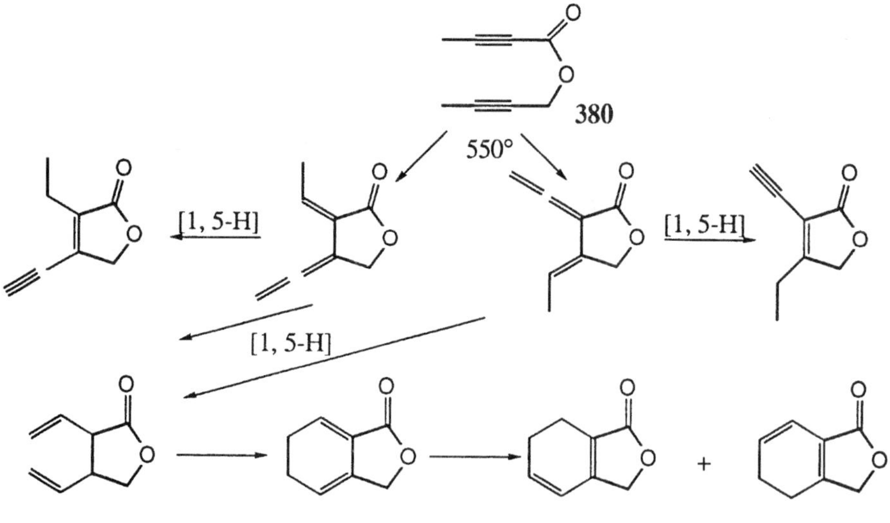

Scheme 45

Flow pyrolysis of propargyl propiolate esters does not proceed by carbene formation but by an initial ene-reaction and results in the formation of butenolides.[126] Pyrolysis of 2-butynyl propynoate (**374**) at 550° gave 3-ethynyl-2-methyl-2-buten-4-olide (**376**) (85%) and of propynyl but-2-ynoate (**377**) gave 2-ethynyl-3-methyl-2-buten-4-olide (**379**) (80%). These butenolides arise from the intermediate lactones (**375**) and (**378**) by [1,5]H shifts. However, if the flow pyrolysis is done at 440° these intermediates are converted by Diels-Alder additions into dimers which are isolated as the major products.

Pyrolysis of but-2-ynyl but-2-ynoate (**380**) at 550° gave four products arising from the two possible *ene*-reactions followed by [1,5-H] shifts and cyclisation (Scheme 45).

14. THERMAL CLEAVAGE OF CYCLOBUTANES

Mehta and his coworkers[127,128] have developed a simple procedure for the synthesis of triquinane bis-enones (**383** a–f) using photochemical [2+2] cycloaddition followed by regiospecific thermal fragmentation of the cyclobutane under FVP conditions to promote metathesis of the double bonds. Beginning with the formation of the 1:1 adducts (**381**) of cyclopentadiene and benzoquinones, this provides a short route into the triquinane system. Examples are shown in Scheme 46 together with the reaction conditions and yields.

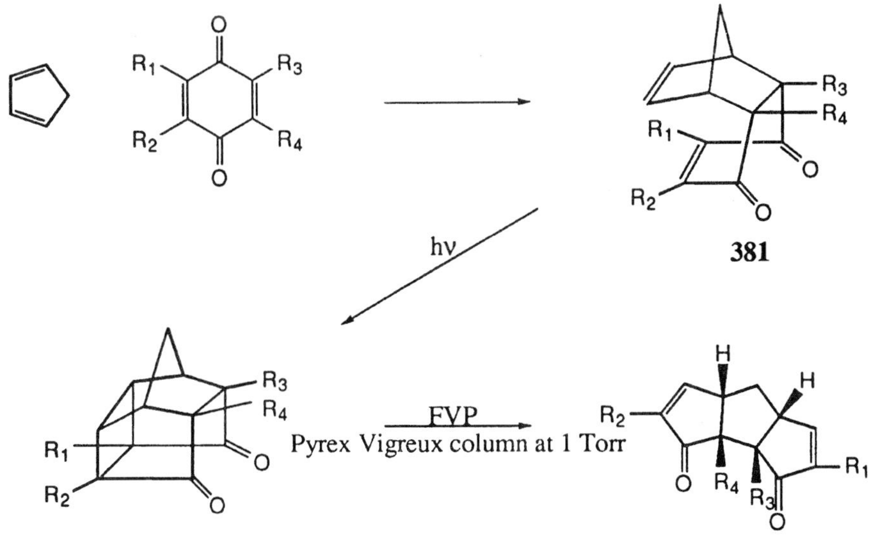

382

383

	R₁	R₂	R₃	R₄		
a	H	H	H	H	560°	96%
b	CH₃	CH₃	H	H	450°	100%
c	Cl	Cl	H	H	450°	20%
d	CH₃	H	H	CH₃	500°	100%
e	Cl	H	H	Cl	500°	65%
f	Br	H	H	Br	500°	60%

Scheme 46

Kanematsu, Osawa and coworkers[129] have confirmed the necessity for the use of FVP for the conversion of (382) into (383), but by adding substituents have sought to decrease the temperature required for cycloreversion. With a variety of substituents but particularly with compound (282, $R_1=R_2=OMe$, $R_3=R_4=H$), the methoxy substituents allowed reversion at a temperature of less than 100°. These effects were attributed to stabilization of the initially formed radical through cleavage of the σ-bond.[129-131]

Mehta and Nair developed this chemistry first to obtain triquinane systems bridged with an ether or carbonyl[132] and then "roofed" with a cyclopentene bridge.[133] Conversion of the diketone (382a) into the ether (384a) by established procedures followed by FVP at 620°/7 Torr gave the ether bridged triquinane (385a) in 95% yield. The ether-bridged ketone (384b) did not decarbonylate under the FVP conditions and gave the ketone (385b) in 40%. Synthesis of the diketone (384c) followed by FVP at 580°/5 Torr gave the tetraquinanedione (385c) in 70% yield.

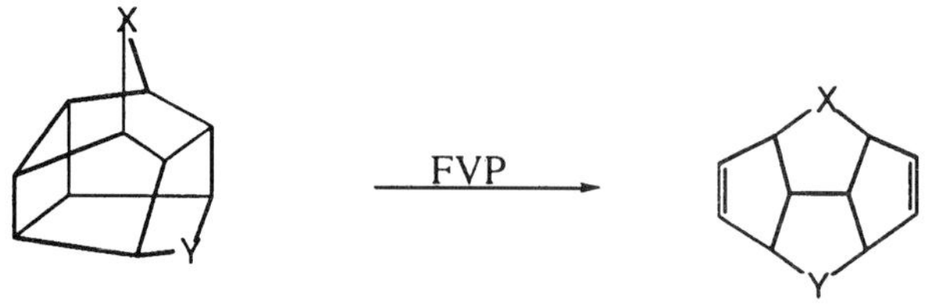

	X	Y			
384a	CH$_2$	O	620°/7 Torr	385a	95%
384b	CO	O	620°/5 Torr	385b	40%
384c	CO	CO	580°/5 Torr	385c	70%

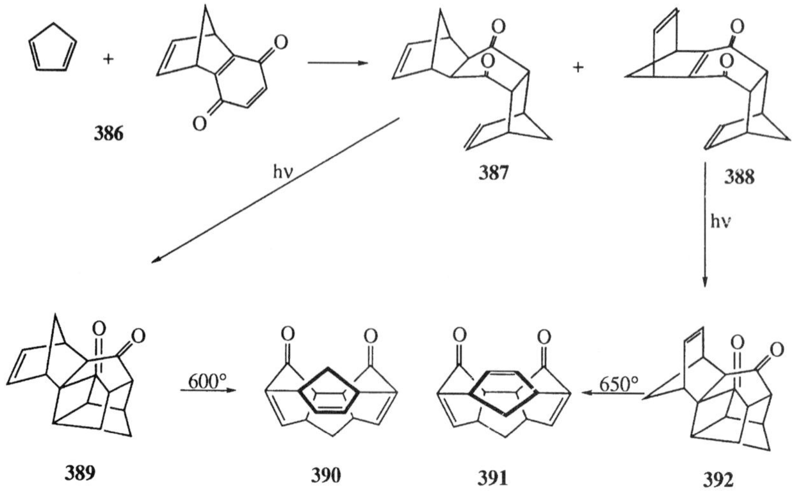

This approach has been extended to incorporate a cyclopentene ring and this provides entry to a segment of the dodecahedrane surface.[133] The sequence begins with addition of cyclopentadiene to norborneno-p-benzoquinone (386) to give the isomers (387) and (388) which on separation and irradiation yield (389) and (392) respectively. FVP of (389) at 600°/0.02 Torr gave the roofed product (390) in 76% yield. Similarly FVP of (392) at 650° gave the isomeric compound (391) in 50% yield.

15. SYNTHESIS OF ALKENES AND ALKYNES BY CYCLOREVERSION

Lasne and Ripoll have comprehensively reviewed the use of cycloreversion as a synthetic procedure for the formation of a double bond.[134] They demonstrated that FVP has been a very successful procedure for achieving this transformation and is often the method of choice. These authors have investigated the use of bis-adducts for the formation of triple bonds.[135] They found that in some cases after the loss of one diene, the intermediate mono-adducts may undergo non-selective retro-Diels Alder reactions and give rise to a number of products.

Pyrolysis of the cyclopentadiene di-adduct (393, X=CH$_2$) at 870°/10^{-6} Torr gave methyl propiolate (395) (20%) and cyclopentadiene. The alternative mode of cycloreversion of the intermediate (394, X=CH$_2$) gave acetylene and the ester (396, X=CH$_2$) (25%). Methyl 2-methylbenzoate (8%) and methyl 3-methylbenzoate (12%) were also obtained from this reaction. Pyrolysis of the double furan adduct (393, X=O) at 500°/10^{-6} Torr gave methyl propiolate (395) (30%) and furan and the intermediate (394, X=O) also broke down to acetylene and methyl 3-furoate (396, X=O) (50%).

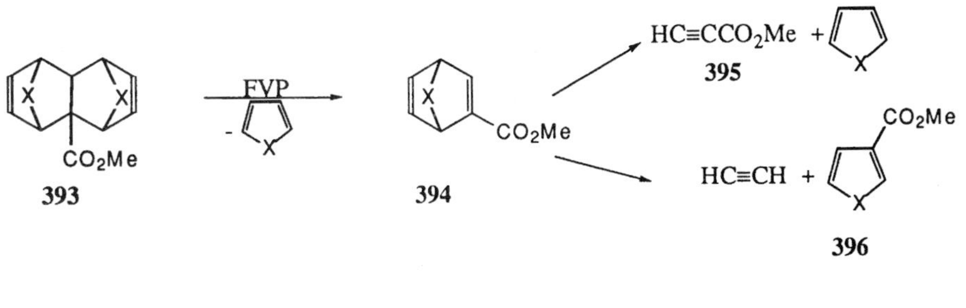

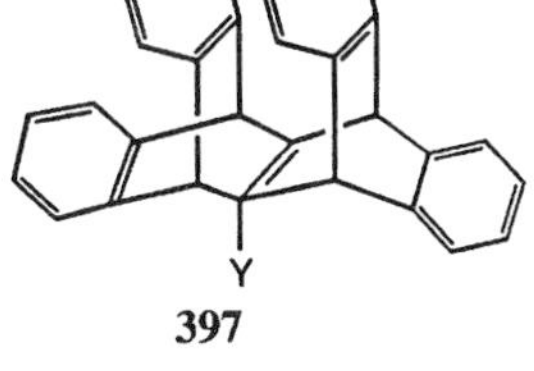

$$HC \equiv CCO_2Me \quad + \quad 395$$

$$HC \equiv CH \quad + \quad CO_2Me \quad 396$$

393   394

	Y
a	= H
b	= CO$_2$CH$_3$
c	= OCH$_3$
d	= OH

397

Cycloreversion of the double anthracene adducts (397a–d) where X=H, CO$_2$Me, OMe and OH has also been studied using FVP at 10^{-6} Torr. The hydrocarbon (**397a**) at 870° gave anthracene and acetylene and the ester (**397b**) at 890° gave methyl propiolate (**395**) (20%). The ether (**397e**) gave anthracene and ketene, CH_2CO, which is the known fragmentation product of methoxyacetylene. On pyrolysis of the alcohol (**397d**) some of the intermediate enol tautomerised to 9,10-ethanoanthracene-11-one (< 10%). The remaining products were anthracene (60%) and ketene, characterised as the anilide, which arose from tautomerisation of the expected hydroxyacetylene.

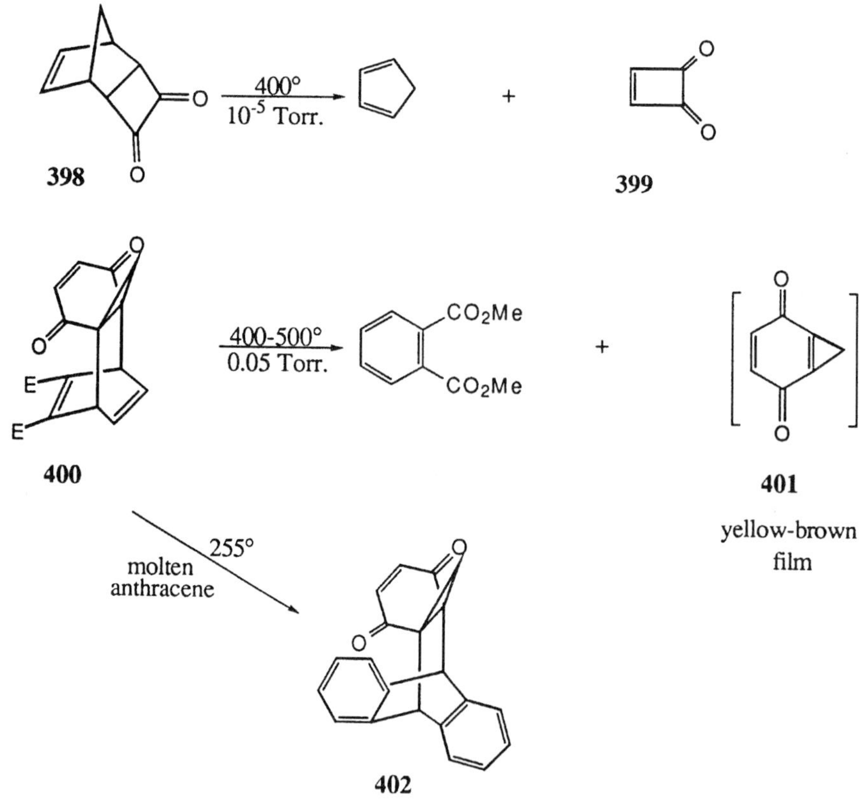

Martin and coworkers[136] have shown that cyclobut-3-ene-1,2-dione (**399**) can be prepared by cycloreversion of the cyclopentadiene and anthracene adducts. FVP of (**398**) through a quartz tube with glass packing at $400°/10^{-5}$ Torr gave the diketone (**399**) in 50–60% yield together with cyclopentadiene. The diketone could be selectively condensed from the pyrolysate at −45°. The corresponding anthracene adduct on FVP at 400° gave the diketone (**399**) in 15% yield.

In an unsuccessful attempt to isolate the cyclopropaquinone (**401**) the precursor (**400**) was pyrolysed through a packed quartz tube at 400–500°/0.05 Torr. Dimethyl phthalate was formed but the remaining material was deposited as a film on the surface of the tube and did not reach the cold trap. Formation of the required cyclopropaquinone (**401**) was demonstrated by fusion of the precursor with anthracene when the adduct (**402**) was formed.

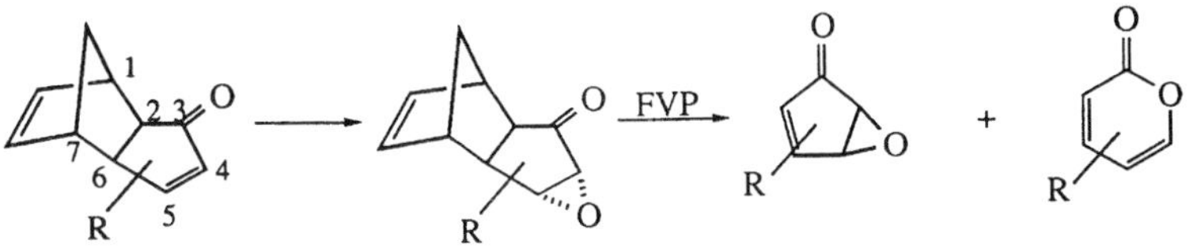

403   404   405   406

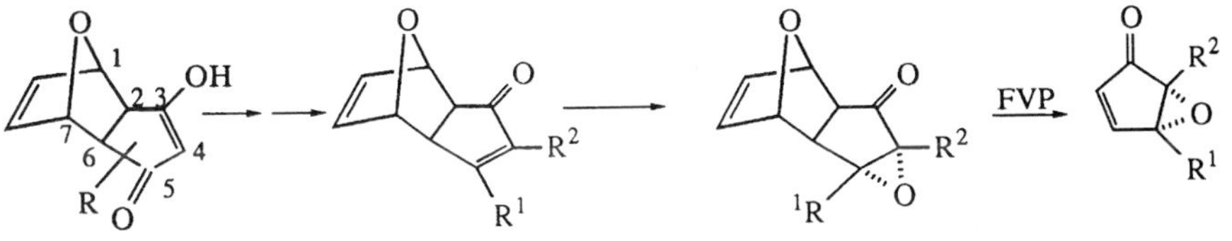

407   408   409   410

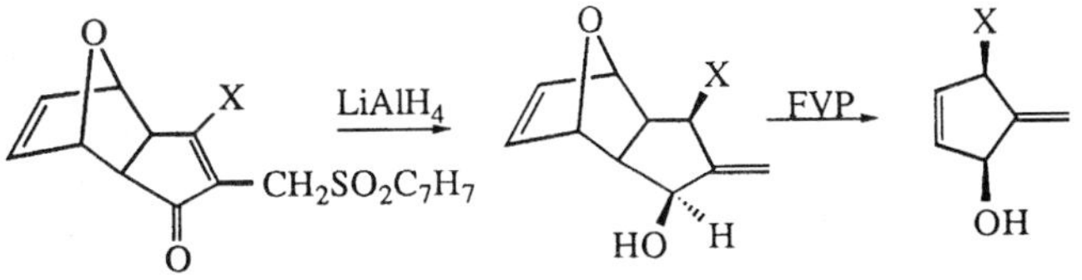

| 411a | X = OEt | 412a | X = OEt | 413a | X = OEt | 81%, 400°/0.076 Torr. |
| 411b | X = H | 412b | X = H | 413b | X = H | 86%, 420°/0.046 Torr. |

| 414a | X = OEt | 415a | X = OEt | 73%, 400°/0.023 Torr. |
| 414b | X = H | 415b | X = H | 99%, 310°/0.23 Torr. |

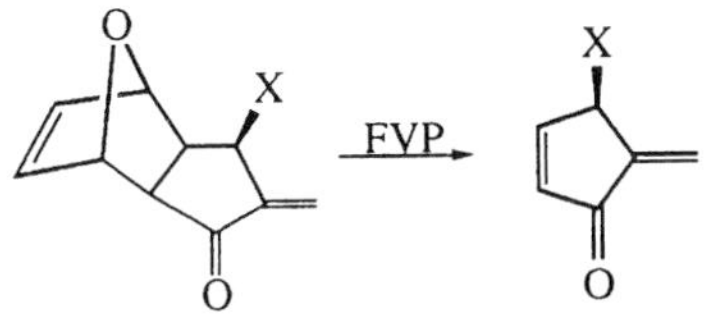

Zwanenburg and his group have shown that selective transformation of the enone system in the tricyclo[5.2.1.0$^{2.6}$]decadienones (403) followed by cyclorever-sion under FVP conditions leads to the formation of a variety of functionalised cyclopentenones.[138] The process can be used for the synthesis of racemic or optically active intermediates suitable for further elaboration. Attempts to pre-pare the cyclopentenone epoxides (405) as precursors for highly oxygenated cyclopentenoids by pyrolysis of epoxides (404) was unsatisfactory as considerable proportions of the epoxides (405) were lost through rearrangement to the α-pyrones (406) because of the high temperatures required for fragmentation.

The furan adduct (408) was known to fragment at a lower temperature than the corresponding cyclopentadiene adduct and an efficient synthesis of cyclopentenone epoxides based on FVP of furan derivatives has been developed.[139]

Reaction of furan with cyclopentene-2,4-dione over a period of 1–2 weeks yields the exo-enolone (407) in 95% yield. 4-Alkoxymethyl groups can be added via the 4-tosylmethyl derivative (411) formed by acid-catalysed reaction of (407) with formaldehyde and toluenesulfinic acid. Formation of the enol ether by ethylation with Meerwein's salt and addition of methyl lithium or diisobutyl aluminium hydride gave the 5-methyl or unsubstituted ketone. The resulting enones (408) can be epoxidised to precursors (409) and fragmented. FVP of a series of epoxides (409) where R^1=H or Me, and R^2=H or CH_2OR, at temperatures and pressures in the range 310–375°/0.04–0.1 Torr gave the cyclopentenone epoxides (410) in yields around 90%.

Lithium aluminium hydride reduction of the 4-tosylmethyl intermediates (411a) and (411b) unexpectedly gave the exocyclic methylene alcohols (412a) and (412b). On pyrolysis these derivatives gave the dienols (413a) and (413b) in good yield. Oxidation of the alcohols (412a) and (412b) with manganese dioxide gave the corresponding ketones (414a) and (414b) which underwent cycloreversion on FVP to form the dienones (415a) and (415b) in 73 and 99% yields respectively.[141]

Gas phase conditions have been used as a convenient method of cleaving a Diels Alder dimer (416) isolated from a preparative reaction in solution.[142] FVP of (416) at 500°/0.02 Torr gave the unsaturated ketone (417) in quantitative yield. The monomer was trapped at liquid nitrogen temperature and could be used directly. When stored at room temperature it reverted to the dimer (416).

416   417

16.   CYCLOREVERSION AND "TRANSFER TECHNOLOGY"

Warrener and Russell[143–146] have developed a concept of transfer technology, an operation previously used in synthesis but not recognised as a general procedure. Defined by the authors as "The transfer reagent is caused to undergo a cycloaddition (or other) reaction with the substrate and the resulting intermediate (or its chemically modified counterpart) subjected to the Alder-Rickert (or other) process in the final deprotection step. In this way the chemical entity is transferred from the transfer agent to the substrate without ever being a free

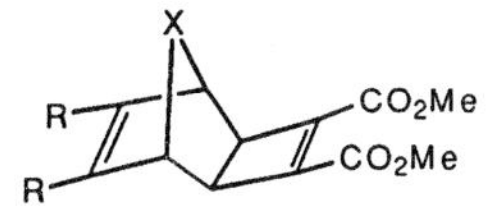

418   R=H, X=CH$_2$
419   R=CF$_3$, X=O
420   R=H, X=O

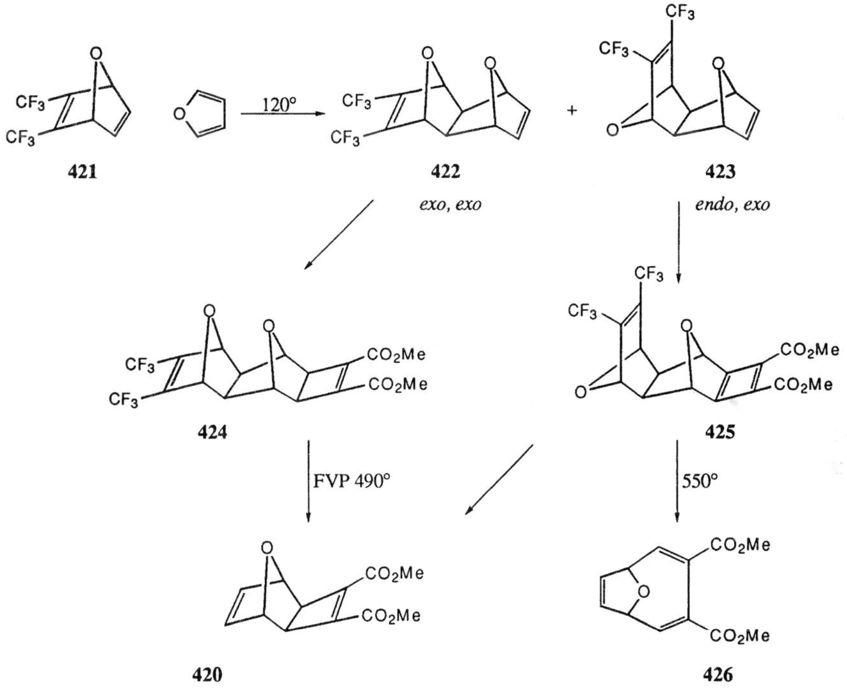

molecule", the procedure has been used for the transfer of dimethyl cyclobuta-diene-1,2-dicarboxylate. FVP reactions are used in these syntheses in two ways; first in the formation of one of the transfer reagents and, second, in the final [4π + 2π] cycloreversion leading to a cyclobutene, the Alder-Rickert cycloreversion.

Three transfer reagents (418), (419) and (420) have been investigated. The first (418) has the disadvantage that the temperatures required for FVP in the final cycloreversion are high and may result in decomposition of the desired products. The second (419) cannot be used in sequences requiring lithium aluminium hydride reduction as this leads to cleavage of the ether bridge. The third (420) overcomes, these difficulties.

The preparation of (420)[147] involved synthesis of the principal precursor (424) and concomitant formation of the less productive isomer (425). Bis(trifluoro-methyl)oxanorbornadiene (421) reacted with furan at 120° to give two thermo-dynamically stable adducts. These were separated into the *exo, exo* isomer (422) (20% yield) and the *endo, exo* isomer (423) (38%). Addition of dimethyl butynedioate catalysed by Ru(CO)H$_2$(PPh$_3$)$_3$ gave the cyclobutane diesters (424) and (425). FVP of the *exo, exo, exo* isomer (424) at $490°/10^{-3}$ Torr gave the required reagent (420) in 90% yield. FVP of the *endo, exo, exo* isomer (425) under the same conditions gave only 20% cleavage with formation of some (420) together with the bridged cyclooctatriene (426) as a by-product. FVP of (425) at 550° gave (426) as the exclusive product.

It is known that the facile cyclooctatetraene–bicyclo[4,2,0]octa-2,4,7-triene equilibrium can be suppressed and the bicyclo form stabilized by bridging the 1,6-positions. Transfer technology has been used for the synthesis of such compounds.[148] Addition of cyclobutadiene (427), formed *in situ* from the iron tricarbonyl complex, to the transfer agent (420) gave the isomers (428) and (429). These were separated and each was reduced with lithium aluminium hydride and the resulting diols cyclised using triphenylphosphene/diethyl azodicarboxylate to give the ethers (430) and (431). FVP of (430) at 450°/0.001 Torr converted the bicyclohexene into the cyclohexadiene (432). At 490°/0.04 Torr the diene (432) and the two precursors (430) and (431) gave the Alder-Rickert cycloreversion to yield 8-oxatricyclo[4,3,2,0]undeca-2,4,10-triene (433).

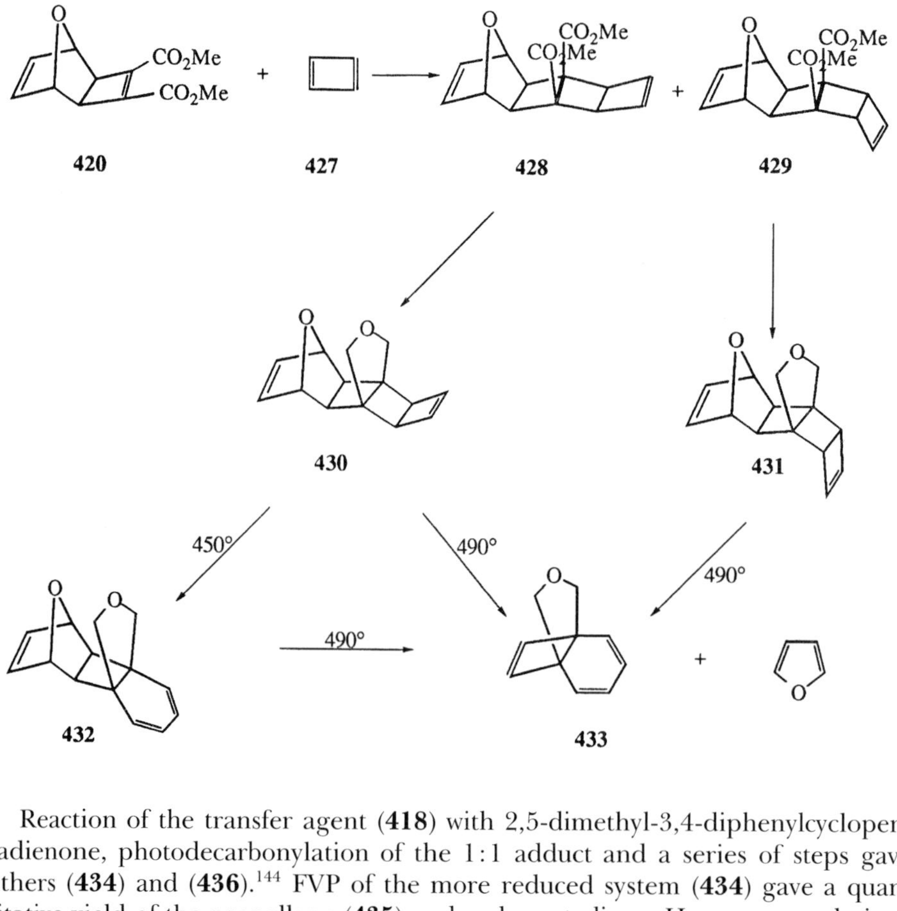

Reaction of the transfer agent (418) with 2,5-dimethyl-3,4-diphenylcyclopentadienone, photodecarbonylation of the 1:1 adduct and a series of steps gave ethers (434) and (436).[144] FVP of the more reduced system (434) gave a quantitative yield of the propellane (435) and cyclopentadiene. However, pyrolysis of

(**436**) gave a mixture of products arising from the Alder-Rickert cleavage, triene (**437**) and from a second route considered to involve a non-concerted mechanism leading to norbornadiene (**438**) and the aromatic product (**439**). The tendency to proceed along this non-concerted route was equated to stabilization of the intervening radical (**440**).

The known photodimer[149] (**441**) of the transfer agent (**418**) on FVP gave detectable amounts of polyene (**442**) but also a number of products arising from retro-Diels-Alder processes.[145] Epoxidation of the double bonds in (**441**) produced the bisepoxide (**443**) which fragmented without these further problems. FVP of (**443**) at 500°/0.003 Torr yielded the epoxypolyene (**444**). Other nitrogen containing compounds have been synthesised using these methods.

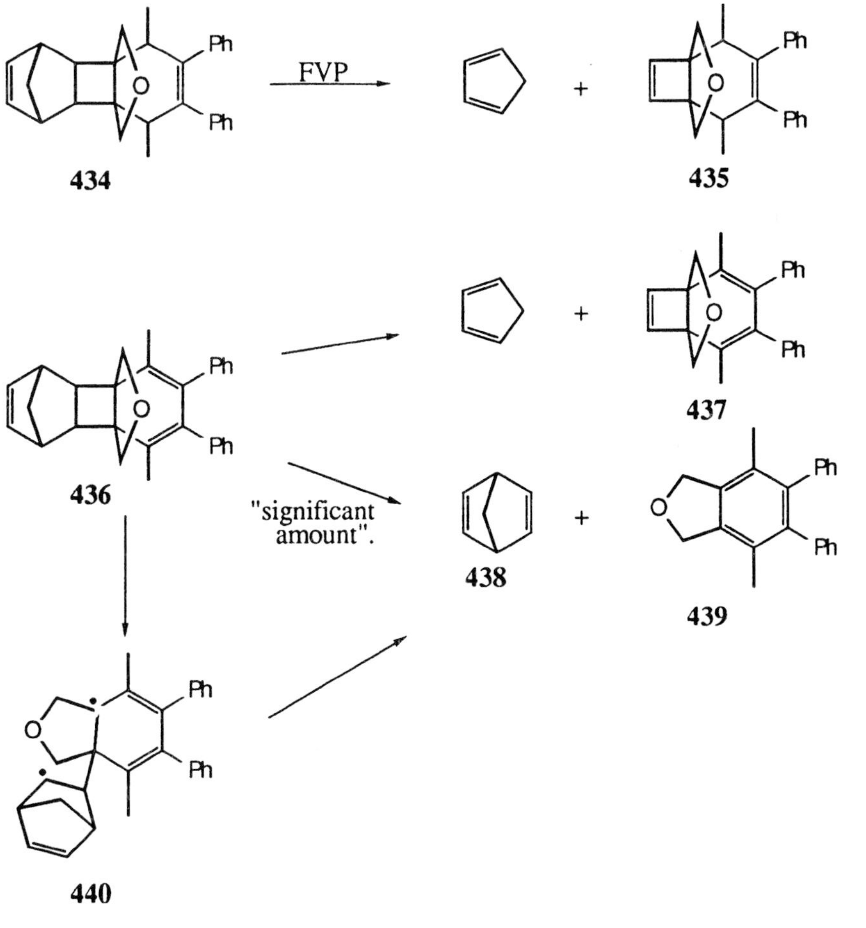

The use of these reagents and FVP for fragmentation for the synthesis of simpler systems has been described.[146] Addition of diazomethane to reagent (**418**) followed by elimination of dinitrogen by ultraviolet irradiation gave the cyclopropane derivative (**445**). FVP at 500–550°/0.01 Torr caused cleavage of the σ bond

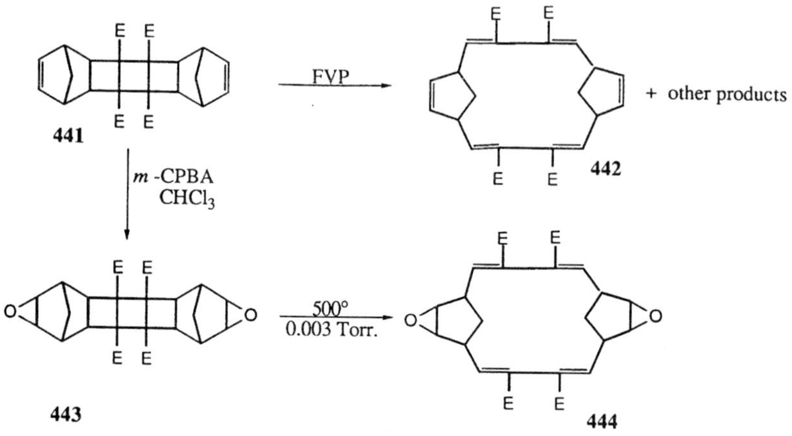
E E
E E
441
FVP
E E
E E
442
+ other products
m -CPBA
CHCl3
E E
O
O
E E
443
500°
0.003 Torr.
E E
O
O
E E
444

E = CO_2Me

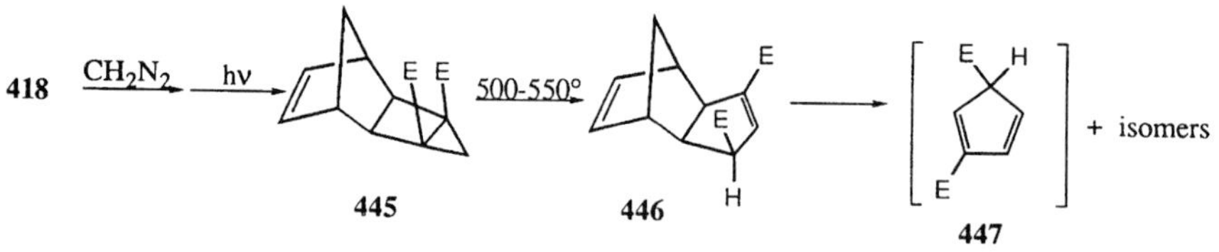
418
CH2N2
hν
E E
445
500-550°
E
E
H
446
E H
E
447
+ isomers

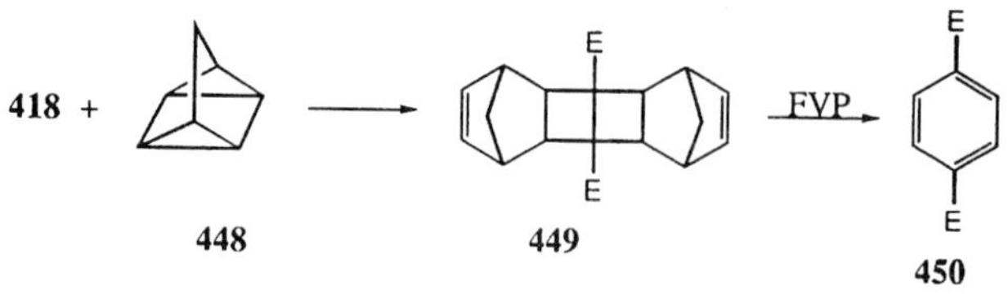
418 +
448
E
E
449
FVP
E
E
450

E = CO_2Me

to yield a mixture of epimeric *exo*-fused cyclopentenes (**446**) which eliminated cyclopentadiene to give dimethyl cyclopentadiene-1,3-dicarboxylate (**447**) together with other isomers arising from hydrogen shifts and prototropic rearrangement.

Transfer agent (**418**) reacts with a molecule of quadricyclane (**448**) to yield a formal cyclopentadiene/Dewar benzene adduct (**449**).[143] Pyrolysis of this yielded the terephthalate ester (**450**). Other electron-withdrawing substituents can be incorporated giving a variety of *p*-disubstituted benzenes.

17. DECARBONYLATION LEADING TO HYDROCARBONS AND KETONES

Diederich and coworkers have developed a useful synthetic method for dialkyl, diaryl and disilyl polyacetylenes (poliynes) by decarbonylation of cyclobutene-1,2-diones.[150]

3,4-Dichlorocyclobutene-1,2-dione (**451**), which had been shown by Bock et al.[151] to yield dichloroethyne on FVP, undergoes substitution of the chlorine by (tri-n-butylstannyl)alkynes in the presence of Pd(PPh$_3$)$_4$ or by copper (I) acetylides to give the diethynyl derivatives (**452**).[152] Introduction into the pyrolysis apparatus was achieved using solution spray FVP whereby a benzene solution was passed through a capillary into the evacuated heated tube. Under these conditions at 650° the dione was decarbonylated to give the triacetylene (**453**).

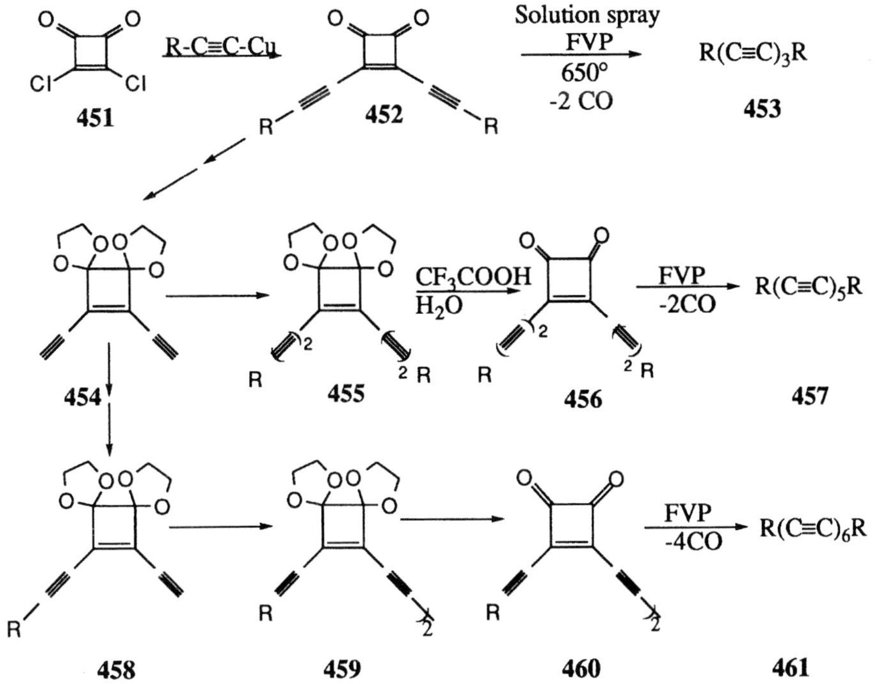

R-C≡C-Cu

Solution spray
FVP
650°
-2 CO

R(C≡C)$_3$R

451   **452**   **453**

CF$_3$COOH
H$_2$O

FVP
-2CO

R(C≡C)$_5$R

454   **455**   **456**   **457**

FVP
-4CO

R(C≡C)$_6$R

458   **459**   **460**   **461**

To develop this method further, the diethynyl compound (**452**, R=t-BuMe$_2$Si) was heated with bis(trimethylsilyloxy)ethane and trimethylsilyl triflate to give the bis ketal which was converted to the diethynyl derivative (**454**) by reaction with potassium hydroxide in methanol. Compound (**454**) could be extended through oxidative coupling to (**455**) which on hydrolysis to (**456**) and solution spray FVP gave the pentaacetylene (**457**). Formation of the unsymmetrical intermediate (**458**) allowed oxidative coupling to (**459**). Hydrolysis yielded the intermediate (**460**) and this on pyrolysis gave the hexayne (**461**).

Vogtle and his group have shown that FVP of cyclophane diones, triones and tetraones at 600–650°/10^{-5} Torr can be controlled so that carbon monoxide is lost successively to give intermediate ketones and finally the cyclophane. The twelve-membered ring diketone (**463**) was selectively converted at 610° to the monoketone (**462**) and at 650° to the cyclophane (**464**).[153] Similarly, pyrolysis of the [3.3.3] cyclophanetrione (**465**) gave the di- and monoketones (**466**) and (**467**) and the hydrocarbon (**468**) with increasing temperature.[153]

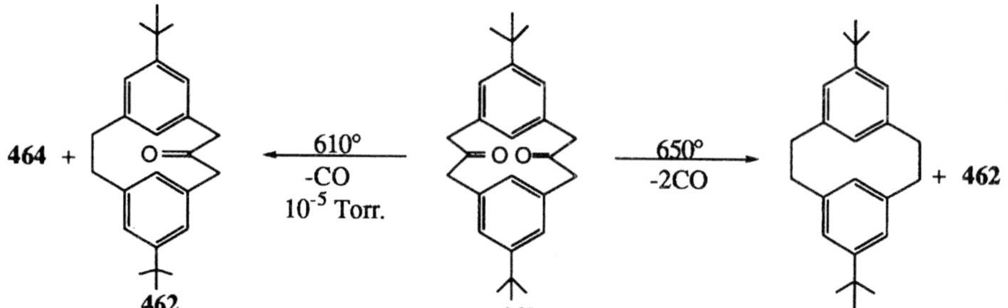

462

463

464

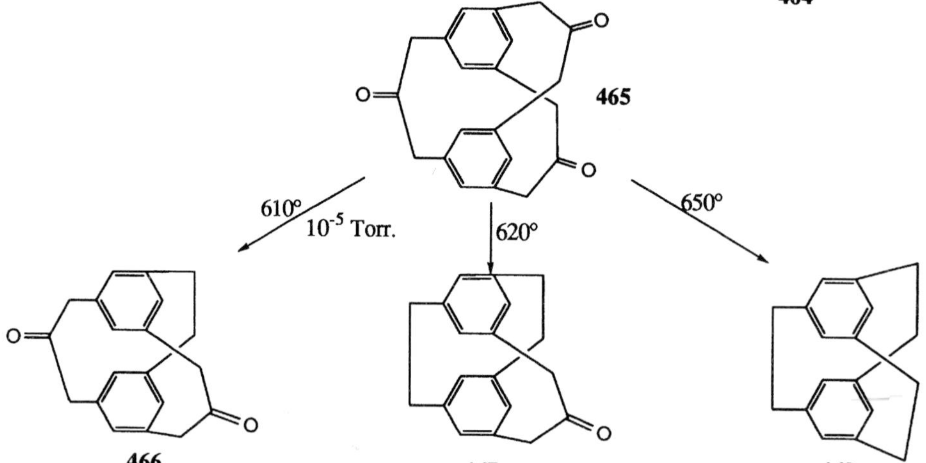

465

466

467

468

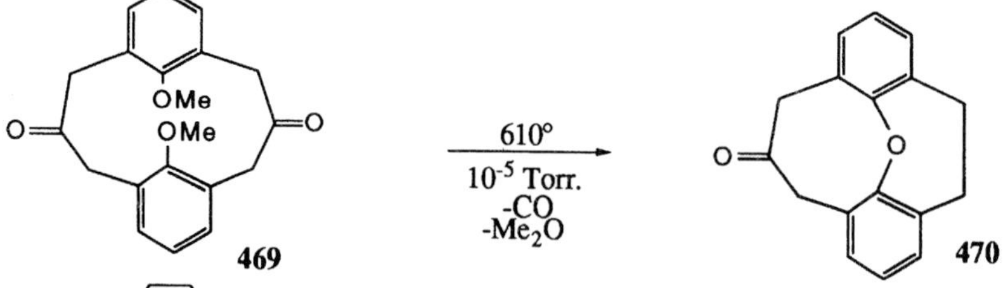

469

470

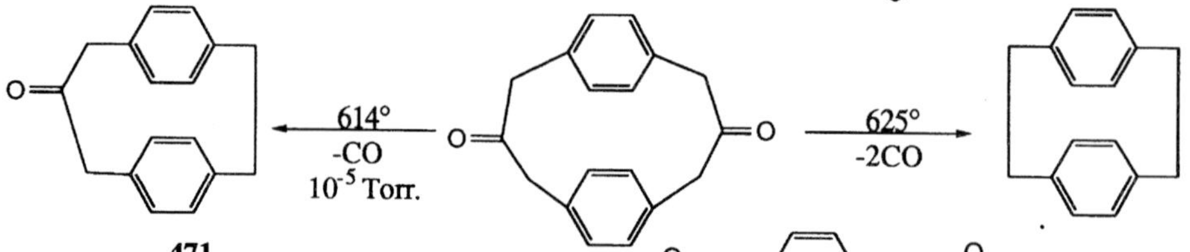

471

472

473

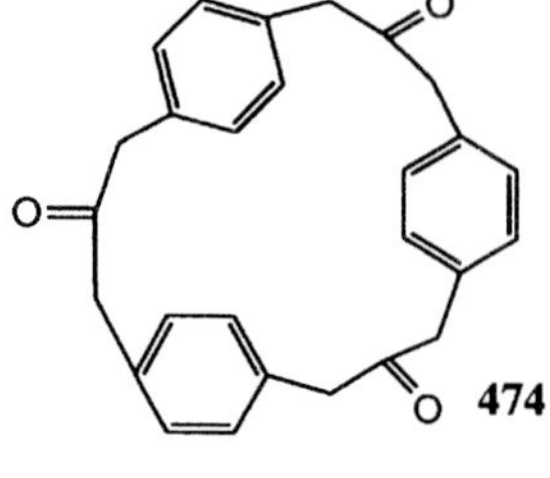

474

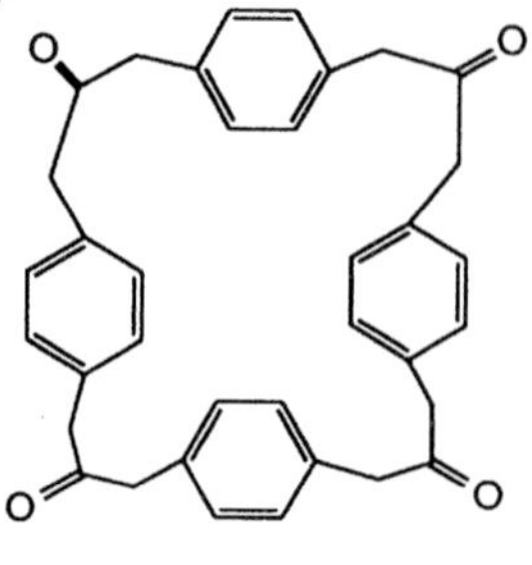

475

The dimethoxy[3.3]metacyclophane-2,11-dione (**469**) on FVP at $614°/10^{-5}$ Torr eliminated carbon monoxide and dimethyl ether to give the bridged cyclophane (**470**).[154] [3.3] Paracyclophane-2,12-dione (**472**) was converted into the monoketone (**471**) (32%) at 614° and into [2.2]-paracyclophane (**473**) (30%) at 625°.[155] The larger ring system in [3.3.3] (1,4) cyclophane-2,22,20-trione (**474**) on FVP at 614° gave the intermediate di- and monoketones and [2.2.2] 1,4-cyclophane. [3.3.3.3] (1,4) Cyclophane-2,11,20,29-tetraone (**475**) on FVP at 670° gave a trione, two diones, a monoketone and [2.2.2.2] 1,4-cyclophane. The reaction has been extended to include 1,4-naphthalene and 2,2"-(1,1':3',1"-terphenyl)cyclophane ketones.[155a]

The reaction could be applied to the aliphatic ketone cyclododecanone which at 800° gave cycloundecane (26%) and to the open chain analogue of the cyclophane ketones, dibenzyl ketone which at 700° was converted into 1,2-diphenylethane (50%).[155]

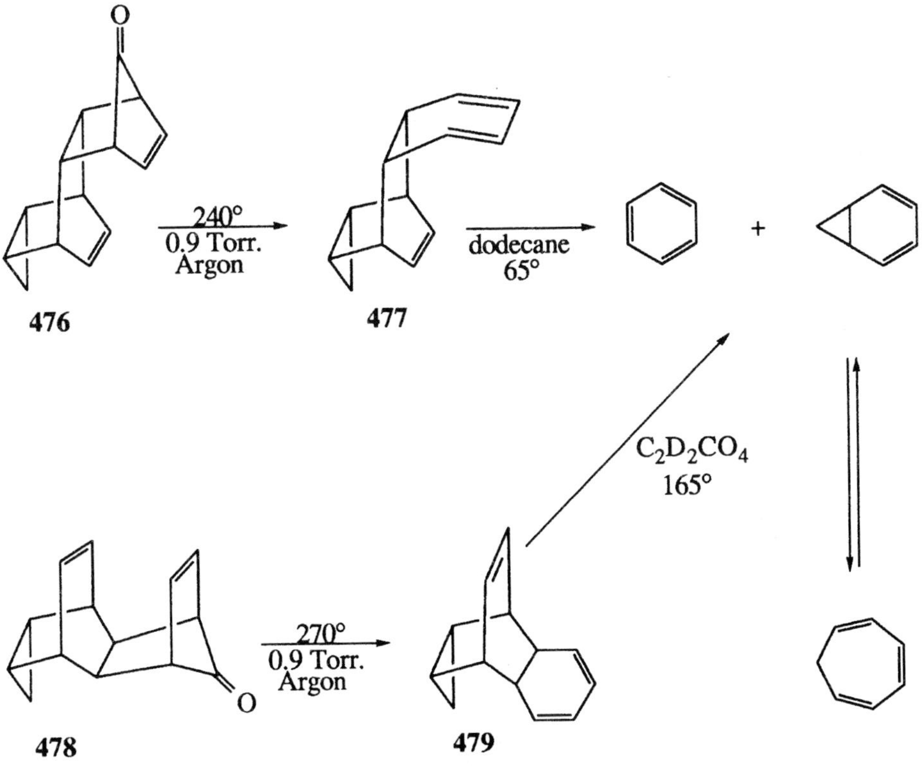

Grimme, Warner and coworkers[156] have used decarbonylation under FVP conditions to prepare two thermally sensitive triene intermediates (**477**) and (**479**) that underwent cycloreversion to benzene and norcaradiene/cyclohepta-triene. The intermediate ketones (**476**) and (**478**) were obtained by hydrolysis of ketals formed by cycloaddition. The pyrolyses were conducted at 0.9 Torr in a flow of argon. Pyrolysis of (476) at 240° gave the *anti*-tetracyclotriene (477) in 77% yield and of (478) at 270° gave the *syn*-tetracyclotriene (479) in 85% yield. The rate ratio of (477) to (479) for the extrusion of benzene was 8.9×10^3 which was attributed to homoaromatic interactions in the cycloreversion transition state of (477).

18. MECHANISMS OF DECARBOXYLATION OF CARBOXYLIC ACIDS

Back and coworkers have examined the gas-phase decomposition of oxalic,[157] pyruvic,[158] glyoxylic[159] and malonic acids in a static quartz system and have identified the products by VPC. Malonic acid on decomposition at 92–151°/0.1 Torr gave acetic acid and CO_2 as major products in approximately equal amounts together with smaller amounts, expressed as a percentage of the amount of CO_2, of CO (5%), acetone (3%), ethane (0.4%) and methane (0.24%). It was proposed that the major products are formed by intramolecular hydrogen atom transfer through a four centre transition state. The alternative unimolecular six centre transition state leading to 1,1-dihydroxyethene has previously been considered in solution and remains a possibility in the gas-phase (Scheme 47). Minor pathways involving radicals occurred in the quartz apparatus even at the low temperatures used.

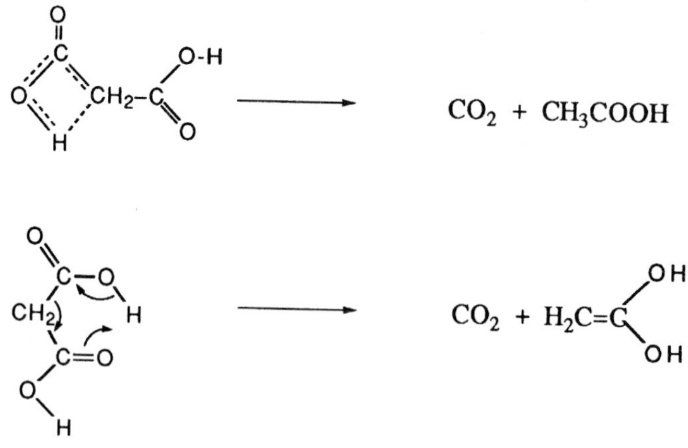

$$CO_2 + CH_3COOH$$

$$CO_2 + H_2C=C\begin{smallmatrix}OH\\OH\end{smallmatrix}$$

Scheme 47

$$CO_2 + CH_3-C\begin{smallmatrix}H\\O\end{smallmatrix}$$

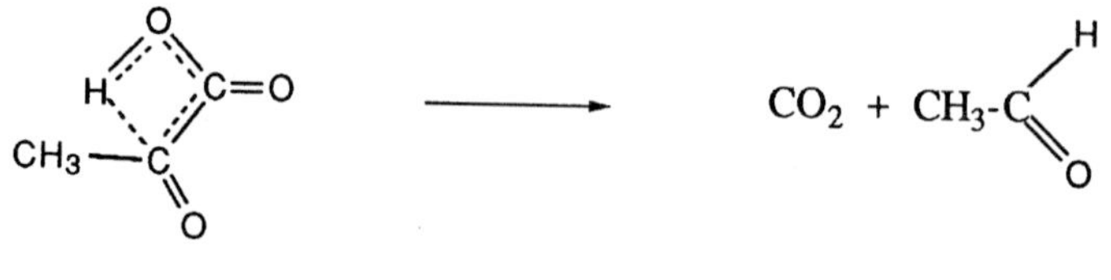

Scheme 48

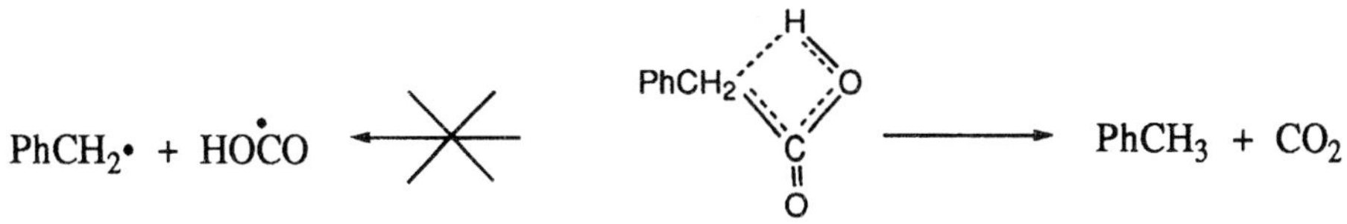

$$PhCH_2\bullet + H\dot{O}CO \quad \xmapsto{} \quad \begin{array}{c} H \\ PhCH_2 \cdots \diamond \cdots O \\ C \\ \| \\ O \end{array} \quad \longrightarrow \quad PhCH_3 + CO_2$$

Scheme 49

The experimental results obtained for the conversion of pyruvic acid[158] into acetaldehyde and carbon dioxide appeared anomalous. The reaction has been studied further by Taylor in a stainless steel apparatus.[161] From a linear Arhenius plot there was obtained an activation energy of 41.25 kcal/mole and a value of log $(A/s^{-1})=13.53$. Four or six-centre transition states generally have log (A/s^{-1}) values around 14. Pyruvic acid was consequently regarded as fragmenting unimolecularly *via* a four-membered transition state (Scheme 48).

Colussi and coworkers[162] have shown that passage of phenylacetic acid through a capillary restriction at a flow rate of 5×10^{13} to 5×10^{14} molecules s^{-1} into a silica flow reactor at very low pressure and heated to 600–750° gave toluene and carbon dioxide as determined by modulated beam mass spectrometry. They concluded that the molecular decomposition to the products involved a concerted four centre process and the mechanism did not involve the benzyl and hydroxy-carbonyl radicals (Scheme 49).

19. 1,2-ELIMINATION OF CARBON DIOXIDE

Following the successful synthesis of the pyramidalized alkene (**481**), which was isolated as the [2+2] dimer (**482**), by thermolysis of the dimethylamino dioxolan (**480**) in tetraglyme containing 1 equivalent of acetic acid,[163] the unsubstituted tricyclo-[3.3.2.0^{3,7}] dec-3(7)-ene (**484**) was sought. Preparation of the β-lactone (**483**) was achieved by ruthenium oxide oxidation of the oxetane.[164]

FVP of (**483**) at 410°/10^{-3} Torr resulted in 50% decarboxylation and a pyroly-sate which consisted of CO$_2$ and the alkene (**484**) which was analysed by IR spec-troscopy at 10 K.[165] When allowed to warm to room temperature the alkene (**484**) gave the [2+2] dimer (**487**). At 440° the pyrolysate on warming consisted of 90% of the dimer and 10% of (**485**). At 530° the hydrocarbon (**485**) was the major pro-duct, the dimer constituted only 10% of the mixture, and the further rearrange-ment product of (**485**), the diene (**486**) was also present. Compound (**485**) was shown to be partially rearranged to 2,6-dimethylenebicyclo[2.2.2]octane (**486**) at 500°.[166]

The β-lactone (**488**), the precursor of the more highly pyramidalized alkene (**491**), was resistant to loss of CO$_2$, and FVP at 550°/10^{-3} Torr gave 50% decomposition but without loss of CO$_2$.[167] The product was the keto ketene (**489**). This mode of β-lactone cleavage was unprecedented and was indicative of the high transition state energy leading to the alkene (**491**). On FVP at temperatures above 550° the β-lactone (**488**) lost some CO$_2$ and a small amount of the [2+2]

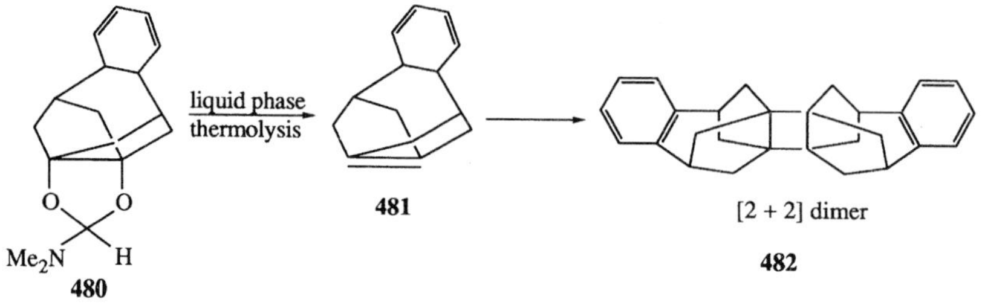

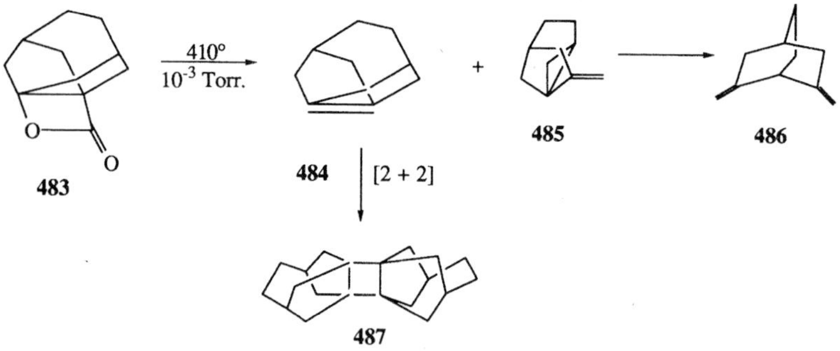

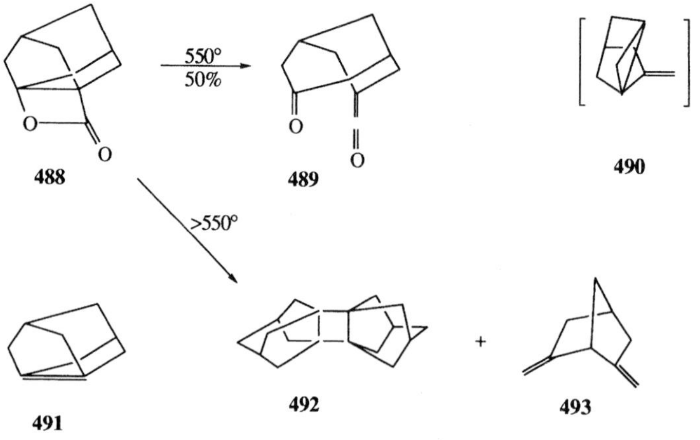

liquid phase thermolysis

480

481

[2 + 2] dimer

482

483

410°
10⁻³ Torr.

484 | [2 + 2]

485

486

487

488

550°
50%

489

490

>550°

491

492

493

dimer (**492**) was obtained. The major product (>95%) was 2,6-dimethylene-bicyclo[2.2.1]heptane (**493**) which may have been formed *via* the vinyl-cyclo-propane structure (**490**) but this was not detected.

Taylor has studied the kinetics of thermal elimination reactions at tempera-tures up to 500° in a gold plated stainless steel reactor.[168] Volatile compounds were studied directly and involatile solids were dissolved in chlorobenzene. Attention has been drawn to the similarities of the kinetic parameters obtained from Arrhenius plots of first order unimolecular reactions for a number of thermal eliminations and these have been compared with a body of data from the literature.[168] 1,4-Elimination of CO_2 from isochroman-3-one (**494**) takes place on pyrolysis to give benzocyclobutene with E=221.1 kJ/mole and log $(A/s^{-1})=$ 14.305, a value characteristic of elimination from cyclic compounds. Pyrolysis of benzyl methyl ether (**495**) (231.7 kJ/mole, 12.92) and 2-phenylethanol (**496**) (200.5 kJ/mole, 11.21) each gave toluene and formaldehyde.

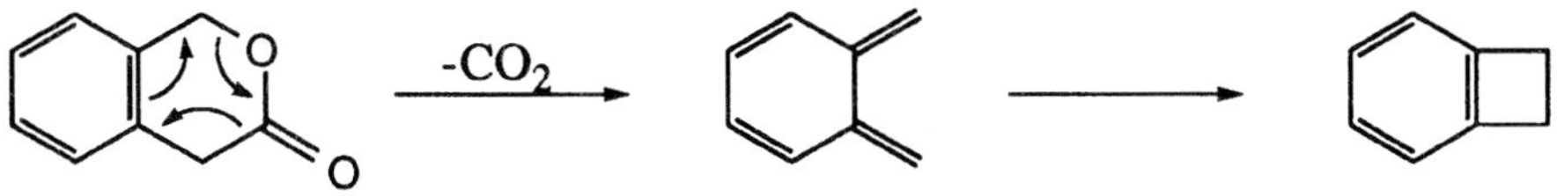

494

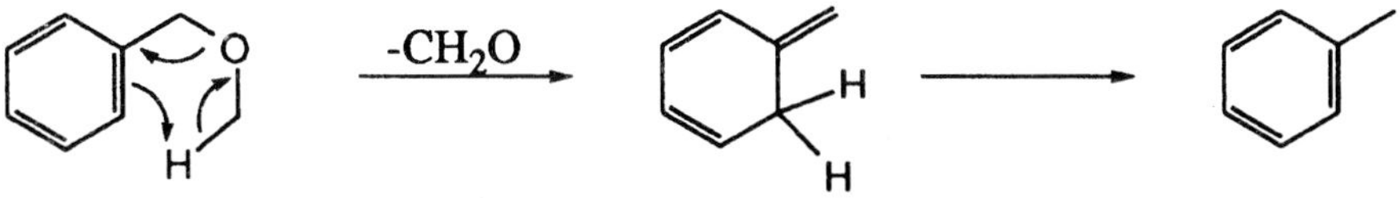

495

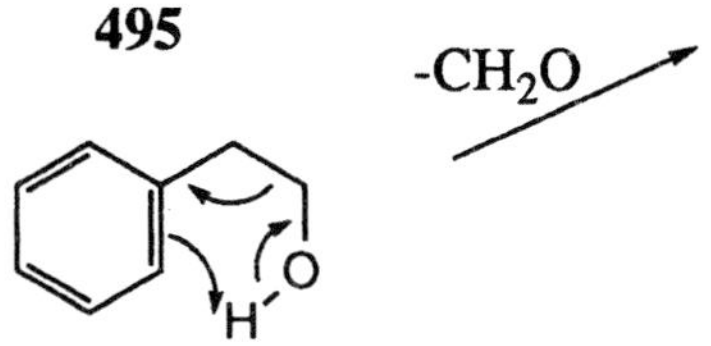

496

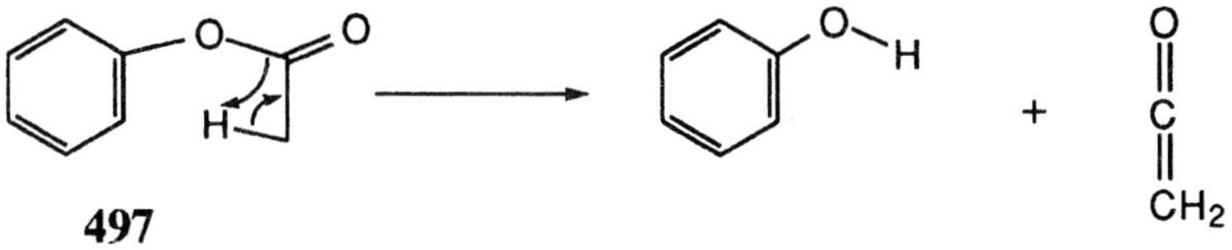

497

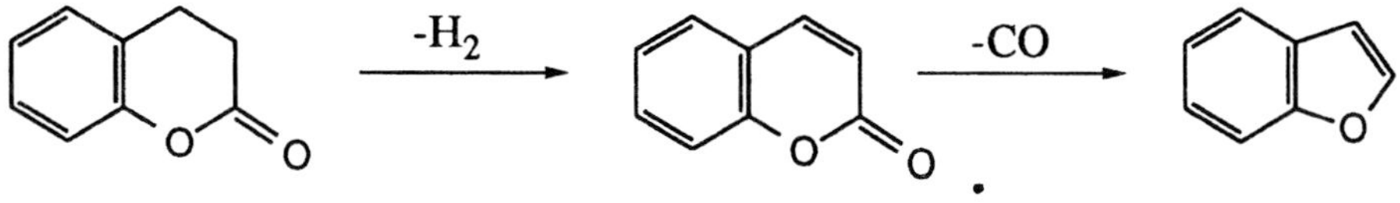

498   **499**   **500**

These reactions took place by a semi-concerted mechanism involving a six-membered transition state. The conversion of phenyl acetate (**497**) to phenol and ketene (221.8 kJ/mole, 11.78) proceeded *via* a four centre transition state. 3,4-Dihydro-2*H*-pyran fragmented to ethylene and propenol by cycloreversion but ethyl phenyl ether decomposed by more than one mechanism. Pyrolysis of chroman-1-one (**498**) gave coumarin (**499**) which led to benzofuran (**500**) by loss of CO. Similar kinetic studies on thermal eliminations in substituted vinyl ethers have led to proposals concerning the non-planarity of the six-membered transition state in these particular reactions.[169]

20. ELIMINATION OF CARBOXYLIC ACIDS FROM ESTERS

In a theoretical study on the 1,2-elimination of acids from esters on gas phase pyrolysis it was found that the experimental order of reactivity was reproduced correctly by AM1 semiempirical M.O. calculations but not by MNDO.[170] The six-membered transition state was confirmed as the pathway of minimum energy and the most important interactions were shown to be the π-donating ability of the π-HOMO of the C=O group and the accepting ability of the σ^* LUMO of the C—H acceptor. Substituents increasing these HOMO–LUMO interactions increase the reactivity.

The effect of ring size in cycloalkyl acetates on the gas-phase elimination has been determined.[171] The relative rates of the unimolecular reactions measured in a static system at 280–370°/35–234 Torr can be compared by setting the cyclohexyl acetate as standard and using $\log K - \log K_6$ to contrast the rates of elimination. The values obtained for the cycloalkyl rings containing n carbons were for n=5 (0.75), 6 (0.0), 7 (0.82), 8 (1.07), 10 (1.58), 12 (1.05) and 15 (0.76). Elimination in each case was faster than in cyclohexyl acetate and this reflects the relative ease of formation of the required planar six-membered transition state.

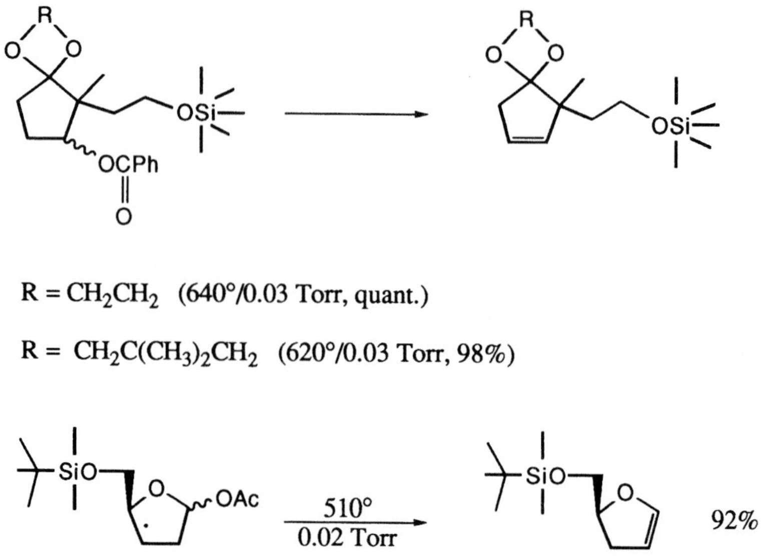
R
O
O
OSi
OCPh
O
R
O
O
OSi
R = CH2CH2 (640°/0.03 Torr, quant.)
R = CH2C(CH3)2CH2 (620°/0.03 Torr, 98%)
SiO
O
OAc
510°
0.02 Torr
SiO
O
92%

Scheme 50

Preparative gas phase ester pyrolyses are well known and widely used. Three examples of the use of FVP for preparative reactions which demonstrate the thermal stability of ketal and *t*-butyldimethylsilyl protecting groups are shown in Scheme 50.[172–174]

1,4-Elimination of carboxylic acids from esters is less common. Trahanovsky and coworkers developed a preparation of 2,3-dimethylene-2,3-dihydrofuran (**502**) by FVP of (2-methyl-3-furyl)methyl benzoate (**501**) at 620–640°/10⁻⁴ Torr.[175] The dimethylene compound dissolved in carbon disulfide and warmed above −30° yielded the [4+ 4] head-to-head dimer (**504**) quantitatively. This dimerization has been shown to proceed stepwise through a diradical intermediate (**503**).[176] A study of the benzo analogue has shown that 2,3-dimethylene-2,3-dihydrobenzofuran (**506**) dimerises by both [4+ 4] and [4+ 2] processes.[177] FVP of (2-methyl-3-benzofuryl)methyl benzoate (**505**, R= CH₃) at 620–640°/10⁻⁴ Torr gave a colourless pyrolysate in the cold trap at −196°. The ¹H NMR spectrum of this pyrolysate measured at −60° showed the presence of the dimethylene compound (**506**) formed in 35% yield. In carbon disulfide solution at room temperature this compound dimerised to give a mixture of the [2+ 4] dimer (**509**) and the [4+ 4] dimer (**510**) in a ratio of 4.4:1. These findings were confirmed and the ¹H NMR spectra were assigned with the aid of the corresponding deuterated compounds arising from the dideuteromethylene intermediate formed by pyrolysis of **505** (R= CD₃). The only [4+ 2] dimer formed was that expected to arise from the alternative mode of cyclisation of the diradical intermediate (**507**). This diradical mechanism was considered more likely than concerted [4+ 2] addition which would be expected to give a mixture of isomers. In a reaction with methyl acrylate the dimethylene compound (**506**) behaved as a diene to give the two expected [2+ 4] regioisomeric esters. The elimination of benzoic acid from (**505**) may involve direct 1,4-elimination or a [3,3] rearrangement to (**508**) followed by 1,2-elimination.

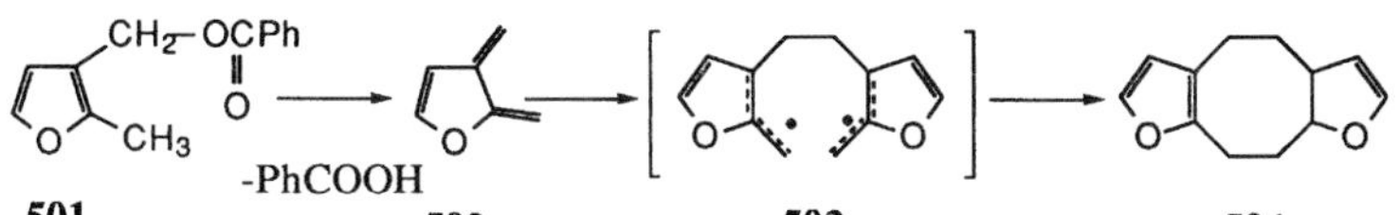

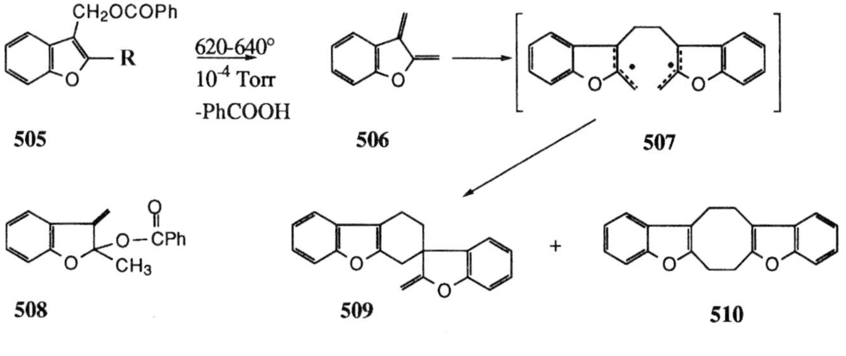

21.   REDUCTIVE ELIMINATION AND OTHER ELIMINATIONS

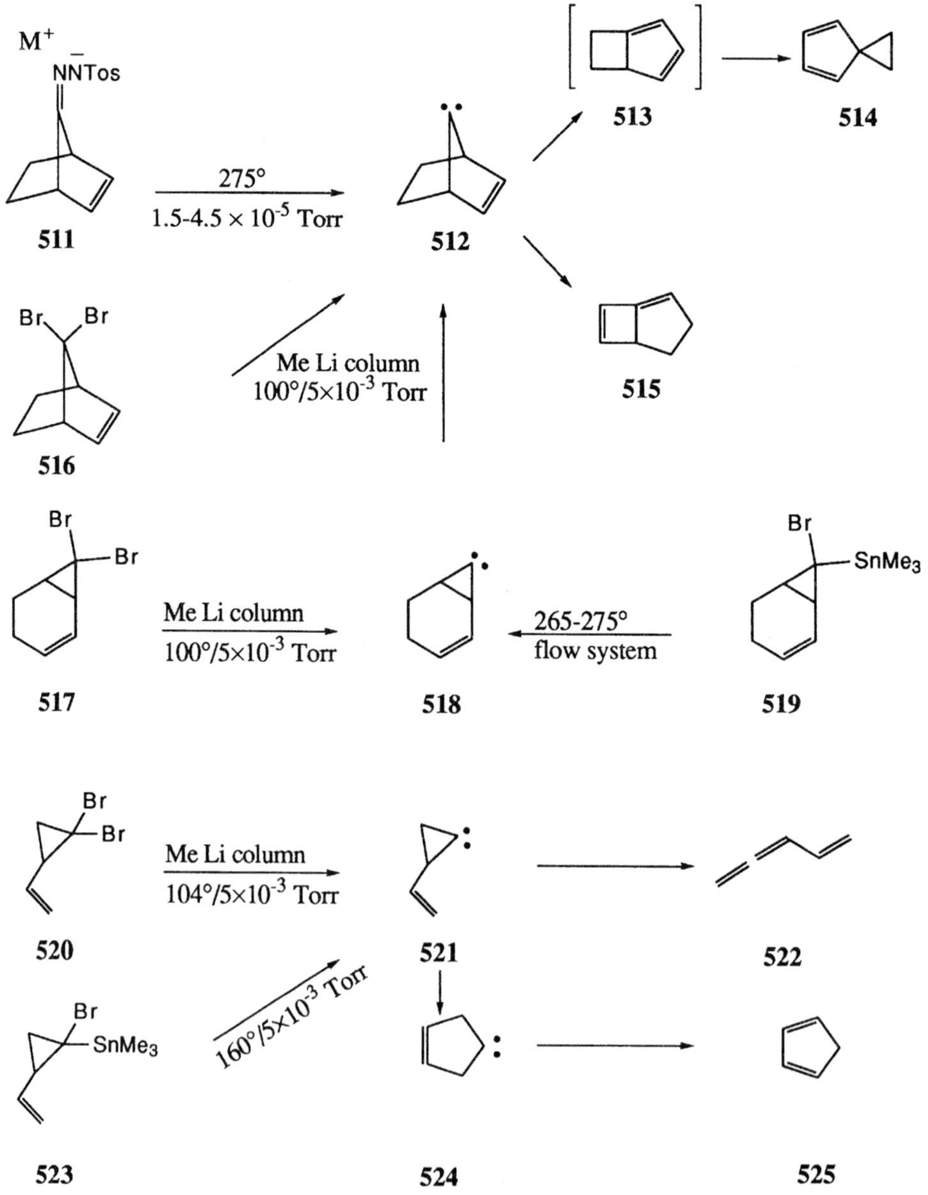

Brinker and Ritzer have explored the rearrangement of 7-norcar-2-enylidene (**518**) to 7-norbornenylidene (**512**) and its conversion to bicyclo[2,3,0]hepta-1,6-diene (**515**) and spiro[2,4]hepta-4,6-diene (**514**).[178] They showed that formation of

7-norbornenylidene (**512**) in the gas phase by pyrolysis at $275°/1.5–4.5 \times 10^{-4}$ Torr of the lithium salt and of the sodium salt of norbornen-7-one tosylhydrazone (**511**) gave (**515**) and (**514**) in 62.4% and 5.2% and in 51% and 4.9% yields respectively. Formation of the carbene (**512**) by reductive debromination of 7,7-dibromonorbornene (**516**) by passage through a bed of glass turnings carrying adsorbed methyl lithium at $100°/5 \times 10^{-3}$ Torr gave the same products (**515**) and (**514**) in 47.8% and 12.5%. Similar passage of the 7,7-dibromonorcar-2-ene (**517**) through a methyl lithium column at $100°/5 \times 10^{-3}$ Torr gave (**515**) and (**514**) in yields of 25% and 4.8% respectively. This provided new evidence for the conversion of 7-norcar-2-enylidene (**518**) to norbornenylidene (**512**) which rearranged to (**515**) and (**514**). As an alternative approach to the carbene (**518**), 7-bromo-7-(trimethyltin)norcar-2-ene (**519**) was pyrolysed at 265–275° in a flow system but gave only traces of the products (**514**) and (**515**).

The rearrangement of vinylcyclopropanylidene (**521**) was also explored in this manner. Passage of 1,1-dibromo-2-vinylcyclopropane (**520**) through a methyl lithium column at $104°/5 \times 10^{-3}$ Torr gave penta-1,2,4-triene (**522**) and cyclopentadiene (**525**) *via* the carbene (**524**) in a combined yield of 85%. An alternative approach to the carbene through pyrolysis of 1-bromo-1-(trimethyltin)-2-vinylcyclopropane (**523**) at $260°/5 \times 10^{-3}$ Torr gave principally buta-1,2,4-triene (**522**).

Columns packed with potassium *t*-butoxide on Chromosorb W also provide an efficient method for the synthesis of sensitive alkenes by elimination of hydrogen halide from halides in the gas phase. Thus, passage of 2-chloromethylenecyclopropane (**526**) through such a column at 240°/0.01 Torr gave nearly pure methylenecyclopropene (**527**) which could be trapped at −196° but which decomposed above −75°.[179] Similar observations were made when 2-bromomethylenecyclopropene (**528**) was passed over the same reagent at room temperature and 0.02–0.03 Torr.[180]

Passage of 2,4-dichlorospiro[2,2]pentane (**529**) through potassium *t*-butoxide on Chromosorb W at 320°/0.01 Torr gave 1-vinylcyclopropene (**531**).[181] This compound was also formed by passage of 2-chlorovinylcyclopropene (**530**) through the column under the same conditions. At −60° the vinylcyclopropene formed a dimer (**532**) which at room temperature formed the valence isomer 1,2-divinylcyclohexa-1,4-diene (**533**).

Reductive elimination from 1,2-dihalo-compounds by passage of the vapour through methyl lithium adsorbed on glass was also successful for the synthesis of a cyclopropene.[181] 1-Bromo-1-trimethylsilyl-2,2-dichlorocyclopropane (**535**) on passage through a methyl lithium column at $25°/10^{-2}$ Torr gave an 85% yield of 1-chloro-2-trimethylsilylcyclopropene (**534**).

Fluoride ion induced elimination of trimethylsilyl fluoride and chloride ion from 1-trimethylsilyl-2-chloro compounds can also be achieved by passage of the substrate in the gas phase through tetrabutylammonium fluoride adsorbed onto glass helices. Passage of (**535**) through such a column at $25°/10^{-2}$ Torr gave 1-bromo-2-chlorocyclopropene (**536**).[181] Similarly, passage of (**537**) through a "fluoride column" at $25°/10^{-2}$ Torr gave the alkene (**538**) which dimerized rapidly even at −120° but could be trapped by cyclopentadiene to give (**539**).

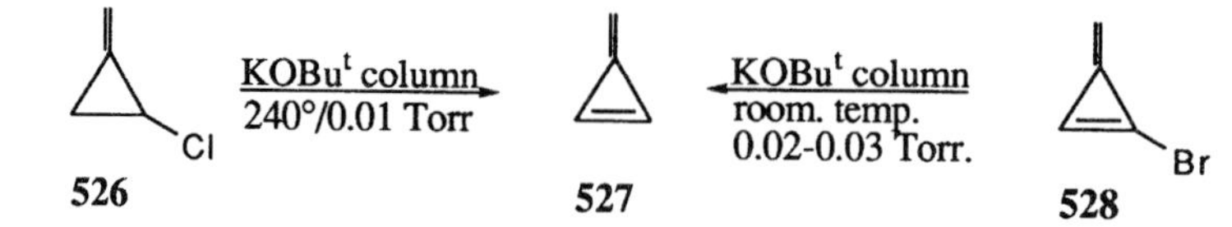

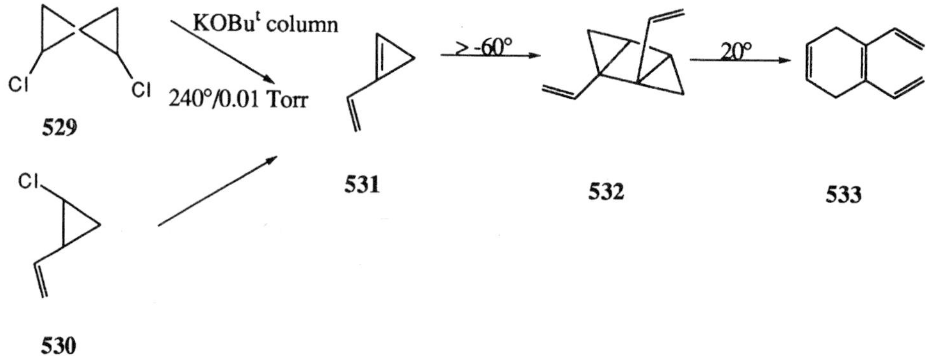

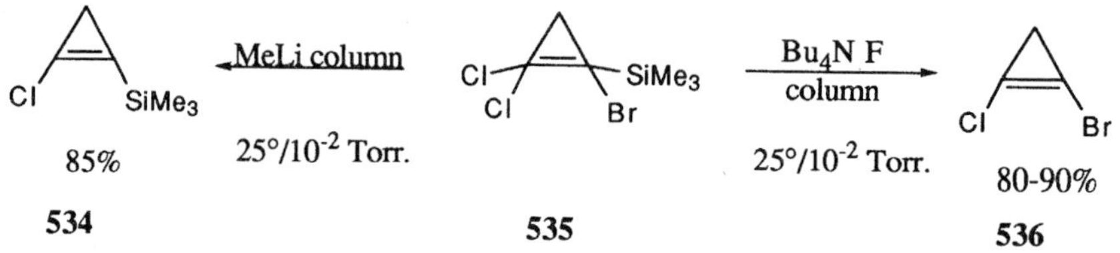

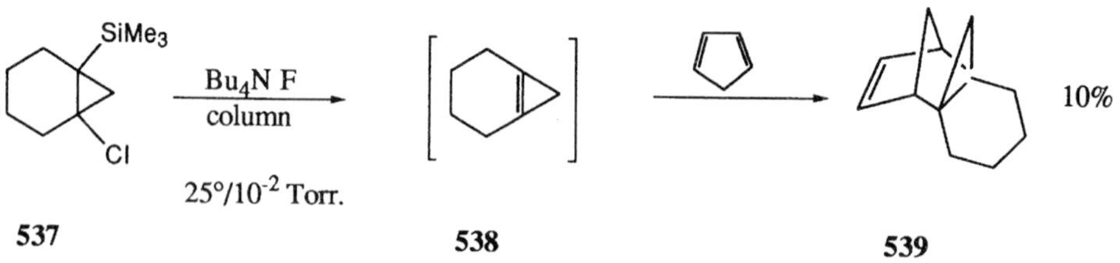

526   KOBu^t column
            240°/0.01 Torr   **527**   KOBu^t column
                                                room. temp.
                                                0.02-0.03 Torr.   **528**

529   KOBu^t column
            240°/0.01 Torr

530   **531**   ≥ -60°   **532**   20°   **533**

534   MeLi column   **535**   Bu$_4$N F   **536**
85%   25°/10^{-2} Torr.   column
                                             25°/10^{-2} Torr.   80-90%

537   Bu$_4$N F   **538**   **539**   10%
            column
            25°/10^{-2} Torr.

22. FORMATION OF ALKYNES BY ELIMINATION
OF TRIPHENYLPHOSPHINE OXIDE

Pyrolytic elimination of triphenylphosphine oxide from oxoylides leads to the formation of acetylenes. Aitken and coworkers showed that reaction of acid chlorides with two equivalents of methylenetriphenylphosphorane gave the intermediate oxoylides (**540**) which on FVP at 750°/10^{-1}–10^{-2} Torr gave the terminal alkynes (**541**).[182] However, the procedure could be simplified by acylation of the available stabilized ylide ethoxycarbonylmethylenetriphenylphosphorane in the presence of triethylamine. The resulting intermediate ylides (542) on FVP at 500° gave the terminal acetylenic esters (543) in good yield. Increasing the temperature of pyrolysis to 750° caused loss of the ethoxycarbonyl group, most probably by ester elimination followed by decarboxylation, and formation of the terminal acetylene (541).[183]

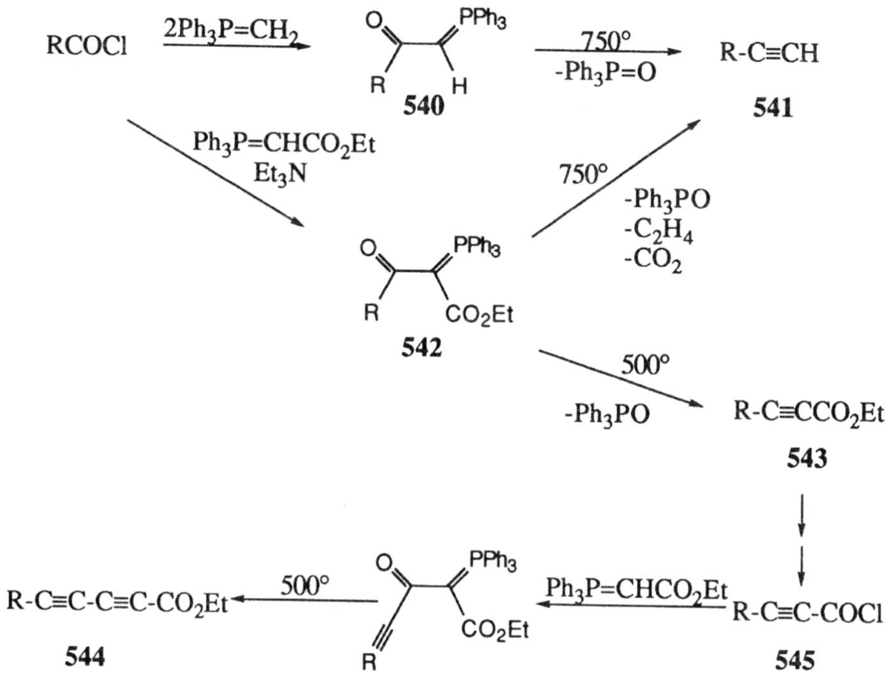

The availability of the acetylenic ester (543) provided the basis for an iterative procedure.[184] Hydrolysis of the ester and conversion of the acid into the acid chloride (545) allowed repetition of the cycle and the preparation of the diacetylenic ester (544). Further extension of the chain by this method was achieved.

23. ARYL, ARYLOXYL AND BENZYL RADICALS IN THE GAS PHASE

Gas phase pyrolysis of phenol at 790–890°/760 Torr with an initial phenol concentration of 500–2016 ppm in nitrogen carrier gas gave carbon monoxide, cyclopentadiene and benzene as the principal products.[185] The rate of production of carbon monoxide was greater than that of cyclopentadiene and a radical combination reaction of cyclopentadienyl and phenoxyl radicals provided a mechanism that reproduced the experimental findings.

Pyrolysis of phenyl methyl ether or phenyl ethyl ether at 600–720°/10⁻⁴ Torr yields phenoxyl radicals which decompose unimolecularly to give carbon monoxide and cyclopentadienyl radicals which give cyclopentadienyl dimers

($C_{10}H_{10}$). These processes were demonstrated using on line mass spectrometry.[186] At temperatures above 1000°, naphthalene is also formed. Pyrolysis of *m*-methylphenyl methyl ether yields the *m*-methylphenoxyl radical which fragments to carbon monoxide and methylcyclopentadienyl radicals which lose a hydrogen atom to give fulvene which rearranges to benzene.[186]

The chemistry of phenyl, benzyl, aryloxyl and other radicals ArX˙ formed in short contact time gas phase reactions was reviewed in 1986.[187] Cadogan, McNab and their coworkers have taken this starting point and have explored the formation of such radicals in the gas phase and examined their behaviour in intramolecular reactions.

Coupling of aryl radicals to yield biphenyls proceeds in poor yields and is accompanied by hydrogen abstraction processes. However, the intramolecular attack by an aryl radical on a radicophilic centre is not hindered by large activation energy requirements or the development of charge separation and the resulting radical can form a stable molecule by loss of hydrogen or other atom. FVP of allyl benzoates at 900° results in loss of the stabilized allyl radical and decarboxylation of the resulting benzoyloxyl radicals to form aryl radicals. A series of *ortho*-substituted allyl benzoates having a phenyl or substituted phenyl attached through C,N,O or S atoms has been investigated.[188] Three of these, (**546a**), (**546b**) and (**546c**) on pyrolysis at 900° gave dibenzofuran (**548a**) (49%), fluorene (**548b**) (40%), and a mixture of fluorenone (**548c**) and benzophenone (**549**), respectively. Formation of benzophenone shows the propensity of such an *o*-benzoylphenyl radical (**547c**) to abstract hydrogen intermolecularly. To determine whether intramolecular hydrogen abstractions were occurring, the ester (**550**) was pyrolysed at 900°. The pyrolysate consisted of a mixture of (**552**) and (**554**) in a ratio of 75:25. This ratio was independent of temperature and indicative of complete equilibration between radicals (**551**) and (**553**).

4'-Substituents in (**546a**) led to 2-substituted dibenzofurans and 3'-substituents gave mixtures of 1- and 3-substituted dibenzofurans. The 2'-methyl compound gave 4-methyldibenzofuran, dibenzofuran (by loss of a methyl radical) and, through formation of the *o*-phenoxybenzyl radical, xanthene and fluoren-1-ol (see below).

Under FVP conditions aryloxy radicals can be formed efficiently from allyl and benzyl aryl ethers and benzyl radicals can be formed from dibenzyl oxalates. Using these methods, Cadogan *et al.*[189] showed that the 2-benzylphenoxyl radical (**556**, X= CH_2, R= H) and the 2-phenoxybenzyl radical (**562**, X= CH_2, R= H) both gave mixtures of 2-benzylphenol (**557**, X= CH_2, R= H), fluoren-1-ol (**560**, X= CH_2, R= H) and xanthene (**563**, X= CH_2, R= H). The yields obtained from pyrolysis of (**555**, X= CH_2, R= H), (**558**) and (**561**, X= CH_2, R= H) at 750°/0.02 Torr were precursor dependent and are shown tabulated in Scheme 51. Pyrolysis of the oxalate (**561**, X= CH_2, R= H) had previously been described but only xanthene had been isolated.[190] Formation of the same products from the three precursors provided evidence for a common radical intermediate (**559**, X= CH_2, R= H). A similar requirement for a common intermediate of this type had been found with other elements[191] but the oxygen compounds led to the most complex results.

Application of the same procedure to the methyl substituted compounds (**555**, X= CH_2, R=Me) and (**561**, X= CH_2, R=Me) gave the corresponding products

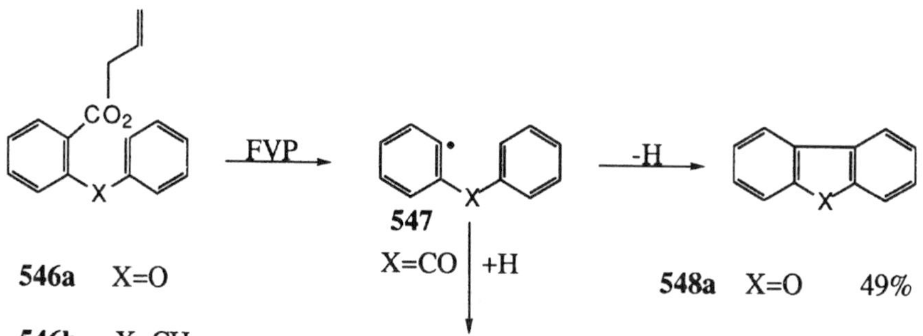

546a X=O

546b X=CH$_2$

546c X=CO

547

X=CO | +H

PhCOPh

549

548a X=O   49%

548b X=CH$_2$ 40%

548c X=CO

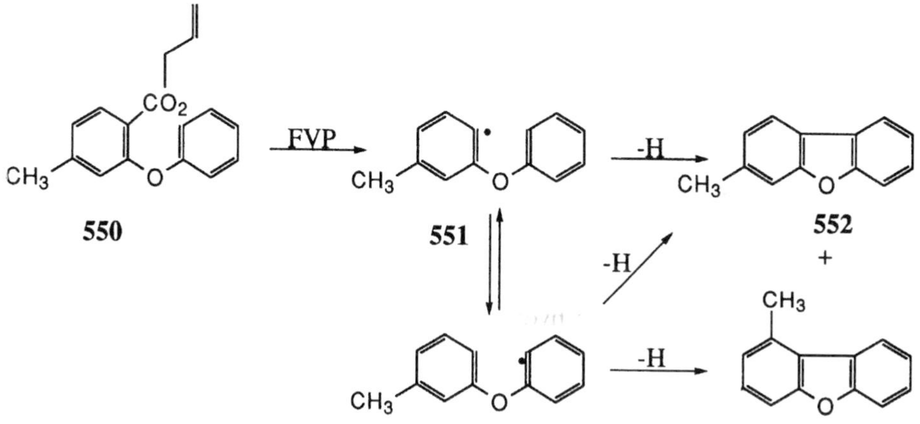

550

551

552
+

553

554

(**557**, X=CH$_2$, R=Me), (**560**, X=CH$_2$, R=Me), (**563**, X=CH$_2$, R=Me) and (**564**, X=CH$_2$, R=Me). The formation of the two methylxanthenes, (**563**, X=CH$_2$, R=Me) and (**564**, X=CH$_2$, R=Me), provided a test for the migratory aptitude of the O and CH$_2$ group in the common radical intermediate (**559**, X=CH$_2$, R=Me). Pyrolysis of (**555**, X=CH$_2$, R=Me) gave (**563**, X=CH$_2$, R=Me) and (**564**, X=CH$_2$, R=Me) in a 1:1 ratio and pyrolysis of (**561**, X=CH$_2$, R=Me) gave a ratio of 1:2.5. These two products were formed in low yield but the findings suggested that, rather than complete equilibration to the intermediate (**559**, X=CH$_2$, R=Me) occurring from both precursors, there was some direct cyclisation of the benzyl radical (**562**, X=CH$_2$, R=Me) to form (**564**, X=CH$_2$, R=Me).

The formation of fluoren-1-ol (**560**, X=CH$_2$, R=H) and 6-methylfluoren-1-ol (**560**, X=CH$_2$, R=Me) in these reactions suggested that the phenoxyl radical may abstract hydrogen from the adjacent phenyl group. Cyclisation of the resulting radical and loss of a hydrogen atom would yield the fluoren-1-ol. Route A in Scheme 52 which involves such a hydrogen abstraction would be expected to be strongly endothermic and an alternative mechanism, route B, involving

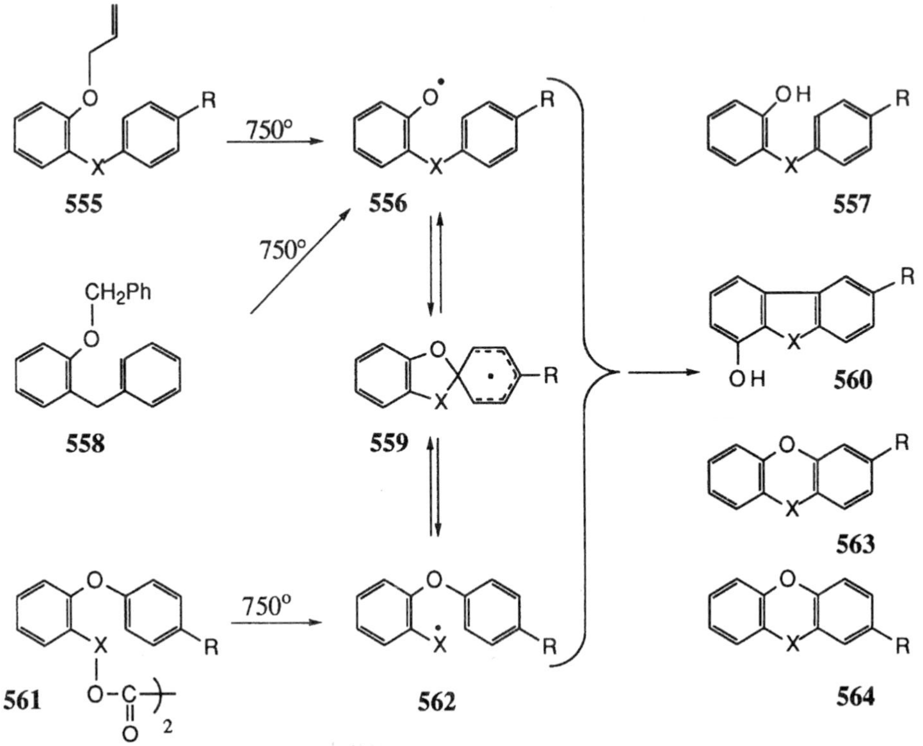

When X=CH$_2$ and R=H the yields of products were

	557	560	563 ≈ 564
from 555	33%	2%	30%
558	18%	13%	63%
561	5%	9%	36%

Scheme 51

intramolecular transfer of hydrogen from the benzylic carbon to the phenoxyl was proposed. To test this possibility, the dideutero compound (**555**, X=CH$_2$, R=H) was pyrolysed at $750°/1 \times 10^{-3}$ Torr to give a mixture of [9-^{2}H$_2$]-, [9-^{2}H][9-^{1}H]- and [9-^{1}H$_2$]-fluoroen-1-ol in a ratio of 4.6:4.4:1 (Scheme 52). This showed that not all the reaction proceeded by transfer of hydrogen from the benzylic position and that route A was a probable pathway.

Pyrolysis of 2-(allyloxy)benzophenone (**555**, X=CO, R=H) gave 2-benzoyl-phenol (**531**, X=CO, R=H), 1-hydroxyfluorenone (**560**, X=CO, R=H), xanthone (**563**, X=CO, R=H) and dibenzofuran (**565**, R=H). Formation of 1-hydroxyfluorenone (**560**, X=CO, R=H) confirmed that the presence of benzylic hydrogens in the intermediate radical was not necessary for this type of product

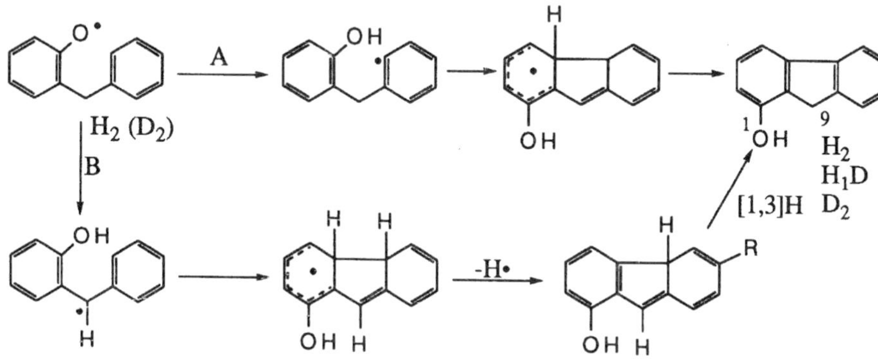

Scheme 52

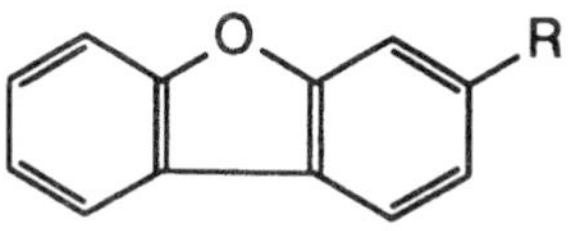

565

to be formed. FVP of the xanthone (**563**, X=CO, R=H) did not yield dibenzo-furan (**565**, R=H) and this compound most probably arises by decarbonylation of 2-phenoxybenzoyl (**562**, X=CO, R=H) followed by cyclisation of the aryl radical and loss of a hydrogen atom.

One further aspect of the mechanism was explored. FVP of (**555**, X=CO, R=Me) gave the four products (**557**, X=CO, R=Me), (**560**, X=CO, R=Me) (**563**, X=CO, R=Me) and (**564**, X=CO, R=Me) as well as 3-methyldibenzofuran (**565** R=Me). It was shown that with increasing temperature the ratio of 2-methylxan-thone (**564**, X=CO, R=Me) to 3-methylxanthone (**563**, X=CO, R=Me) remained as 1:1 while the yield of 3-methyldibenzofuran (**565**, R–Me) increased through decarbonylation of the acyl radical (**562**, X=CO, R=Me). This demonstrated that the acyl radical was not on the pathway to the xanthones and therefore these compounds must arise by direct sigmatropic migration from the spiro-radical (**559**, X=CO, R=Me), the migratory aptitude of O and CO groups being equal.

Intramolecular coupling of an alkyl radical with a phenoxyl radical may take place at the *ortho* position. In an extension of their work on cyclophanes, Bickelhaupt and his group pyrolysed spiro[5,7]trideca-1,4-dien-3-one (**566**) and obtained 9-hydroxy[7]metacyclophane (**567**) together with 4-(6-heptenyl)phenol (**568**).[192] Pyrolysis through an alumina tube at 0.03 Torr gave 90% recovery of material with varying proportions of products depending on temperature. At 520°, 560°, 600° and 650° the ratios of (**566**):(**567**):(**568**) were 38:29:23, 14:23:53, 0:15:75 and 0:0:90 respectively. The open chain phenol formed by hydrogen transfer was formed exclusively at high temperatues. The formation of a phenoxyl radical carrying a radical side chain also arises in pyrolysis of

coumaran, 2,3-dihydrobenzofuran (**569**). When a 1 mole % solution of coumaran
in toluene was pyrolysed in nitrogen carrier gas at 700°/760 Torr, 2-vinylphenol
(**570**) was formed as the major product (~50%) together with benzofuran
(**571**).[193] At 750° the proportion of the fully aromatic benzofuran increased and
these two products were formed in nearly equal amounts (~40%). Pyrolysis of
2,3-dihydro-2,3-dimethylbenzofuran (**572**) at 695° under the same conditions
gave benzofuran (**571**) (~54%) and some monomethylbenzofuran. These reactions could be rationalised by hydrogen atom transfer or in the second example
by ready loss of methyl radicals.

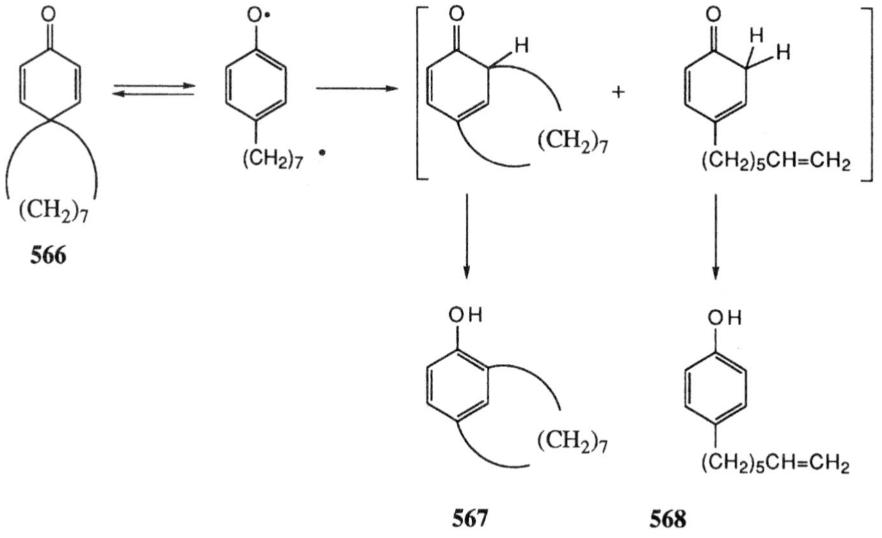

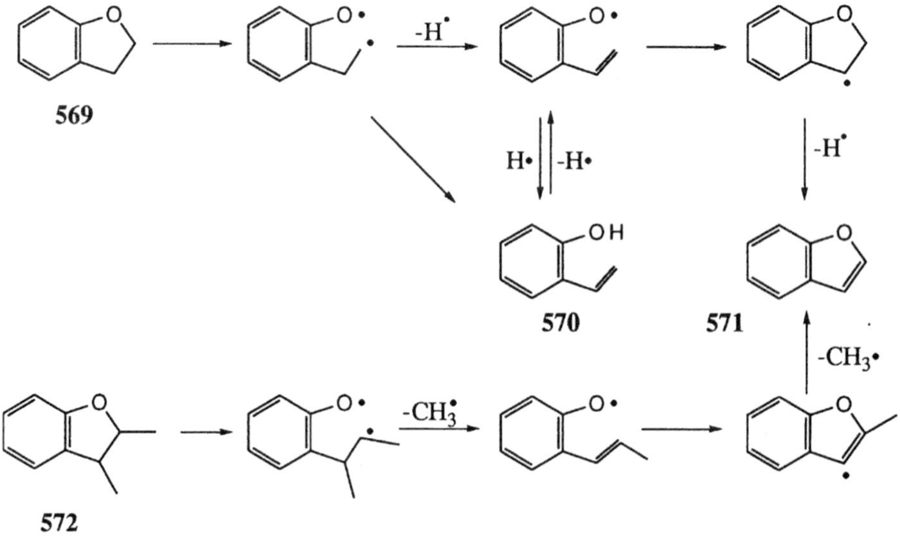

Benzofuran (**571**) is also the end product of FVP of 2-ethynylphenol (**576**) at 800°/0.02 Torr. At this temperature the conversion is quantitative. This fact provided support for the mechanism proposed to account for the quantitative conversion of 3-methylenebenzo[*b*]furan-2-one (3-methylene-2-coumaranone) (**573**) (initially protected as the cyclopentadiene adduct) to benzofuran (**571**) on FVP at 1000°/0.02 Torr.[194] The initial cleavage was considered to involve homolysis of the ester to the diradical (**574**) which underwent decarbonylation and a hydrogen shift to give 2-ethynylphenol (**576**) which was converted into benzofuran (**571**) *via* the carbene (**575**).

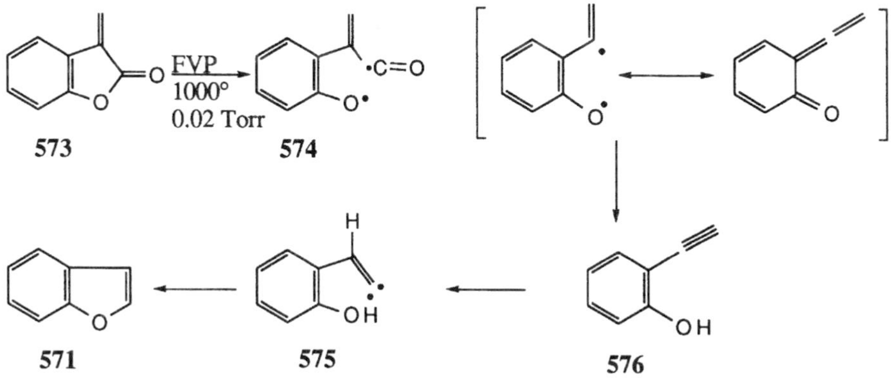

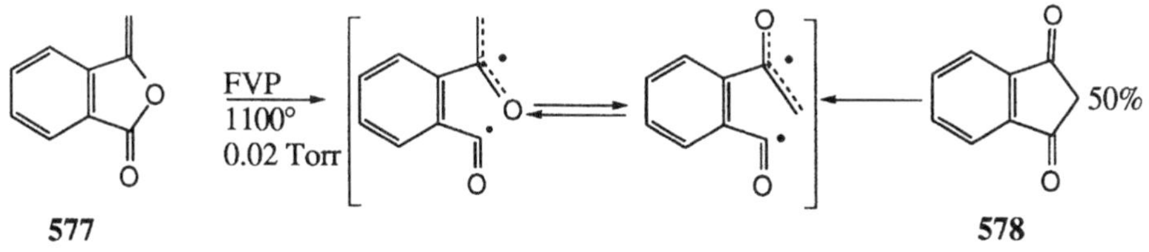

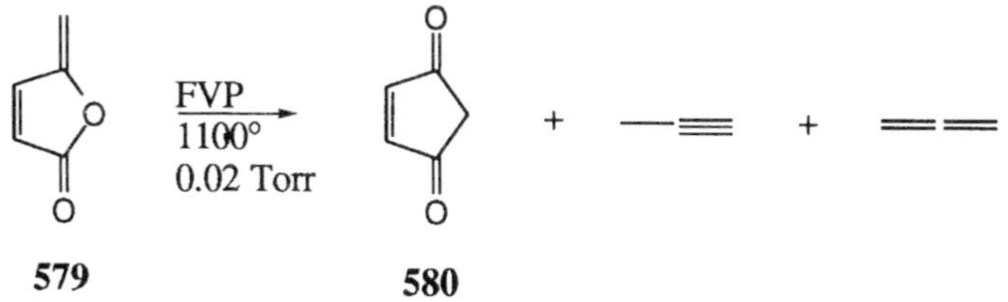

Radical cleavage of enollactones requires high temperatures. FVP of 3-methyl-enephthalide (**577**) at 1100°/0.02 Torr gave indane-1,3-dione (**578**) (50%) by radical cleavage and rearrangement with no evidence of decarbonylation. On the other hand, FVP of 5-methylene-2(5*H*)-furanone (protoanemonin) (**579**) at 1100°/0.02 Torr gave cyclopentene-3,5-dione (**580**) (60%), propyne (25%) and allene (10%).[194]

24. SYNTHESIS OF KETENES

The chemistry of ethenones, the simplest ketenes, has burgeoned in recent years but has almost exclusively been pursued using solution chemistry.[195] Gas phase methods have been used in this area principally in conjunction with spectroscopic methods for defining the properties of substituted ketenes.[196] On the other hand, the higher ketenes, the cumulenones, have been prepared and studied largely by gas phase methods.[197] In a recent review[1j] the development of the chemistry of higher ketenes was described and the relationship between ketenes, carbenes, acetylenes and arynes in gas phase chemistry was outlined. In this section the findings not considered in that review are discussed.

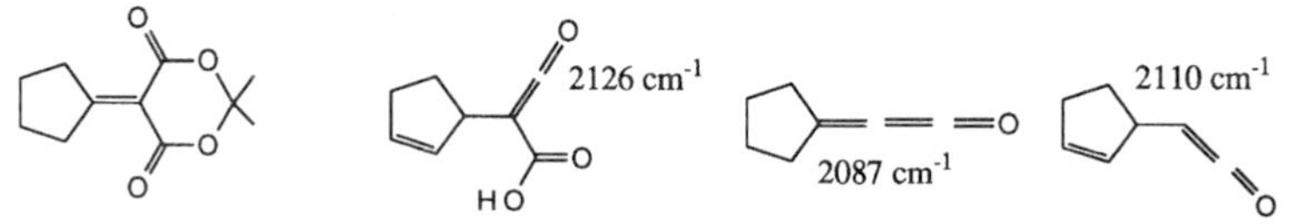

581   582   583   584

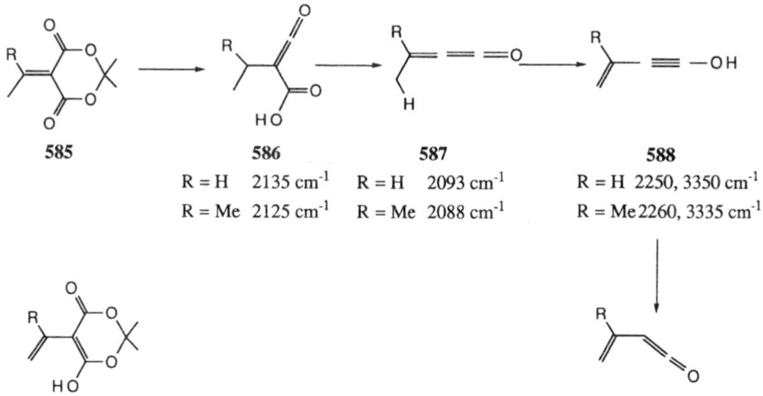

                          2126 cm⁻¹   2110 cm⁻¹
                                        2087 cm⁻¹

585   586   587   588

              R = H   2135 cm⁻¹   R = H   2093 cm⁻¹   R = H 2250, 3350 cm⁻¹
              R = Me 2125 cm⁻¹   R = Me 2088 cm⁻¹   R = Me 2260, 3335 cm⁻¹

589   590

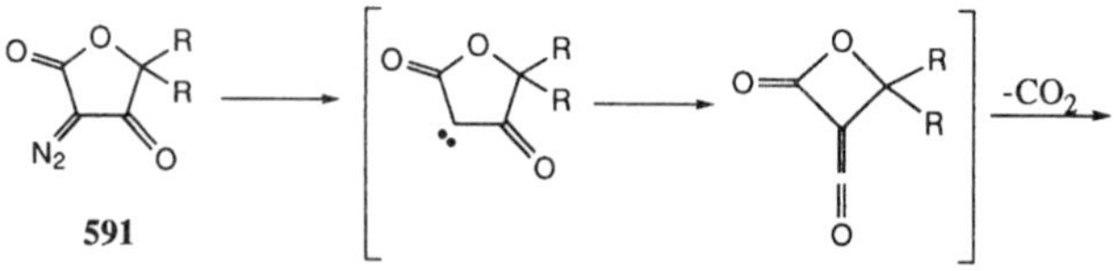

591   —   —   — $-CO_2$ $R_2C=C=C=O$
                                                          592

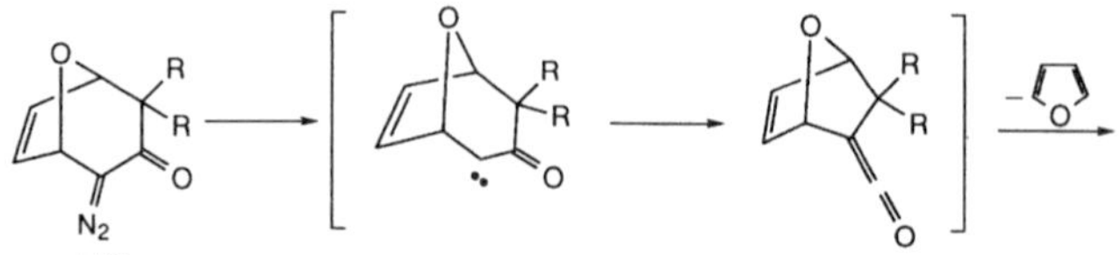

593

R = H
R = Me

                                                          592

Propadienones

The synthesis of propadienones was initially achieved by FVP of the products of condensation of aldehydes with Meldrum's acid, 2,2-dimethyl-1,3-dioxan-4,6-dione. This method was successful in those compounds in which there was no possibility of tautomerism. Pyrolysis of enolizable 5-alkylidene Meldrum's acid derivatives did yield alkyl substituted propadienones under particular conditions, but the fragmentation pathways were later shown to be complex and the processes involved have been elucidated by Wentrup and his coworkers in a series of studies.[198–200]

5-Cyclopentylidene-2,2-dimethyl-1,3-dioxan-4,6-dione (**581**) was shown to fragment to carboxycyclopenten-3-ylethenone (**582**), cyclopentylideneethenone (**583**) and cyclopenten-3-ylethenone (**584**) but the precise sequence of generation of these ketenes was not established.[198] Further studies on the 5-ethylidene- (**585**, R=H) and 5-(1-methylethylidene)- (**585**, R=Me) derivatives of Meldrum's acid using a double pyrolysis system and infrared detection allowed the sequence of formation of the intermediates to be determined.[199,200] In each case the compound initially formed was carboxyethenone (**586**) presumeably through fragmentation of the enolic form (**589**), although this intermediate could not be detected. Decarboxylation of (**586**) led to the alkylideneethenone (**587**) which at higher temperatures gave the alkenylethenone (**590**) through the intermediacy of the detectable hydroxyacetylene (**588**).[200] The complexity of these reactions threw doubt on earlier spectroscopic investigations of pyrolysates of precursors of this type.

Alternative methods of synthesis of propadienones based on the rearrangement of diazoketones to ketenes have been developed. Chapman and coworkers[201] have shown that FVP of the diazoketone (**591**) at 680°/10^{-5}Torr resulted in loss of nitrogen and CO_2 and the formation of propadienones (**592**). Brahms and Dailey[202] synthesised the diazoketones (**593**, R=H) and (**593**, R=Me) and these on FVP at 430°/10^{-4}Torr underwent rearrangement and loss of furan by a reverse Diels-Alder reaction to yield the propadienones (**592**, R=H) and (**592**, R=Me). Alkylpropadienones are highly reactive species which are difficult to study because of their lability. Introduction of a heteroatom at the 3-position, particularly nitrogen or sulfur carrying an available lone pair of electrons, markedly increases their stability.[203,204] Here we are concerned with the methoxy substituent which similarly, stabilizes the system.

Early in the development of this chemistry, methoxymethylene Meldrum's acid (**594**, R=H) was pyrolysed unexceptionally to yield methoxypropadienone[205] (**595**, R=H). With the development of understanding concerning the stabilisation of these species, Pommelet, Chuche and coworkers have shown that methoxypropadienone (**595**, R=H) prepared in this way has a stability in solution comparable with dimethylaminopropadienone.[206] The ^{1}H and ^{13}C NMR spectra of (**595**, R=H) could be measured in CD_2Cl_2 between −70° and 28°. The compound was stable below −20° but polymerised above this temperature. Pyrolysis of 1-methoxyethylidene Meldrum's acid (**594**, R=Me) gave 3-methoxybuta-1,2-dien-1-one (**595**, R=Me) which was stable below 0° and at room temperature dimerised slowly to yield a mixture of the two stereoisomers of 2,4-bis(methoxyethylidene)cyclobutane-1,3-dione (**596**).[206] The methoxypropadienones were studied using photoelectron spectroscopy.[207] The findings obtained on (**594**, R=H) were consistent with simple fragmentation while the additional methyl group in (**594**, R=Me) resulted in fragmentation through the enolic form and formation of 3-methoxybuta-1,3-dien-1-one.[207]

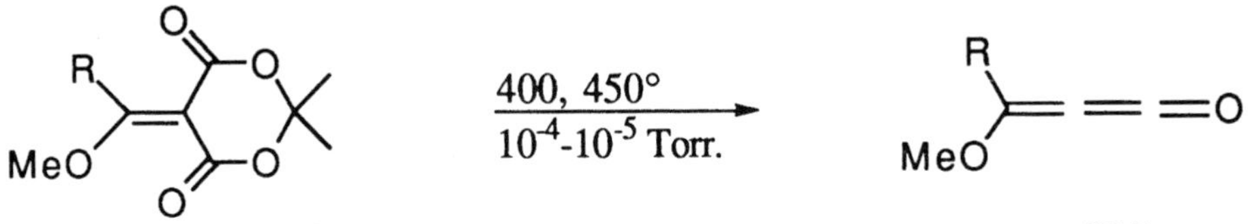

$$\frac{400,\ 450°}{10^{-4}\text{-}10^{-5}\ \text{Torr.}}$$

594   **595**

R = H   2030 cm^{-1}

R = Me   2060 cm^{-1}

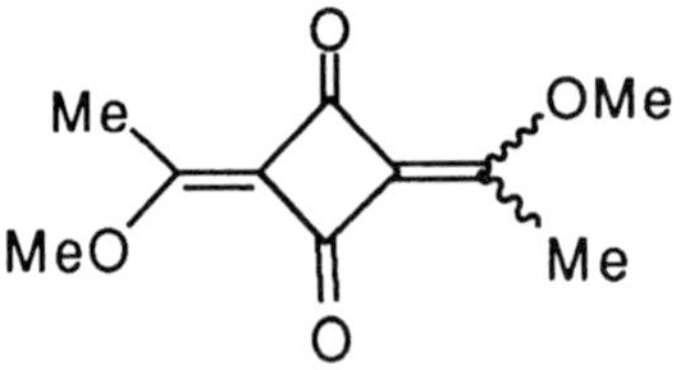

596

Butatrienones

The chemistry of butatrienone (**601**) is limited to its formation by FVP and detection by argon matrix infrared spectroscopy. The ketene has been formed as the major product of FVP of six precursors (**597** a, b; **598** a, b; **599** c, d) and as a minor product from the pyrolysis of (**600**, R=H).[208] The infrared spectrum was characterised by a ketene band (v_2) at 2242 cm^{-1} which in the spectrum of [1-^{13}C]-butatrienone was shifted to 2200 cm^{-1}. Analysis of the microwave spectrum of butatrienone showed it to have C_{2v} symmetry with either a shallow single-minimum potential or a very low barrier to planarity.[209]

An attempt to make diphenylbutatrienone by pyrolysis of (**600**, R=Ph) failed[210] and formation of dimethylbutatrienone (**606**) required development of alternative methods.[211] Pyrolysis of the mixed anhydride (**602a**) and of the acid chloride (**602b**) at 550–650°/0.05 Torr gave the ketene (**603**) which decarbonylated to the carbene (**604**). Formation of the cycloalkyne **607** followed by cycloreversion gave the ketene (**606**) which was detected by argon matrix infrared spectroscopy (v_{max} 2224, 2216 cm^{-1}). In [1-^{13}C]-dimethylbutatrienone these bands shifted to 2183 and 2177 cm^{-1}. Reaction of the pyrolysate with methanol gave the allenic ester (**605**, X=OMe). Ketene (**606**) was also formed in poorer yield by pyrolysis of the allenic mixed anhydride (**605a**) or the acid chloride (**605b**) at 600°/0.05 Torr.

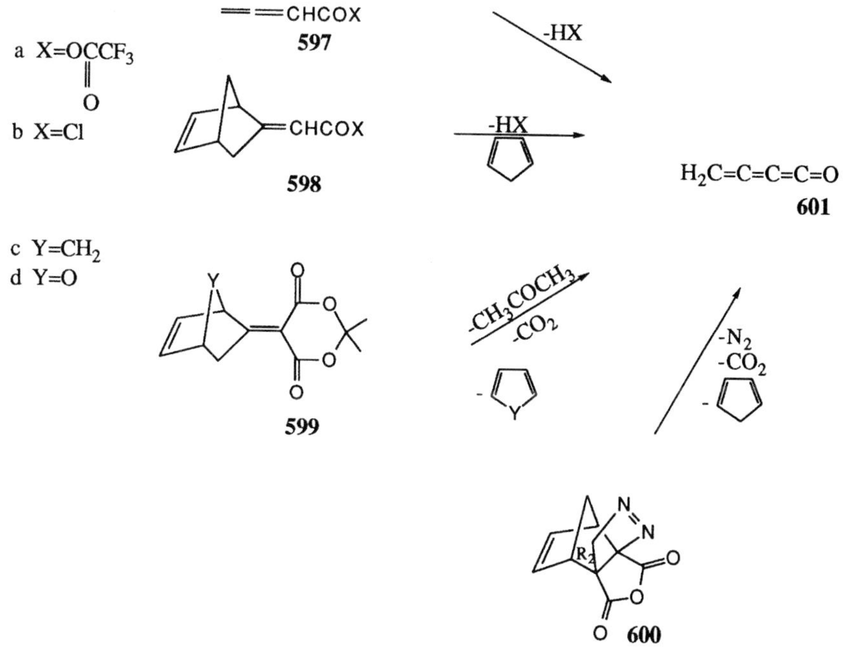

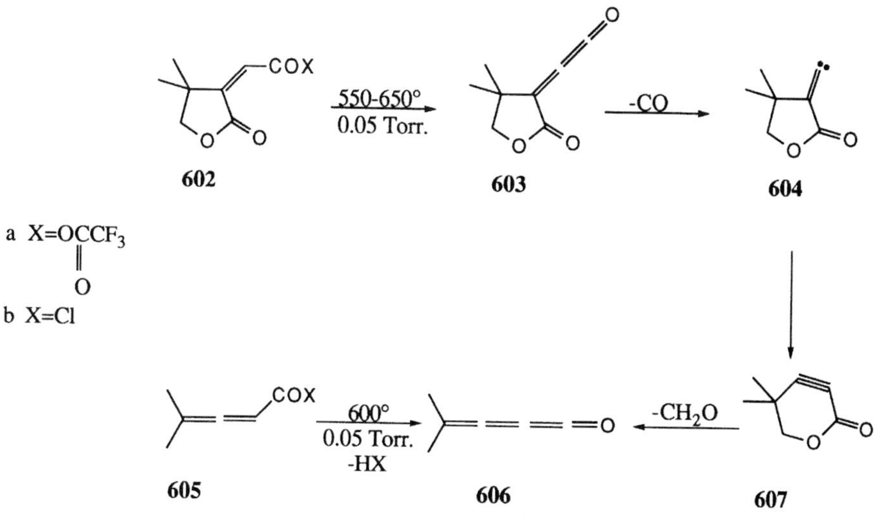

a X=OCCF₃

==CHCOX
597

-HX

b X=Cl

=CHCOX
598

-HX

H₂C=C=C=C=O
601

c Y=CH₂
d Y=O

599

-CH₃COCH₃
-CO₂

-N₂
-CO₂

600

-COX
602

550-650°
0.05 Torr.

603

-CO

604

a X=OCCF₃

b X=Cl

COX
605

600°
0.05 Torr.
-HX

606

-CH₂O

607

Pentatetraenone

Evidence for the formation of pentatetraenone (**611**) has been obtained from FVP of the five precursors (**608–610**).[212] Infrared spectroscopy on the pyrolysates in argon matrix showed the ketene band attributed to pentatetraenone at $2207\,\text{cm}^{-1}$ shifting to $2168\,\text{cm}^{-1}$ in the $[1\text{-}^{13}\text{C}]$ isotopomer.

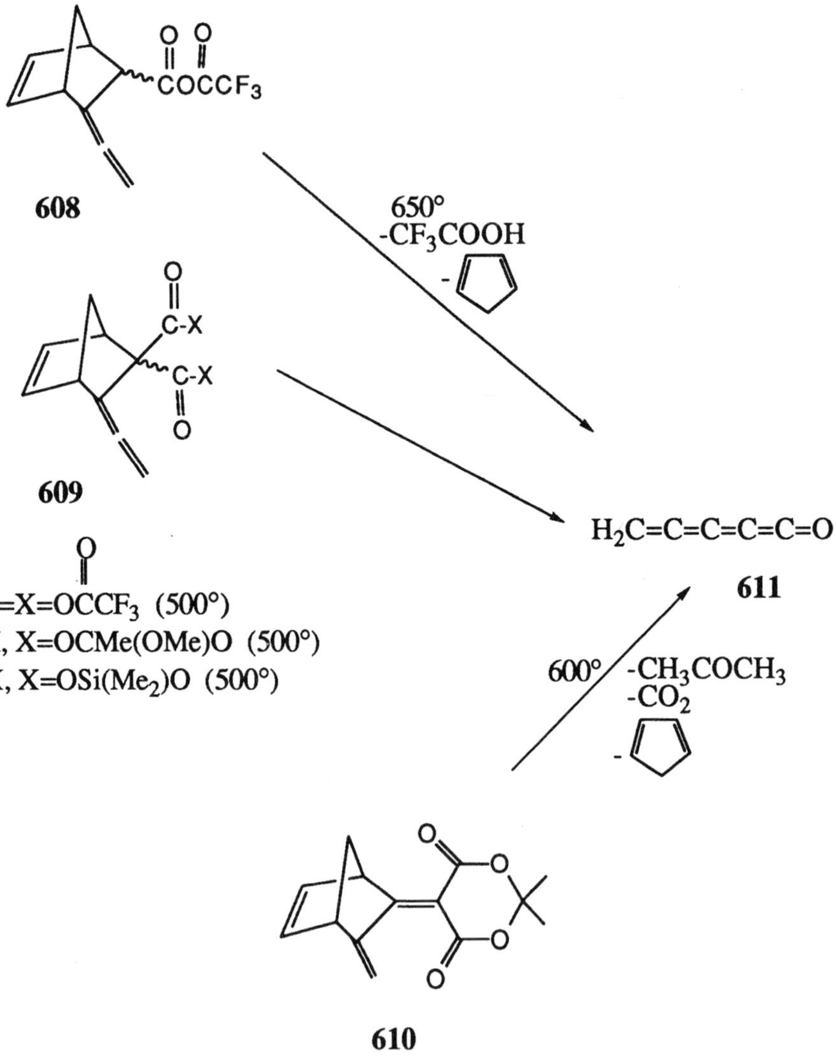

608

650°
-CF$_3$COOH

609

X=X=OCCF$_3$ (500°)
X, X=OCMe(OMe)O (500°)
X, X=OSi(Me$_2$)O (500°)

H$_2$C=C=C=C=C=O

611

600° -CH$_3$COCH$_3$
-CO$_2$

610

25. OXOKETENES

3-Oxoketenes (acylketenes) (**616**) are intermediates occurring in a variety of thermolytic and photolytic reactions in solution and in gas phase pyrolytic fragmentations.[213] Thermolytic decarbonylation of 2,3-dihydrofuran-2,3-diones (**612**), thermolytic or photolytic elimination of a carbonyl compound from 1,3-dioxin-4-ones (**613**) or of nitrogen from 2-diazo-1,3-dicarbonyl compounds (**614**) and FVP of enolizable 3-oxo-esters (**615**) yield acylketene intermediates (**616**) which are useful for 3-oxoacylation or for [2+4] cycloaddition reactions.[214]

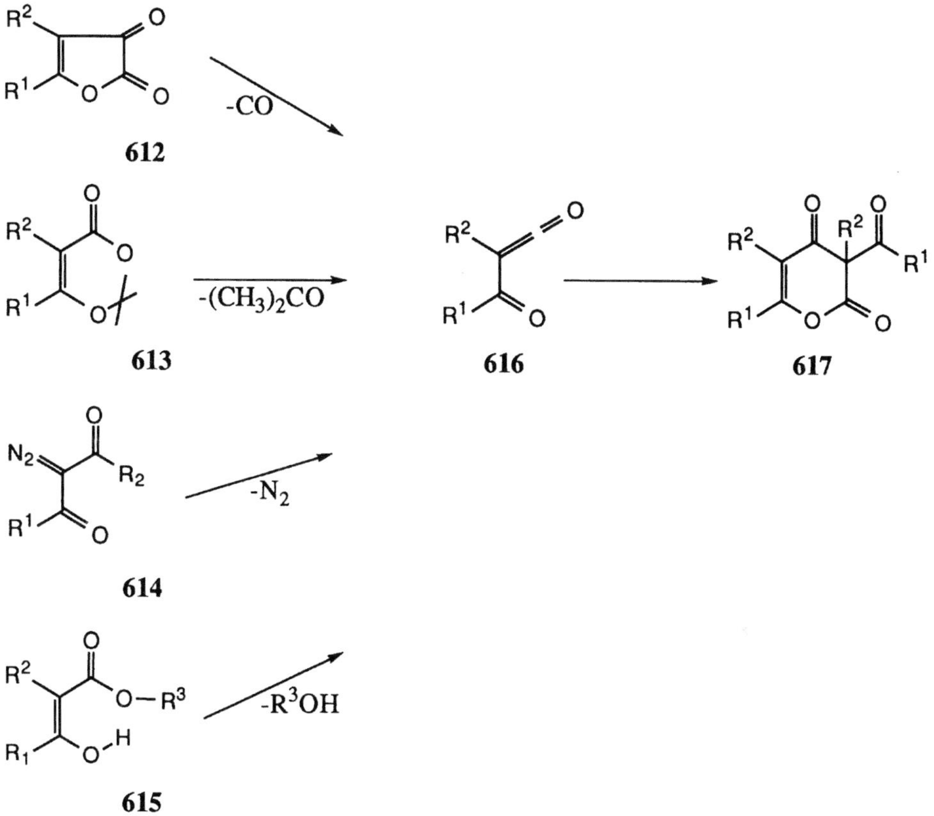

612

613   −(CH₃)₂CO

616

617

−CO

614   −N₂

615   −R³OH

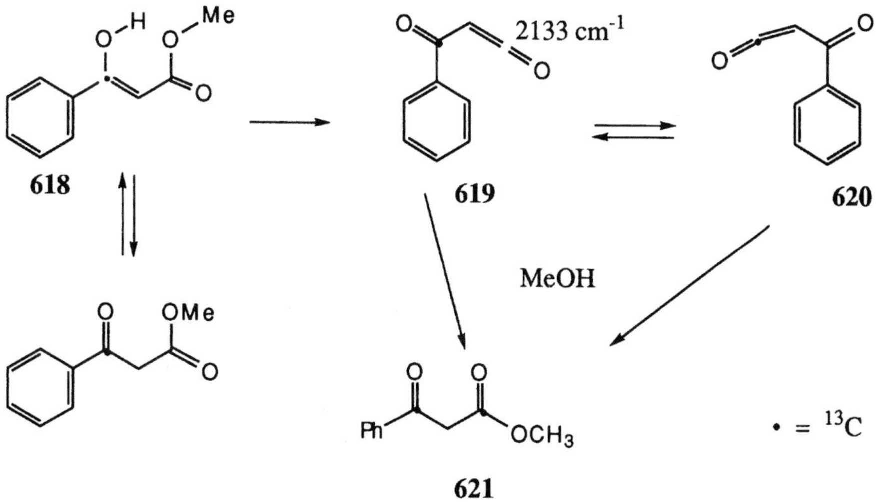

618

619   2133 cm⁻¹

620

MeOH

621

• = ¹³C

The reactivity of the intermediate, however, may lead to side reactions such as dimerization to the pyran derivative (617).

The acylketene intermediates cannot normally be observed under the reaction conditions generally employed and evidence for their existence has come either from pyrolytic formation in gas phase reactions and deposition in a matrix or by irradiation of a suitable precursor in a matrix and examination of the trapped products by infrared spectroscopy. Kinetic evidence for their intermediacy has also been described.[215]

FVP of methyl benzoylacetate (methyl 3-oxo-3-phenylpropanoate) (618) at $700°/10^{-4}$ Torr gave benzoylketene (619, unlabelled) which was trapped as a film at 82 K (v_{max} 2133 cm^{-1}, $^{12}C=^{12}C=O$).[216] FVP of [3-^{13}C]-3-oxo-3-phenyl-propanoate (619) (90% ^{13}C) under the same conditions gave a pyrolysate with v_{max} 2133 ($^{12}C=^{12}C=O$) and 2080 cm^{-1} ($^{12}C=^{13}C=O$) (intensity ratio, 55:45). These findings were interpreted as being due to a 1,3- phenyl shift, (619) $\rightleftarrows$ (620), involving the orthogonal ketene orbitals. At 400° no rearrangement occurred (v_{max} 2133 cm^{-1}) and at 550° there was 50% rearrangement (intensity ratio, 75:25). The fully rearranged labelled benzoylketene on reaction with methanol gave the doubly labelled ester (621).[216]

Aroylketenes were obtained in better yields than the method outlined above by FVP of 5-aryl-2,3-dihydrofuran-2,3-diones (612, R^1=Ph, R^2=H) at 400–500°/ 7.6×10^{-5} Torr. The same conversion could be achieved by irradiating the dione with unfiltered light from a high-pressure Xe–Hg lamp.[217]

Clemens and Witzeman[215a] have examined the rates of reaction of a series of nucleophiles (RNH$_2$, ROH, ArOH) with 6-methyl-1,3-dioxin-4-one (622) at 91.7° and 106.7°. They found the rate of reaction to be first order and independent of the nature and concentration of the nucleophile. This provided evidence for the intermediacy of the strongly electrophilic acetylketene (623). Passage of the dione (622) through a gas chromatrograph/matrix isolation apparatus (argon, 5–12 K) with the transfer line from the gas chromatograph at 180–240° gave a matrix showing distinct absorption bands at 2135 and 2142 cm^{-1} due to s-E and s-Z conformers of acetylketene. The carbonyl band of the acetyl group was observed at 1676 cm^{-1}. Kinetic studies on thermally induced trans-acetoacetylation pro-vided further evidence for the intermediacy of the acetylketene and t-butyl acetoacetate was found to be a useful source of the ketene because elimination is facile and the reverse reaction is strongly hindered.[215b]

Friermuth and Wentrup[218] have found that FVP of enolizable 3-oxoesters results in fragmentation of the enolic form with production of the alcohol and the s-E and s-Z conformers of 3-oxoketenes. These products were trapped in an argon matrix at 18 K for spectroscopic examination. On warming, the mixtures regenerated the enolic 3-hydroxy-2,3-unsaturated esters. Pyrolysis of methyl or ethyl acetoacetate gave the s-E conformer (v_{max} 2147w, 2143s (C=C=O) and 1681m cm^{-1} (C=O)) of acetylketene (623). Eight other esters were pyrolysed under similar conditions to yield the corresponding 3-oxoketenes. These were found to polymerize in the cold, but at elevated temperatures the 3-oxoketenes formed dimers.

The infrared spectra of three of these acylketenes were measured in argon matrices by Leung-Toung and Wentrup and used for comparison with specta

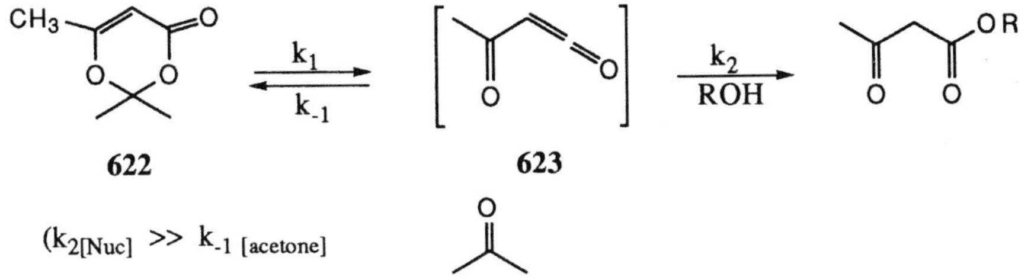

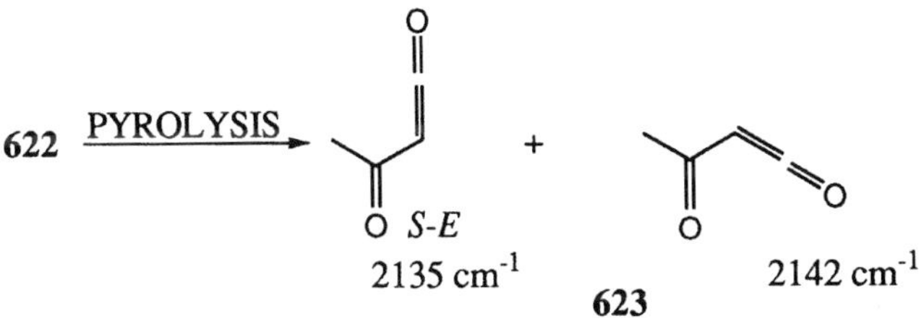

CH3

622

$\xrightarrow[k_{-1}]{k_1}$

623

$\xrightarrow[\text{ROH}]{k_2}$

OR

$(k_{2[\text{Nuc}]} \gg k_{-1\,[\text{acetone}]})$

622 $\xrightarrow{\text{PYROLYSIS}}$

O S-E
2135 cm^{-1}

+

2142 cm^{-1}

623

obtained from products of irradiation of 2-diazo-1,3-diketones.[219] FVP of methyl 4,4-dimethyl-3-oxopentanoate (**624**) at 475°/7.6 × 10^{-6} Torr gave pivaloylketene (**625**) (*s-E*, v_{max} 2134, 1680 cm^{-1}; *s-Z*, v_{max} 2142, 1668 cm^{-1}) and of methyl 2-methyl-3-oxobutanoate (**626**) gave acetylmethylketene (**627**) (*s-E*, v_{max} 2123, 2120, 1684 cm^{-1}; *s-Z*, 2131, 1666 cm^{-1}). Methyl 2-oxocyclopentanecarboxylate (**628**) at 650°/0.076 Torr gave the cyclic acylketene (**629**) (v_{max} 2133, 1708 cm^{-1}). This ketene was also formed on pyrolysis of 4,5,6,7-tetrahydropenta-1,3-dioxin-4-one (**630**) at 400°/7.6 × 10^{-6} Torr.

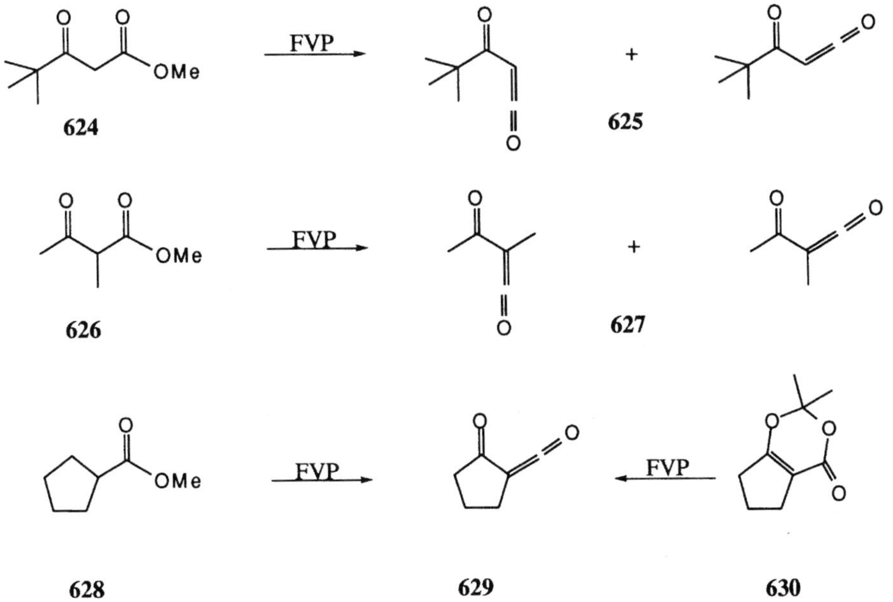

OMe

624

$\xrightarrow{\text{FVP}}$

+

625

OMe

626

$\xrightarrow{\text{FVP}}$

+

627

OMe

628

$\xrightarrow{\text{FVP}}$

629

$\xleftarrow{\text{FVP}}$

630

In a study of the FVP of cyclic 3-oxoesters incorporating a vinyl-cyclopropane system, De Meijere and coworkers showed that work up with ethanol of the pyrolysate of (**631**) gave an ethyl ester which contained a cyclopentene ring (**632**).[220] The rearrangement was regarded as taking place in the 3-oxoketene intermediate. In another example (**633**), the cyclopropane interacted with the ketene to give a phenolic product (**634**).[220]

3-Oxoketenes were also formed by fragmentation of 2,4-dioxoesters in a GC–FTIR instrument with loss of carbon monoxide and the esterifying alcohol.[221] The injection port of the GC was heated in the range 170–280° and the tube transferring the effluent from the GC column to the FTIR instrument was at 240–280°. It was established that fragmentation principally occurred during the brief period after the compound emerged from the GC column. Ketenes were identified by infrared bands in the range 2120–2140 cm^{-1}.

Passage of ethyl 4,4-dimethyl-3-oxopentanoate through the GC–FTIR gave ethanol and pivaloylketene (**625**) with ν_{max} 2138 cm^{-1}. Passage of ethyl 5,5-dimethyl-2,4-dioxohexanoate (**635**) through the instrument gave ethanol, carbon monoxide and pivaloylketene (**625**) (ν_{max} 2140 cm^{-1}). The mechanism proposed involved formation of the enolic forms (**636**) and (**637**) followed by fragmentation with loss of CO.

Dibenzoylketene (**639**) was formed by pyrolysis of 4-benzoyl-2,3-dihydro-5-phenyl-2,3-dioxofuran (**638**) at 250°/10^{-3} Torr.[222] It had ν_{max} 2140 cm^{-1} (at −194°) and this band disappeared above −75°. At higher temperatures (>120°) dibenzoylketene formed a dimer (**641**) and another product (**642**) in which two molecules combined with loss of CO$_2$. The [2+4] cyclodimerization product (**640**) was proposed as the key intermediate.

Dipivaloylketene (**644**) was prepared by Wentrup and coworkers by FVP of 5-*t*-butyl-4-pivaloyl-2,3-dihydrofuran-2,3-dione (**643**) at 500°/7.6 × 10^{-4} Torr.[213,223] The ketene (**644**) when stored in CD$_2$Cl$_2$ at −20° was stable indefinitely and it was stable for brief periods at room temperature. At room temperature the liquid product over a period of 48 h gave a crystalline dimer the structure of which included a highly hindered acylketene (**646**).[223] Further study of dipivaloylketene (**644**) showed that in non-polar and in some polar solvents it formed the [2+4] dimer (**646**). However, on dimerization in the presence of dimethylsulfoxide, triphenylphosphine oxide or pyridine it formed a different dimer by addition across the C=O bond of the ketene to give the dioxinone (**645**).[213] FVP of either (**645**) or (**646**) at 400°/7.6 × 10^{-4} Torr gave the ketene (**644**). The dipivaloylketene (**644**) reacted with heterocumulenes to give [2+4] cycloaddition products.[213] The dimer (**646**), which retained the pivaloylketene structure, reacted with carbodimides to yield [2+2] adducts[213] and with primary aromatic amines to give bridged bis-dioxines.[224]

FVP of methyl 3-methoxy-3-trimethylsilyloxypro*p*-2-enoate (**647**) at 500° gave the ketene ester methyl 3-oxoprop-2-enoate (**648**) which formed the enolisable dimer (**649**) at −80°. At temperatures >850° (**647**) gave C$_3$O$_2$.[225]

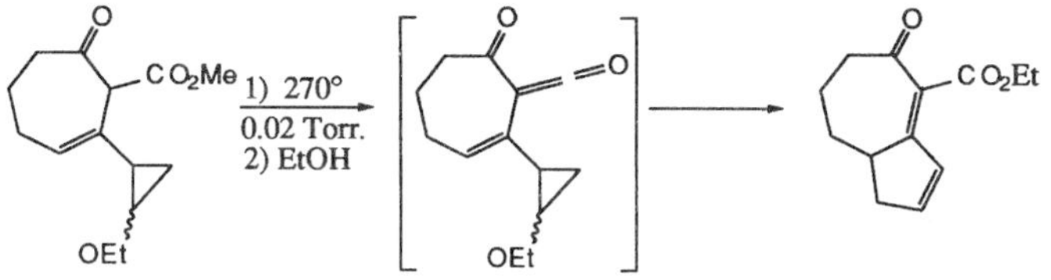

631 → **632**

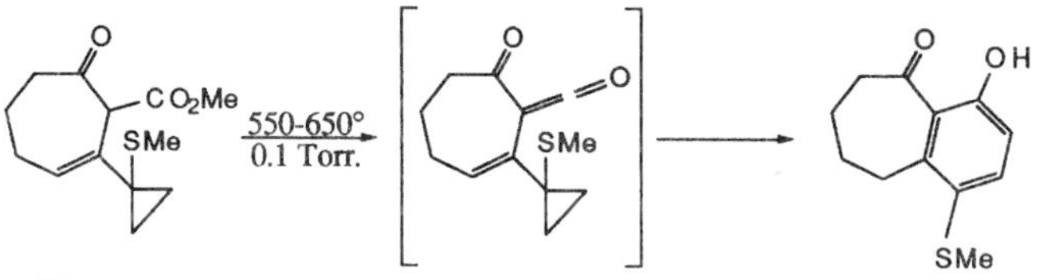

633 → **634**

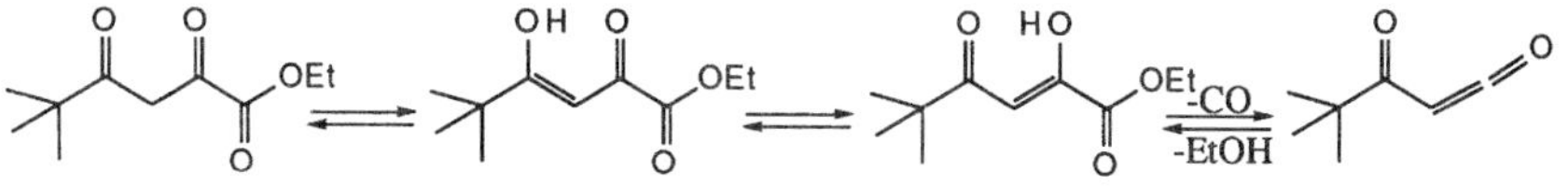

635 ⇌ **636** ⇌ **637** → **625**

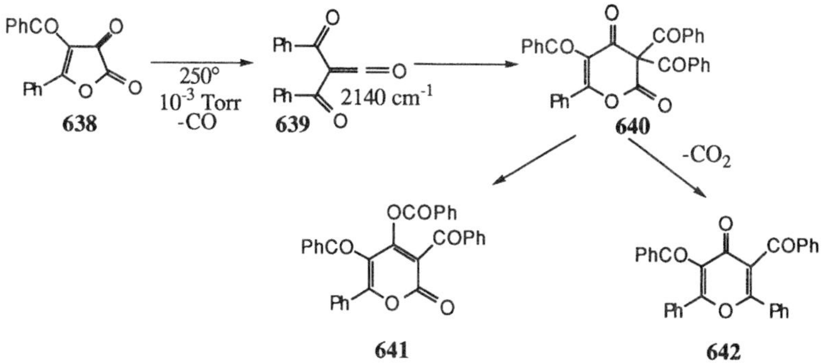

638 → **639** → **640**

641 **642**

26. OXIDES OF CARBON

The carbon oxides C_3O, C_3O_2, C_5O_2 and C_7O_2 have been prepared using gas phase pyrolytic methods. Other oxides, C_4O, C_4O_2 and C_6O, have been obtained in associated photolytic experiments.

In 1971 DeKock and Weltner detected infrared absorption at $2244 \, \text{cm}^{-1}$ in the reaction of carbon atoms with CO in an argon matrix and attributed this band to C_3O (**651**).[226] Pyrolysis of the Meldrum's acid derivative (**650**) through a quartz tube at 600° into a quadrupole mass spectrometer and microwave spectrometer cell gave the first spectroscopic evidence of C_3O in the gas phase.[227] Later refinements of the microwave spectrum were achieved using fumaroyl chloride (652) as precursor[228] and deposition of C_3O in an argon matrix and measurement of the infrared spectrum confirmed the strong band at $2243\,cm^{-1}$.[229] There was no evidence in any of these experiments for the intermediacy of C_4O_2 (653) and in a later study of the high resolution infrared spectrum of C_3O in the gas phase the ν_1 band was centred at $2257\,cm^{-1}$ and there were no bands attributable to C_4O_2.[230] Tricarbon monoxide was also obtained by FVP of 3,5-dimethyl-1-propynoylpyrazole (654) and detected by microwave spectroscopy.[231] The diazoketones (655) and (656) have also been reported as giving C_3O on FVP.[232]

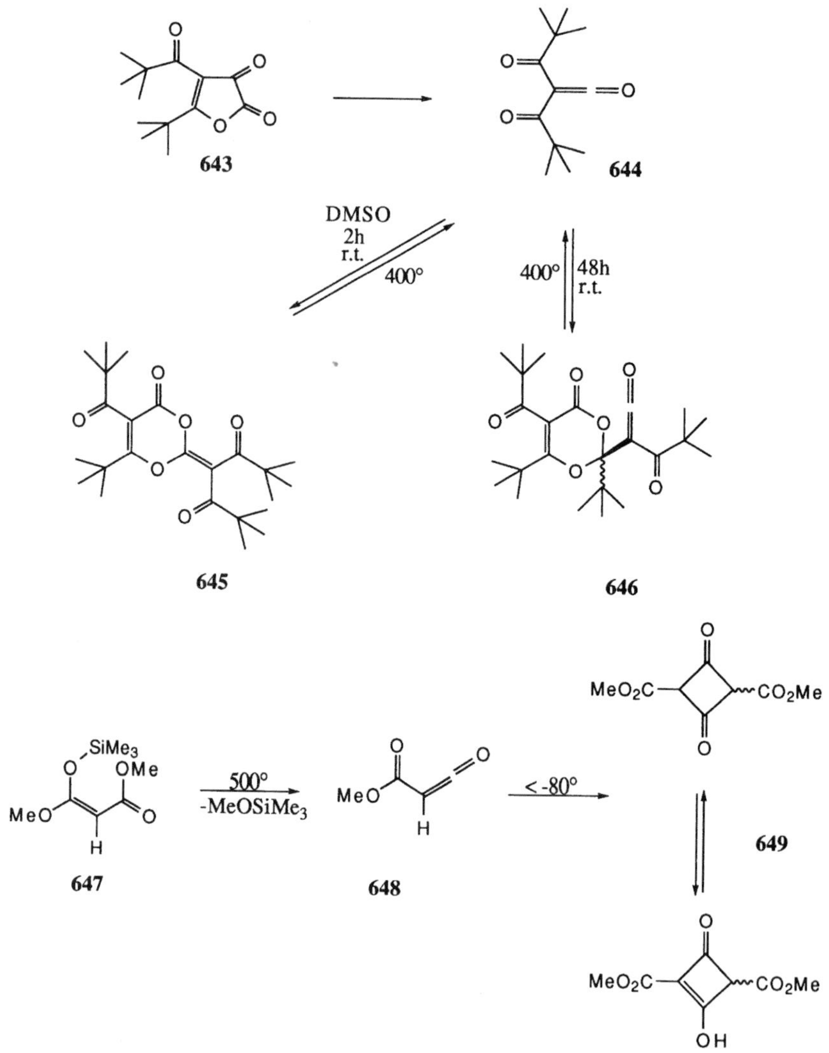

In these cases, however, it was shown that irradiation of (655) or (656) with ultraviolet light (254 nm) gave C_4O_2 (653) (v_{max} 2130 cm^{-1}) which on further irradiation with light of wavelength >300 nm gave C_3O. Tricarbon dioxide (658), carbon suboxide C_3O_2, has long been known and in the same study it was obtained from FVP or irradiation of the diazoketone (657).[232]

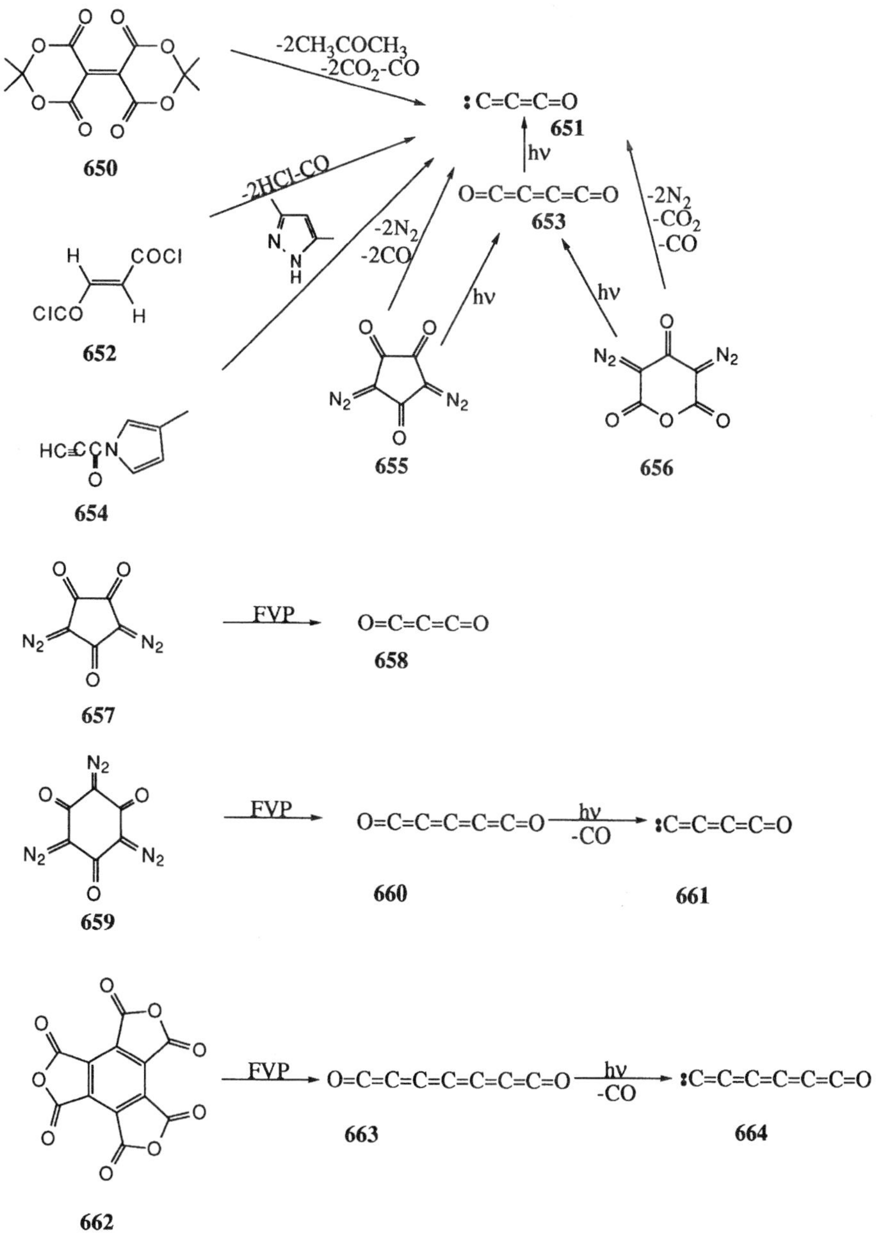

651

653

655

656

657

O=C=C=C=O

658

659

O=C=C=C=C=C=O
660

$\xrightarrow[-CO]{hv}$

:C=C=C=C=O
661

662

O=C=C=C=C=C=C=C=O
663

$\xrightarrow[-CO]{hv}$

:C=C=C=C=C=C=O
664

652

654

Pentacarbon dioxide, C_5O_2 (**660**), has been prepared by FVP of the diazo-ketone (**659**) at $700°/10^{-3}$ Torr.[233] In an argon matrix at 12 K the principal infrared absorption band was at $2213\,cm^{-1}$. The high resolution spectrum in the gas phase has been reported.[234] In the condensed phase C_5O_2 is only stable to $-90°$ when it polymerises, but when trapped in a frozen solvent and thawed it can be obtained as a solution which is stable at room temperature. Irradiation of C_5O_2 yields C_4O (**661**) and CO.[233,236]

FVP at $900–1000°$ or matrix photolysis of mellitic anhydride (**662**) yields a number of carbon oxides.[235] The infrared spectrum of the matrix at 12 K was interpreted in terms of the formation of C_7O_2 (**663**) (ν_{max} $2118\,cm^{-1}$) which on irradiation was converted into C_6O (**664**) and CO.[235,236]

REFERENCES

1a. R.F.C. Brown, *Pyrolytic Methods in Organic Chemistry* (Academic Press, New York, 1980).
1b. J.J. Gajewski, *Hydrocarbon Thermal Isomerizations* (Academic Press, New York, 1981).
1c. A. Viola, J.J. Collins and N. Filipp, *Tetrahedron* **37**, 3765 (1981).
1d. U.E. Wiersum, *Recl. Trav. Chim. Pays-Bas* **101**, 317, 365 (1982); *Aldrichichimica Acta* **17**, 31 (1984).
1e. C. Wentrup, *Reactive Molecules* (Wiley, New York, 1984).
1f. P. Schiess and S. Rutschmann, *Chimia* **39**, 213 (1985).
1g. M. Karpf, *Angew. Chem. Int. Ed. Engl.* **25**, 414 (1986).
1h. R.F.C. Brown, *Rec. Trav. Chim. Pays-Bas* **107**, 655 (1988).
1i. R.F.C. Brown, *Pure and Applied Chem.* **62**, 1981 (1990).
1j. R.F.C. Brown and F.W. Eastwood, *Synlett 1993*, 9.
1k. R.F.C. Brown and F.W. Eastwood, *Pure and Appl. Chem.* **68**, 261 (1996).
1l. L.T. Scott, *Pure and Appl. Chem.* **68**, 291 (1996).
2a. L.Shevelkova, L. Gaselnikov, G. Bach and G. Zimmermann, *Russian Chem. Rev.* **61**, 433 (1992).
2b. S.E. Stein, *Acc. Chem. Res.* **24**, 350 (1991).
2c. U.E. Wiersum, *Janssen Chimica Acta* **10**(3), 3 (1992).
 3. R.F.C. Brown, N.R. Browne, K.F. Coulston, L.B. Danem, F.W. Eastwood, M.J. Irvine and A.D.E. Pullin, *Tetrahedron Lett.* **27**, 1075 (1986).
 4. R.F.C. Brown, N.R. Browne, K.J. Coulston, F.W. Eastwood, M.J. Irvine, A.D.E. Pullin and U.E. Wiersum, *Aust. J. Chem.* **42**, 1321 (1989).
 5. M. Barry, R.F.C. Brown, F.W. Eastwood, D.A. Gunawardana and C. Vogel, *Aust. J. Chem.* **37**, 1643 (1984).
 6. C. Wentrup, R. Blanch, H. Briehl and G.J. Gross, *J. Am. Chem. Soc.* **110**, 1874 (1988).
 7. R.F.C. Brown, K.J. Coulston, F.W. Eastwood and C. Vogel, *Aust. J. Chem.* **41**, 1687 (1988).
 8. R.F.C. Brown, K.J. Coulston and F.W. Eastwood, *Aust. J. Chem.* **47**, 47 (1994).
 9. R.F.C. Brown, K.J. Coulston, F.W. Eastwood and S. Saminathan, *Aust. J. Chem.* **41**, 107 (1987).
 10. R.F.C. Brown, K.J. Coulston, B.J. Dobney, F.W. Eastwood and G.D. Fallon, *Aust. J. Chem.* **40**, 1687 (1987).
 11. R.F.C. Brown, K.J. Coulston, F.W. Eastwood and T. Korakis, *Tetrahedron Lett.* **29**, 6791 (1988).
 12. R.F.C. Brown, N. Choi and F.W. Eastwood, *Aust. J. Chem.* **48**, 185 (1995).
 13. R.F.C. Brown, F.W. Eastwood and C.J. Smith, *Aust. J. Chem.* **45**, 1315 (1992).
 14. M.R. Anderson, R.F.C. Brown, K.J. Coulston, F.W. Eastwood and A. Ward, *Aust. J. Chem.* **43**, 967 (1990).
 15. M. Adeney, R.F.C. Brown, K.J. Coulston, F.W. Eastwood and I.W. James, *Aust. J. Chem.* **44**, 967 (1991).
16a. U.E. Wiersum and L.W. Jenneskens, *Tetrahedron Lett.* **34**, 6615 (1993).
16b. R.F.C. Brown, N. Choi, K.J. Coulston, F.W. Eastwood, U.E. Wiersum and L.W. Jenneskens, *Tetrahedron Lett.* **35**, 4405 (1994).

17. R.R. Jones and R.G. Bergman, *J. Am. Chem. Soc.* **94**, 660 (1972).
18. T.P. Lockhart, C.B. Mallon and R.G. Bergman, *J. Am. Chem. Soc.* **102**, 5976 (1980).
19. T.P. Lockhart, P.B. Comita and R.G. Bergman, *J. Am. Chem. Soc.* **103**, 4082 (1981).
20. T.P. Lockhart and R.G. Bergman, *J. Am. Chem. Soc.* **103**, 4091 (1981).
21. R.F.C. Brown, D.V. Gardner, J.F.W. McOmie and R.K. Solly, *Aust. J. Chem.* **20**, 139 (1967).
22. M.D. Banciu, R.F.C. Brown, K.J. Coulston, F.W. Eastwood, C. Jurss, I. Mavropoulos, M. Stanescu and U.E. Wiersum, *Aust. J. Chem.* **49**, 965 (1996).
23. M.C. Saver and L.M. Dorfman, *J. Chem. Phys.* **35**, 497 (1961); H. Okabe and J.R. McNesby, *J. Chem. Phys.* **36**, 601 (1962).
24. P.S. Skell and J.H. Plonka, *J. Am. Chem. Soc.* **92**, 5620 (1970).
25. R.F.C. Brown, F.W. Eastwood and G.P. Jackman, *Aust. J. Chem.* **31**, 579 (1978).
26a. Y. Osamura, H.F. Schaefer, S.K. Gray and W.H. Miller, *J. Am. Chem. Soc.* **103**, 1904 (1981).
26b. A.C. Scheiner and H.F. Schaefer, *J. Am. Chem. Soc.* **107**, 4451 (1985).
27a. R.P. Duran, V.T. Amorebieta and A.J. Colussi, *J. Phys. Chem.* **92**, 636 (1988).
27b. A.M. Wodke and Y.T. Lee, *J. Phys. Chem.* **89**, 4744 (1985).
28. R.P. Duran, V.T. Amorebieta and A.J. Colussi, *J. Am. Chem. Soc.* **109**, 3154 (1987).
29. M.A. Grela, V.T. Amorebieta and A.J. Colussi, *J. Phys. Chem.* **96**, 9861 (1992).
30. H.-D. Martin, T. Urbanek, R. Braun and R. Walsh, *Int. J. Chem. Kinet.* **16**, 117 (1984).
31. M. Remmler, G. Zimmerman, F.-D. Kopinke, B. Ondruschka, M. Findeisen and U. Böhme, *Z. Chem.* **28**, 246 (1988).
32. R.F.C. Brown, F.W. Eastwood, K.F. Harrington and G.L. McMullen, *Aust. J. Chem.* **27**, 2393 (1974).
33. R.F.C. Brown, F.W. Eastwood and G.P. Jackman, *Aust. J. Chem.* **30**, 1757 (1977).
34. R.F.C. Brown, F.W. Eastwood and N.R. Wong, *Tetrahedron Lett.* **34**, 1223 (1993).
35. M. Adeney, R.F.C. Brown, K.J. Coulston, F.W. Eastwood and I.W. James, *Aust. J. Chem.* **44**, 967 (1991).
36. L.T. Scott, M.M. Hashemi, D.T. Meyer and H.B. Warren, *J. Am. Chem. Soc.* **113**, 7082 (1991).
37. L.T. Scott, M.M. Hashemi and M.S. Bratcher, *J. Am. Chem. Soc.* **114**, 1920 (1992).
38. A. Borchardt, A. Fuchicello, K.V. Kilway, K.K. Baldridge and J.S. Siegel, *J. Am. Chem. Soc.* **114**, 1921 (1992).
39. G. Metha, S.R. Shah and K. Ravikumar, *J. Chem. Soc., Chem. Commun. 1993*, 1006.
40. R.F.C. Brown, F.W. Eastwood and N.R. Wong, *Tetrahedron Lett.* **34**, 3607 (1993).
41. L.T. Scott and N.H. Roelofs, *J. Am. Chem. Soc.* **109**, 5461 (1987).
42. P.R. West, O.L. Chapman and J.-P. Le Roux *J. Am. Chem. Soc.* **104**, 1779 (1982).
43. W.J. Baron, M. Jones Jr. and P.P. Gaspar, *J. Am. Chem. Soc.* **92**, 4739 (1970).
44. P.P. Gaspar, J.-P. Hsu, S. Chari and M. Jones Jr., *Tetrahedron* **41**, 1479 (1985).
45. J.M. Fox, J.E. Gillen Scacheri, K.G.L. Jones, M. Jones Jr., P.B. Shevlin, B. Armstrong and R. Sztyrbicka, *Tetrahedron Lett.* **33**, 5021 (1992).
46. R.W. Hoffmann, I.H. Loof and C. Wentrup, *Liebigs Ann. Chem. 1980*, 1198.
47. W.T. Brown and W.M. Jones, *J. Org. Chem.* **44**, 3090 (1979).
48. M.W. Baum, J.L. Font, M.E. Meislich, C. Wentrup and M. Jones Jr., *J. Am. Chem. Soc.* **109**, 2534 (1987).
49. J. Becker and C. Wentrup, *J. Chem. Soc., Chem Commun. 1980*, 190.
50. T.A. Engler and H. Shechter, *Tetrahedron Lett.* **23**, 2715 (1982).
51. P.R. West, A.M. Mooring, R.J. McMahon and O.L. Chapman, *J. Org. Chem.* **51**, 1316 (1986).
52. M. Balci, W.R. Winchester and W.M. Jones, *J. Org. Chem.* **47**, 5180 (1982).
53. K.A. Horn and J.E. Chateauneuf, *Tetrahedron* **41**, 1465 (1985).
54. C. Wentrup and J. Benedikt, *J. Org. Chem.* **45**, 1407, 1980.
55. C. Wentrup and J. Becker, *J. Am. Chem. Soc.* **106**, 3705 (1984).
56. E. Heilbronner, P.A. Plattner and K. Weiland, *Experientia* **3**, 70 (1947).
57. E. Heilbronner and K. Wieland, *Helv. Chim. Acta.* **30**, 947 (1947).
58. L.T. Scott, N.H. Roelofs and T.-H. Tsang, *J. Am. Chem. Soc.* **109**, 5456 (1987).
59. K. Merz and L.T. Scott, *J. Chem. Soc., Chem Commun. 1993*, 412.
60. N.J. Turro, C.A. Renner, J.J. Katz, K.B. Wiberg and H.A. Connon, *Tetrahedron Lett. 1976*, 4133.
61. M.L. Heffernan and A.J. Jones, *Chem. Commun. 1966*, 120.
62. B.J. Gaynor, R.G. Gilbert, K.D. King and P.J. Harman, *Aust. J. Chem.* **34**, 449 (1981).

63. L.T. Scott and G.K. Agopian, *J. Am. Chem. Soc.* **99**, 4506 (1977).

64. R.W. Alder and G. Whittacker, *J. Chem. Soc., Perkin Trans. 2 1975*, 2.

65. R.W. Alder and C. Wilshire, *J. Chem. Soc., Perkin Trans. 2 1975*, 1464.

66. R.W. Alder, R.W. Whiteside, G. Whittaker and C. Wilshire, *J. Am. Chem. Soc.* **101**, 629 (1979).

67. J. Becker, C. Wentrup, E. Katz and K.-P. Zeller, *J. Am. Chem. Soc.* **102**, 5110 (1980).

68. L.T. Scott and M.A. Kirms, *J. Am. Chem. Soc.* **103**, 5875 (1981).

68a. K.-P. Zeller, *Angew. Chem. Int. Ed. Engl.* **21**, 440 (1982).

69. R.F.C. Brown, G.E. Gream, D.E. Peters and R.K. Solly, *Aust. J. Chem.* **21**, 2223 (1968).

70. L.T. Scott, M.M. Hashemi, T.H. Schultz and M.B. Wallace, *J. Am. Chem. Soc.* **113**, 9692 (1991).

70a. U. Burger, Y. Mentha and P.J. Thorel, *Helv. Chim. Acta.* **69**, 670 (1986).

71. G. Gross, R. Schulz, A. Schweig and C. Wentrup, *Angew. Chem. Int. Ed. Engl.* **20**, 1021 (1981).

72. L.T. Scott and N.H. Roelofs, *Tetrahedron Lett.* **29**, 6857 (1988).

73. A.J. Stone and D.J. Wales, *Chem. Phys. Lett.* **128**, 501 (1986).

74. P.W. Fowler, D.E. Manolopoulos and R.P. Ryan, *J. Chem. Soc., Chem. Commun. 1992*, 408.

75. L.T. Scott, The 3rd Tohwa University International Symposium on Synthetic Mechanistic Hydrocarbon Chemistry, Fukuoka, Japan (1993).

76. L.T. Scott, M.A. Kirms, A. Berg and P.E. Hansen, *Tetrahedron Lett.* **23**, 1859 (1982).

77. A.G. Anderson, E.D. Daugs, L.G. Kao and J.-F. Wang, *J. Org. Chem.* **51**, 2961 (1986).

78. A.G. Anderson and R.D. Haddock, *J. Org. Chem.* **56**, 550 (1991).

79. L.T. Scott, T.-H. Tsang and L.A. Levy, *Tetrahedron Lett.* **25**, 1661 (1984).

80. L.T. Scott, *Acc. Chem. Res.* **15**, 52 (1982).

81. K.M. Merz and L.T. Scott, *J. Chem. Soc., Chem. Commun. 1993*, 412.

82. M. Remmler, G. Zimmermann, B. Ondruschka, H. Bader and H. Hopf, *Chem. Ber.* **123**, 1375 (1990).

83. B. Ondruschka, G. Zimmermann, M. Remmler, U. Ziegler, F.D. Dopinke, B. Olk and M. Findeisen, *Chem. Ber.* **122**, 715 (1989).

84. H. Hopf, *Angew. Chem. Int. Ed. Engl.* **23**, 948 (1984).

85. K. Hassenück, H.-D. Martin and R. Walsh, *Chem. Rev.* **89**, 1125 (1989).

85a. H.-D. Martin, T. Urbanek, R. Braun and R. Walsh, *Int. J. Chem. Kinet.* **16**, 117 (1984).

86. A.G. Griesbeck, *J. Org. Chem.* **54**, 4981 (1989).

87. J.J. Gajewski and C.J. Cavender, *Tetrahedron Lett.* 1057 (1971).

88. H. Meier, A. Pauli, H. Kolshorn and P. Kochhan, *Chem. Ber.* **120**, 1607 (1987).

89. H. Pauli, H. Kolshorn and H. Meier, *Chem. Ber.* **120**, 1611 (1987).

90. H. Meier, T. Echter and H. Petersen, *Angew. Chem. Int. Ed. Engl.* **17**, 942 (1978).

91. H. Meier, N. Hanold and H. Kolshorn, *Angew. Chem. Int. Ed. Engl.* **21**, 66 (1982).

92. N. Hanold, T. Molz and H. Meier, *Angew. Chem. Int. Ed. Engl.* **21**, 917 (1982).

93. H. Meier and T. Echter, *Angew. Chem. Int. Ed. Engl.* **21**, 67 (1982).

94. J.J.M. Lamberts and W.H. Laarhoven, *J. Org. Chem.* **49**, 100 (1984).

95. H. McNab, *J. Chem. Res. (S) 1985*, 320.

96. M.D. Bancui, M. Pop, A. Petride and M.D. Stanescu, *Rev. Roumaine Chim.* **36**, 907 (1991).

97. M.D. Banciu, *Rev. Roumaine Chim.* **21**, 1083 (1976).

98a. W.-D. Fessner, B.A.R.C. Murty and H. Prinzbach, *Angew. Chem. Int. Ed. Engl.* **26**, 451 (1987).

98b. W.-D. Fessher, B.A.R.C. Murty, J. Wörth, D. Hunkler, H. Fritz, H. Prinzback, W.D. Roth, P. von R. Schleyer, A.B. Metwen and W.F. Maier, *Angew. Chem. Int. Ed. Engl.* **26**, 452 (1987).

99. H. Meier and M. Schmidt, *Tetrahedron Lett.* **30**, 5873 (1989).

100. J.W. van Straten, W.H. De Wolf and F. Bickelhaupt, *Recl. Trav. Chim. Pays-Bas* **96**, 88 (1977).

101. L.W. Jenneskens, W.H. De Wolf and F. Bickelhaupt, *Tetrahedron* **42**, 1571 (1986).

102. P.A. Kraakman, E.T.J. Nibbering, W.H. De Wolf and F. Bickelhaupt, *Tetrahedron* **43**, 5109 (1987).

103. L.W. Jenneskens, A.H.P. Krul, P.A. Kraakman, W. Moene, W.H. de Wolf and F. Bickelhaupt, *J. Org. Chem.* **51**, 2162 (1986).

104. D.F. Murray, M.W. Baum and M. Jones, *J. Org. Chem.* **51**, 1 (1986).

105. T. H. Peterson ad B.K. Carpenter, *J. Am. Chem. Soc.* **114**, 1496 (1992).

106. R.P. Johnson, *Chem. Rev.* **89**, 1111 (1989).

107. W. Adam, C. Cadiz and F. Mazenod, *Tetrahedron Lett.* **22**, 1203 (1981).

108. J.D. Price and R.P. Johnson, *Tetrahedron Lett.* **26**, 2499 (1985).

109. K.J. Shea, L.D. Burke and W.P. England, *Tetrahedron Lett.* **29**, 407 (1988).

110. C. Wentrup, G. Gross, A. Maquestiau and R. Flammang, *Angew. Chem. Int. Ed. Engl.* **22**, 542 (1983).
111. A. Runge and W. Sander, *Tetrahedron Lett.* **27**, 5835 (1986).
112. W. Sander and O.L. Chapman, *Angew. Chem. Int. Ed. Engl.* **27**, 398 (1988).
113. A. Bertsch, W. Grimme, G. Reinhardt, H. Rose and P.M. Warner, *J. Am. Chem. Soc.* **110**, 5112 (1988).
114. K.B. Wiberg and F.H. Walker, *J. Am. Chem. Soc.* **104**, 5239 (1982).
115. J. Belzner and G. Szeimies, *Tetrahedron Lett.* **27**, 5839 (1986).
116. T. Hudlicky, T.M. Kutchan and S.M. Naqvi, *Org. React.* **33**, 247 (1985).
117. T. Hudlicky, R. Fan, J.W. Reed and K.G. Gadamasetti, *Org. React.* **41**, 1 (1992).
118. H.N.C. Wong, M.-Y. Hon, C.-W. Tse, Y.-C. Yip, J. Tanko and T. Hudlicky, *Chem. Rev.* **89**, 165 (1989).
119. R.P. Short, B.C. Ranu, J.M. Revol and T. Hudlicky, *J. Org. Chem.* **48**, 4453 (1983). T. Hudlicky and F.J. Koszyk, *Tetrahedron Lett.* **21**, 2487 (1980).
120. T. Hudlicky, G. Sinai-Zingde, M.G. Natchus, B.C. Ranu and P. Papadopoulos, *Tetrahedron* **43**, 5685 (1987).
121. A. Fleming, G. Sinai-Zingde, M.G. Natchus and T. Hudlicky, *Tetrahedron Lett.* **28**, 167 (1987).
122. T. Hudlicky, A. Fleming and T.C. Lovelace, *Tetrahedron* **45**, 3037 (1989).
123. M. Karpf and A.S. Dreiding, *Helv. Chim. Acta.* **62**, 852 (1979).
124. G.G.G. Manzardo, M. Karpf and A.S. Drieding, *Helv. Chim. Acta.* **69**, 659 (1986).
125. M. Koller, M. Karpf and A.S. Dreiding, *Helv. Chim. Acta.* **69**, 560 (1986).
126. V. Bilinski, M. Karpf and A.S. Dreiding, *Helv. Chim. Acta.* **69**, 1734 (1986).
127. G. Mehta, A.V. Reddy and A. Srikrishna, *Tetrahedron Lett.* **20**, 4863 (1979).
128. G. Mehta, A. Srikrishna, A.V. Reddy and M.S. Nair, *Tetrahedron* **37**, 4543 (1981).
129. Y. Okamoto, K. Kanematsu, T. Fujiyoshi and E. Osawa, *Tetrahedron Lett.* **24**, 5645 (1983).
130. G. Mehta, D.S. Reddy and A.V. Reddy, *Tetrahedron Lett.* **25**, 2275 (1984).
131. Y. Okamoto, K. Senokuchi and K. Kanematsu, *Chem. Pharm. Bull. Japan.* **33**, 3074 (1985).
132. G. Mehta and M.S. Nair, *J. Am. Chem. Soc.* **107**, 7519 (1985).
133. G. Mehta, K.R. Reddy, R. Gleiter, S. Lalitha and J. Chandrasekhar, *J. Org. Chem.* **56**, 7048 (1991).
134. M.-C. Lasne and J.-L. Ripoll, *Synthesis* **1985**, 121.
135. M.-C. Lasne and J.-L. Ripoll, *Bull. Soc. Chim. France* **1986**, 766.
136. B. Albert, C. Heller, R. Iden, G. Martin, H.D. Martin, B. Mayer and A. Oftring, *Isr. J. Chem.* **25**, 74 (1985).
137. T. Watabe, K. Okada and M. Oda, *J. Org. Chem.* **53**, 216 (1988).
138. A.J.H. Klunder, W. Bos, J.M.J. Verlaak and B. Zwanenburg, *Tetrahedron Lett.* **22**, 4553 (1981).
     A.J.H. Klunder, W. Bos and B. Zwanenburg, *Tetrahedron Lett.* **22**, 4557 (1981).
     J.M.J. Verlaak, A.J.H. Klunder and B. Zwanenburg, *Tetrahedron Lett.* **23**, 5463 (1982).
     A.J.H. Klunder, W.B. Huizinga, P.J.M. Sessink and B. Zwanenburg, *Tetrahedron Lett.* **28**, 357 (1987).
     A.J.H. Klunder, M.J.F.M. Crul, A.A.M. Houwen-Claassen, M.G. Kooy and B. Zwanenburg, *Tetrahedron Lett.* **28**, 3147 (1987).
     J.H.M. Lange, A.J.H. Klunder and B. Zwanenburg, *Tetrahedron Lett.* **30**, 127 (1989).
139. A.A.M. Houwen-Claassen, A.J.H. Klunder, M.G. Kooy, J. Steffann and B. Zwanenburg, *Tetrahedron* **45**, 7109 (1989).
140. A.A.M. Houwen-Claassen, A.J.H. Klunder and B. Zwanenburg, *Tetrahedron* **45**, 7134 (1989).
141. A.A.M. Houwen-Claassen, A.J.H. Klunder and B. Zwanenburg, *Tetrahedron* **45**, 7149 (1989).
142. D.J. Collins, G.D. Fallon and C.E. Skene, *Aust. J. Chem.* **45**, 71 (1992).
143. R.N. Warrener, R.A. Russell, R. Solomon, I.G. Pitt and D.N. Butler, *Tetrahedron Lett.* **28**, 6503 (1987).
144. D.N. Butler, D.L. Officer, I.G. Pitt, R.A. Russell and R.N. Warrener, *Tetrahedron Lett.* **28**, 6507 (1987).
145. R.N. Warrener, J.-M. Wang, K.D.V. Weerasuria and R.A. Russell, *Tetrahedron Lett.* **31**, 7069, (1990).
146. R.N. Warrener, S. Bina, D.N. Butler, I.G. Pitt and R.A. Russell, *Tetrahedron Lett.* **31**, 7073 (1990).
147. R.A. Russell, R.W. Longmore, K.D.V. Weerasuria and R.N. Warrener, *Aust. J. Chem.* **44**, 1341 (1991).
148. R.N. Warrener, I.G. Pitt, K.D.V. Weerasuria and R.A. Russell, *Aust. J. Chem.* **45**, 155 (1992).
149. W. Eberbach, *Chem. Ber.* **107**, 3287 (1974).

150. Y. Rubin, S.S. Lin, C.B. Knobler, T.J. Anthony, A.M. Boldi and F. Diederich, *J. Am. Chem. Soc.* **113**, 6943 (1991).

151. H. Bock, W. Reid and V. Stein, *Chem. Ber.* **114**, 673 (1981).

152. Y. Rubin, C.B. Knobler and F. Diederich, *J. Am. Chem. Soc.* **112**, 1607 (1990).

153. T. Brietenbach, F. Ott and F. Vögtle, *Angew. Chem. Int. Ed. Engl.* **31**, 307 (1992).

154. T. Brietenbach, F. Ott, M. Nieger and F. Vögtle, *Chem. Ber.* **125**, 1283 (1992).

155. F. Ott, T. Brietenbach, M. Nieger and F. Vögtle, *Chem. Ber.* **126**, 97 (1993).

155a. E. Schmohel, F. Ott, J. Breitenbach, M. Nieger and F. Vögtle, *Chem. Ber.* **126**, 2477 (1993).

156. A. Bertsch, W. Grimme, G. Reinhardt, H. Rose and P.M. Warner, *J. Am. Chem. Soc.* **110**, 5112 (1988).

157. S. Yamamoto and R.A. Back, *J. Phys. Chem.* **89**, 622 (1985).

158. S. Yamamoto and R.A. Back, *Can. J. Chem.* **63**, 549 (1985).

159. R.A. Back and S. Tamamoto, *Can. J. Chem.* **63**, 542 (1985).

160. J.-R. Cao and R.A. Back, *Can. J. Chem.* **64**, 967 (1986).

161. R. Taylor, *Int. J. Chem. Kinet.* **19**, 709 (1987).

162. J. Colussi, V.T. Amorebieta and M.A. Grela, *J. Chem. Soc., Faraday Trans.* **88**, 2125 (1992).

163. R. Greenhouse, W.T. Borden, K. Hirotsu and J. Clardy, *J. Am. Chem. Soc.* **99**, 1664 (1977).

164. G.E. Renzoni, T.-K. Yin, F. Miyake and W.T. Borden, *Tetrahedron* **42**, 1581 (1986).

165. J.G. Radziszewski, T.-K. Yin, F. Miyake, G.E. Renzoni, W.T. Borden and J. Michl, *J. Am. Chem. Soc.* **108**, 3544 (1986).

166. T.-K. Yin, J.G. Radziszewski, G.E. Renzoni, J.W. Downing, J. Michl and W.T. Borden, *J. Am. Chem. Soc.* **109**, 820 (1987).

167. G.E. Renzoni, T.-K. Yin and W.T. Borden, *J. Am. Chem. Soc.* **108**, 7121 (1986).

168. R. Taylor, *J. Chem. Soc., Perkin Trans. II*, **1988**, 183.

169. R. Taylor, *J. Chem. Soc., Perkin Trans. II*, **1988**, 737.

170. I. Lee, O.J. Cha and B.-S. Lee, *J. Phys. Chem.* **94**, 3926 (1990).

171. G. Chuchani, I. Martin, A. Rotinov, R.M. Dominguez, D.G. Morris and A.G. Shepherd, *Int. J. Chem. Kinet.* **20**, 145 (1988).

172. M. Campbell, D.J. Collins and A.M. James, *Aust. J. Chem.* **42**, 17 (1989).

173. D.J. Collins, G.D. Fallon and R.P. McGeary, *Aust. J. Chem.* **47**, 739 (1994).

174. R.F.C. Brown and R.P. McGeary, *Aust. J. Chem.* **46**, 1321 (1993).

175. W.S. Trahanovsky, T.J. Cassady and T.L. Woods, *J. Am. Chem. Soc.* **103**, 6691 (1981).

176. C.-H. Chou and W.S. Trahanovsky, *J. Am. Chem. Soc.* **108**, 4138 (1986).

177. C.-H. Chou and W.S. Trahanovsky, *J. Org. Chem.* **51**, 4208 (1986).

178. U.H. Brinker and J. Ritzer, *J. Am. Chem. Soc.* **103**, 2116 (1981).

179. W.E. Billups, L.-J. Lin and E.W. Casserly, *J. Am. Chem. Soc.* **106**, 3698 (1984).

180. S.W. Staley and T.D. Norden, *J. Am. Chem. Soc.* **106**, 3699 (1984).

181. W.E. Billups and L.-J. Lin, *Tetrahedron* **42**, 1575 (1986).

182. R.A. Aitken and J.I. Atherton, *J. Chem. Soc., Chem. Commun.* **1985**, 1140.

183. R.A. Aitken and S. Seth, *Synlett.* **1990**, 211.

184. R.A. Aitken and S. Seth, *Synlett.* **1990**, 212.

185. A.B. Lovell, K. Brezinsky and I. Glassman, *Int. J. Chem. Kinet.* **21**, 547 (1989).

186. B. Ondruschka, V. Engelstädter, D. Vorwerk and G. Zimmermann, *J. prakt. Chem.* **331**, 923 (1989).

187. J.I.G. Cadogan, C.L. Hickson and H. McNab, *Tetrahedron* **42**, 2135 (1986).

188. J.I.G. Cadogan, H.S. Hutchinson and H. McNab, *Tetrahedron* **48**, 7747 (1992).

189. J.I.G. Cadogan, H.S. Hutchinson and H. McNab, *J. Chem. Soc., Perkin Trans. 1*, **1991**, 385.

190. W.S. Trahanovsky and C.C. Ong, *J. Am. Chem. Soc.* **92**, 7174 (1970).

191. J.I.G. Cadogan, H.S. Hutchinson and H. McNab, *J. Chem. Soc., Perkin Trans. I* **1987**, 1407; **1988**, 2875; J.I.G. Cadogan, C.L. Hickson, H.S. Hutchinson and H. McNab, *J. Chem. Soc., Perkin Trans. I* **1991**, 385.

192. L.W. Jenneskens, J.W. Van Straten, W.H. De Wolf and F. Bickelhaupt, *Tetrahedron* **40**, 3117 (1984).

193. M. van Meurs, I.W.C.E. Arends, R. Louw and P. Mulder, *Rec. Trav. Chim. Pays-Bas* **110**, 475 (1991).

194. R. Bloch and P. Orvanne, *Tetrahedron Lett.* **22**, 3597 (1981).

195. T.T. Tidwell, *Acc. Chem. Res.* **23**, 273 (1990); *Ketenes* (John Wiley & Sons, Inc., New York, 1995).

196. H. Bock, T. Hirabayashi and S. Mohmand, *Chem. Ber.* **114**, 2595 (1981).

197. R.F.C. Brown and F.W. Eastwood, *The Chemistry of Ketenes, Allenes and Related Compounds* (Ed. S. Patai, Wiley, Chichester, 1980).
198. C. Wentrup, G. Gross, H.-M. Berstermann and P. Lorencak, *J. Org. Chem.* **50**, 2877 (1985).
199. A. Maquestiau, P. Pauwels, R. Flammang, P. Lorencak and C. Wentrup, *Org. Mass. Spectrom.* **21**, 259 (1986).
200. C. Wentrup and P. Lorencak, *J. Am. Chem. Soc.* **110**, 1880 (1988).
201. O.L. Chapman, M.D. Miller and S.M. Pitzenberger, *J. Am. Chem. Soc.* **109**, 6867 (1987).
202. J.C. Brahms and W.P. Dailey, *Tetrahedron Lett.* **31**, 1381 (1990).
203. E. Schaumann, *Tetrahedron*, **44**, 1827 (1988).
204. For references to propadienones carrying heteroatom substitutents see references 1j and 206.
205. G.J. Baxter, R.F.C. Brown, F.W. Eastwood and K.J. Harrington, *Aust. J. Chem.* **30**, 459 (1977).
206. A. Ben Cheikh, H. Dhimane, J.C. Pommelet and J. Chuche, *Tetrahedron Letts.* **29**, 5919 (1988).
207. F. Chuburu, S. Lacombe, G. Pfister-Guillozo, A. Ben Cheikh, J. Chuche and J.C. Pommelet, *J. Am. Chem. Soc.* **113**, 1954 (1991).
208. R.F.C. Brown, K.J. Coulston, F.W. Eastwood, A.D.E. Pullin and A.C. Staffa, *Aust. J. Chem.* **43**, 561 (1990).
209. R.D. Brown, P.D. Godfrey, M.J. Ball, S. Godfrey, D. McNaughton, M. Rodler, B. Kleibomer and R. Champion, *J. Am. Chem. Soc.* **108**, 6534 (1986).
210. M.R. Anderson, R.F.C. Brown, N.R. Browne, F.W. Eastwood, G.D. Fallon, D.P.C. Gan, A.D.E. Pullin and A.C. Staffa, *Aust. J. Chem.* **43**, 549 (1990).
211. R.F.C. Brown, K.J. Coulston, F.W. Eastwood and M.J. Irvine, *Aust. J. Chem.* **44**, 87 (1991).
212. R.F.C. Brown, K.J. Coulston, F.W. Eastwood, M.J. Irvine and A.D.E. Pullin, *Aust. J. Chem.* **41**, 225 (1988).
213. For key references see C.O. Kappe, G. Farber, C. Wentrup and G. Kollenz, *J. Org. Chem.* **57**, 7078 (1992).
214. C. Kaneko, M. Sato, J.-i. Sasaki and Y. Abe, *J. Heterocycl. Chem.* **27**, 25 (1990).
215a. R.J. Clemens and J.S. Witzeman, *J. Am. Chem. Soc.* **111**, 2186 (1989).
215b. J.S. Witzeman, *Tetrahedron Lett.* **31**, 1401 (1990).
216. C. Wentrup and K.-P. Netsch, *Angew. Chem. Int. Ed. Engl.* **23**, 802 (1984).
217. Y.S. Andreichikov, G. Kollenz, C.O. Kappe, R. Leung-Toung and C. Wentrup, *Acta. Chem. Scand.* **46**, 683 (1992).
218. B. Friermuth and C. Wentrup, *J. Org. Chem.* **56**, 2286 (1991).
219. R. Leung-Toung and C. Wentrup, *J. Org. Chem.* **57**, 4850 (1992).
220. G. Bengston, S. Keyaniyan and A. de Meijere, *Chem. Ber.* **119**, 3607 (1986).
221. D.W. Emerson, R.L. Titus and R.M. González, *J. Org. Chem.* **56**, 5301 (1991).
222. C. Wentrup, H.-W. Winter, G. Gross, K.-P. Netsch, G. Kollenz, W. Ott and A.G. Bierdermann, *Angew. Chem. Int. Ed. Engl.* **23**, 800 (1984).
223. C.O. Kappe, R.A. Evans, C.H.L. Kennard and C. Wentrup, *J. Am. Chem. Soc.* **113**, 4234 (1991).
224. C.O. Kappe, G. Farber, C. Wentrup and G. Kollenz, *Tetrahedron Lett.* **33**, 4553 (1992).
225. J. Jullien, J.M. Pechine and F. Perez, *Tetrahedron Lett.* **24**, 5525 (1983).
226. R.L. DeKock and W. Weltner, *J. Am. Chem. Soc.* **93**, 7106 (1971).
227. R.D. Brown, F.W. Eastwood, P.S. Elmes and P.D. Godfrey, *J. Am. Chem. Soc.* **105**, 6496 (1983).
228. R.D. Brown, P.D. Godfrey, P.S. Elmes, M. Rodler and L.M. Tack, *J. Am. Chem. Soc.* **107**, 4112 (1985).
229. R.D. Brown, A.D.E. Pullin, E.H.N. Rice and M. Rodler, *J. Am. Chem. Soc.* **107**, 7877 (1985).
230. D. McNaughton, D. McGilvery and F. Shanks, *J. Mol. Spectrosc.* **149**, 458 (1991).
231. R.F.C. Brown, P.D. Godfrey and S.C. Lee, *Tetrahedron Lett.* **26**, 6373 (1985).
232. G. Maier, H.P. Reisenauer, H. Balli, W. Brandt and R. Janoschek, *Angew. Chem. Int. Ed. Engl.* **29**, 905 (1990).
233. G. Maier, H.P. Reisenauer, U. Schäfer and H. Balli, *Angew. Chem. Int. Ed. Engl.* **27**, 566 (1988).
234. F. Holland, M. Winnewisser, G. Maier, H.P. Reisenauer and A. Ulrich, *J. Mol. Spectrosc.* **130**, 470 (1988).
235. G. Maier, H.P. Reisenauer and A. Ulrich, *Tetrahedron Lett.* **32**, 4469 (1991).
236. R.J. Van Zee, G.R. Smith and W. Weltner, *J. Am. Chem. Soc.* **110**, 609 (1988).

2. GAS PHASE THERMOLYSIS IN NATURAL PRODUCT SYNTHESIS

ANTONIUS J.H. KLUNDER and BINNE ZWANENBURG

Department of Organic Chemistry, NSR Center for Molecular Structure, Design and Synthesis, University of Nijmegen, Toernooiveld, 6525 ED Nijmegen, The Netherlands

1. INTRODUCTION

The thermal degradation of natural products, both organic and inorganic, is undoubtedly one of the oldest preparative techniques in chemical synthetic history. It is a convenient method which does not require sophisticated apparatus, and, moreover it can be applied with a minimum of creative imagination. This simple preparative technique allowed the earliest chemists to discover the first principles of chemistry. By degradative distillation of natural matter, either solid or liquid, new unknown products were found which occassionally had great social impact on human society and therefore became landmarks in human history. A most illustrative example is the anearobic heating of bituminous coal at temperatures between 1000–1300°C that leads to coke, coal gas and coal tar. Whereas the gas was welcomed for heating and lighting, the formation of coal tar triggered the organic chemistry of aromatic hydrocarbons.[1] A most intriguing account on the large contribution of pyrolysis to the development of organic chemistry is given in Hurd's classic book entitled: "The Pyrolysis of Carbon Compounds" which appeared in 1929.[2]

With the development of more sophisticated methods to isolate or synthesize new compounds and unravel their structures, pyrolysis lost its analytical and synthetic importance in organic chemistry. As molecules became bigger and more complicated, the pyrolytic methodology is usually no longer suitable for their structure determination or to achieve a synthetic conversion.

A revival of the pyrolytic methodology in organic synthesis took place in the fifties and the sixties with the development of modern chromatographic and spectroscopic techniques, which allowed the separation and identification of the products formed. At the same time theoretical organic chemistry provided mechanistic answers for reaction patterns observed for thermal processes. The application of molecular orbital theory and the formulation of the Woodward-Hoffmann rules led numerous organic chemists to explore the thermal behavior of a great variety of structures. New experimental techniques were developed which allowed more subtle adaptation of the thermal conditions to the specific features of the substrates.

Thermal reactions involving a single substrate can be defined as a static or as a dynamic process. In a static process the substrate is heated, either as such or in a solvent, in an open or closed vessel *e.g.* a sealed tube. Eliminations, fragmentations and pericyclic rearrangements are typical reaction types that can be accomplished successfully in this manner. However, it is clear that this method is not suitable for the preparation of thermally labile products as they will undergo further reactions intra- as well as intermolecularly, ultimately leading to complex mixtures. In a dynamic process the substrate is distilled or sublimed through a hot zone, $e.g.$ a quartz tube heated in an oven, and the products formed are trapped at low temperatures (-78 to $-190°C$) immediately after they leave the hot zone. By careful selection of the dimensions of the hot zone (length and width) and choosing the appropriate temperature, flow and pressure, the substrate can now be subjected to just the amount of thermal energy required to undergo the desired conversion. A factor which plays a crucial role in determining the ultimate product formation is the residence time of the substrate in the hot zone. The shorter the contact time the smaller the probability that primary product molecules can undergo secondary reactions. In order to realize satisfactorily short contact times, dynamic or gas flow pyrolysis is usually best run under low or medium pressure conditions (10^{-1}–10^{-3} torr). Under these conditions and using standard equipment, typical contact times are 10^{-3}–10^{-1} sec. Here, thermal excitation of the substrate molecules mainly occurs by collision with the hot wall of the reactor. Because very short reaction times are involved, this pyrolytic methodology is commonly known as Flash Vacuum Thermolysis (FVT) or Pyrolysis (FVP) technique. In some cases, it may be advantageous to use an inert carrier gas, $e.g.$ nitrogen, at pressures between 10^{-1} and 10 torr. Thermal excitation of the substrate molecules is now primarily achieved by intermolecular energy transfer. Although contact times are longer and undesirable secondary reactions may therefore be more likely, this method is usually preferred for those reactions which do not involve very sensitive products because larger amounts of substrate can be processed per unit of time.

The most impressive advantage of gas flow pyrolysis over the static process is the ability to produce highly reactive compounds which may be isolated as such or be trapped in a matrix at extreme low temperature and identified by spectroscopic or chemical means. The use of moderate or high vacuum is essential here as it isolates the individual product molecule from its surroundings and therefore completely blocks intermolecular secondary reactions which otherwise could take place in the hot zone. Also for the synthesis of more stable compounds gas flow pyrolysis is generally the technique of choice when relatively high temperatures are required. In static systems either the temperature required to provide the necessary activation energy for the primary reaction cannot be reached or considerable decomposition of either the substrate or products may take place at this temperature. Other major advantages of using a dynamic system over a static system are (i) the absence of any solvent which may interfere in the reaction or complicate workup and (ii) the continuous character of the process which not only allows easy scaling up but also makes the technique suitable for substrates which have a relatively low volatility and therefore need more time to be processed.

There are also restrictions to this dynamic pyrolysis technique. In order to make gas flow pyrolysis a practical synthetic method the substrate must have a sufficient volatility. In fundamental studies this limitation is usually less important as only small quantities of material are needed. However, if a thermal conversion constitutes one of the steps in a total synthesis of a natural product, the efficiency of this reaction, both in yield, time and energy, is crucial. By applying special experimental techniques such as "solvent-assisted sublimation"[3] volatility problems can sometimes be satisfactorily overcome. A most promising solution has recently been introduced by Meth-Cohn et al.[4] who developed the "spray-pyrolysis" technique as a modification of the conventional flash vacuum pyrolysis methodology. Instead of using the distillation or sublimation technique, a liquid or molten substrate is now directly sprayed into the hot zone. As this technique is not suited for high melting substrates the "solution spray flash vacuum thermolysis" appeared the appropriate solution to this problem. Diederich et al.[5] showed that direct introduction of a solution of the substrate in a solvent e.g. benzene as a sprayed aerosol inside a hot quartz tube maintained under a moderate vacuum (1–2 torr) gives excellent results.

Notwithstanding these solutions, poor or low volatility of the substrate is the main reason for choosing a static pyrolysis technique. Fortunately, for many reaction types and substrates this method leads to satisfactory results. In particular the use of an inert high boiling solvent as the energy transfer agent is the preferred method to accomplish thermal reactions involving thermally stable primary products. It should, however, be realized that under these conditions the substrate molecules are completely solvated and therefore solvent mediated thermolysis may lead to a different reaction pattern as compared to gas phase reactions, as the result of solvent–substrate interaction. In this chapter we will compare static pyrolysis approaches with gas phase pyrolysis when appropriate.

For many gas flow pyrolysis experiments simple equipment is usually adequate to obtain the desired preparative results. Only when highly unstable intermediates are formed which must be trapped in a matrix or be detected by dynamic analysis, more sophisticated apparatus may be necessary. Several types of apparatus designed for a variety of applications are extensively described in literature.[6,7]

In our laboratory we successfully use the conventional pyrolysis set-up which is schematically depicted in Figure 1. The heart of the apparatus is a horizontal quartz tube heated with an external electric furnace, the temperature of which is controlled by a thermocouple positioned on the outer wall of the tube. An electronic control unit allows accuracies up to 10°C over a range of 300–700°C. To achieve good reproducibility of experiments it is essential that the rate of transfer of the substrate into the gas phase can be well controlled. Uneven heating of the sample should be avoided as this may lead to local decomposition. In order to realize these conditions we applied a Büchi air bath. After introduction of the substrate this air bath is slided over the sample flask against the side wall of the oven. In this way the temperature gradient in the preheating zone is minimal. Electronic control of the air bath allows accuracies up to 1–2°C over a temperature range of 50–250°C. The distance between sample flask and oven is kept as short as possible. At the exit of the oven at a minimal distance of about 2 cm the first cold trap is connected. In most cases studied, the products nicely condensed on the inner wall of the cold trap. Less volatile products may condense just after the glass joint but this never led to isolation problems. Due to the extremely high

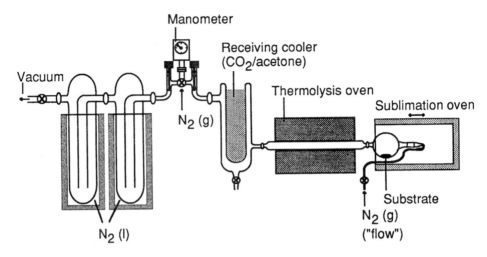

Figure 1 Simple effective apparatus for flash vacuum thermolysis.

temperature of the joints close to the oven, viton O-rings are used to ensure leak free operation. A normal vacuum system protected by a second cold trap completes our set-up. Pressure and flow can be modulated by leaking in gas, *e.g.* nitrogen or argon, at different places. Standard pressure is generally 10^{-2} torr which is adequate for most pyrolysis experiments.

In general, pyrolytic reactions in organic chemistry can be divided into two main categories.

(i) Symmetry-allowed pericyclic processes. Examples are the Diels-Alder and the ene reaction, sigmatropic rearrangements and eliminations. This category is extremely useful for preparative purposes as these reactions in principle proceed with high stereoselectivity. Moreover, combinations of different pericyclic conversions in a tandem fashion are conceivable and may lead to complex molecules in one thermolysis operation.[8] It is therefore not surprising that most of the applications of pyrolysis in natural product synthesis belong to this category.

(ii) Homolytic processes involving the formation of radical intermediates. Unless pericyclic processes are allowed, thermolysis of organic substrates will generally lead to fragmentation of the weakest bond in the substrate molecule to generate either two radicals or a biradical intermediate. Such reactions often lead to rather complex product mixtures as the result of competitive cleavages of different bonds with similar bond strengths. To obtain synthetically useful results, careful optimization of the pyrolysis conditions and chemical adaptation of the substrate are necessary in almost all cases. Only a few examples of pyrolytic radical reactions in natural product synthesis have been reported.

The everincreasing number of highly reactive and sometimes exotic molecules that can be generated using flash vacuum thermolysis, has initiated an growing interest in applying this technique for the preparation of precursors or intermediates for natural product synthesis. Both the efficiency and product selectivity usually observed in this thermal procedure are particularly attractive features. Moreover this technique is not limited to the generation of small quantities of product but can usually be easily scaled up to preparative useful amounts. Exceptions are those cases for which the volatility of the starting substrates is too low to allow reasonably effective conversion rates. Here static pyrolysis methods are sometimes to be preferred.

The purpose of this chapter is to demonstrate that thermal conversions constitute an important synthetic tool in natural product synthesis. The effectiveness of this tool is sometimes considerably enhanced by using dynamic gas flow pyrolysis instead of some static method. The examples selected will clearly illustrate this point. In this chapter we have limited ourselves to fragmentation reactions dealing mainly with the retro-Diels-Alder reaction. The broad scope of this reaction already convincingly demonstrates the wide applicability of gas flow pyrolysis for the synthesis of wide variety of natural products. Obviously, we have restricted us to representative cases which have some relevance to natural product synthesis. Many other examples are to be found in a comprehensive monograph and in several extensive reviews on flash vacuum thermolysis.[6-21]

2. ELIMINATION REACTIONS

The β-elimination of carboxylic acids from esters to form alkenes is one of the most well-known pyrolytic reactions which frequently is performed under gas phase conditions.[6,22] Stereochemically it is a *cis*-elimination reaction and its regiochemistry is mainly controlled by the number of β-hydrogen atoms available for each mode.

The synthesis of tricyclic marine natural product sinularene **6** is a nice illustration of the application of elimination by gas phase thermolysis as here this methodology has been utilized twice.[23] Both the introduction of the methylene function at C_7 and the *endo*-isopropyl group at C_5 in **6** are accomplished in this way (Scheme 1). Tricyclic dimethylcarbinol **1** was acylated and the acetate **2** was then subjected to flow pyrolysis. The gas phase elimination reaction was carried out

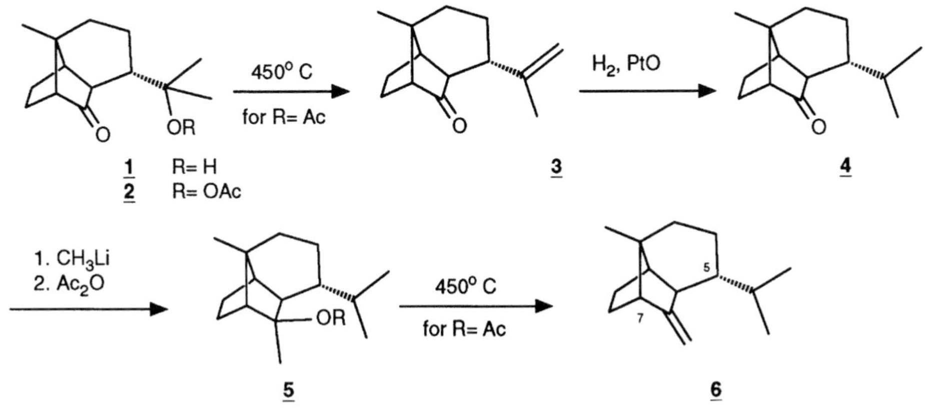

Scheme 1

at atmospheric pressure by slowly injecting a benzene solution of the acetate into a vertically mounted pyrex column (350 × 14 mm) heated in a tubular furnace at 450°C, while a slow stream of nitrogen was passed through the column. The pyrolysate was collected at the column exit in a flask immersed in dry ice. This set-up is the most commonly used one for such ester eliminations and works generally quite satisfactorily for substrates which lead to relative stable products. In order to achieve a more efficient thermal excitation the tube may be packed with pyrex or quartz beads. In the case of **2** pyrolysis is very effective already at 450°C, to give, without packing terminal alkene **3** in 95% yield along with 5% of the undesired internal alkene. Pure **3** could readily be obtained by preparative gas chromatography. Catalytic hydrogenation of **3** over reduced platinum oxide in methanol gave stereoselectively ketone **4** which, on reaction with methyllithium followed by acylation, gave acetate **5**. In a similar way, as decribed above, gas phase pyrolysis of **5** at 450°C resulted in a fast elimination to give *rac*-sinularene **6** in 63% yield, after distillation.

A recent example of pyrolytic acetate elimination using the same pyrolysis technique is encountered in the latest total synthesis of the sesquiterpene (+)-longifolene **8** (Scheme 2).[24] Using an intramolecular Diels-Alder cycloaddition reaction to obtain the basic tricyclic skeleton of this much studied sesquiterpene,

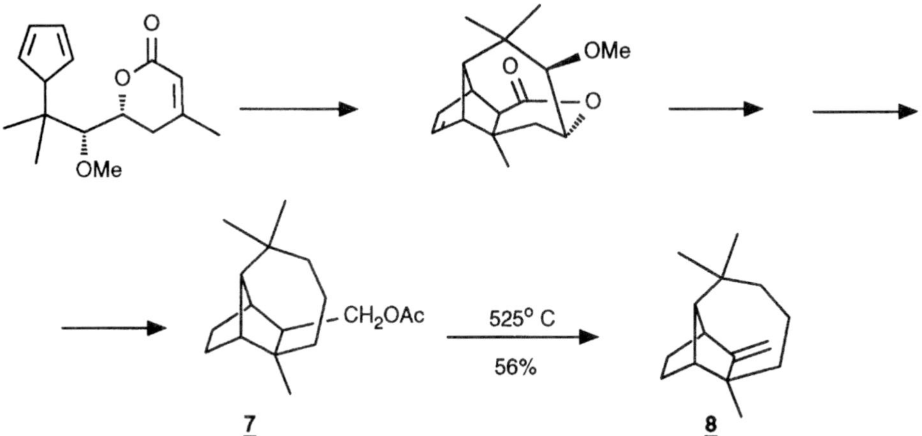

Scheme 2

the synthesis of **8** was completed by dropping a solution of acetate **7** in benzene into a heated quartz tube (525°C) packed with quartz-wool under a flow of nitrogen. A moderate yield of 56% was obtained for this atmospheric gas phase pyrolysis.

The modest yields obtained here may be caused either by an incomplete elimination reaction or by secondary reactions in the gas phase. Under these atmospheric conditions contact times are relatively long and intermolecular reactions are quite conceivable. Moreover, using the set-up as described above, acetic acid is condensed together with the product which on warming up may lead to undesired rearrangement products. Flash vacuum thermolysis would have avoided these undesired reactions and would probably have led to a cleaner product and a higher yield.[7,10]

The advantage of the flash vacuum thermolysis technique in ester elimination reactions is nicely demonstrated in the preparation of menthene from (−)-menthyl acetate **9**. When carried out in a similar way as described above at 450°C at atmospheric pressure over a column packed with pyrex beads, a mixture of menth-1-ene **10** and menth-2-ene **11** was obtained in 86% total yield[25] (Scheme 3). Alkaline extraction and subsequent distillation are necessary to obtain a pure menthene mixture. However, if the elimination is carried out under flash vacuum

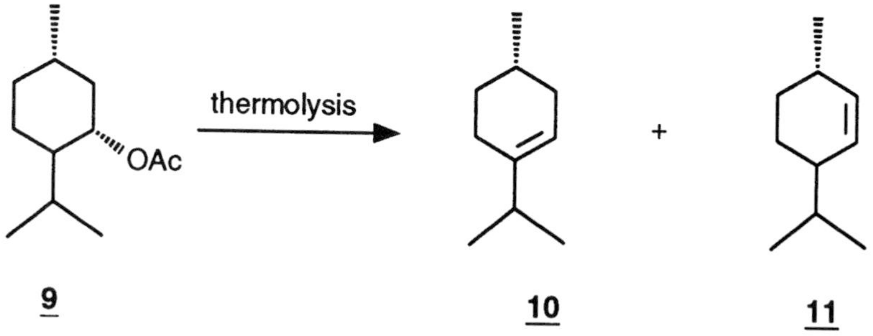

Scheme 3

thermolysis conditions using the apparatus as depicted in Figure 1 applying 600°C at 10^{-3} torr and trapping the pyrolysate at −78°C a clean mixture of menthenes is obtained in almost quantitative yield without distillation.[26]

A striking example of the preparative significance of flash vacuum thermolysis in the synthesis of sensitive biological structures is encountered in a route to the furo-pyran ring system.[27] This structural moiety is found in azadirachtin **12** which is responsible for the anti-feedant activity against the larva of *spodoptera littoralis*. To test this hypothesis, the sensitive cyclic enol ether systems, dihydrofuro-2,3-pyrans **16** and **17** were prepared from tetrahydropyrane **13** using the thermal ester elimination reaction as the final step (Scheme 4). Using a sequence of protection/deprotection reactions acetates **14** and **15** were prepared. Distillation of **14** and **15** through a quartz tube heated at 550°C at 2×10^{-1} torr afforded **16** and **17** in 70 and 30% yield, respectively. The considerably lower yield observed in the pyrolysis of **15** as compared to methyl ether **14** is most likely due to the lower volatility of the former substrate. The higher temperatures needed to transfer sufficient starting material into the gas phase to achieve a reasonable conversion rate, may lead to substantial decomposition already at this preheating stage. The presence of an alcohol function generally leads to lower yields in gas phase pyrolyses due to this volatility effect combined with the increased acidity of the compound. By appropriate masking of the alcohol function with a thermally stable and apolar protecting group, *e.g.* an ether or ester, the conversion rate as

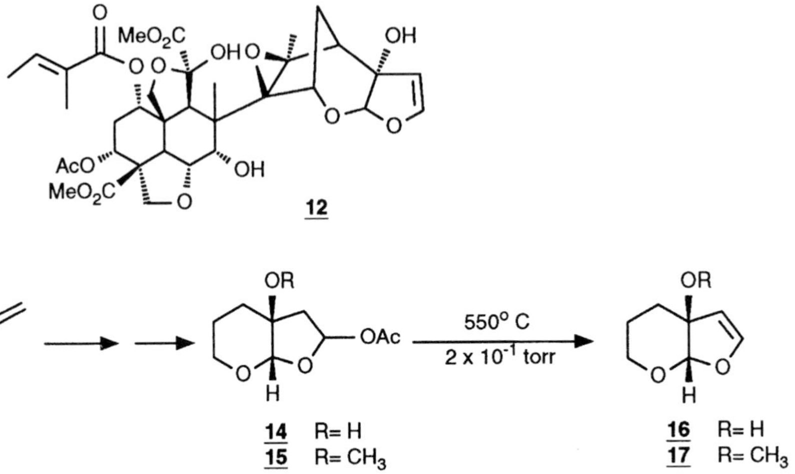

Scheme 4

well as the yield may be considerably increased. Of course, removal of such a masking group must be feasible after the pyrolysis step.

In many cases the pyrolytic elimination methodology may be the only method possible to introduce an olefinic bond in a sensitive structure. En route to the *Aristotelia* alkaloid penduncularine **18**, Speckamp et al.[28] subjected bicyclic alcohol **19** and derivatives to a variety of anionic and cationic elimination procedures to introduce the required endocyclic C_3–C_4 double bond (Scheme 5). However, all attempts to prepare **21** were in vain due to rapid rearrangement or retroconden-sation. Therefore, the neutral pyrolytic elimination of acetate **20** was attempted using flash vacuum conditions. However, at pyrolysis temperatures between 400

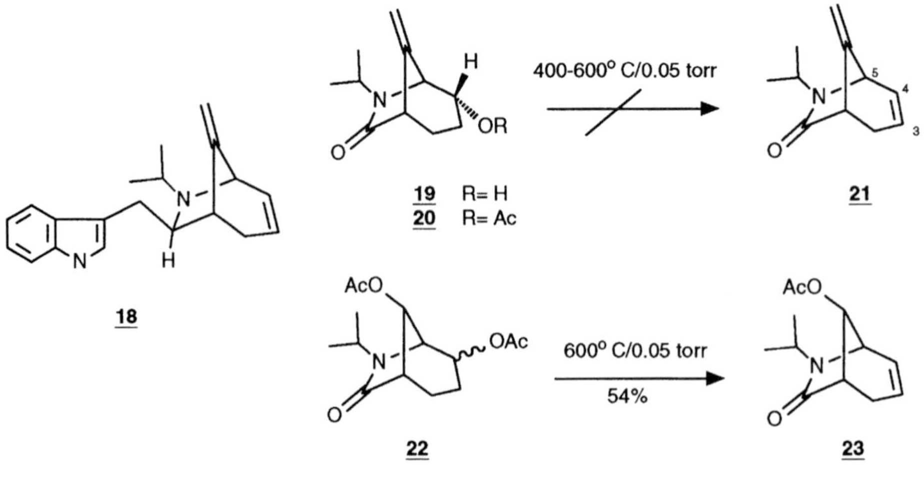

Scheme 5

and 600°C at $5 \cdot 10^{-2}$ torr either starting material or decomposition products were isolated. This failure was ascribed to the homolytic weakness of the (allylic) N–C$_5$ bond in either the starting acetate or more likely in the desired product **21**. To alleviate this problem the flash vacuum thermolysis technique was applied to diacetate **22**, which was readily obtained from an earlier intermediate in this synthetic sequence. When **22** was subjected to pyrolysis at 600°C smooth elimination of the C$_4$ acetate function was observed to give the desired alkene **23** in an acceptable yield of 54%. The remaining steps to the basic bicyclic skeleton of **18** all proceeded smoothly and in high yields. Besides the advantageous use of a neutral pericyclic process to accomplish the desired elimination, this example also convincingly demonstrates that complications due to labile groups or bonds can sometimes be readily solved by minor adaptations of the structure.

Another possibility to circumvent undesired thermal decomposition of substrate or product is lowering the activation energy required for the thermal process. For thermal elimination reactions a considerable temperature drop can be achieved by using the S-methyl xanthate group instead of an ester function.[22] Recently, this well-known xanthate elimination has been successfully applied for the multigram synthesis of 6-chlorohexene **25** from 6-chlorohexanol using the flash vacuum thermolysis technique[29] (Scheme 6). Chlorohexene **25** is an inter-

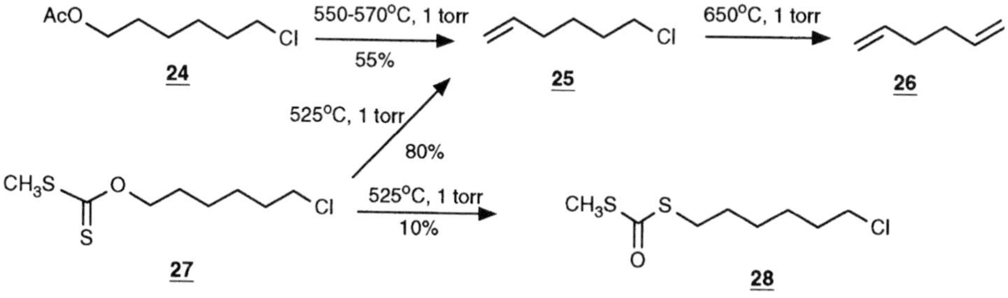

Scheme 6

esting starting material for natural product synthesis.[30,31] Using acetate **24** as the starting substrate complete conversion was observed at 650°C at a pressure of 1 torr. However, at this temperature there was already excessive elimination of hydrogen chloride from **25**, leading to the predominant formation of 1,5-hexadiene **26** as a consecutive product. Optimum yields of 55% for **25** could be obtained from runs at 570–575°C. A much better result was obtained when the S-methyl xanthate **27** was used. Complete conversion was already observed at 525°C to afford 80% of the desired chlorohexene **25**. Although no elimination of hydrogen chloride took place under these conditions, a surprisingly facile Schönberg reaction was observed leading to dithiocarbonate **28** in 10% yield.

Gas phase thermolysis may also be the method of choice when expensive isotopically labeled compounds are required. A convenient lab-scale synthesis of selectively [^{13}C]-labeled acrylonitrile **30** has been realized from readily available and relatively cheap 2-[^{13}C]-cyanoacetate **29** (Scheme 7).[32] Flash vacuum

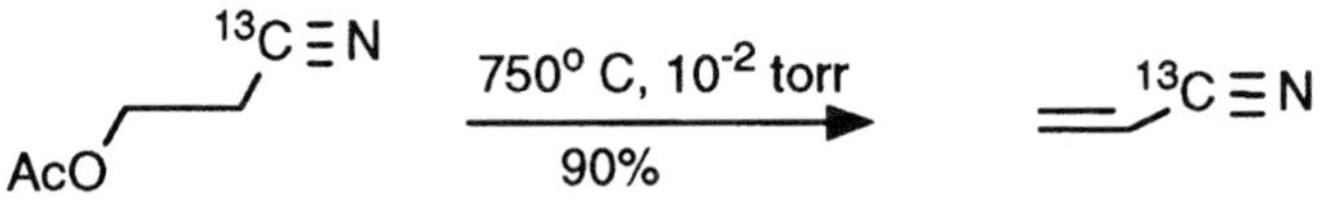

Scheme 7

pyrolysis of **29** at 750°C and 10^{-2} torr afforded the desired acrylonitrile in 90% yield with 99.9% purity after preparative gas chromatography. This labeled acrylonitrile has been utilized for the synthesis of [^{13}C]-labeled bacteriorhodopsin to enable detailed ^{13}CNMR-studies.[33]

Compounds of type **31**, where **X** is a purine or pyrimidine base, have been investigated intensively because of their potent activity against human immuno-deficiency virus (**HIV**). Dihydrofurans **32** have proved valuable in the synthesis of nucleosides and dideoxynucleosides. A simple route to dihydrofuran **35**, a potentially useful synthon for these compounds, is the flash vacuum thermolysis of lactol acetate **34**, which is readily available from lactone **33** (Scheme 8).[34] At 510°C at 0.02 torr dihydrofuran **35** was obtained in 92%.

Scheme 8

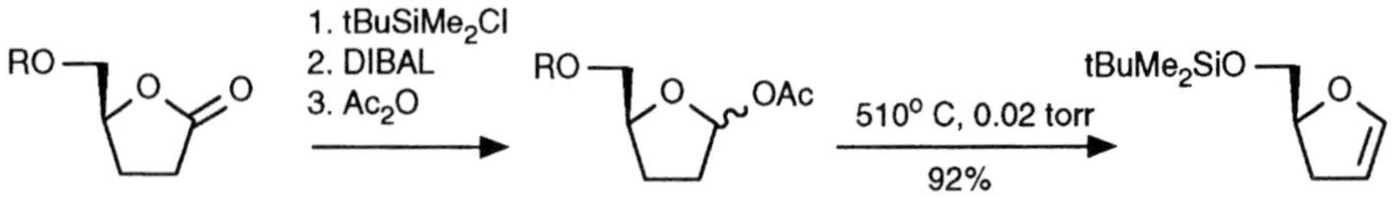

An illustrative example of the gas flow elimination of a sulfoxide function to produce a sensitive olefinic system was recently reported by Pohmakotr and Popaung[35] in their synthesis of racemic pentenomycin I and *epi*-pentenomycin I **40**. These highly oxygenated cyclopentenoids are of interest because of their antibiotic activity. Starting from readily available acetal ester **36** sulfoxide **37** was prepared (Scheme 9). Treatment of **37** with lithium diisopropyl amide resulted in the intramolecular acylation of the transient α-sulfinyl carbanion to give a diastereomeric mixture of spiro-ketosulfoxides **38**. Flash vacuum thermolysis of **38** at the relatively low temperature of 340°C (0.1 torr) afforded a mixture of

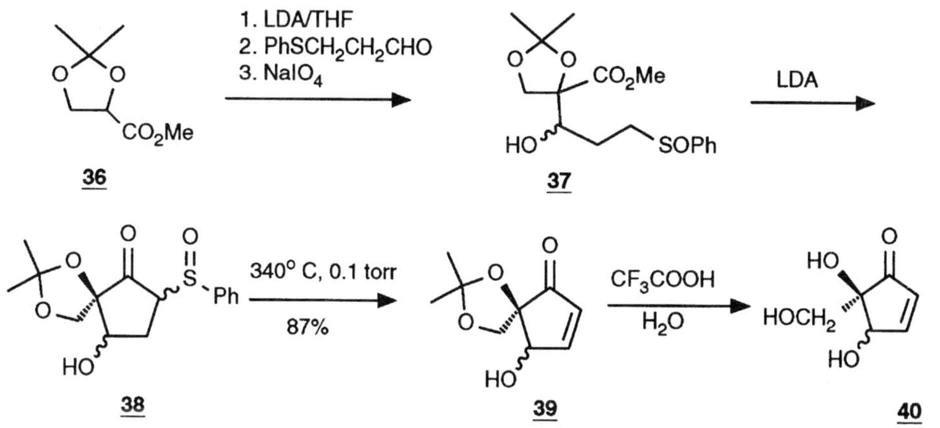

Scheme 9

pentenomycin I and *epi*-pentenomycin I acetonides **39** in an excellent total yield of 87%. In contrast pyrolysis of **38** under conventional conditions, either neat or in refluxing toluene, led to a very low yield of pentenomycins **39** due to considerable decomposition. Separation of the epimers and subsequent hydrolysis of **39** led to the pentenomycins **40**.

3.   RETRO-DIELS-ALDER REACTIONS

The retro-Diels-Alder strategy for the access to natural products has been recognized as a powerful synthetic tool only since the mid 60s. A first review paper by Kwart and King[36] already indicated the enormous potential of this methodology especially to protect an olefinic or an alkyne bond in a Diels-Alder adduct to enable conversions which would otherwise lead to extensive decomposition of the starting substrate or product. Most of the cycloreversions described in this review make use of static thermolysis procedures. Solids and liquids are simply heated at normal or reduced pressures and the volatile products collected. Due to the drawbacks of these static methods a limited number of applications of this methodology for the synthesis of natural products are mentioned. Much of the earlier work was actually aimed at the generation of transient substrates or the study of the mechanism of the Diels-Alder reaction. An interesting and very early application relevant to natural product chemistry, however, is the structure proof for cantharidin, a vesicant in the terpene series isolated from *Cantharis vesicatoria*. Heating cantharidin **41** together with palladium asbestos at 280°C in order to accomplish dehydrogenation to give **42** afforded furan and dimethylmaleic anhydride as the only products (Scheme 10).[37]

The synthesis of *rac*-ar-turmerone **47** is one of the first examples of a retro-Diels-Alder application in natural product synthesis.[38] Complications accompanying the introduction of a methyl group onto the benzylic position of the

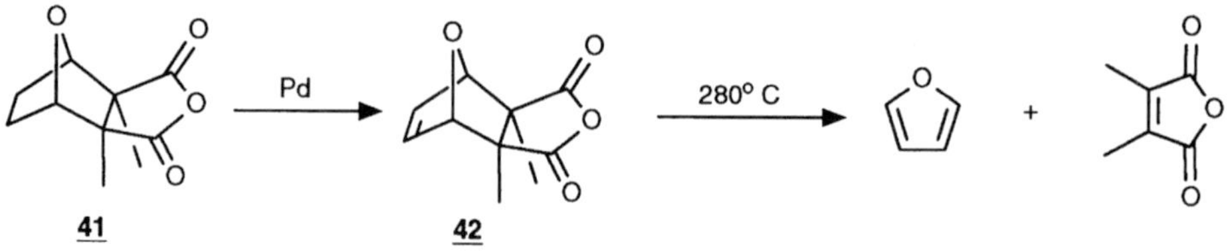

Scheme 10

cross-conjugated dienone **43** brought Ho to design a novel route in which the dimethyl enone unit was masked (Scheme 11). To this end the Diels-Alder adduct **44** of cyclopentadiene and mesityl oxide was prepared and subjected to aldolization with p-tolualdehyde to give the α, β-unsaturated aryl enone **45** in 70% yield.

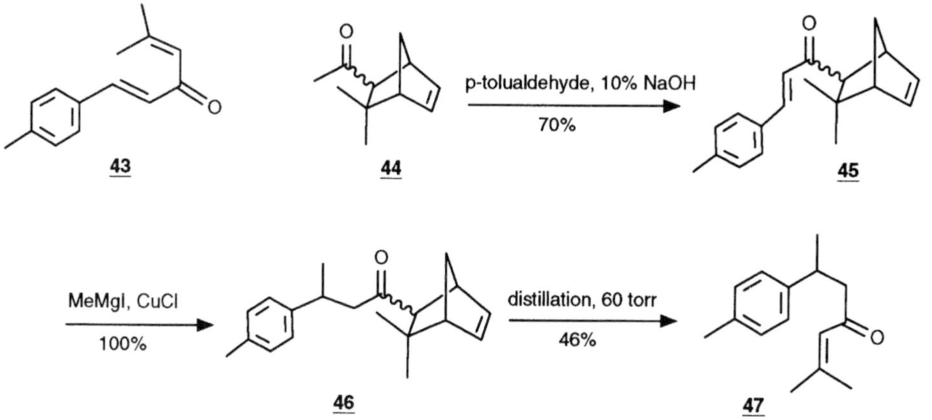

Scheme 11

Michael addition of methylmagnesium iodide in presence of cuprous chloride then led to ketone **46** in almost quantitative yield. This ketone appeared to be thermally quite unstable as it undergoes cycloreversion already on distillation at 60 torr to give rac-ar-turmerone **47** in 46% yield, after column chromatography.

This relatively simple synthetic sequence clearly illustrates the synthetic strategy which generally underlies the intelligent use of the retro-Diels-Alder methodology. Such strategy involves (i) masking the olefinic or acetylenic bond in a Diels-Alder adduct; (ii) appropriate group transformations in the adduct and (iii) retro-Diels-Alder reaction to restore the unsaturated moiety.

Stork and Rouessac[39] were among the first synthetic organic chemists to realize the importance of this cycloreversion approach for the synthesis of sensitive compounds, e.g. cyclopentenoids. In 1971 they published the synthesis of cis-jasmone essentially starting from dicyclopentadiene and applying the thermal cracking of appropriately functionalized tricyclodecanone **50** which formally is a

Diels-Alder adduct of cyclopentenone **51** and cyclopentadiene, as the key step (Scheme 12).[39] Allylic oxidation of dicyclopentadiene followed by Jones oxidation gave tricyclodecadienone **48**. Stereoselective Michael addition employing methylmagnesium iodide in the presence of copper chloride led to *exo-β-methyl-* tricyclodecane **49** in excellent yield. In order to prevent dialkylation, the

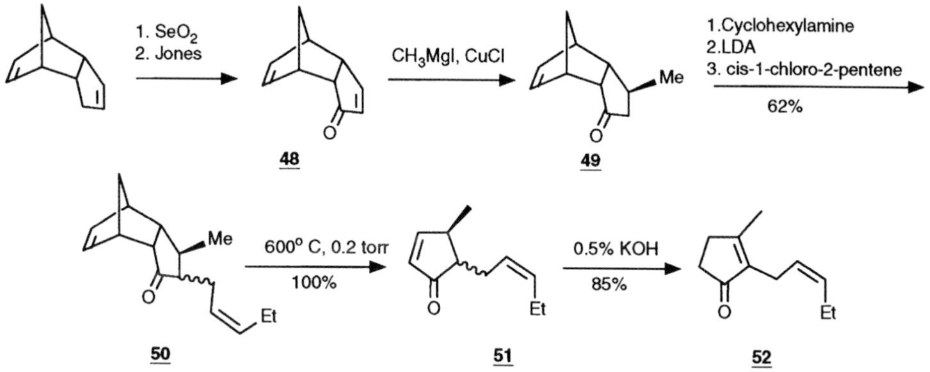

Scheme 12

introduction of the *cis*-cyclopentene group was accomplished using the metal-loenamine procedure. Due to base catalyzed epimerization a mixture of *cis*- and *trans*-**50** was obtained. The cyclopentenone synthesis was completed by the slow addition of **50** at the top of a quartz column filled with quartz chips and maintained at 600°C at 0.2 torr. Only under these conditions a quantitative yield of cyclopentenone **51** was obtained, free from the isomeric **52**. Atmospheric pressure distillation or heating in sealed quartz or pyrex tubes not only reduced the yield but also led to mixtures of isomeric cyclopentenones, *i.e.* **51** and **52**. Complete conversion of **51** to the desired *cis*-jasmone **52** was accomplished by refluxing with 0.5% potassium hydroxide for 2 hr.

Applying essentially the same approach methyl jasmonate and jasmonic acid **56** were synthesized from tricyclodecadienone **48** (Scheme 13).[40] Stereoselective addition of dimethyl malonate to **48** followed by hydrolysis and decarboxylation afforded carboxylic acid **53** in 50% overall yield. The stereoselective introduction of the required propenyl chain to give **54** was realized by the following sequence of reactions: (i) esterification of **53** followed by (ii) carbomethoxylation at C_4 using sodium hydride and dimethyl carbonate; (iii) alkylation at C_4 with 1-bromopent-2-yne and subsequent decarbomethoxylation with lithium iodide in dimethyl formamide. *Trans*-pentynyl ester **54**, thus obtained in 53% overall yield, was now subjected to gas flow pyrolysis at 430°C to give a mixture of isomeric cyclopentenones from which **55** was isolated in 80% yield. The synthesis of *rac*-jasmonate **56** was completed by stereoselective reduction of the alkyne moiety and a two-step reductive/oxidative removal of the enone moiety.

The strategy depicted in Scheme 12 was also successfully applied for the synthesis of a variety of prostaglandine PGA_2 derivatives.[41,42] Gas phase pyrolysis of the

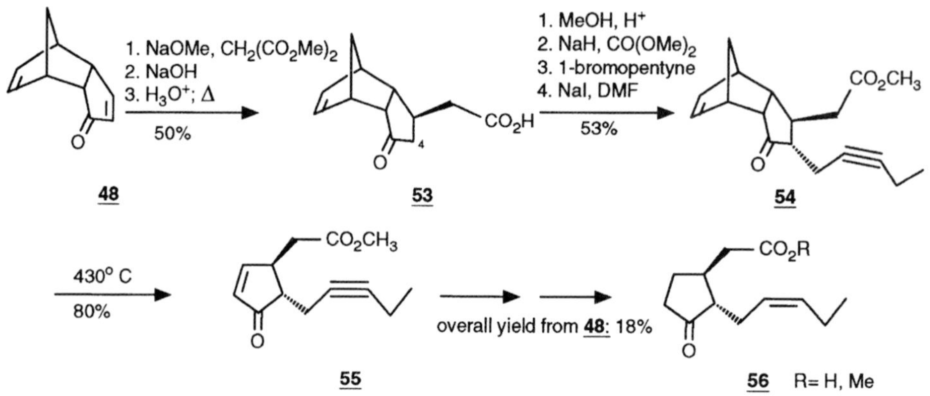

Scheme 13

intermediate *trans*-dialkylated tricyclodecenones **57** at 500°C and 0.01 torr led to complete cycloreversion to give the expected *trans*-2,3-dialkylated cyclopentenones **58** in quantitative yield without any double bond isomerization (Scheme 14).

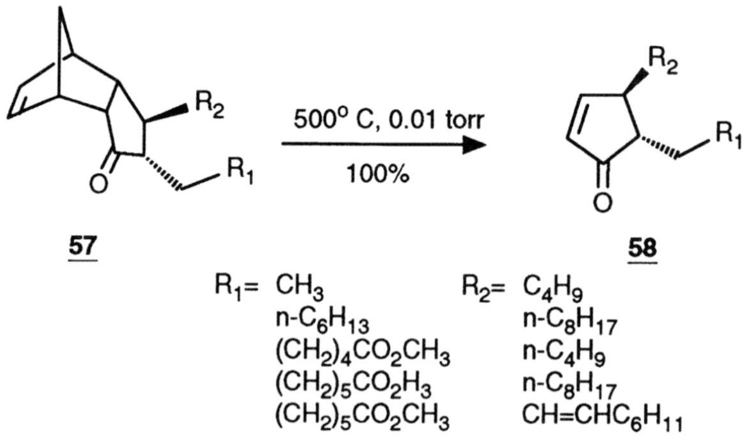

$R_1= CH_3$ $R_2= C_4H_9$
$n\text{-}C_6H_{13}$ $n\text{-}C_8H_{17}$
$(CH_2)_4CO_2CH_3$ $n\text{-}C_4H_9$
$(CH_2)_5CO_2H_3$ $n\text{-}C_8H_{17}$
$(CH_2)_5CO_2CH_3$ $CH=CHC_6H_{11}$

Scheme 14

An interesting and effective approach to prostanoid precursors involves the cycloaddition of *endo*-dicyclopentadiene with dichloroketene (Scheme 15).[43] A mixture of two regioisomeric cycloaddition was obtained with **59** predominating. Reduction with zinc in acetic acid to remove the chlorine substituents, followed by Baeyer-Villiger oxidation with hydrogen peroxide in acetic acid, gave lactone **60**, which was reduced with dibutylaluminum hydride to afford the corresponding lactol. Wittig reaction followed by oxidation and esterification gave tricyclic ketone **61** which on flash vacuum thermolysis at 460°C and 0.01 torr smoothly

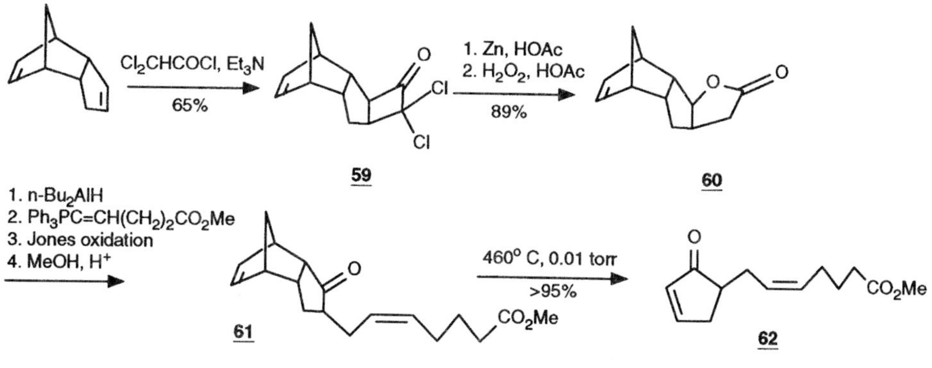

Scheme 15

underwent a retro-Diels-Alder reaction to produce prostanoid precursor **62** in nearly quantitative yield.

The [2 + 2]-cycloaddition of dichloroketene and *endo*-dicyclopentadiene also constituted the starting reaction for the synthesis of *rac*-multifidene and related structures.[44] Multifidene, a highly unsaturated cyclopentene derivative with a *cis*-vicinal substitution pattern, is the major constituent of the essential oil of the Mediterranean seaweed *Cutleria multifida* which functions as a chemical signal in the control of the reproduction. Dicyclopentadiene was selected as the starting compound as it allows vicinal *cis*-substitution by initial ketene cycloaddition and already contains, although masked, the cyclopentene double bond.

Dichloro ketene addition product **59** was ring opened with sodium hydroxide in water/dioxane to give a mixture of carboxylic acids which could be separated by crystallization to give **63** as the major product in 52% yield (Scheme 16). Hydride reduction followed by silver catalyzed hydrolysis of the dichloromethyl group led to lactol **64**, which was subjected to a Wittig reaction to introduce the vinyl side chain, then oxidized to the *cis*-vinyl aldehyde and again reacted in a Wittig reaction with triphenylpropylidene phosphorane to give the desired

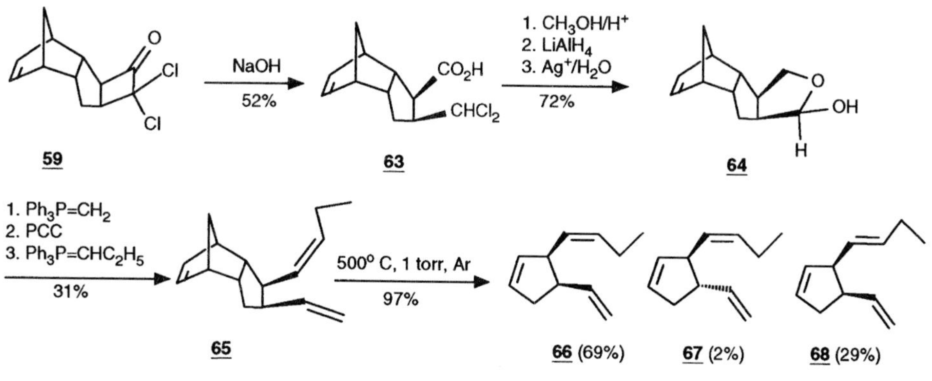

Scheme 16

cis-disubstituted tricyclodecene **65**. Flash vacuum thermolysis of **65** at 500°C and 1 torr using a small gas flow of argon led to complete cycloreversion which, however, was accompanied by substantial rearrangement of the initially formed multifidene **66**. Under the thermal conditions applied **66** was found to undergo a double Cope rearrangement to give considerable quantities of stereoisomers **67** and **68**. Lowering the pyrolysis temperature did increase the relative yield of **66**, however, the cycloreversion was then far from complete. In order to avoid this undesired [3 + 3]-rearrangement an alternative sequence was chosen, *viz* introduction of the vinyl group after the thermal generation of the cyclopentene system (Scheme 17).[45] For this purpose *cis*-methanol **69** was prepared and subjected to gas flow pyrolysis. In addition to its very low volatility, which is often

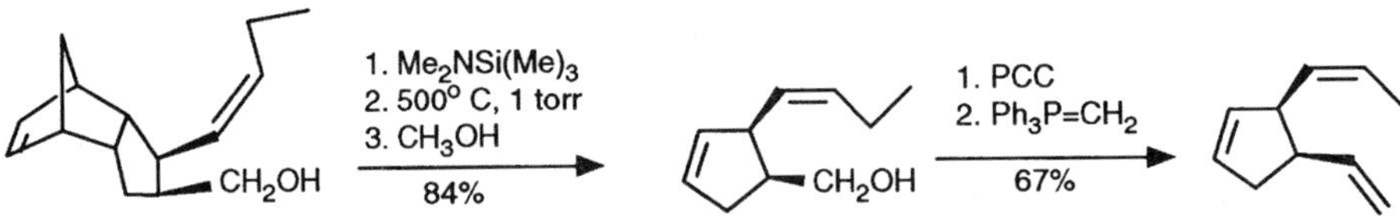

Scheme 17

a problem with alcohols, it turned out that neither the free alcohol nor its corresponding aldehyde or ester gave satisfactory results. Considerable amounts of *trans*-substituted disubstituted cyclopentenes were obtained, most likely as a result of the intermediate formation of biradicals or enolization of the carbonyl moiety. An elegant solution to problem appeared to be the use of the corresponding trimethylsilyl ether of **69**. Flash pyrolysis at 500°C and 1 torr applying an argon flow now proceeded smoothly to give, after removal of the silyl group, cyclopentene **70** in an excellent yield with only minor amounts of *trans*-impurities present. Subsequent oxidation and Wittig olefination afforded multifidene **66** in an acceptable overall yield and with high stereochemical purity. This sequence of reactions has also been used for the synthesis of a series of congeners of multifidene.

Although the pioneering work of Rouessac and coworkers already indicated the synthetic potential of the tricyclodecadienone system as synthon for cyclopentenoid natural products, its broad synthetic applicability was only recognized after the successful stereoselective syntheses of the naturally occurring antibiotics terrein and pentenomycin. Inspired by the attempts of Chapman and Hess to prepare cyclopentadienone epoxide **72** by flash vacuum thermolysis of tricyclodecadienone epoxide **71**[46] (Scheme 18), we considered both epoxides as interesting synthons for highly oxygenated cyclopentenoids. Unfortunately, the results of Chapman and Hess were not particularly encouraging as no cyclopentadienone epoxide was isolated when **71** was subjected to flash vacuum thermolysis at 600°C and 0.01 torr. Instead, a quantitative yield of α-pyrone was isolated which is the result of a fast intramolecular [4π$_a$ + 2π$_a$] cycloreversion reaction of the initially formed cyclopentadiene epoxide **72**. However, we reasoned that appropriate

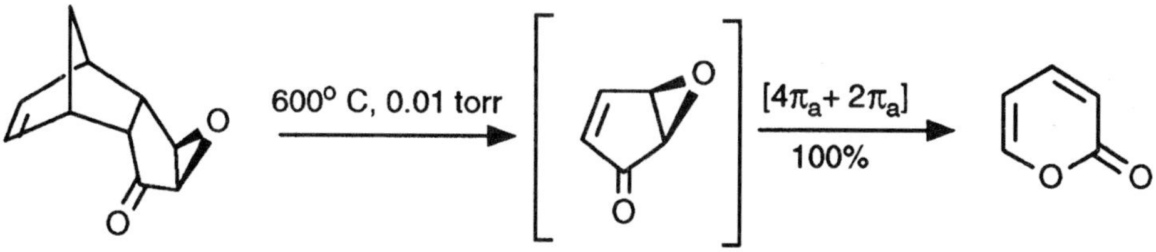

Scheme 18

functionalization could considerably increase the thermal stability of the cyclopentadienone epoxide system, thus making the subsequent thermal rearrangement less favorable.

A practical synthesis of 3-functionalized cyclopentadienone epoxides and their
dimethyl acetals was accomplished starting from ethyl tricyclodecadienone
6-carboxylate **73**, which is readily accessible from benzoquinone and cyclopentadiene[47] (Schemes 19 and 20). Alkaline epoxidation of **73** with hydrogen peroxide
occurred exclusively from the sterically less hindered *convex* face of the tricyclo-

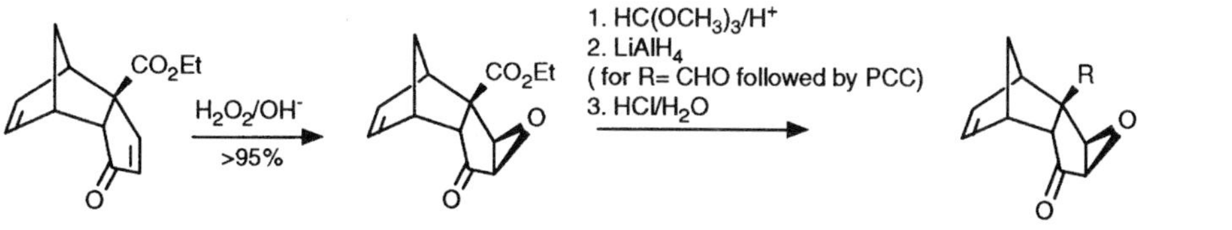

Scheme 19

decadienone skeleton to give *exo*-epoxide ester **74** in quantitative yield. In order
to achieve selective conversion of the ester function, the cyclopentanone carbonyl
group was protected as the dimethyl acetal. Hydride reduction and subsequent
oxidation with pyridinium chlorochromate, followed by deprotection of the
ketone function with aqueous hydrogen chloride, led to tricyclic alcohol **75** and
aldehyde **76**, respectively, in excellent overall yields. All three tricyclic epoxy
ketones **74**, **75** and **76** were subjected to flash vacuum thermolysis (Scheme 20).
Around 500°C and 0.02 torr almost complete cycloreversion was observed for all
epoxy ketones. At that temperature both ester **74** and aldehyde **76** produced
mixtures of cyclopentadienone epoxides **77** and **79**, and α-pyrones **80** and **82**,
respectively. Optimum yields for cyclopentadienone epoxides **77** and **79** were
obtained at 430°C and 420°C, respectively. Only in case of the aldehyde **76**
the pyrolysis could be controlled in such a way that epoxide **79** was obtained
as the exclusive product. Thermal reaction of alcohol **75** did not lead to any

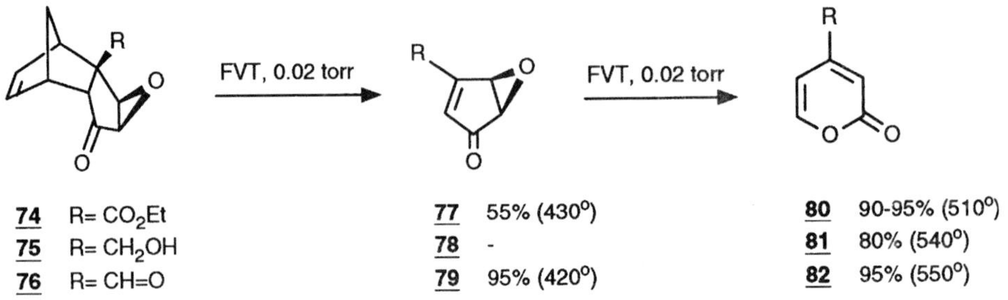

74	R= CO$_2$Et		**77**	55% (430°)	**80**	90-95% (510°)
75	R= CH$_2$OH		**78**	-	**81**	80% (540°)
76	R= CH=O		**79**	95% (420°)	**82**	95% (550°)

Scheme 20

cyclopentadienone epoxide **78**. At 400–450°C only starting alcohol and α-pyrone **81** were obtained. When the pyrolysis temperatures over 500°C are applied complete and almost quantitative conversion into the corresponding α-pyrones is observed in all three cases. These results convincingly demonstrate that the thermal stability of cyclopentadienone epoxides is considerably enhanced by extending the conjugation of the cycloenone system. As a consequence, cyclopentadienone epoxides such as **77** and **79** can be efficiently generated by gas phase pyrolysis of appropriate tricyclodecenones. When such an extended π-system is lacking, the cyclopentadienone epoxides rapidly rearrange to the corresponding α-pyrones under the conditions of the thermal fragmentation. For preparative purposes this cycloreversion constitutes a useful route to functionalized cyclopentadienone epoxides and α-pyrones. Both classes of compounds are important structural entities in natural product synthesis.

The high yield generation of epoxy aldehyde **79** is a mere demonstration of the great potential of the flash vacuum thermolysis technique. By choosing short reaction times (oven length only 16 cm and unpacked) and carefully selecting the pyrolysis temperature, formation of α-pyrone **82** could be completely avoided. It should be noted that a relatively small increase in temperature causes a subsequent rearrangement of the cyclopentadienone epoxides to the corresponding α-pyrones. These strained highly functionalized five membered ring compounds are not attainable at all applying the conventional 'wet' chemistry as they rapidly react with both nucleophilic and electrophilic reagents. The many attempts to isolate or synthesize cyclopentadienone epoxides in solution have always been in vain.[48]

A clear demonstration of the synthetic applicability of cyclopentadienone epoxides in natural product chemistry is given by the stereoselective synthesis of rac-terrein **88**, a mould metabolite from Aspergillus terreus.[49] Although already isolated in 1935, its sensitivity to acid and base precluded an efficient synthesis. Tricyclic aldehyde **83** was the starting material of choice (Scheme 21). In the same efficient manner as described for aldehyde ketone **76** flash vacuum thermolysis (475°C, 0.1 torr) of **83** led to a quantitative formation of **85**. Subsequent Wittig-ethenylidation afforded the trans-propenyl derivative **86** in 55% yield. This compound can also be prepared by ethenylidation of tricyclic aldehyde **83** to give alkene **84**, which was followed by thermolysis (450°C,

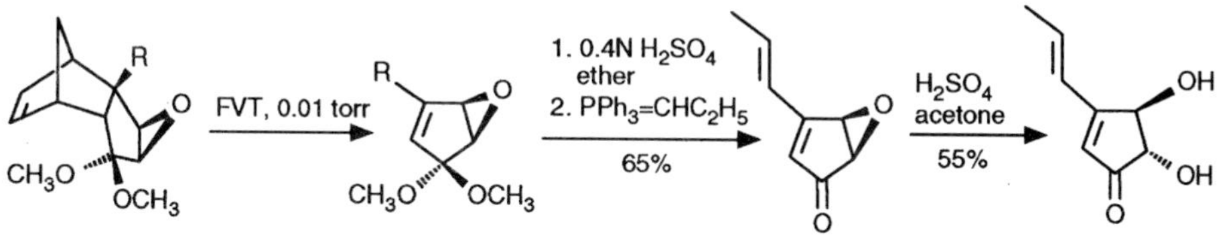

83 R= CH=O
84 R= trans-C=CCH₃

85 R= CH=O
86 R= trans-C=CCH₃

87

88

Scheme 21

0.1 torr) to yield **86**. However, this sequence of steps is less efficient as **86** is now obtained in only 35% yield after a tedious column chromatographic separation step. Selective hydrolysis of the dimethyl ketal function of **86** was accomplished with 0.4 N sulfuric acid in ether at room temperature to give cyclopentadienone epoxide **87** in 65% yield. Using 1% of an aqueous 5N sulfuric acid solution in acetone led after 4 days of stirring at room temperature to *rac*-terrein **88** in 55% yield.

Having successfully completed the terrein synthesis, a route to *epi*-penteno-mycin, containing also a vicinal *trans*-diol moiety, was devised. Applying the same strategy, it proved necessary to use the 10-oxa-analog of tricyclodecadienone epoxide **89**, *viz* **90**. These furan derived Diels-Alder adducts undergo cyclorever-sion at such a temperature that no competitive rearrangement of the initially

89

90

formed cyclopentadienone epoxides to α-pyrones takes place.[50–52] Starting from the Diels-Alder adduct of furan and cyclopentene-1,4-dione, **91**, a series of 3-alkoxymethyloxatricyclodecadienones **92** was synthesized following the se-quence depicted in Scheme 22.[50,52] Alkaline epoxidation of **92** led stereoselec-tively and quantitatively to tricyclic epoxides **93** which on flash vacuum thermolysis at temperatures as low as 350°C (0.05 mbar) afforded the cyclopen-tadienone epoxides **94** in yields of 90%. Only traces of α-pyrones were detected. Subsequent hydrolysis of **94** with 1% 5N sulfuric acid in acetone led to a smooth

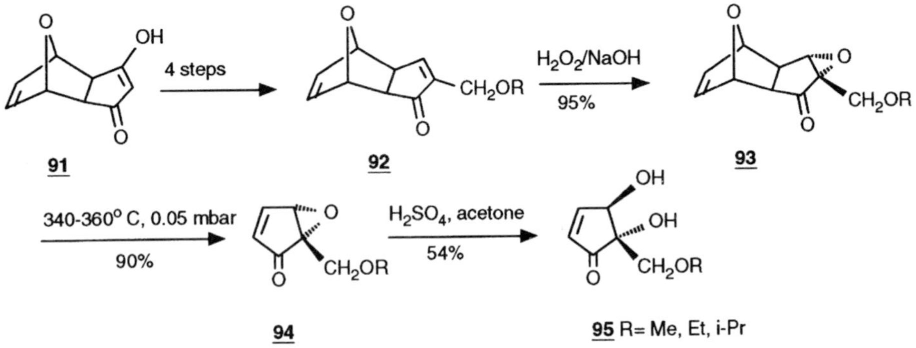

Scheme 22

cleavage of the epoxide ring affording the alkyl protected *epi*-pentenomycins **95** in yields of 54%.

The synthesis of *rac*-pentenomycin **100**, an antibiotic isolated from *Streptomyces eurythermus*, required a somewhat different strategy as it contains, in contrast to *epi*-pentenomycin, a vicinal diol moiety with the *cis*-configuration. Since simple hydrolysis of cyclopentadienone epoxide would lead to a *trans*-diol, *cis*-hydroxylation of an appropriately substituted tricyclodecadienone, *viz.* hydroxy-protected 4-hydroxymethyl-*exo*-tricyclodecadienone **96** was considered.[53] It is obvious that direct *cis*-hydroxylation using electrophilic reagents would not meet with success as the strained norbornene double bond would react preferentially. Therefore, a multistep synthesis starting from *exo*-epoxide **97** was devised (Scheme 23). Alkaline epoxidation of **96** gave stereoselectively the desired epoxide **97** which on treatment with concentrated hydrobromic acid in methanol was regioselectively converted into a *trans*-bromohydrin in 60–80% yield. Demethylation with

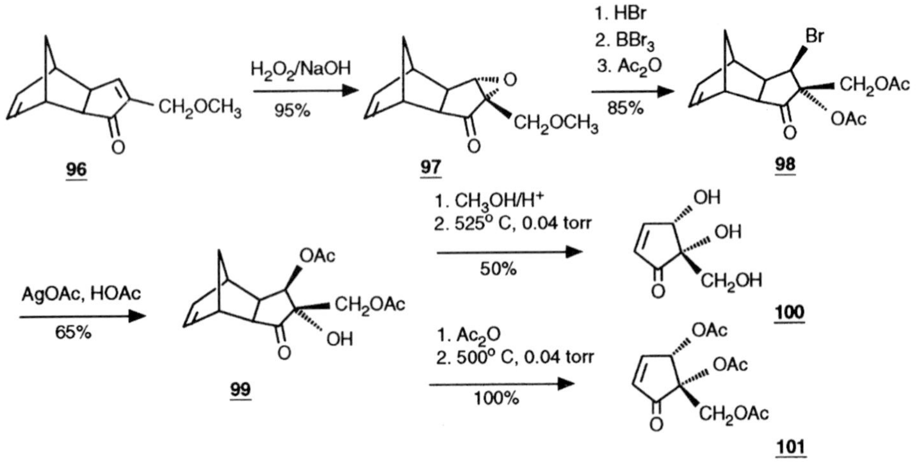

Scheme 23

borontribromide followed by acylation afforded diacetate **98** in excellent yield. Silver assisted replacement of the bromine by an acetate group gave acetoxy alcohol **99** with the desired *cis*-configuration. Acid-catalyzed methanolysis of **99** led quantitatively to the corresponding triol which on flash vacuum thermolysis at 525°C and 0.04 torr was smoothly transformed into *rac*-pentenomycin **100** in 50% yield. The relative low yield obtained in this thermolysis is mainly due to the low volatility of the triol which requires higher preheating temperatures and consequently leads to partial decomposition of the starting triol. This effect of poor volatility is nicely demonstrated by the result of the flash vacuum thermolysis of the corresponding tricyclic triacetate. Being much more volatile, a lower preheating temperature was needed resulting in a quantitative formation of pentenomycin triacetate **101**.

As was shown above the synthetic potential of the tricyclodecadienone system as a synthon for naturally occurring or biologically interesting cyclopentenoids lies undoubtedly in its tricyclic structure which enforces chemical transformations to occur in a highly stereoselective manner, ultimately leading to cyclopentenones with a well-defined stereochemistry. Our finding that tricyclodecadienone ester **73** is most readily and efficiently resolved by enzymatic hydrolysis using pig's liver esterase[54] has added extra value to this synthon as a variety of optically active cyclopentenoids are now readily accessible. Recent examples from our group are the stereospecific enantioselective syntheses of clavulones I and II **102** and **103**,[55,56] and (−)-kjellmanianone[57] (Schemes 24 and 28).

Clavulones I and II, **102** and **103**, isolated from the Japanese soft coral *Clavularia viridis*, have interesting antitumor activity. Several groups have synthesized these marine prostanoids.[58] The approach used by the groups of Corey and Yamada is based on racemic or only partly resolved 4-hydroxycyclopen-

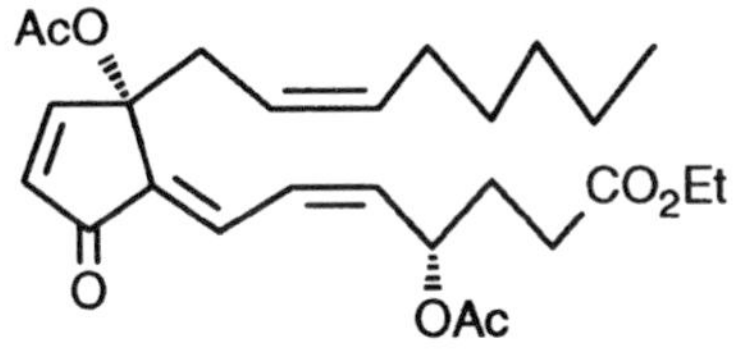

102 Clavulone I 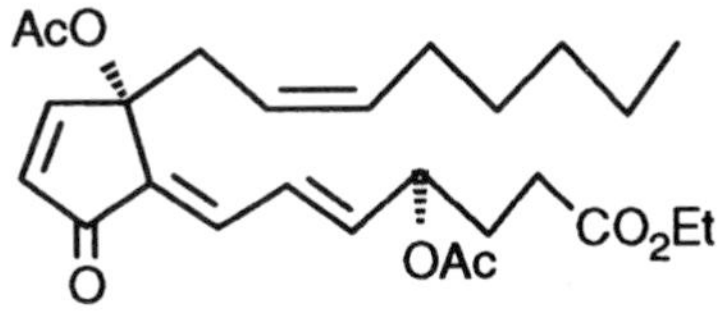   **103** Clavulone II

tenone **107** as the key intermediate. Our alternative completely enantioselective synthesis of this intermediate is again based on a tricyclodecadienone epoxide, *viz.* homochiral (+)-**71** (Scheme 24).[55,56] Addition of zinc octynyl bromide to (+)-**71** gave a chemo- and stereoselective reaction to produce alcohol **104** in nearly 90% yield. Subsequent regio- and stereoselective reductive opening of the epoxide function in **104** with lithium aluminum hydride in tetrahydrofuran led to the desired *trans*-1,3-diol **105** in good yield. Flash vacuum thermolysis of **105** at 550°C and 0.05 torr smoothly afforded cyclopentene-diol **106** in 72% yield.

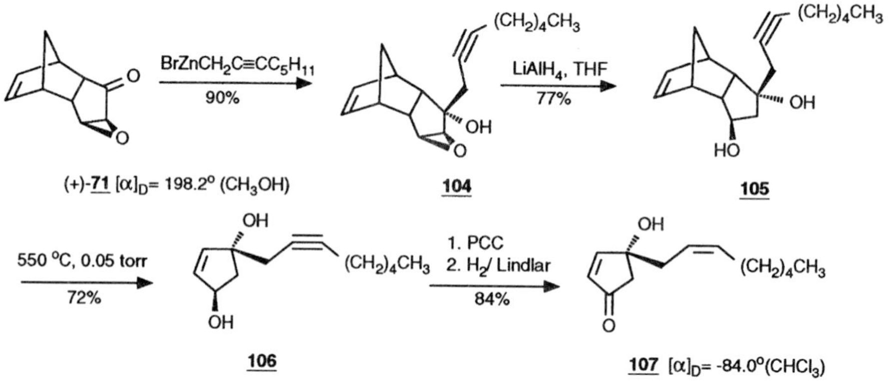

(+)-**71** [α]$_D$= 198.2° (CH$_3$OH)   **104**   **105**

Scheme 24

Subsequent oxidation of **106** with pyridinium chlorochromate and stereoselective hydrogenation of the alkyne function using Lindlar catalyst produced optically pure (−)-γ-hydroxycyclopentenone **107** in almost quantitative yield, thereby completing the formal synthesis of clavulones **102** and **103**.

Very recently, Liu *et al.* reported the synthesis of chromomoric acid D I methyl ester using the same methodology.[59] Chromomoric acids, isolated from *Chromolaens morii* and *C. chaslae*, are metabolites of linolenic acid. Their structures are very similar to the marine prostanoids clavulones. Starting from racemic **71** triol **108** was prepared similarly to **105** (Scheme 25).[59] Oxidation to cyclopentanone followed by condensation with 2-pentynal gave enone **109** terminal alcohol of which was then transformed into a methyl ester function. Gas flow thermolysis of this tricyclic ester gave cyclopentenone **110** in 64% yield. Stereoselective hydrogenation afforded *rac*-chromomoric acid D I methyl ester **111**.

rac-Chromomoric acid C I methyl ester **114** which lacks the γ-hydroxylic function was conveniently prepared[60] following the synthetic procedure already

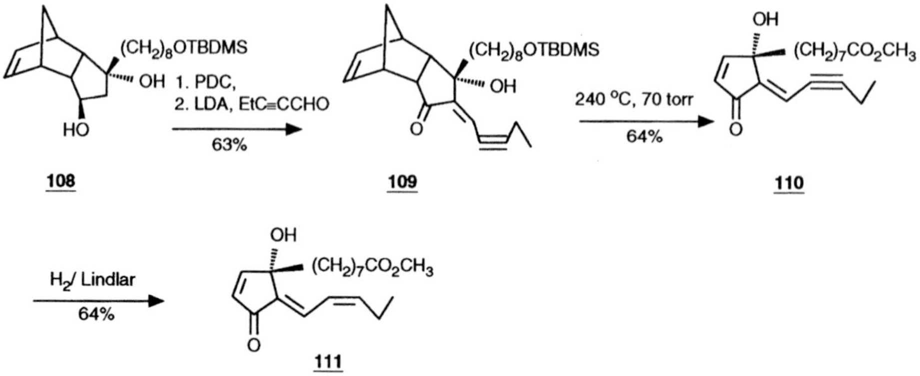

Scheme 25

described by Wiel and Rouessac for the synthesis of prostanoids.[41] Conjugate addition to racemic tricyclodecadienone **73** followed by aldol condensation led to enone **112** (Scheme 26). Gas flow pyrolysis at 254°C using a nitrogen flow at a reduced pressure of 200 torr afforded cyclopentenone **113** in 73%. Stereoselective hydrogenation of the triple bond completed the synthesis of the natural product **114**.

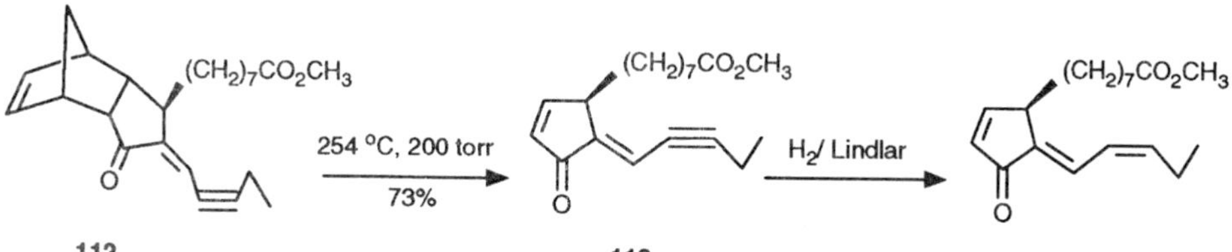

Scheme 26

In a similar way the queen recognition pheromone invictolide **119** was synthesized starting from tricyclic exocyclic enone **115** (Scheme 27).[61] Stereoselective conjugate addition of dimethylcopper lithium gave **116** which on gas flow pyrolysis at 410°C and 310 torr afforded cyclopentenone **117** in high yield. Reduction of the enone double bond and Baeyer-Villiger oxidation afforded δ-lactone **118** which can be transformed in invictolide **119** by known procedures.

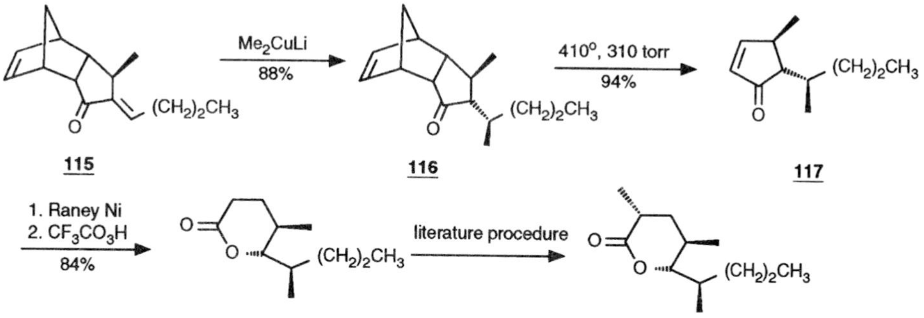

Scheme 27

Kjellmanianone **125** is a highly oxygenated cyclopentenoid isolated from the brown algae *Sargassum kjellmanianum* and shown to posses moderate activity against gram positive bacteria, such as *E. Coli* K12 and *Bacillus subtilis var niger*. The optical rotation of the natural product as isolated by Nakayama was remarkably low, *viz* $[\alpha]_D = +1.6°$ (CHCl₃).[62] Soon after its isolation Smith *et al.*[63]

achieved an enantioselective synthesis of (+)-kjellmanianone by asymmetric hydroxylation of 3-methoxy-5-methoxycarbonyl-cyclopent-2-enone using enantiopure N-sulfonyloxaziridines in ee's up to 68.5%. Although no enantiopure (+)-kjellmanianone was obtained, its optical rotation was calculated to be $[\alpha]_D = \sim 100°$ showing that the natural product is largely racemic. Our route to enantiopure kjellmanianone **125** started from homochiral ethyl tricyclodeca-dienone 2-carboxylate (+)-**73** (Scheme 28).[57] Selective reduction of the enone double bond in ester (+)-**73** at $-78°C$ followed by alkaline hydrolysis gave an

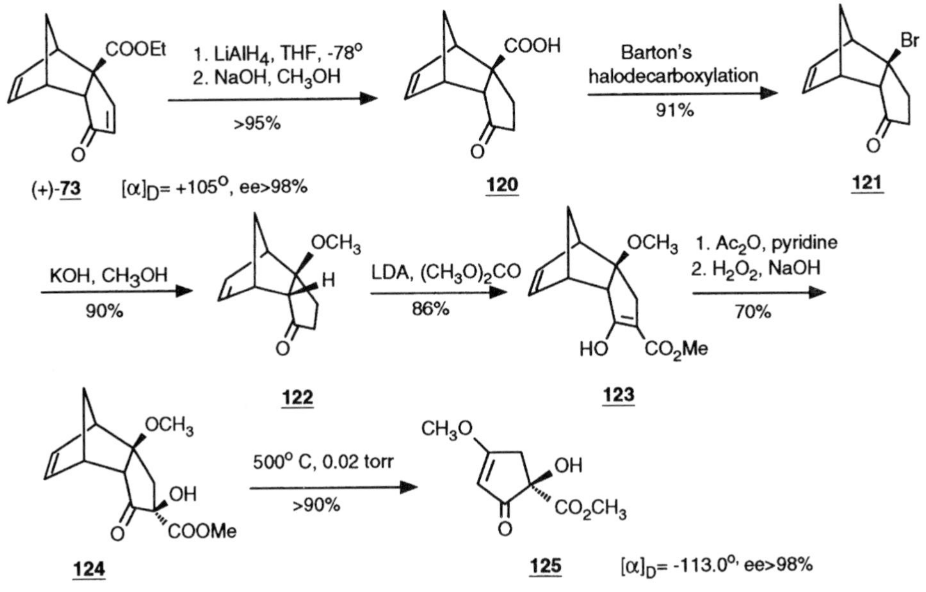

Scheme 28

almost quantitative yield of carboxylic acid **120**. Applying Barton's halodecar-boxylation procedure led to bridgehead bromide **121** in 91% yield. Replacement of the bromine function in **121** by a methoxy group was readily accomplished by reaction with potassium hydroxide in methanol to give **122** in 90% yield. Subsequent condensation of **122** with diethyl carbonate gave β-keto ester **123** which appeared to be completely enolized. Acylation of **123** followed by peroxi-dation of the enol acetate with alkaline hydrogen peroxide gave the desired α-hydroxy-β-keto ester **124** in 70% yield. Although epoxidation of enol acetate **123** is expected to occur exclusively from the sterically less hindered *exo*-face of the molecule to give **124**, this structure was unambiguously ascertained by an X-ray analysis. In the final step, enantiopure (−)-kjellmanianone **125** was produced in high yield by thermal cycloreversion by flash vacuum thermolysis of **124** at 500°C and 0.02 torr.

The general synthetic applicability of the tricyclodecadienone system **48** for the stereo- and enantioselective synthesis of cyclopentenones has induced some other research groups to explore this rigid enone as building block for natural product synthesis.

Liu et al.[64] presented an alternative synthesis for rac-sarkomycin **130**, a cyclopentenoid with interesting antitumor activity. Starting from cyanotricyclodecadienone **126**, a compound already prepared earlier by Rouessac et al.,[65] sarkomycin precursor **129** was obtained from tetracyclic lactone **128** by gas flow pyrolysis at 220°C and 30 torr (Scheme 29). Lactone **128** was readily prepared from **126** by double hydroxymethylation to give **127**, followed by decarboxylation.

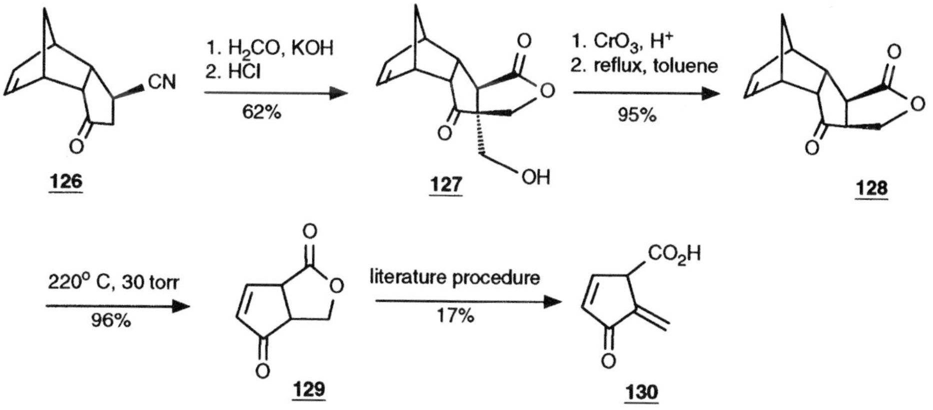

Scheme 29

The research group of Takano has published some elegant syntheses of a series of natural products using this retro-Diels-Alder methodology. Examples are the cyclopentanoid (+)-α-cuparenone,[66] the steroids (+)-equilenin[67] and (+)-estrone[68] and the alkaloids (−)-aphanorphin,[69] (−)-goniomitin,[70] (−)-physovenin and (−)-physostigmin.[71] In all cases reported no gas flow pyrolysis was applied to accomplish the cycloreversion reaction. Instead high boiling solvents such as o-dichlorobenzene and diphenyl ether in open or closed systems were used. The good yields that are generally obtained show that the static thermolysis technique may be a good alternative when rather stable and large molecules with a relatively low volatility are handled.

Ichihara et al. showed that static methods also work for more labile molecules, such as quinone epoxides and epoxycyclohexenones, provided that the temperature of cycloreversion is low enough to avoid decomposition or polymerisation of the substrates involved.[72] Using dimethylfulvene instead of cyclopentadiene as the diene component, cycloreversion of modified p-benzoquinone adducts could readily be accomplished in high yields in either high boiling solvents or in low boiling solvents using a sealed tube (Scheme 30). A series of bioactive naturally occurring cyclohexene epoxides has been synthesized using this approach. Examples are (±)-senepoxyde,[73] phyllostin, epoxydione, epiepoxydone, epiepoformin and epoformin.[74]

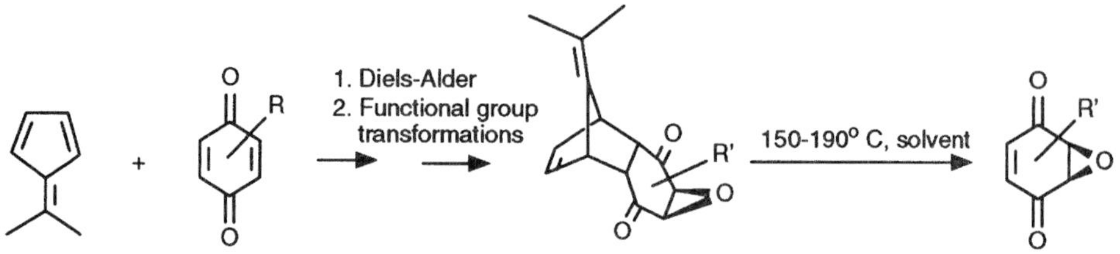

Scheme 30

The flash vacuum thermolysis technique was the method of choice for the synthesis of macrolide *rac*-pyrenolide B **135**. (Scheme 31).[75] Starting from the cyclopentadiene/benzoquinone adduct **131** appropriate functional group transformations led to tricyclic diol **132** which on treatment with lead tetraacetate gave the required macrolide skeleton **133** constrained in the tricyclic system. Another

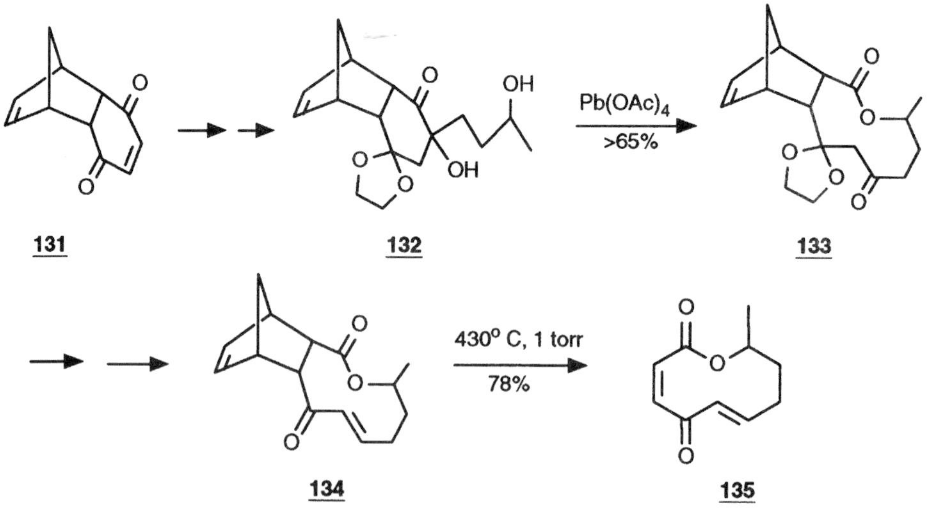

Scheme 31

series of standard operations gave **134** which smoothly underwent cycloreversion upon gas flow pyrolysis at 430°C and 1 torr to give *rac*-pyrenolide B **135** in 78% yield.

A stereoselective synthesis of conduritols F and A from **131** was recently accomplished in our laboratory (Schemes 32 and 33).[76] These highly oxygenated monocyclic cyclohexenes, which occur in *marsedenia condurago*, are active site directed inhibitors of glycosidases. Stereoselective alkaline epoxidation of **131**, subsequent sodium borohydride reduction followed by acylation afforded tricyclic epoxydiacetate **136** in 82% overall yield (Scheme 32). Flash vacuum thermolysis at 500°C and 2–10 torr led to cyclohexene epoxide **137**. Hydrolytic opening of the epoxide ring followed by careful aminolysis gave conduritol F **138** in an excellent overall yield of 44% from **131**.

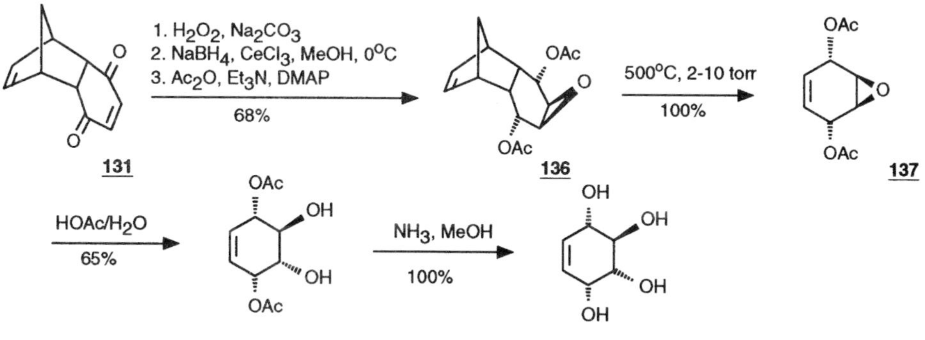

Scheme 32

Conduritol A was synthesized as depicted in Scheme 33.[76] Tricyclic diacetate **139** gave the hitherto unknown 1,4-diacetoxy-cyclohexa-1,4-diene **140** (together with some acetoxybenzene which is the result of aromatization of **140**) when subjected to flash vacuum thermolysis at 500°C (10^{-1} torr). Hydroxylation of **140** with osmium tetroxide followed by aminolysis gave conduritol A **141** in excellent overall yield.

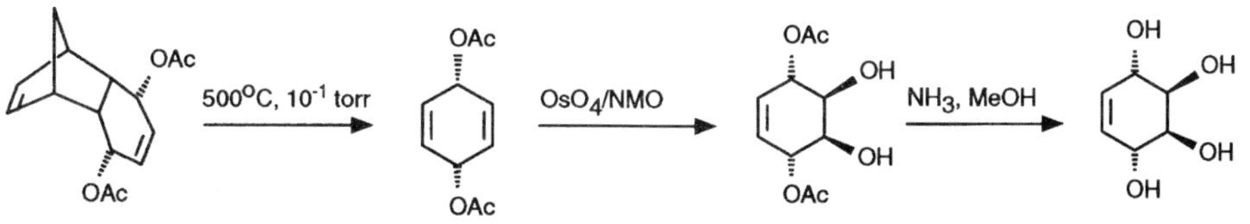

Scheme 33

The γ-lactone structural unit occurs in quite a number of natural products. Effective access to such γ-lactones has been realized applying the retro-Diels-Alder methodology.[77] A notable example is the synthesis of tulipalin A **144**, a constituent of tulip bulbs, from anthracene adduct **142** (Scheme 34).[78] Deprotonation of ester **142** with lithium diisopropylamide followed by substitution with ethylene oxide gave lactone **143** in 72% yield. Flash vacuum thermolysis at 250–300°C and 0.1 torr afforded tulipalin A in 77% yield.

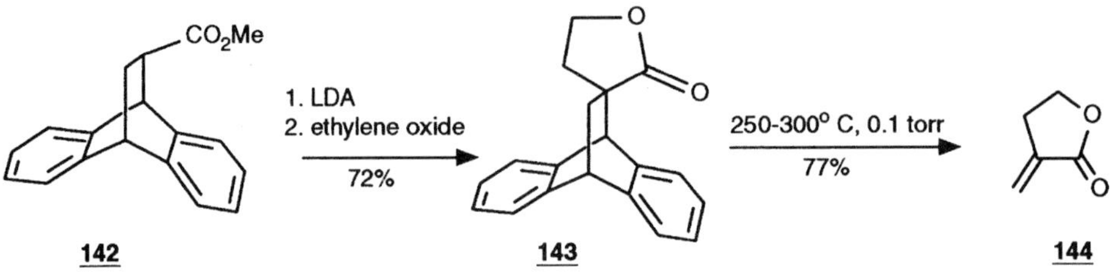

Scheme 34

The anthracene adduct **142** is also the starting material for the syntheses of the cyclopentenoid antibiotics methylenomycin B **146** and sarkomycine methyl ester **148** (Scheme 35).[79,80] Via a three-carbon annelation reaction spirocyclopentenones **145** are obtained in high yields. Vacuum thermolysis of **145** (R = Me) at 400–450°C and 0.05 torr gave methylenomycin B **146** in almost quantitative yield. The introduction of a γ-carboxylic ester function in **145** (R = H) afforded the precursor **147** for sarkomycin in 65% overall yield. Flash

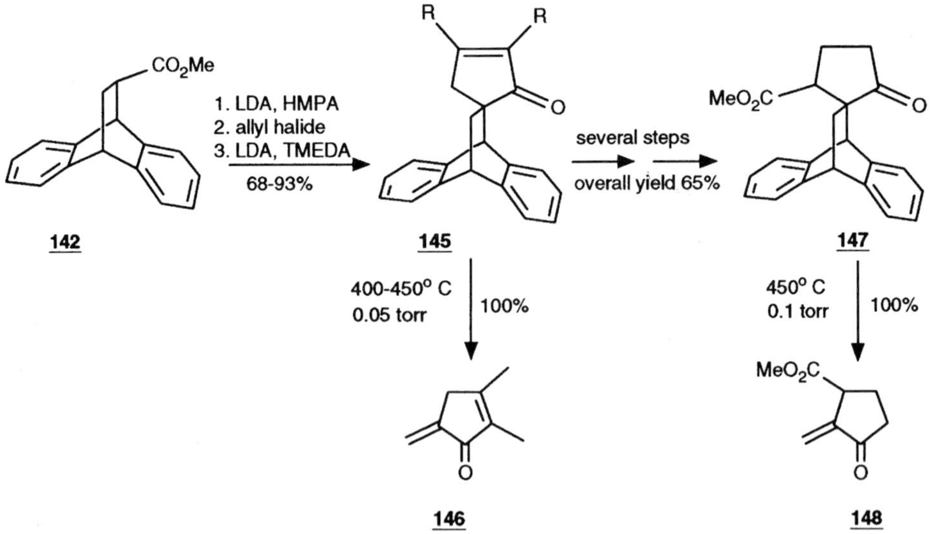

Scheme 35

vacuum thermolysis of **147** at 450°C and 0.1 torr led to a quantitative yield of *rac*-sarkomycine methyl ester **148**, which upon hydrolysis gave the natural product.

An efficient and simple synthesis of naturally occurring dihydrofuran (+)-eldanolide **153**, the sex pheromone of *Eldana saccharine*, was accomplished starting from enantiopure oxatricyclic lactol **149** (Scheme 36).[81] A stereoselective tandem Wittig-Horner olefination/intramolecular Michael addition afforded **150**

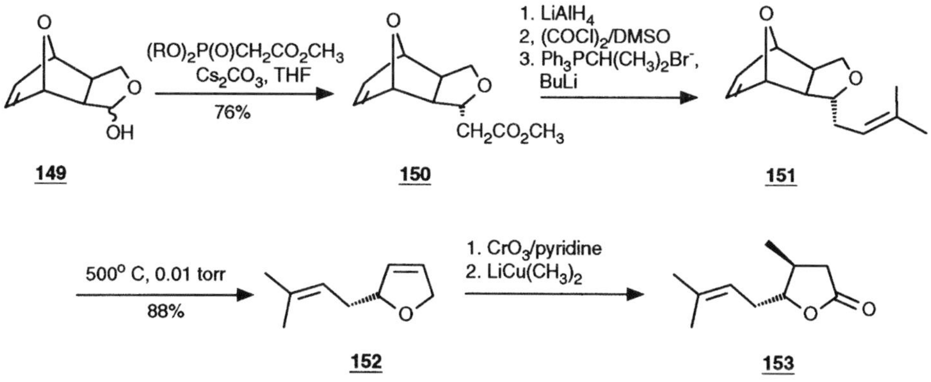

Scheme 36

in 76% yield and with excellent optical purity (ee > 95%). Reduction of the ester function, Swern oxidation and Wittig olefination led to tricyclic alkene **151**. Flash vacuum thermolysis of **151** gave dihydrofuran **152** in excellent yield. Oxidation to the corresponding lactone and subsequent conjugate addition of lithium-dimethyl cuprate completes this enantioselective synthesis of (+)-eldanolide **153**.

Oxatricyclic lactol **149** has also been applied[82] as chiron for hydroxyeicosatet-raenoic acids (HETEs), biological important metabolites of the lipoxygenation of arachidonic acid in mammalian tissues, (+)-coriolic acid,[83] a metabolite of linoleic acid, and a sex attractant for the beetle *Hylecoetus dermestoides*.[84] Here the respective cycloreversion steps were effected by heating in xylene and toluene at temperatures not exceeding 130°C. The high yield formation of these relative sensitive compounds under the static conditions applied, shows that Diels-Alder adducts derived from furan fragmentate at such low temperatures that gas phase conditions are generally not required.

Gas phase pyrolysis was, however, used for the synthesis for the host specific substance **156** for the ambrosis beetle *Try podendron lineatum oliv.*,[85,86] (Scheme 37). Enantiopure lactol **149** was transformed into alcohol **154** using essentially the same procedure as depicted in Scheme 36. Flash vacuum thermolysis at 500°C gave rise to dihydrofuran **155** in 82% yield. Reduction of the olefinic bond, regioselective opening of the hydrofuran ring, alkylation and deprotection ultimately led to (−)-**156** in 28% overall yield from **155**.

(+)-Lipoic acid, a biologically interesting compound which is widely distribu-ted in animal and plant tissue, was synthesized from tricyclic ester **150** using essentially the same strategy as depicted in the Schemes 36 and 37.[86]

Starting from enantiopure norbornene annulated lactam **157** the synthesis of a series of bicyclic alkaloids has been realized[87,88] (Scheme 38). Key steps in these syntheses involve diastereoselective addition to an acyliminium ion generated from **157** and the retro-Diels-Alder reaction. The synthesis of (+)-indolizidine **160** was accomplished using alkene **157a**.[87] Acyliminium cyclization of **157a** using formic acid gave formate **158** as the only product in excellent yield. Hydrolysis of **158** followed by flash vacuum pyrolysis at 450°C and 0.5 Pa gave

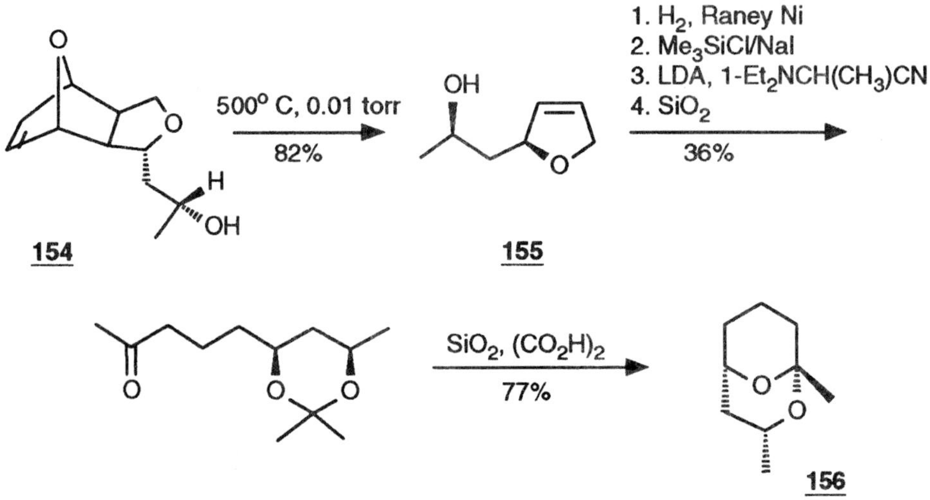

Scheme 37

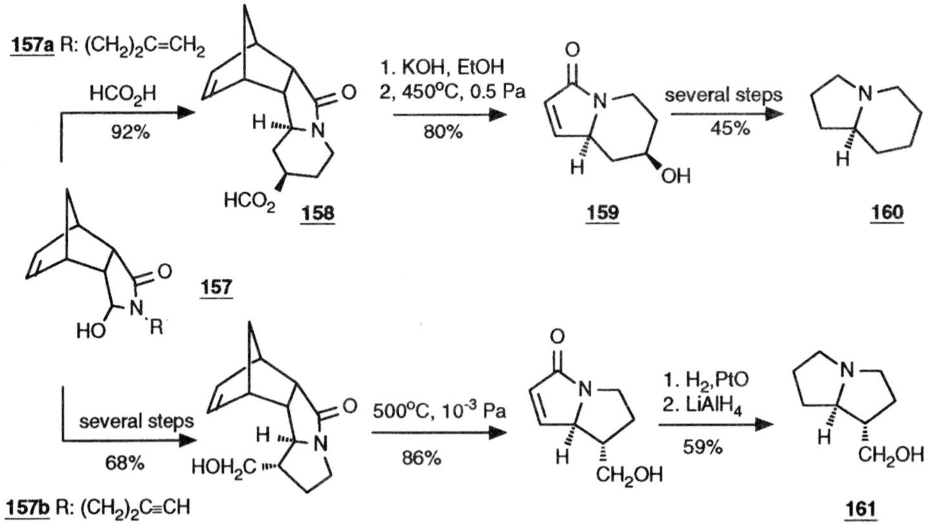

Scheme 38

bicyclic alcohol **159** in 80% yield. Reduction of the double bond and removal of the alcohol function afforded the natural product **160** in good overall yield. Enantiopure (+)-laburnine **161** was synthesized using essentially the same procedure starting from alkyne **157b**.[87]

An asymmetric synthesis of *Elaeocarpus* alkaloids (+)-elaeokanine A and (+)-elaeokanine C was accomplished by intermolecular addition of 2-(trimethyl-silyloxy)-pent-l-ene in presence of ethereal boron trifluoride to the bicyclic acyliminium ion derived from **162** to give **163** (Scheme 39).[88] Subjecting **163** to intramolecular aldol condensation under acidic conditions afforded the tetracyclic lactam **164**. Cycloreversion of **164** using the flash vacuum thermolysis technique at 435°C and 0.004 torr gave a mixture of two pyrrolidinones in a ratio of 3:1 and

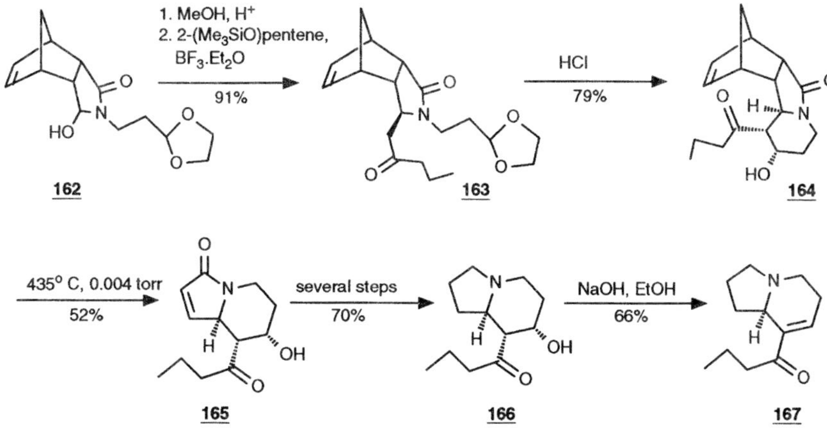

Scheme 39

in high yield. The major product **165** (52% yield) could readily be separated from its isomer in which the olefinic bond had migrated from the 2,3- to the 3,4 position under the thermal conditions used. Reduction of the double bond and removal of the ring carbonyl function afforded (+)-elaeokanine C **166**. Its unsaturated analog, (+)-elaeokanine A **167** was readily obtained by dehydration of **166** under alkaline conditions.

Relative simple norbornenes are excellent precursors for a variety of open chain natural products. An early example is the synthesis of *rac*-turmerone **170** which is most efficiently prepared from cyclopentadiene and 4-methylpent-3-en-2-one (Scheme 40).[89] Aldol condensation of adduct **168** with p-tolualdehyde

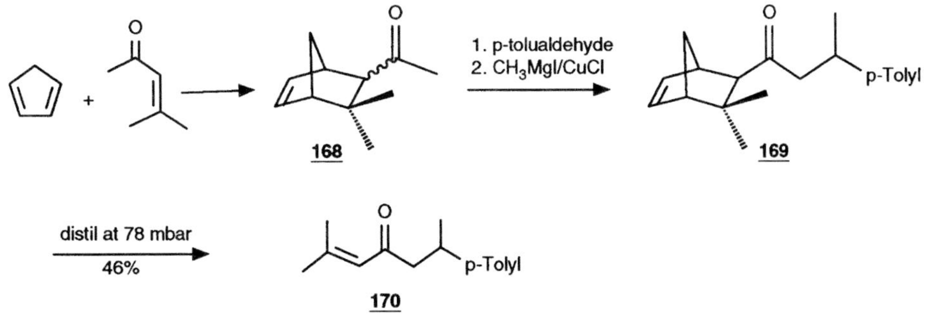

Scheme 40

followed by Michael addition gave **169**. Distillation at 78 mbar gave *rac*-tur-merone in 46% yield.

High yield syntheses of linalool **174** and nerolidol **175** were achieved from norbornenyl ketones **171** (Scheme 41).[90] Addition of methylmagnesium iodide gave the corresponding bicyclic alcohols **172** which on gas phase thermolysis at 450°C and 0.01 torr led amongst others to linalool **174** and nerolidol **175** in 75%

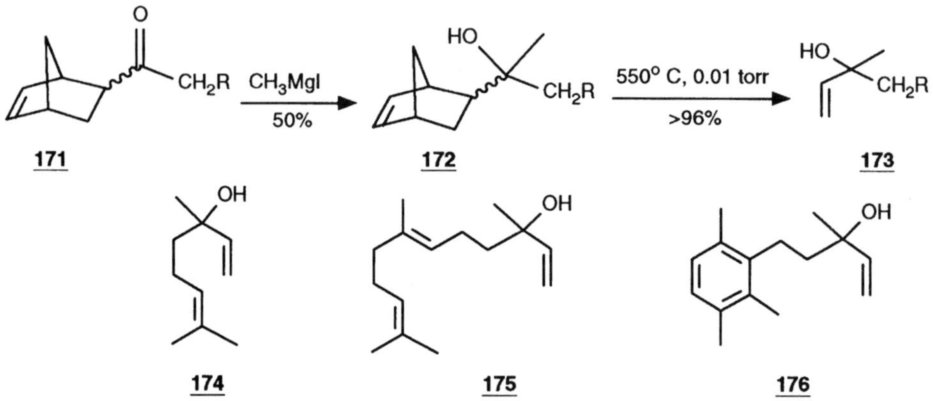

Scheme 41

overall yields (calculated on **171**). Using the same approach sesquiterpene **176**, a constituent of the alga *Laurencia nidifica*, was also prepared.

Stereocontrolled syntheses of the C(1)–C(7) fragments of deoxyerythronolide **179** and erythronolide seco acids have been realized by stereoselective aldol condensation of 2-arylpropanal and the lithium enolate of 2-methyl-5-norbornenyl ethyl ketone **177**, followed by flash vacuum thermolysis of **178** at 500°C and 0.001 torr to give **179** in quantitative yield (Scheme 42).[91] After syn-selective reduction of the ketone group in **179**, the configuration required at carbon C(6)

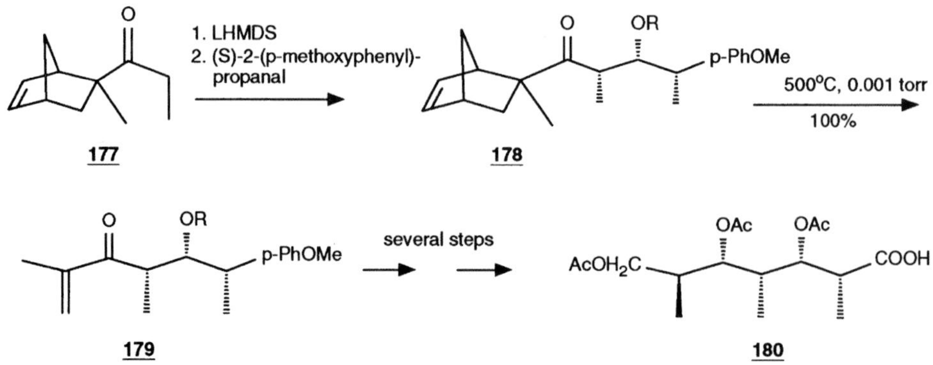

Scheme 42

was introduced by selective reactions carried out on the double bond regenerated by the thermal cycloreversion to give **180**. The C(1)–C(7) fragment of erythonolide was made using essentially the same methodology.

An elegant route to naturally occurring 1,3-dienes **185** is based on the tandem [4 + 2]cycloreversion/chelotropic elimination of tricyclic sulfones **183** (Scheme 43).[92, 93] The initially formed dihydrothiophenes **184** are not stable under the

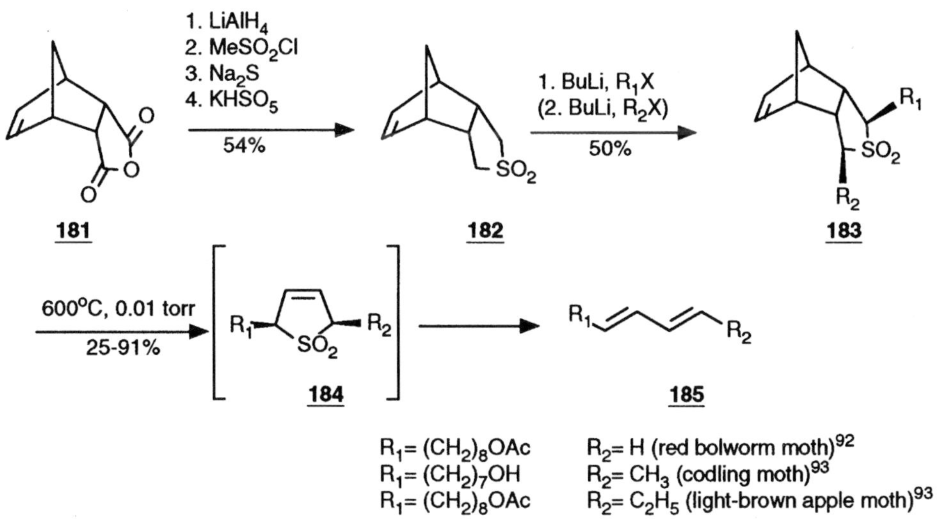

1. LiAlH$_4$
2. MeSO$_2$Cl
3. Na$_2$S
4. KHSO$_5$

54%

1. BuLi, R$_1$X
(2. BuLi, R$_2$X)

50%

181   **182**   **183**

600°C, 0.01 torr

25-91%

184   **185**

R$_1$= (CH$_2$)$_8$OAc   R$_2$= H (red bolworm moth)[92]
R$_1$= (CH$_2$)$_7$OH   R$_2$= CH$_3$ (codling moth)[93]
R$_1$= (CH$_2$)$_8$OAc   R$_2$= C$_2$H$_5$ (light-brown apple moth)[93]

Scheme 43

thermal conditions applied, but immediately undergo sulfur dioxide extrusion to give the corresponding 1,3-dienes. Starting from maleic anhydride adduct **181** these sulfones are readily available by mono or bis-alkylation of parent structure **182**. By applying the flash vacuum thermolysis technique (600°C, 0.01 torr) a variety of sex pheromones **185** was obtained. Using the same strategy some naturally occurring trienes were also prepared.[94–96]

The retro-Diels-Alder/Diels-Alder tandem reaction is an interesting possibility to prepare polycyclic naturally products in one thermal step. An illustrative example, employing gas phase thermolysis, is shown in Scheme 44. At 450°C and

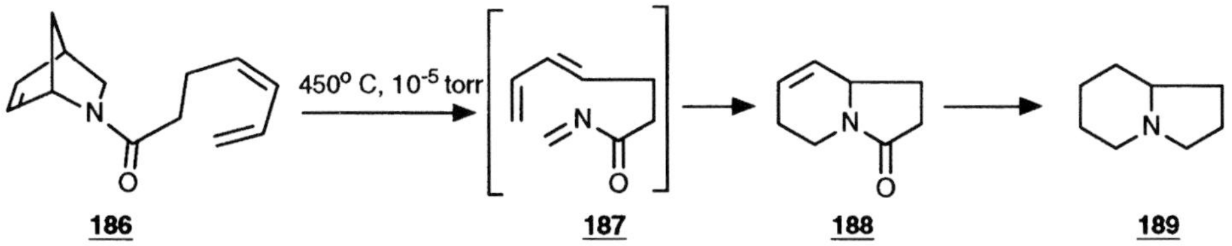

450° C, 10^{-5} torr

186   **187**   **188**   **189**

Scheme 44

a relative low pressure of 10^{-5} torr N-acyl-2-azabicyclo[2.2.1]hept-2-ene **186** gives lactam **188**, a precursor of δ-coniceine **189**, via intramolecular Diels-Alder reaction of transient acylimine **187**.[97] More examples of this synthetically useful methodology are given in Ho's excellent book on tandem organic reactions.[8]

4. CONCLUSIONS

Gas flow pyrolysis constitutes a powerful tool in organic synthetic methodology. The many examples presented in this chapter illustrate that this synthetic technique is not only applicable for the generation of transient intermediates but can also be quite successfully used for the synthesis of rather complex natural products. Although volatility may seem an obstacle in some cases, intelligent derivatization usually eliminates this problem and allows satisfactory results.

In our laboratory the technique of gas flow pyrolysis now holds an indispensable position not only as a crucial intermediate process in the synthesis of cyclopentanoid natural products but even more so as an efficient means to effect thermal reactions in all fields of chemistry. The availibility of this technique in our laboratory and its simplicity both in equipment and operation has undoubtedly helped us to solve sometimes complicated synthetic problems by considering thermal approaches. Therefore, together with the godfather of modern flash vacuum thermolysis, prof. Roger Brown, whose enthusiasm stimulated us to start in this field, we feel that this technique should be standard in any modern synthetic organic laboratory. To stimulate this is the main purpose of this contribution.

5. REFERENCES

1. Fieser, L.F. and Fieser, M.L. 'Advanced Organic Chemistry' p. 640, Rheinhold Publishing Corporation, New York (1961).
2. Hurd, C.D. 'The Pyrolysis of Carbon Compounds', Organic Chemistry Monograph Series, Chem. Catalog Co., New York (1929).
3. Magrath, J. and Fowler, F.W. *Tetrahedron Lett.* 1988, **29**, 2171–2174.
4. Clancy, M.G., Hawkins, D.G., Hesabi, M.M., Meth-Cohn, O. and Rhouati, S. *J. Chem. Res. Synop.* 1982, 78; Clancy, M.G., Hesabi, M.M. and Meth-Cohn, O. *J. Chem. Soc., Perkin Trans. I* 1984, 429–434.
5. Rubin, Y., Lin, S.S., Knobler, C.B., Anthony, J., Boldi, A.M. and Diederich, F. *J. Amer. Chem. Soc.* 1991, **113**, 6943–6949.
6. Brown, R.F.C. 'Pyrolytic methods in Organic Chemistry', Chapter 2, Academic Press, New York (1980).
7. Wiersum, U.E. *Aldrichimica Acta* **17**, 2, 31 (1984).
8. Lok, T.-L. 'Tandem Organic Reactions', Wiley, New York (1992).
9. Seybold, G. *Angew. Chem.* 1977, **89**, 377; *Angew. Chem., Int. Ed. Engl.* 1977, **16**, 365.
10. Wiersum, U.E. *Recl. Trav. Chim. Pays-Bas* 1982, **101**, 317; Wiersum, U.E. *Recl. Trav. Chim. Pays-Bas* 1982, **101**, 365.
11. Schiess, P. and Rutschmann, S. *Chimia* 1970, **39**, 313.
12. Karpf, M. *Angew. Chem.* 1986, **98**, 413; *Angew. Chem., Int. Ed. Engl.* 1986, **25**, 414.
13. Schiess, P. *Thermochimica Acta* 1987, **112**, 31.
14. Griesbeck, A.G. *J. Prakt. Chem.* 1993, **335**, 498.

15. Wentrup, C. *Chimia* 1977, **31**, 258.
16. Wentrup, C. *Lectures in Heterocyclic Chem.* 1984, 91.
17. Schiess, P. *Chimia* 1985, **39**, 213.
18. Cadogan, J.I.G., Hickson, C.L. and McNab, H. *Tetrahedron* 1986, 2135.
19. Brown, R.F.C. *Recl. Trav. Chim. Pays-Bas* 1988, **107**, 655.
20. Ripoll, J.-L. *Inform. Chimie* 1988, **295**, 221.
21. Brown, R.F.C. *Pure and Appl. Chem.* 1990, **62**, 1981.
22. DePuy, C.H. and King, R.W. *Chem. Reviews* 1960, **60**, 431.
23. Collins, P.A. and Wege, D. *Aust. J. Chem.* 1979, **32**, 1819.
24. Lei, B. and Fallis, G. *J. Org. Chem.* 1993, **58**, 2186.
25. Wibaut, J.P., Beyerman, H.C. and Leeuwen, H.B. *Rec. Trav. Chim. Pays-Bas* 1952, **71**, 1027; McNiven, N.L. and Read, J. *J. Chem. Soc.* 1952, 2067.
26. Van der Waals, A.C.L.M., Klunder, A.J.H. and Zwanenburg, B. Unpublished results.
27. Pflieger, D., Muckensturm, B., Robert, C.P., Simonis, M.-T. and Kienlen, J.-C. *Tetrahedron Letters* 1987, **28**, 1519.
28. Klaver, W.J., Hiemstra, H. and Speckamp, W.N. *J. Am. Chem. Soc.* 1989, **111**, 2588.
29. Jenneskens, L.W., Hoefs, A.M. and Wiersum, U.E. *J. Org. Chem.* 1989, **54**, 5811.
30. Miahara, S., Tateba, H., Nishimura, O., Macchi, Y. and Kishino, K. *J. Agric. Food Chem.* 1987, **35**, 532.
31. Black, H.J. and Weedon, B.C.L. *J. Chem. Soc.*, 1953, 1785.
32. Van den Berg, E.M.M., Richardson, E.E., Lugtenburg, J. and Jenneskens, L.W. *Synth. Commun.* 1987, **17**, 1189.
33. Van den Berg, E.M.M., Baldew, A.U., De Goede, A.T.J.W., Raap, J. and Lugtenburg, J. *Rec. Trav. Chim. Pays-Bas* 1988, **107**, 73.
34. Brown, R.F.C. and McGeary, R.P. *Aust. J. Chem.* 1993, **46**, 1321.
35. Pohmakotr, M. and Popuang, S. *Tetrahedron Letters* 1991, **32**, 275.
36. Kwart, H. and King, K. *Chem. Rev.* 1968, **68**, 415.
37. Von Bruchhausen, F. and Bersch, H.W. *Arch. Pharm.* 1928, **266**, 697.
38. Ho, T.-L. *Synth. Commun.* 1974, **4**, 189.
39. Stork, G., Nelson, G.L., Rouessac, F. and Gringore, O. *J. Am. Chem. Soc.* 1971, **93**, 3091.
40. Ducos, P. and Rouessac, F. *Tetrahedron* 1973, **9**, 3233.
41. Wiel, J.B. and Rouessac, F. *J. Chem. Soc. Chem. Commun.* 1976, 446.
42. Wiel, J.B. and Rouessac, F. *Bull. Soc. Chem. Fr.* 1979, *II*-273.
43. Wiel, J.B. and Rouessac, F. *J. Chem Soc. Chem. Commun.* 1975, 180.
44. Boland, W. and Jaenicke, L. *Chem. Ber.* 1978, **111**, 3262.
45. Boland, W. and Jaenicke, L. *J. Org. Chem.* 1979, **44**, 4817.
46. Chapman, O.L. and Hess, T.C. *J. Org. Chem.* 1979, **44**, 962.
47. Klunder, A.J.H., Bos, W., Verlaak, J.M.M. and Zwanenburg, B. *Tetrahedron Letters* 1981, **22**, 4553.
48. Baltrop, J.A., Barrett, J.C., Carder, R.W., Day, A.C., Harding, J.R., Long, W.E. and Samuel, C.J. *J. Am. Chem. Soc.* 1979, **101**, 7510; Baltrop, J.A., Day, A.C. and Samuel, C. J. *J. Am. Chem. Soc.* 1979, **101**, 7521.
49. Klunder. A.J.H., Bos. W. and Zwanenburg, B. *Tetrahedron Letters* 1981, **22**, 4557.
50. Klunder, A.J.H., Houwen-Claassen, A.A.M., Kooy, M.G. and Zwanenburg, B. *Tetrahedron Letters* 1987, **28**, 1329.
51. Houwen-Claassen, A.A.M., Klunder, A.J.H., Kooy, M.G., Steffan, J. and Zwanenburg, B. *Tetrahedron* 1989, **45**, 7109.
52. Houwen-Claassen, A.A.M., Klunder, A.J.H. and Zwanenburg, B. *Tetrahedron* 1989, **45**, 7134.
53. Verlaak, J.M.J., Klunder, A.J.H. and Zwanenburg, B. *Tetrahedron Letters* 1982, **23**, 5463.
54. Klunder, A.J.H., Huizinga, A.J.M., Hulshof, A.J.M. and Zwanenburg, B. *Tetrahedron Letters* 1986, **27**, 2543.
55. Klunder, A.J.H., Zwanenburg, B. and Liu, Z.Y. *Tetrahedron Letters* 1991, **26**, 3131.
56. Zhu, J., Yang, J.-Y, Klunder, A.J.H., Liu, Z.-Y. and Zwanenburg, B. *Tetrahedron* 1995, **51**, 5847.
57. Zhu, J., Klunder, A.J.H. and Zwanenburg, B. *Tetrahedron Letters* 1994, **35**, 2787.
58. Corey, E.J. and Mehrotra, M.M. *J. Am. Chem. Soc.* 1984, **106**, 3384; Nagoka, H., Miyakoshi, T. and Yamada, Y. *Tetrahedron Letters* 1984, **33**, 3621; Hashimoto, S., Arai, Y. and Hamanaka, N. *Ibid.* 1985, **26**, 2679; Shibasaki, M. and Ogawa, Y. *Ibid.* 1985, **26**, 3841.

59. Liu, Z.-Y. and Chu, X.-J. *Tetrahedron Letters* 1993, **34**, 349.
60. Liu, Z.-Y. and Chu, X.-J. *Tetrahedron Letters* 1993, **34**, 3885.
61. Liu, Z.-Y., He, L. and Zheng, H. *Syn. Lett.* 1993, 191
62. Nakayama, M., Fukuoka, Y., Nozaki, H., Matsuo, A. and Hayashi, S. *Chem. Letters* 1980, 1243.
63. Boschelli, D., Smith, III, A.B., Stringer, O.D., Jenkins, R.H. and Davis, F.A. *Tetrahedron Letters* 1981, **22**, 4385; Chen, B.-C., Weismiller, M.C., Davis, F.A., Boschelli, D., Emfield, J.R. and Smith III, A.B. *Tetrahedron* 1991, **47**, 173.
64. Liu, Z.-Y., Shi, W. and Zhang, L. *Synthesis* 1990, 235.
65. Bugel, J.-P., Ducos, P., Gringore, O., and Rouessac, F. *Bull. Soc. Chem. Fr.* 1972, 4371.
66. Takano, S., Inomata, K. and Ogasawara, K. *J. Chem. Soc. Chem. Commun.* 1989, 271.
67. Takano, S., Inomata, K. and Ogasawara, K. *J. Chem. Soc. Chem. Commun.* 1990, 1544.
68. Takano, S., Moriya, M. and Ogasawara, K. *Tetrahedron Letters*, 1992, **33**, 1909.
69. Takano, S., Inomata, K., Sato, T., Takahashi, M. and Ogasawara, K. *J. Chem. Soc. Chem. Commun.* 1990, 290.
70. Takano, S., Sato, T., Inomata, K. and Ogasawara, K. *J. Chem. Soc. Chem. Commun.* 1991, 462.
71. Takano, S., Moriya, M. and Ogasawara, K. *J. Org. Chem.* 1991, **56**, 5982.
72. Ichihara, A., Kobayashi, M., Oda, K., Sakamura, S. and Sakai, R. *Tetrahedron* 1979, **35**, 2861.
73. Ichihara, A., Oda, K., Kobayashi, M. and Sakamura, S. *Tetrahedron* 1980, **36**, 183.
74. Ichihara, A., Kimura, R., Oda, K., Moriyasu, K. and Sakamura, S. *Agric. Biol. Chem.* 1982, **46**, 1879.
75. Asaoka, M., Naito, S. and Takei, H. *Tetrahedron Letters*, 1985, **26**, 2103.
76. Mgani, Q.A., Klunder, A.J.H., Nkunya, M.H.H. and Zwanenburg, B. *Tetrahedron Letters* 1995, **36**, 4661.
77. Lasne, M.-C. and Ripoll, J.-L. *Synthesis* 1985, 121.
78. Jenkitkasemwong, Y., Thebtaranonth, Y. and Wajirum, N. *Tetrahedron Letters* 1979, **18**, 1615.
79. Siwapinyoyos, T. and Thebtaranonth, Y. *J. Org. Chem.* 1982, **47**, 599.
80. Kodpinid, M., Siwapinyoyos, T. and Thebtaranonth, Y. *J. Am. Chem. Soc.* 1984, **106**, 4862.
81. Bloch, R. and Seck, M. *Tetrahedron* 1989, **45**, 3731.
82. Bloch, R. and Gasparini, G. *J. Org. Chem.* 1989, **54**, 3370.
83. Bloch, R. and Perfetti, M.-T. *Tetrahedron Letters* 1990, **31**, 2577.
84. Bloch, R. and Brillet, C. *Tetrahedron Asymmetry* 1991, **2**, 797.
85. Bloch, R. and Seck, M. *Tetrahedron Asymmetry* 1990, **1**, 855.
86. Bloch, R., Bortulussi, C., Girard, C. and Seck, M. *Tetrahedron* 1992, **48**, 453.
87. Arai, Y., Kontani, T. and Koizumi, T. *Chem. Letters* 1991, 2135.
88. Arai, Y., Kontani, T. and Koizumi, T. *Tetrahedron Asymm.* 1992, **3**, 535.
89. Ho, T.L. *Synth. Commun.* 1974, **4**, 189.
90. Bloch, R. *Tetrahedron* 1983, **39**, 639.
91. Ahmar, M., Romain, I. and Bloch, R. *J. Org. Chem.* 1993, **58**, 2953.
92. Bloch, R. and Abecassis, J. *Tetrahedron Letters* 1982, **23**, 3277; Bloch, R., Abecassis, J. and Hassan, D. *Can. J. Chem.* 1984, **62**, 2019.
93. Bloch, R. and Abecassis, J. *Tetrahedron Letters* 1983, **24**, 1247.
94. Bloch, R. and Abecassis, J. *Synth. Commun.* 1985, **15**, 959.
95. Bloch, R., Benecou, C. and Guibé-Jampel, E. *Tetrahedron Letters* 1985, **26**, 1301.
96. Bloch, R. and Hassan-Gonzales, D. *Tetrahedron* 1986, **42**, 4975.
97. Lasne, M.C., Ripoll, J.L. and Thuillier, A. *J. Chem. Res (S)* 1982, 214.

3. THE FORMATION OF POLYAROMATIC HYDROCARBONS, FULLERENES AND SOOT IN COMBUSTION: PYROLYTIC MECHANISMS AND THE INDUSTRIAL AND ENVIRONMENTAL CONNECTION

ULFERT E. WIERSUM[*,†] and LEONARDUS W. JENNESKENS[**]

*Akzo Nobel Central Research, Velperweg 76, 6800 SB, Arnhem, The Netherlands;
**Debye Institute, Department of Physical Organic Chemistry, Utrecht University,
Padualaan 8, 3584 CH, Utrecht, The Netherlands

1. INTRODUCTION: THE PREVALENCE OF POLYCYCLIC AROMATIC HYDROCARBONS (PAH)

Ever since their isolation from coal tar in the beginning of the nineteenth century, polycyclic aromatic hydrocarbons (PAH) have been significant structures in the development of organic chemistry.[1,2] Recent observations on the formation of PAH in pyrolysis and combustion, to be reviewed in this chapter, corroborates the significance of PAH chemistry.[3] The coal tar industry started as early as 1860 as a spin off from coke production for steelworks, providing naphthalene, anthracene and phenanthrene for production of synthetic dyes. This was the beginning of the organic chemical industry that presently utilizes PAH in a great diversity of applications.[4] Around 1930, coal tar was found to cause skin cancer. Benzo[a]pyrene, isolated as an active component, and dibenz-[ah]anthracene became the first identified carcinogenic chemicals.[2,5] From that time many PAH have been identified as carcinogenic and/or mutagenic. To make things worse, beginning in the 1960's, ever more PAH were analysed in soot, fly ash and engine exhaust, as it is released to the environment from combustion processes. Because of their origin from tar and soot, PAH have always been associated with coal,[6] pyrolysis[7] and combustion.[8] Major efforts[9] are made to assess the environmental threat and the risks of human exposure in connection with carcinogenicity.[10,11] This includes study of the biomechanism of carcinogenesis[5] and synthesis of PAH metabolites.[2,12]

Soot formation in combustion is a different process than tar production during thermal charing of solid materials, since it must involve condensation of small reactive species, present in the hot gas phase.[13] The recent discovery of fullerenes in the high temperature plasma of the carbon arc,[14] as well as in soot from flames, fits in fact in the great number of observations on thermal PAH formation during combustion. In contrast with all analytical studies to monitor their presence,

---

† Current Address: Thermo Chemical Services, Alteveerselaan 6, 6881 AV, Velp, The Netherlands

143

systematic work to find mechanistic rationalisations for thermal PAH build up has remained occasional and unfunded *via* administered governmental programs. Nevertheless, typical mechanistic routes can be formulated, based on flash vacuum pyrolysis, FVP[15,16] or flash vacuum thermolysis, FVT[15,17] experiments with individual PAH. Some PAH were found to interconvert selectively at high temperature. In addition, unknown PAH, or PAH requiring tedious multistep procedures, e.g. cyclopenta-PAH, can conveniently be made *via* preparative FVT in substantial quantities. As PAH become increasingly less available from coal tar sources, FVT makes them available as reference compounds, which is important for environmental and toxicity subjects as well as further PAH research.[18]

2. THE PYROLYTIC ORIGIN AND ENVIRONMENTAL OCCURRENCE OF PAH. THE EPA LIST

The emission of (substituted) PAH from combustion is presently in the forefront of environmental concern. All major industrialized countries have agencies aiming to control the handling of chemicals and to govern waste and toxicity problems created by PAH.[4] In the USA rules are established by the Environmental Protection Agency (EPA) and the National Bureau of Standards[19] and in Europe by the European Commission Joint Research Centre (JRC) institutions.[20] PAH with three to seven rings are treated all alike either as carcinogenics, or at least as suspect or harmful chemicals, although many may by inactive.[21]

The analytical routine of PAH tracing in samples of all kind concerns compounds with three to maximal six and rarely seven rings. Here, many of the possible structures are known and quite a few are commercially available as reference compounds. The majority of the larger PAH containing six and more rings, is unknown. The possible PAH structures include the compounds with five or seven membered rings, in addition to the all six membered ring systems. Because of the rapidly increasing number of structures with additional rings, the larger PAH are inaccessible for systematic study,[22] and they are rarely included in PAH analysis of environmental samples. The number of PAH isomers containing the total of six rings, (molecular weight 302), is already too large to cover all possible structures. When reference compounds are available, Shpol'skii fluorescence spectroscopy can identify structures in this category, where chromatographic separation becomes more difficult.[23]

Sixteen PAH, compounds **1–16**, from naphthalene to compounds with six fused rings, are listed by the EPA,[24] as typical contaminants (Table 1 and Scheme 1). As a consequence the EPA compounds **1–16** are the ones that are most frequently identified PAH in air,[25] water[26] or soil samples.[27] Many analytical papers focus mainly on the EPA series. As a set the EPA compounds **1–16** are well matched, because they are biologically persistent and generally found as major components in environmental samples as well as in coal tar and soots of all kind.[28] An experimental stability ranking of the incinerability of hazardous organic compounds, gave some EPA PAH a top position as thermally stable organics.[29]

Compounds **1–8** are commodities available from the coal tar industry. When not won from coal tar sources, PAH, including compounds **9–16**, are very

Table 1 The PAH of the EPA list

Structure	Naphthalene $C_{10}H_8$ (**1**)	Fluoranthene $C_{16}H_{10}$ (**7**)	Benzo[a]pyrene $C_{20}H_{12}$ (**13**)
CA hits	16891	4667	12803
Structure	Acenaphthylene $C_{12}H_8$ (**2**)	Pyrene $C_{16}H_{10}$ (**8**)	Benzo[ghi]perylene $C_{22}H_{12}$ (**14**)
CA hits	2080	8937	2937
Structure	Acenaphthene $C_{12}H_{10}$ (**3**)	Benz[a]anthracene $C_{18}H_{12}$ (**9**)	Indeno[1,2,3-cd]pyrene $C_{22}H_{12}$ (**15**)
CA hits	2551	4638	1641
Structure	Fluorene $C_{16}H_{10}$ (**4**)	Chrysene $C_{18}H_{12}$ (**10**)	Dibenz[ah]anthracene $C_{22}H_{14}$ (**16**)
CA hits	4408	4624	2285
Structure	Phenanthrene $C_{14}H_{10}$ (**5**)	Benzo[b]fluoranthene $C_{20}H_{12}$ (**11**)	
CA hits	8035	1999	
Structure	Anthracene $C_{14}H_{10}$ (**6**)	Benzo[k]fluoranthene $C_{20}H_{12}$ (**12**)	
CA hits	13044	2095	

expensive and sold by research chemicals suppliers in mg quantities. This indicates that either direct synthesis requires lengthy routes, or that purification from suitable tar fractions involves complex work up procedures. In Table 1 the number of references found by CAS number computer search (June 1995) is given for compounds **1–16** (Scheme 1). These numbers give the total of papers citing the specific PAH after 1967. In the cumulative Chemical Abstracts subject indexes one can see that the very large numbers of references often concern analytical studies for environmental purposes.

The EPA list does not include heterocyclic PAH, although they are generally found in combination with the EPA contaminants. Quinolines, carbazoles, phenols, benzofurans, typically present in coal tar, are also generated in combustion.[28] Benzodioxins, especially 2,3,7,7-tetrachlorodibenzodioxin[30] are much sought in waste incinerators, with chlorinated materials in the feed.[31,32] Benzothiophenes are found after burning fuel that contains sulphur.[28] Nitrogen heterocycles are markedly present in tobacco smoke.[33] Another important group of combustion emittants are nitro substituted PAH, often found in diesel exhaust[20,34] and methyl substituted PAH.[28] Examples of parent PAH, not listed by the EPA, but frequently reported in combustion, are 2,3-benzofluorene (**17**, X= CH$_2$), 2,3-benzocarbazole (**17**, X=NH), benzo[b]naptho[2,3-d]furan (**17**, X=O), benzo[b]-naptho[1,2-d]furan (**18**, X=O), benzo[b]naphtho[2,3-d]thiophene (**17**, X=S), 1,2-benzofluorene (**18**, X= CH$_2$), benzo[b]naphtho[1,2-d]thiophene (**18**, X=S), naphthacene (tetracene, **19**), triphenylene (**20**), benzo[e]pyrene (**21**), picene (benzo-[a]chrysene, **22**), perylene (**23**), dibenz-[ac]anthracene (**24**) and coronene (**25**), all

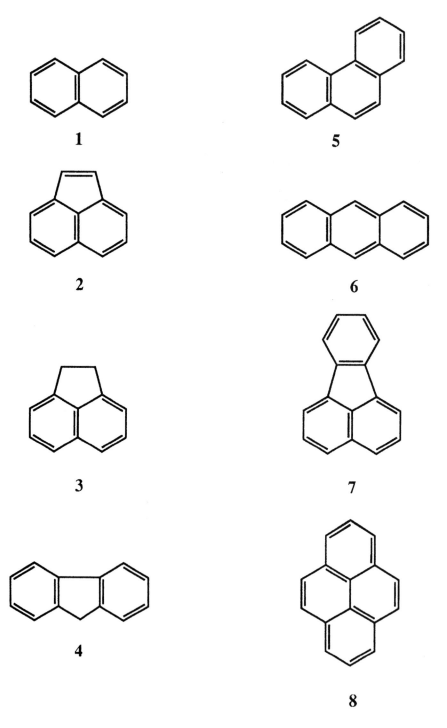

Scheme 1: The PAH listed by the EPA.

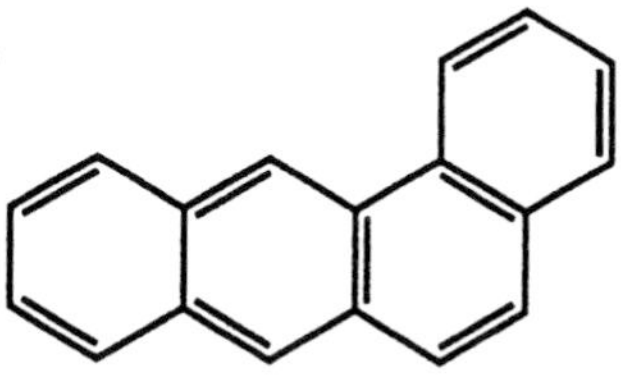

9

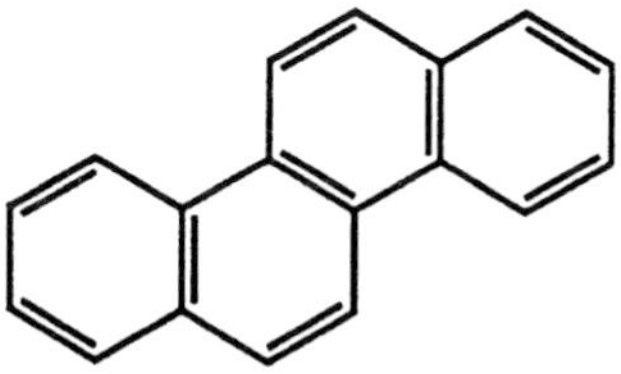

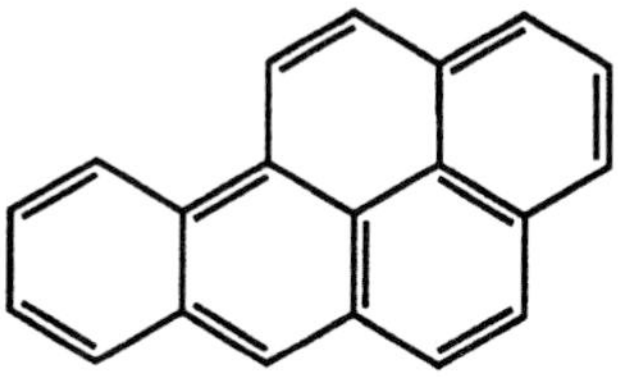

13

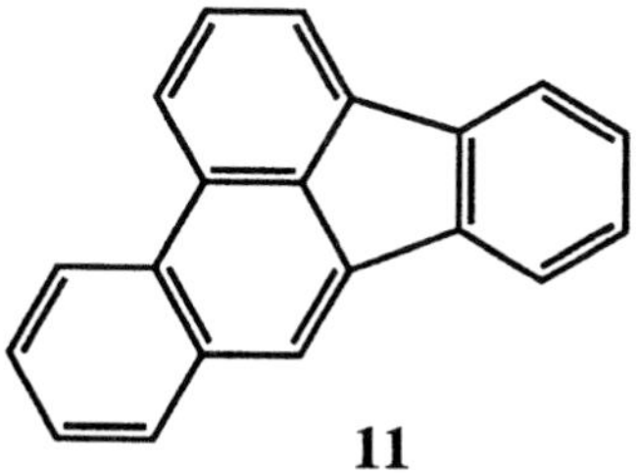

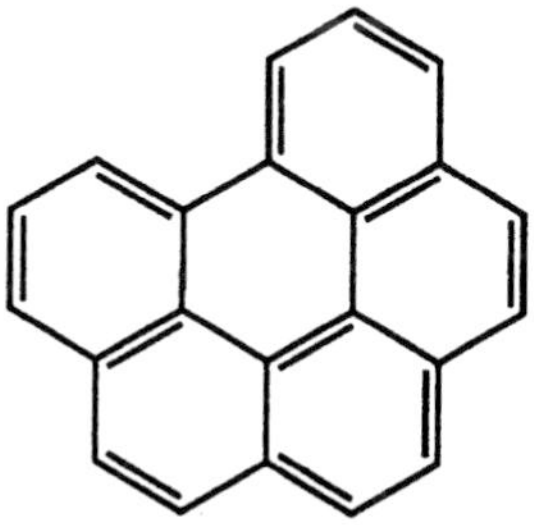

10

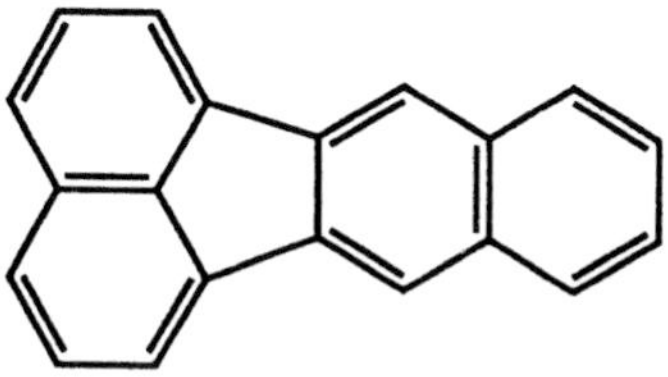

14

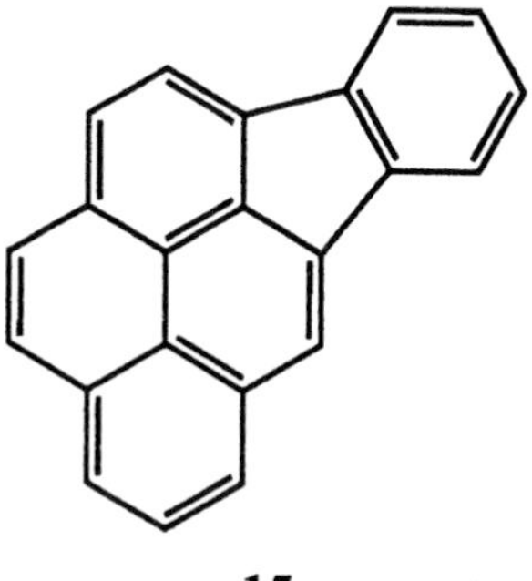

11

15

12

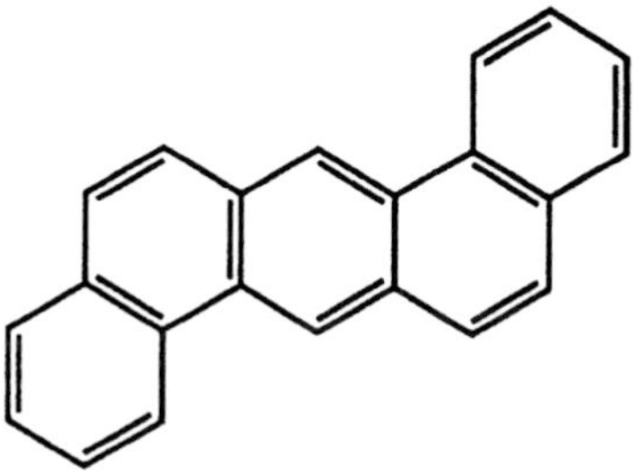

16

Scheme 1 (*Continued*)

commercially available. Compounds **9** and **21** have got much extra attention because they are the work-horses for cancer research.[5] Certain types of PAH, clearly present in soot, are unlikely to survive very long as environmental pollutant due to photoxidation and further decomposition pathways, e.g. **19** and its homologue pentacene (**26**), or the active carcinogen 7,12-dimethyl benz[a]-anthracene (**27**). For detailed information, including environmental data, on a large number of individual PAH, Harvey's book[2] is an indispensable source.

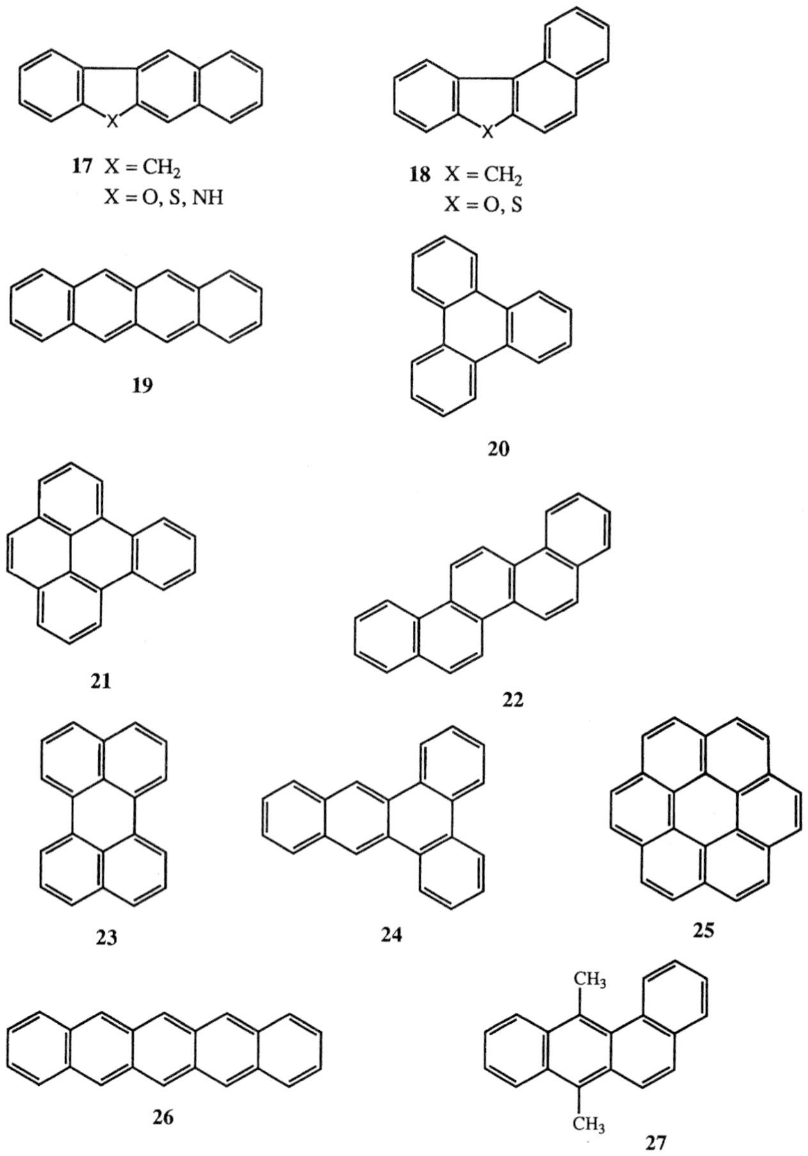

17 X = CH$_2$
      X = O, S, NH

18 X = CH$_2$
      X = O, S

19

20

21

22

23

24

25

26

CH$_3$

CH$_3$
      27

Scheme 2

Analytical work depends strongly on the availability of reference compounds. Over the years some laboratories have gained specialized expertise and utilize non-commercial reference compounds that they have collected.[8,9,20] As rule CAS number search of non-commercial PAH comes up with strikingly less citations on toxicity and environmental studies, although they are clearly proven as combustion emittants. Examples are benzo[c]phenanthrene (**28**), pentaphene (**29**), benzo[a]naphthacene (**30**), benzo[j]fluoranthene (**31**), benzo[ghi]fluoranthene (**32**), naphth[2,3-a]aceanthrylene (**33**) and dibenzo[def,mno]chrysene (anthanthrene, **34**) to mention a few.

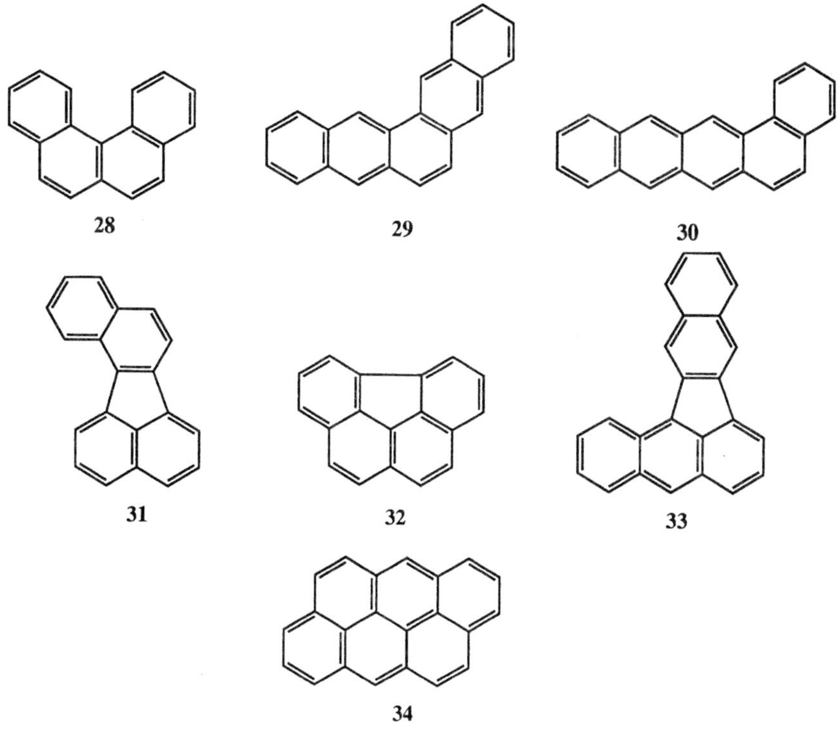

28   29   30

31   32   33

34

Scheme 3

Cyclopenta-PAH (CP-PAH), especially with outer five membered rings are, except for the commodities indene (**35**) and acenaphthylene (**2**), not much studied in connection with coal tar and/or soot. Of the many possible CP-PAH structures many are still unknown[22] and therefore not registered by agencies. Aceanthrylene (**36**) and acephenanthrylene (**37**) were found to equilibrate with fluoranthene (**7**), at high temperature.[35] CP-PAH are potential mutagenics and tumorigenics. Liquid phase synthesis of dicyclopenta[1,2,3-cd:1',2',3'-lm] perylene (**38**),[36,37] and benz[j]aceanthrylene (**39**),[38] was explored to test for mutagenic properties.[37] Cyclopenta[cd]pyrene (CPP, acepyrylene, **40**)[39] is a strong mutagenic and is the CP-PAH citation champion of this moment, much sought in flames, soot and combustion effluents. It has recently become evident from preparative FVT, that many more compounds of the CP-PAH group are present, or are likely to be formed in combustion.[40,41] Examples are cyclopent[hi]aceanthrylene (CPAA, **41**) and its isomer cyclopent[hi]acephenanthrylene (CPAP, **42**),[42,43] cyclopenta[cd]fluoranthene (acefluoranthylene, CPF, **43**),[44] dicyclopentapyrenes, e.g. dicyclopenta[cd,fg]pyrene (cyclopent[fg]acepyrylene, **44**)[45,46] and cyclopentacorannulene (**45**).[47] The mechanistic rationalisation for the process is discussed in Part 3.

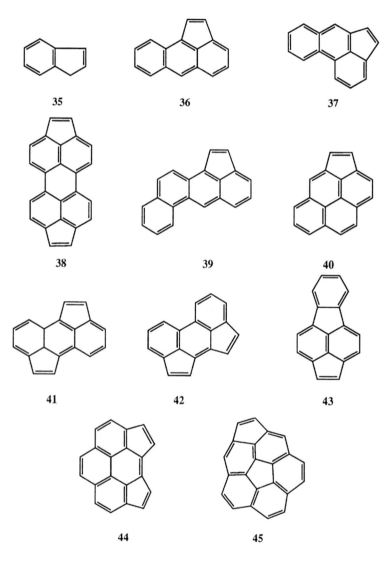

Scheme 4

3. PREPARATIVE FLASH VACUUM THERMOLYSIS (FVT).
   THE ACETYLENE–METHYLENE CARBENE
   REARRANGEMENT TO CYCLOPENTA-PAH

In 1978 it was commented that the development of synthetic methodology for (parent) PAH remained stagnant after the peak in the 1930s under the leadership of Fieser, Cook, Bachmann and Badger.[18] Nevertheless, the systematic work of Clar[1,48] and the coal tar contributions of the Rütgerswerke group[49-51] had revealed unique spectroscopic and physical properties, especially of larger PAH systems. More recently Harvey and others explored new synthetic routes.[2,52] Despite all this, most PAH systems have remained rare and expensive chemicals, because of the tedious and multistep synthetic routes involved. There is still a great need for straightforward synthetic routes towards PAH, to make them more available for research.[18] When one looks through Harvey's recent PAH monograph,[2] pyrolytic methods are not found as synthetic strategy, except for the Elbs cyclization of o-methyldiarylketones. In a number of reviews on preparative FVT, examples of thermal PAH synthesis were collected.[13, 53-56] More pyrolytic reaction possibilities for PAH with review papers are summarized below:

- Elbs reaction[57] and maleic anhydride ring addition-decarboxylation reaction.[1,55]
- Cyclodehydrogenation[55,58,59,60] and thermal catalytic dehydrogenation.[18]
- Eliminations of volatile fragments.[54,61,62]
- Retro-Diels-Alder and retro-ene reaction.[63-65]
- Cyclophane and superphane synthesis,[66-69] pyrenophanes,[70] radialenes.[71]
- Interconversion of polyfluoro-[72] and polychloro aromatics.[73-75]
- FVT induced generation and use of highly reactive intermediates as building blocks, e.g. isobenzofurans and related systems,[17,63] or for spontaneous build up of heterocycles.[53,61,76]
- Dimerisation and polymerisation reactions of o- and p-quinodimethanes,[77,78] benzocyclobutenes[79] and biphenylenes.[69]
- Pyrolysis of sulphones and sulfoxides.[17,80,81]
- Cyclization via arylmethylene–carbenes[40,82,83] and conversion of parent PAH to CP-PAH.[41]

The cyclization of arylmethylene carbenes via rearrangement of ethynyl PAH is an important pyrolytic route for preparation of CP-PAH, e.g. of compounds **40–45**. It deserves closer attention within the scope of this review, because it is operative in combustion and probably relevant in thermal fullerene formation (Part 7). Methylene carbenes result from thermal equilibration around 700°C with acetylenes. This fundamental thermal process, discovered by Brown,[82] is a key mechanism in the build up of CP-PAH. The crucial feature of the reaction is that internal trapping of the carbene intermediate is possible, as shown for the preparation of compounds **40** and **43** (Scheme 5). The synthetic potential[83] and the scope of the reaction for CP-PAH came very recently in focus.[40,41,46] The

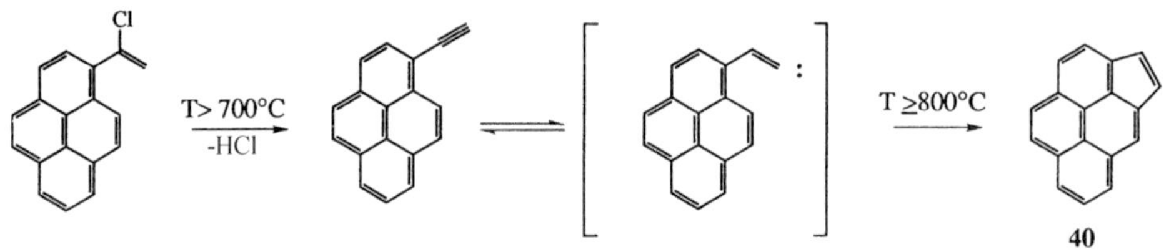

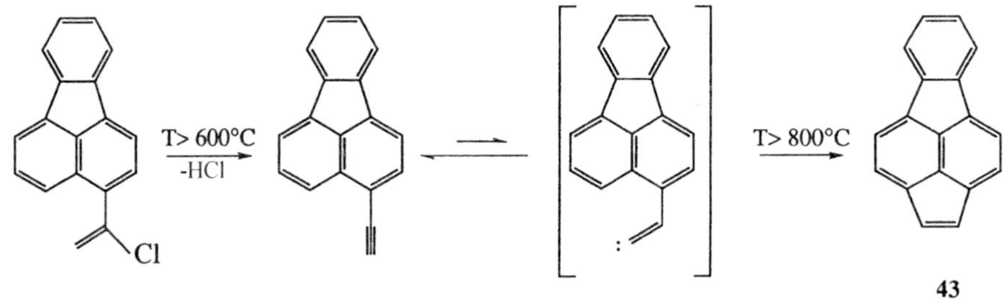

Scheme 5

preparation of CP-PAH by FVT is illustrative both for the mechanism and for the technical procedure. Our FVT apparatus is shown in Figure 1A. The precursor compound is flow-pyrolyzed, or more simply sublimed *in vacuo*, through the hot quartz tube. For preparation of CPP (**40**)[39] and CPF (**43**),[44] the respective (1-chloroethenyl)PAH derivatives are applied as precursors for *in situ* generation of ethynyl derivatives[84] which then undergo internal carbene cyclization as shown (Scheme 5). The products come as crystalline deposits in the bend of the quartz tube.

When a run is finished, the quartz tube is taken out and the bend is filled with solvent for convenient extraction of the products. Hydrogen chloride, eliminated as a volatile fragment, is trapped in the liquid nitrogen trap that is connected to the quartz tube. The cold trap can also be part of a spectroscopic apparatus (Figure 1B). For example a multitude of extended (hetero)cumulenes, inaccessible *via* liquid phase procedures, were trapped at cryostatic temperatures by FVT, directly coupled with spectroscopic techniques,[85,85] or monitored with (UV) photoelectron spectroscopy.[87] Coupled IR and/or mass spectroscopy is especially convenient.[88] When the fragment is a highly reactive molecule, the IR spectrum can be monitored directly at cryostatic temperatures and its sequential condensation or rearrangement behaviour can simply be followed by allowing the IR cell to warm up. An illustrative example is the trapping and characterization of transient 1-bromoethanol.[89]

Multistep liquid phase procedures, for example of CPAA (**41**)[90] and CPAP (**42**)[91] can be replaced by FVT procedures. Scheme 6 shows the formation of both **41** and **42** by FVT of 1,5-bis-(1-chloroethenyl)anthracene.[42]

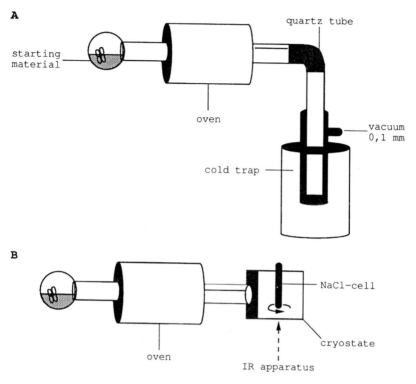

A
starting
material
quartz tube
oven
vacuum
0,1 mm
cold trap
B
NaCl-cell
cryostate
oven
IR apparatus

Figure 1 (A) Preparative FVT apparatus with cold trap. A PAH or a precursor compound is sublimed into the hot quartz tube. The products are, depending on volatility, collected in the bend of the quartz tube and the cold trap. (B) When the cold trap is part of a spectroscopic technique, reactants can be monitored in the cold and their sequential (condensation) reactions can be followed e.g. by the IR cell being warmed up.

Cl
FVT
Cl
Cl
T≥ 700°C
T≥ 800°C
T≥ 900°C

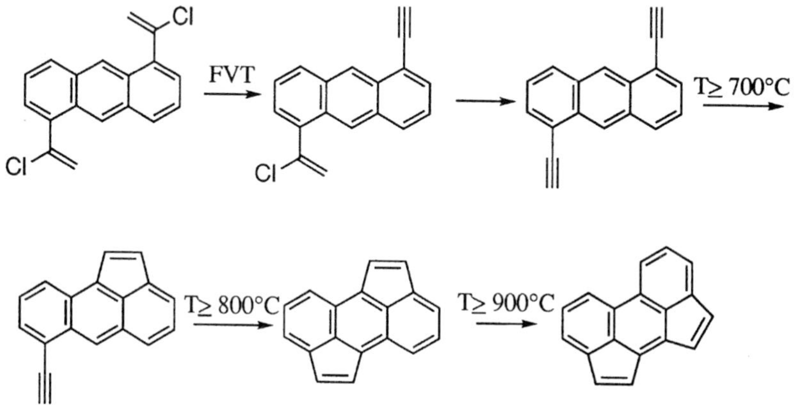

41 **42**

Scheme 6

The temperature of the quartz tube may have influence on the product composition. For example the intermediate ethynyl compounds in Schemes 5 and 6 are found in the pyrolysates when the reactions are run at somewhat lower temperatures than indicated. PAH usually interconvert with isomers of the same molecular formula *via* (open chain) carbene and diradical type intermediates as in the case of **41** and **42**. In this way a series of species of the same molecular formula are the energy minima on a so called thermal energy surface of that molecular formula. Several intermediates, both aromatic and (poly) ene-, yne- and allene-type open chain structures are likely involved, e.g. in thermal Bergman type cycloaromatisations to aryne intermediates.[92] Some thermal inter-conversions do not occur at first sight. FVT with ^{13}C labeled PAH causes automerisation, which proves that transient (open chain) structures come into play that reverse to the same PAH with the label shifted to another position in its skeleton.[93]

An unforeseen reaction, closely related with the acetylene–methylene carbene rearrangement, is that a PAH may loose two hydrogens upon FVT at high temperature, thus entering a new energy surface as a biradical or a benzyne homologue (aryne intermediate). The aryne undergoes ring contraction to its cyclopentadienylidene carbene isomer, followed by intramolecular ringclosure to a CP-PAH.[83] Examples are the formation of CPP (**40**), together with ben-zo[ghi]fluoranthene (**32**) from benzo[c]phenanthrene (**28**)[94] and CPAP (**42**) from triphenylene (**20**)[43] (Scheme 7). In the latter case the primary ring closure pro-duct is thought to engage in a further ring opening-ring closure sequence, similar to the one observed with **41** (Scheme 6), that also rearranges to **42**.[42] Aryne intermediates are also accessible from anhydrides by loss of CO and CO_2 and give rise to CP-PAH derivatives as well, as in FVT of 3-phenylnaphthalene-1,2-carboxylic anhydride to dibenzopentalene (**46**)[95] (Scheme 7), or 4-phenylphenan-threne-2,3-dicarboxylic anhydride to benzo[e]pyrene (**21**).[96] Since hydrogen ab-straction from PAH by OH radicals is a major process in combustion,[97] the aryne-cyclopentadienylidene carbene routes are likely to be common pathways in flames (Part 7). Carbon–hydrogen bond dissociation energies along graphite edges with different shape structure and of different PAH positions were cal-culated.[98]

The preparative strength of FVT has replaced the liquid phase synthesis[99] of corannulene (**47**). In the cyclization of 7,10-diethynylfluoranthene[93] to **47**, the carbene can only engage in formation of six membered rings (Scheme 8). Despite the fact that five membered ring cyclization apparently is more efficient,[39] the double cyclization of 7,12-bis-(1-chloroethenyl)benzo[k]fluoranthene to indeno-[7,7a,1,2,3-lmno]-1,12-etheno-chrysene, (**48**) balances strain effects by selective closure with a five- and a six membered ring (Scheme 8).[100]

For CP-PAH, FVT is the method of choice and for some of them probably an unique route. Even four-fold FVT cyclization to semibuckminsterfullerene (**49**) has been achieved (Scheme 8).[101] Compounds **47–49** are seen as prefullerene structures.[102,103]

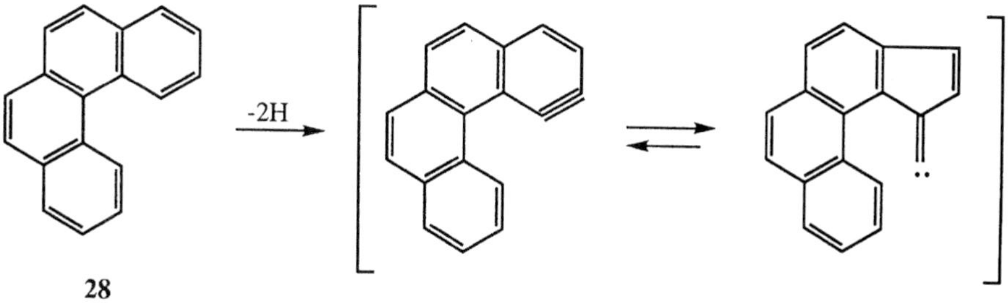

28

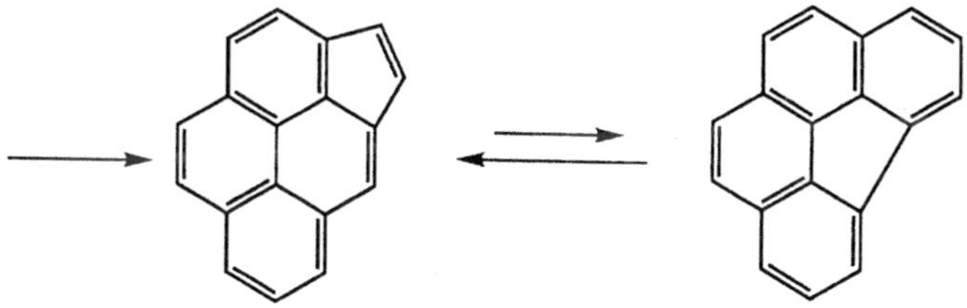

40   **32**

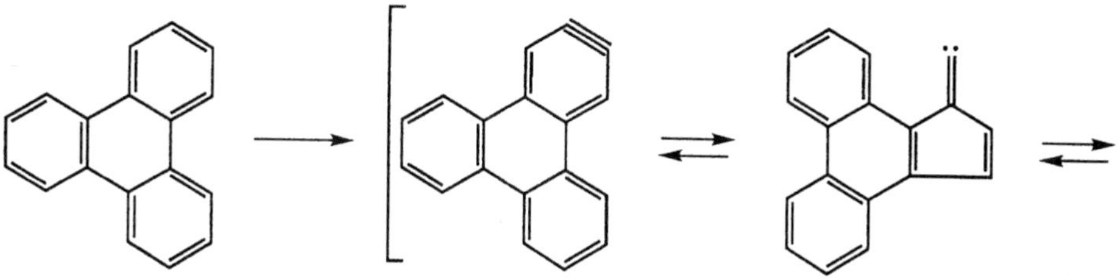

20

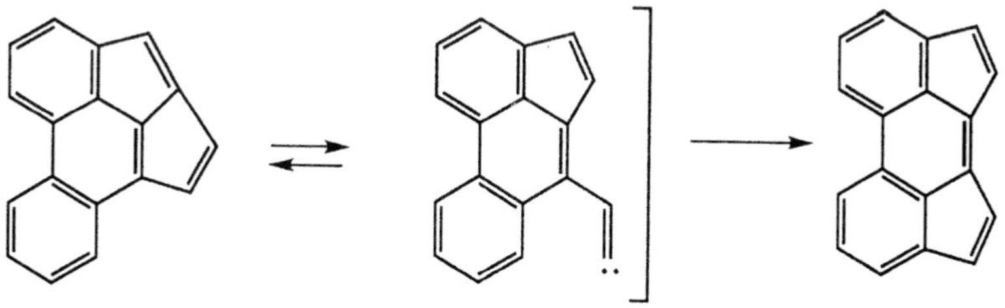

42

Scheme 7

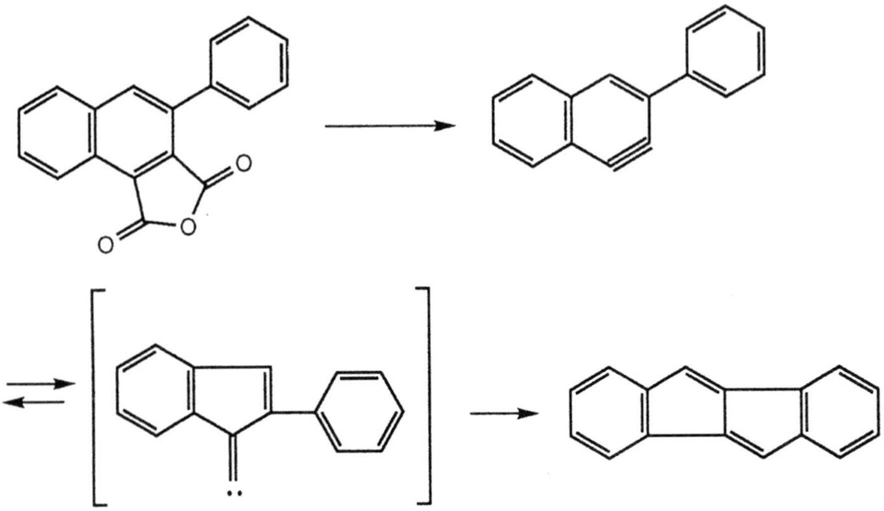

46

Scheme 7   *(Continued)*

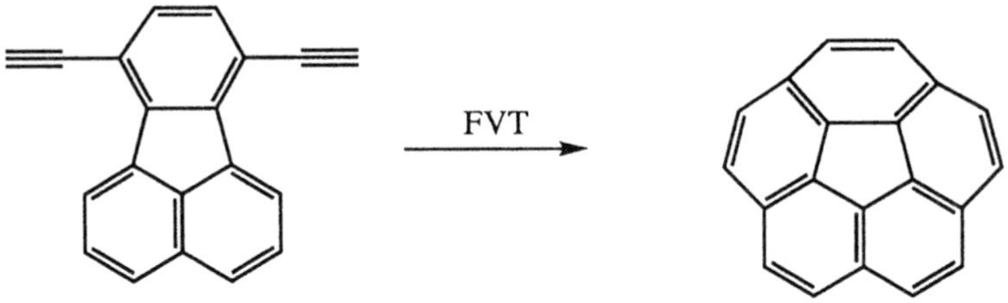

47

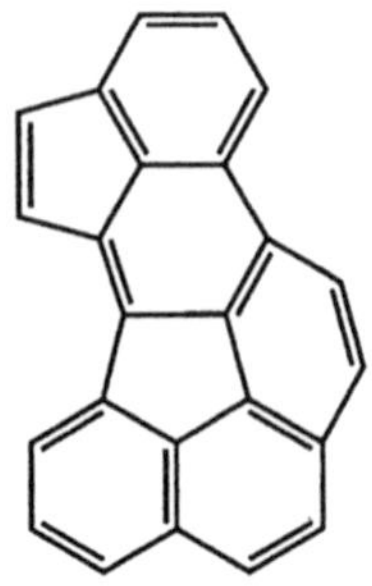

48

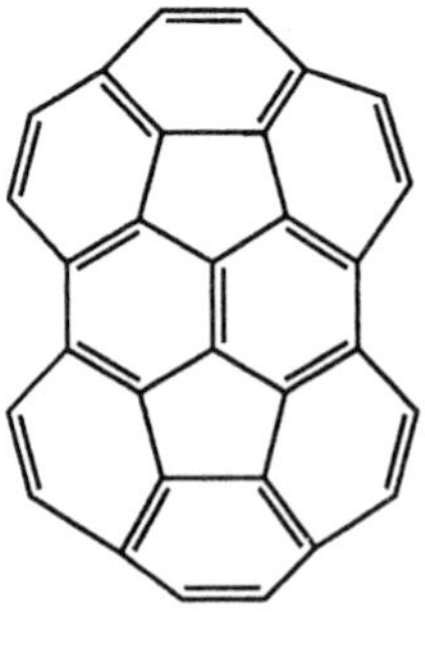

49

Scheme 8

4. SOOT FORMATION AND PAH INTERMEDIATES. PROBING THERMAL ENERGY SURFACES

Our interest in PAH pyrolysis chemistry arose from work at Akzo Nobel to characterize the soot that covers spent zeolite catalysts, taken from fluidized catalytic cracking (FCC) plants. The FCC process[104,105] converts high boiling petroleum fractions (vacuum gas oils) over zeolites[106,107] to gasoline fractions at relatively low temperatures, between 500 and 550°C. On contact with the hot petroleum feed the catalyst is almost immediately covered with soot. Therefore, the zeolite must continuously be recycled and regenerated by burning off the soot.[4] Spent FCC catalysts were found to release PAH when studied with laser ablation techniques coupled to mass spectrometry.[108] Our experience with FVT[17,109] led us to characterize soots by their activity as a hydrogen donor to highly reactive carbene and radical intermediates. To this end, soot or a spent catalyst was placed in the quartz tube of our FVT apparatus. The soots were exposed to the highly reactive interconverting cyclopentadienylidenecarbene–benzyne fragments (C_6H_4 intermediates[82,85]), generated by 5-[7-(2-norbornenylidene)] Meldrum's acid (**50**), or by phthalic anhydride (**51**) pyrolysis. The C_6H_4 gas phase is then exposed to the soot matrix (Scheme 9). Hydrogen transfer to either fulvene (**52**) or benzene (**53**) occurred and the fulvene/benzene ratio was taken as a measure for the donor activity of a particular soot. Benzyne condensates (see below), e.g. biphenylene (**54**), were not observed in pyrolysis over soots.[110]

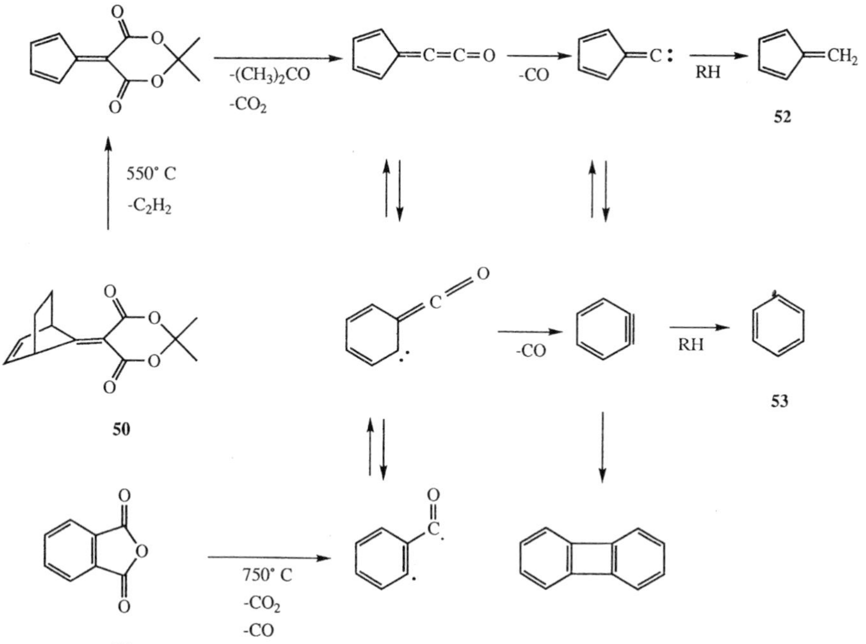

Scheme 9

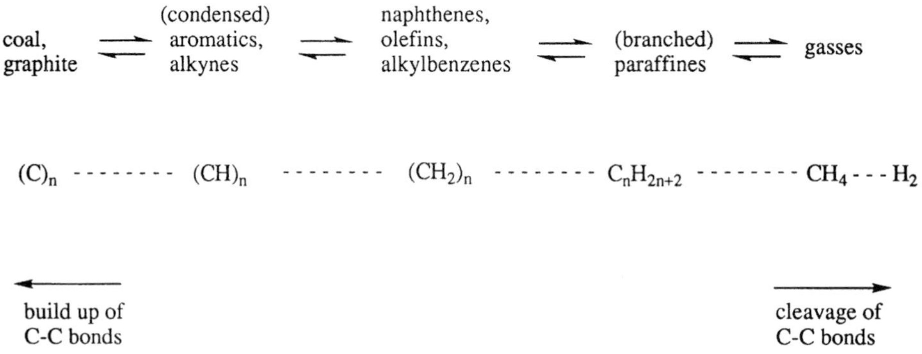

	(condensed)		naphthenes,				
coal, $\longrightarrow$	aromatics, $\longrightarrow$	olefins,	$\longrightarrow$	(branched)	$\longrightarrow$	gasses
graphite $\longleftarrow$	alkynes $\longleftarrow$	alkylbenzenes	$\longleftarrow$	paraffines	$\longleftarrow$	

$(C)_n$ -------- $(CH)_n$ -------- $(CH_2)_n$ -------- C_nH_{2n+2} -------- CH_4 --- H_2

$\longleftarrow$ $\longrightarrow$

build up of   cleavage of
C-C bonds   C-C bonds

Scheme 10

The formation of PAH and soot in thermal decomposition, no matter what variety of organic material it concerns, is a most significant feature of pyrolysis. The pyrolytic process, in a thermodynamical sense, is summarized in Scheme 10. The higher the temperature becomes, the more dissociation into the elements carbon and hydrogen occurs. At very high temperatures, up till 3000°C, for example in the carbon arc, carbon will emerge from graphite in atomic form, or as the more stable C_2 species and C_n oligomers.[111,112] Free carbon species can only exist in very high temperature plasma's, as applied for depositing graphite and synthetic diamond layers[113,114] and fullerenes.[115] These hot small carbon fragments insert immediately into organic substrate molecules.[116,117] The controlled condensation of C_n species by cooling down the plasma of the carbon arc in a helium or argon flow to 1200°C, brings the fullerenes in focus.[118] When C_n is generated in the presence of hydrogen or hydrogen donating sources, a series of PAH and CP-PAH is formed.[119] Fullerenes are also formed in hydrocarbon flames (Part 7). Under controlled conditions carbon fibers can be deposited at 1100°C from the vapor phase of hydrocarbon feedstocks, with iron particles as catalyst.[120] Coking rates from the gas phase at 810°C on different surfaces were measured for pure individual hydrocarbons.[121] Scheme 10 would not be very interesting for organic chemists if no characterizable relatively stable (kinetic) intermediates were involved. They are present in any temperature range during the process. It starts at the lower temperature side above 300°C, with generation of alkyl radicals and ends with the carbon species mentioned above. Our subject molecules, PAH and CP-PAH appear and interconvert in the 700–1200°C temperature range. As rather stable intermediates, they can survive in combustion for some time at that temperature and can escape to the environment. When the initial feed is paraffinic, branched alkanes already start to form radicals and olefines around 300°C.[62] Olefins cleave to (allylic) radicals that further dehydrogenate to more unsaturated olefins and cycloalkenes. Tetralines and phenols are of considerable interest as chain carriers in coal liquefaction and as model compounds to study aromatisation via radical chain hydrogen transfer processes in coal pyrolysis.[62,122] Smaller fragments like ethylene and C_4 fragments may originate from retro-Diels Alder and retro-ene cleavage.[65] When alkanes crack more completely, above 700°C, both ethene and methane as well as aromatics start to form. Despite the intrinsic unselectivity of alkane pyrolysis, the industrial preparation of ethene, propene and other light olefins on industrial bulk scale is uniquely done by thermal cracking of alkane fractions. Detailed technological descriptions of pyrolysis of individual hydrocarbon feedstocks generally mention coal and coke formation, respectively decoking procedures.[6,123,124] Hexadecane is a favoured aliphatic model compound for thermal cracking. The yield of ethene was found to depend on temperature and contact time, but the product distribution from hexadecane essentially follows Scheme 10 with production of PAH fractions.[125] In the context of Scheme 10 all fractions are kinetic products, including methane, since its pyrolysis leads to build up of larger species including PAH.[126] Thermal coupling of methane to form higher hydrocarbons (gasoline)[127,128] and oxidative coupling especially to C_2 fractions, are large subjects (See also Part 7).[129,130] The pyrolysis of waste polyethylene and polypropylene around 800°C (with exclusion of oxygen) produces mixtures that exactly reflect the trend in Scheme 10 with formation of methane, ethene, butadiene, benzene, toluene and PAH.[131] Thermolysis of polyvinyl chloride starts with elimination of hydrogen chloride to highly unsaturated material, that like polystyrene, yields a large variety of PAH.[132] In the production process of vinylchloride by pyrolysis of 1,2-dichloroethane, 2-chlorobutadiene (chloroprene) was found as an effective coke precursor.[121]

Poly-unsaturated and cyclic species are equilibrating with aromatics.[92] When not trapped in condensations, almost any kind of poly-unsaturated species interconverts and cyclizes within its own energy surface. To mention a few, C_7H_6 and C_8H_8,[53] C_9H_{10},[133] $C_{10}H_{10}$,[134] come with a variety of aliphatic or non-aromatic stages that can be subjected to FVT to study their isomerisation pathways. At temperatures above 700°C the non-aromatic intermediates become transitional structures to aromatics.[135] Molecules that are conjugated (aromatic, carbonyl) or cumulated π-systems, including free carbene and/or biradical sites, can survive best in the hot gas phase above this temperature, because they do not have localized single C–C bonds. Resonance energies of this type of species were calculated.[136] Higher temperatures are then needed to force further ionisation or cracking. Rearrangements within the particular energy surface of any conjugated C_nH_m molecule can occur. The acetylene–methylene carbene intermediates and CP-PAH interconversions (Part 3) are examples. In the hot gas phase of combustion and when nitrogen and sulphur are present in the feed, nitrogen-, oxygen- and sulfur centred radicals play part in cyclizations to pyridines, indoles, quinolines, benzofurans and benzothiophenes.[62] Individual thermal stabilities of coal related heterocycles have been studied.[137]

Scheme 11 shows some examples of small species to illustrate the course discussed above. The biradical and carbene intermediates of C_7H_6, undergo ring contraction to fulvenallene and dimerisation to dihydroanthracene and heptafulvalene (Eq. 1).[17,61] Oxocyclohexadienylidene, C_6H_4O, that is isoelectronic with C_7H_6, gives a similar ring contraction to fulvenone (Eq. 2).[138] With chlorine present under combustion conditions, as in the Beilstein copperwire halogen test, dichlorofulvenones will be formed, which then dimerise to dioxins.[139] With benzyne, C_6H_4O will combine to dibenzofurans (Eq. 2). Furthermore, fulvenone

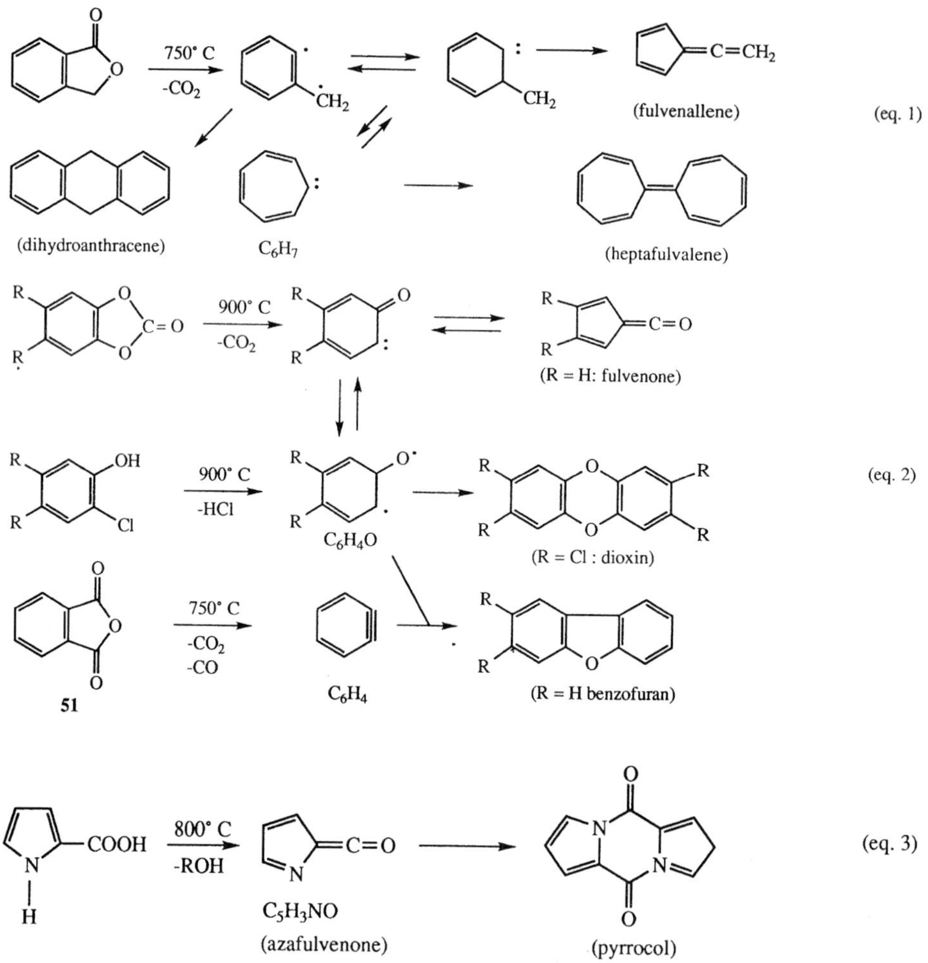

(eq. 1)

(dihydroanthracene) C_6H_7 (heptafulvalene)

(R = H: fulvenone)

(eq. 2)

C_6H_4O (R = Cl : dioxin)

51 C_6H_4 (R = H benzofuran)

(eq. 3)

C_5H_3NO
(azafulvenone)   (pyrrocol)

Scheme 11

decarbonylates at elevated temperature to C_5H_4, that dimerizes to naphthalene. Fulvenone dimer degrades to indene.[138] In general when carbonyl species, e.g. quinones[54] or aryloxy radicals,[140] are formed in combustion of hydrocarbons, they may decarbonylate and again generate C_nH_m species. A product isolated from cigarette tar, pyrrocol[141] is formed in FVT of pyrrole carboxylic acid derivatives *via* dimerisation of azafulvenone (Eq. 3). The carbon atoms of ^{13}C labeled fulvenallene and the nitrogen in azafulvenones scramble at high temperature.[61,76] The FVT reactions in Scheme 11 can be run on a gram scale.[17]

Intermediate benzyl- and phenoxy radicals,[142] carbenes, (cyclo)olefins and (alkyl)aromatics are chain carriers in hydrogen transfer. Phenols become reactive intermediates at relatively low temperature.[62,143,144] Pyrolysis of naphthols in the presence of hydrogen at 450°C yields dinaphthofurans.[145] Formation of benzofurans in coal also involve dehydrocyclizations of phenols.[146] (Aryl)radical- and olefin intermediates form a solid phase of tar and soot condensates, with about 1/1 carbon/hydrogen ratios. These soot particles are connected by benzylic- and oxygen- or sulfur bridges and contain active radical sites.[147] The soot matrix does internally homolyse at a fast rate and interacts with species from the gas phase. Soot material apparently is a radical sponge, that actively engages in hydrogen transfer. Thus, the soot is a kinetic intermediate itself, that dynamically interacts with the gas phase by taking up and releasing volatiles. PAH and methyl-PAH can be formed in the coal sponge and released from it during interaction with the gas phase. We found a similar type of soot present on spent FCC catalysts. The importance of interaction of the soot matrix on zeolites with the hydrocarbon gas phase during the FCC process, is largely underestimated.[104] The soot represents an essential part of the chemistry in the FCC process and it was concluded,[148] that the high aromatic content in the gasoline fraction orig- inates from the soot rather than in the zeolite cavities. The Lewis acid cavities primarily generate reactive olefins, that engage in subsequent non-ionic thermal chemistry, as discussed above. This includes the formation of light ends, e.g. of C_4 by retro-ene reaction. It was recommended that recycling of the light ends with the vacuum gas oil (VGO) feed may result in improved gasoline yields.[148] This representation of the FCC process is substantiated by the thermolysis of polyethy- lene, respectively over silica alumina (zeolite) and activated carbon at 550°C. The zeolite was found to produce preferentially a C_3–C_5 alkane fraction, while C_1–C_5 alkanes and a high yield of C_6–C_8 aromatics are formed over the carbon material.[149] Surface infrared techniques are applied to follow further carboniz- ation of soots and chars as function of temperature.[150]

The identity of soot is sometimes associated with fullerene formation.[151] There are almost as many soots as ways of generation, e.g. for carbon black and in flames (Part 8) or in shock tubes.[152] In our longtime FVT experience, some transient intermediates appeared so reactive that they cause carbonization at relatively low temperature in the quartz tube, despite the short contact time. For example, we observed already strong carbonization around 400°C with the benzyne precursor **50** (Scheme 9). At higher temperature **50** and phthalic anhydride (**51**), above its decomposition temperature of 750°C, gave extreme carbonization in the hot zone. In addition FVT of **51**, run on a 50 g scale, gave large amounts of tar in the bend behind the furnace and in the cold trap.[110] The low molecular products, benzene, naphthalene, acenaphthylene, biphenyl and anthracene, the benzyne dimer biphenylene (**54**), mass number 152 and the benzyne trimer triphenylene (**20**), mass number 228, were found in earlier small scale investigations.[56,153] After extraction of these compounds with hexane, respectively ethanol, a large black tar fraction is left. The mass spectrum of the tar shows large peaks at mass numbers 228, 302, 376, 450, indicating build up of a series of PAH with increments of 74. Shpol'skii fluorescence experiments with the tar solution showed sofar the presence of benz[a]anthracene (**9**), mass number 228, and naphtho[1,2,3,4- def]chrysene (dibenzo[ae]pyrene, **55**), respectively dibenzo[def,p]chrysene (dibenzo[al]pyrene **56**), with mass number 302.[154] (Scheme 12) Compound **56** is suspect as the most carcinogenic PAH.[155]

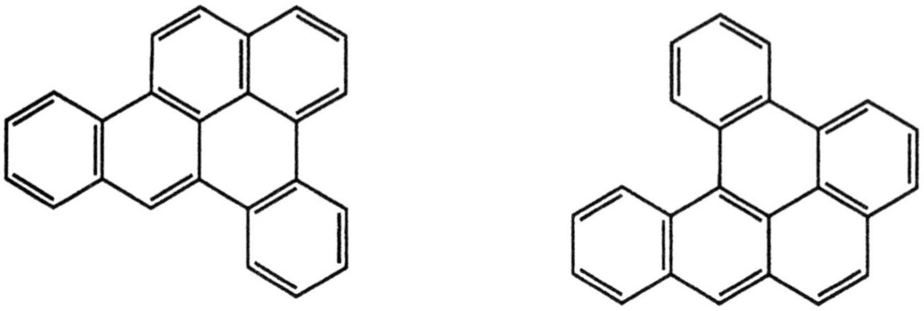

55   **56**

Scheme 12

More PAH with mass number 302 were detected with the Shpol'skii fluorescence technique in developing standard analytical procedures.[23,24] Fetzer has separated and identified larger PAH in petroleum distillate residues and in carbon black.[156,157]

5. A RETROSPECTIVE VIEW ON COAL TAR CHEMISTRY

The discovery of the individual aromatic and heteroaromatic PAH components in pit coal and in brown coal (lignite) tar fractions in the nineteenth century, revealed a richness of organic compounds with unforeseen properties. They were sometimes nicely coloured or fluorescent. In retrospect, the development of aromatic chemistry was of preponderant importance for the understanding of organic structure. The concept of aromaticity, from the early Kekule/Loschmidt theory,[158,159] *via* the Huckel $[4n + 2]$ π-electron approach, alternant and non-alternant PAH,[1,2] cyclophane chemistry[66,160] to MO computational methods for PAH,[161–163] is a continuous topic. The discovery of the fullerenes[164] has created general awareness that PAH are not per definition flat structures. However, the dogma that simple and condensed aromatics are necessarily flat and have rigid frameworks, was much earlier criticized.[165] Cyclophane chemistry has largely contributed to prove the flexibility of aromatics.[160] Incorporation of pentagonal rings in all-hexagon PAH and graphite, causes bended structures as the fullerenes.[14,166] The spectacular [60]fullerene (buckminsterfullerene, C_{60}, **57**) resembles a molecular soccerball. Prefullerene structures are bowl shaped, e.g. corannulene (**47**) with a flexible bowl.[167,168] Semibuckminsterfullerene ($C_{30}H_{12}$, **49**), and its isomers pentacyclopenta[bc,ef,hi,kl,no]corannulene (**58**), triindeno-[4,3,3a,2,1-cdef:4',3',3a',2',1'-ijkl:4",3",3a",2",1"-opqr]triphenylene (**59**) have rigid bowl structures.[101–103] Introduction of seven membered rings allows saddle like curvatures[169,170] as in [7,7]circulene (**60**). Large flat PAH structures e.g. of the naphthalene octamer **61** develop during mesophase stages in the carbonizing of pitch fractions (Scheme 13).[4,171] ESR measurements have shown that high concentrations of stable free radical PAH structures like perinaphthyl (**62**) and naphthanthryl (**63**) are formed during carbonization. The powder diffraction X-ray technique has been an important method to follow the carbonization process.[171] The crystal layering of the flat hexabenzo[bc,ef,hi,kl,no,qr]coronene (**64**) and benzo[1,2,3-bc:4,5,6-b'c']dicoronene (**65**), was used as a model to predict graphitic build up.[172] Larger structures with pentagonal ring connections like in decacyclene (**66**), a trimer of **2**, or in the naphthalene tetramer **67**, also to consider as a dimer of benzo[k]fluoranthene (**12**), are proposed as models for fullerene build up *via* internal ring closure.[102,173,174] When internally crosslinked, as indicated by the arrows, **66** and **67** make up for more than half of a soccer ball (Scheme 13). Stepwise pyrolytic condensation of naphthalene C_{10} fragments has been explored for fullerene synthesis.[175] Decacyclene (**66**) is depicted as a nearly perfect building block for the formation of concentric shells.[166] In the solid state **66** is not flat but has the shape of a three bladed molecular propeller.[176] Mechanisms of fullerene formation may include Stone-Wales rearrangement- and Bergmann cyclization steps.[177] Strategies and phantasies for build up of molecular and polymeric carbon allotropes predict exiting future challenges.[178,179]

Prior to the commercialization of coal tar, distillation of extractable oils from vegetable material, destructive distillation of leaves, bones, wood and resin, were sources of organic solvents and products. The residues were applied as active coal and charcoal. Benzene was first isolated from a sooting flame and toluene and styrene from vegetable balsams, before they became available on industrial scale from coal.[180] The pyrolysis of coal and lignite started around 1820 for the fabrication of lightning gas and lightning oil. The volatiles for lightning, only a few percent of the total mass, were obtained by heating the coal to *ca.* 500°C. After rising the temperature to 700°C, the so called low boiling coal tar fractions, containing alkanes, alkenes, alicyclics, benzene, toluene, xylenes, ethylbenzene, etc., now called the BTX aromatics, are set free. On drastic further heating of the coal, from 900°C to ultimately 1200–1300°C, the high boiling PAH tar fraction is liberated. Modern coal technology, including the product volumes and main uses in a great variety of applications of BTX and PAH, is discussed in monographs.[4,6] The technological encyclopedia[181,182] are also excellent sources. The high temperature treatment of coal was developed to produce a metallurgic coke residue to meet charcoal shortages for the fast expanding steel industry. At the same time this has innovated steelmill technology. Initially the accompanying PAH coal tar fractions gave an early waste problem. However soon an outlet was found in another application, namely wood conservation, especially of railroad sleepers.[183]

Tar is the distillate set free on pyrolysis from molten and solid phases, leaving pitch, coke or graphite residues.[184,185] Apart from coal tar and coke, other specific carbonization technologies are operated e.g. for charcoal, active coal, and carbon black, for application in purification of drinking water, printing inks, rubber vulcanisation, etc.[181,182] Coal tar pitches are important starting materials for making carbon fibers and graphite electrodes for metal production.[186,187] Note that the asphalt residue from petroleum refinery operations is also a pitch fraction. Pitches are thought to contain very large PAH structures formed by condensation of smaller coal tar components in the hot distillation residues.

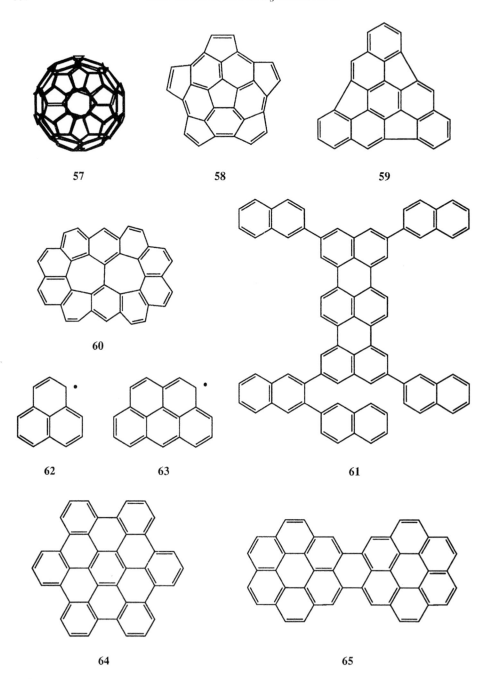

57

58

59

60

62

63

61

64

65

Scheme 13

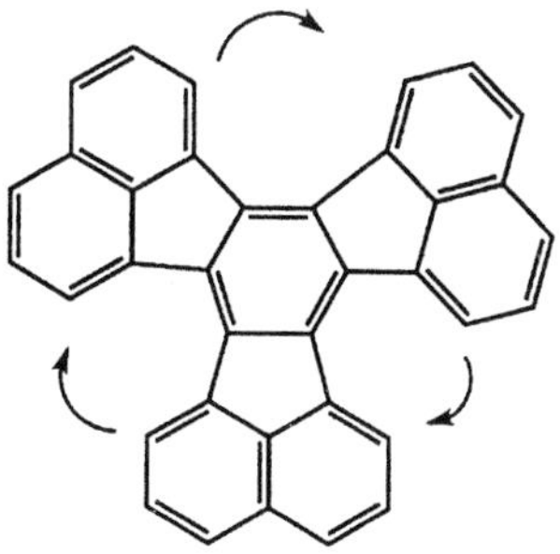

66 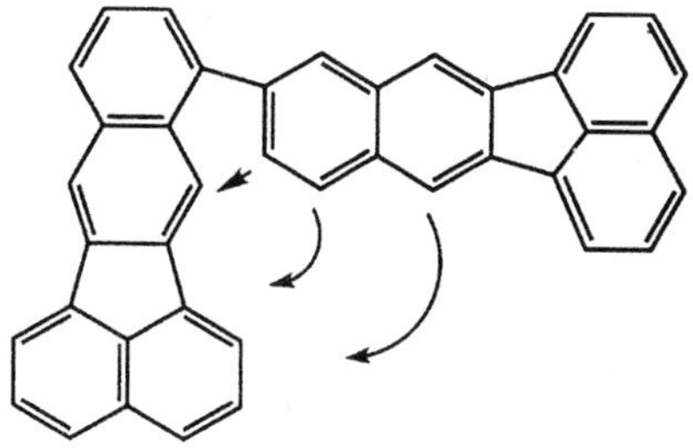   67

Scheme 13   (*Continued*)

They are not volatile enough to vaporize and carbonize on further heating in the molten tar phase.[188] Pitch structure and carbonization,[171,189] coal structure and coal pyrolysis,[147] hydrogen transfer in coal liquefaction,[122] coal gasification,[190] all coal conversion processes, are important subjects of current research.[6]

It is clear, that PAH, naturally present in coal and crude oil, are set free in large quantities by industrial coal pyrolysis and by oil refinery operations. To a large extent the waste streams are controllable,[191] however the exponential growth of these activities makes it also desirable to recycle the consumer products such as man made plastics. Pyrolysis of plastic waste[192,193] is pursued on a large scale and produces ethylene and propylene, together with a BTX fraction. The product mix is comparable with that from steam cracking of oil fractions. In both cases a residue is left that contains most of the PAH found in coal tar.[131,194]

The systematic study of BTX chemistry started around 1840, mainly in England.[4] For example, the nitration of benzene to nitrobenzene (used as a perfume) and its reduction to aniline was already known prior to 1850. The analysis of the high temperature PAH tar fraction, took much longer and required more thorough and diverse studies. These were in particular practiced in Germany. Although naphthalene was already isolated in 1819, and soon was followed by anthracene and phenanthrene, it took well into the second part of the nineteenth century before these compounds became important raw materials for the rising synthetic dye industry.[4,180] Thanks to this industry specialty PAH like acenaphthylene (**2**), fluorene (**4**), dibenzofuran (diphenylen-oxide **68**, X=O), carbazole (**68**, X=NH), dibenzothiophene (**68**, X=S), fluoranthene (**7**), pyrene (**8**), benz[a]anthracene-7,12 dione (**69**), benzanthrone (**70**), produced *via* anthraquinone (Scheme 14, Eq. 4), have become available at relatively low prices. Many interesting products can be made from them by heating them neat, with sulphur, or melting them with sodium, etc.[1] Decacyclene (**66**), by trimerisation of **2** and diindeno[1,2,3-cd: 1',2',3'-lm]perylene (periflanthene, **71**) by dimerisation of **7**, are examples. Indigo, one of the first synthetic dyes, was recently found[195] to undergo selective thermal rearrangement into dibenzo[bg](5H,11H)-1,5-naph-thyridine-6,12-dione (**72**, Scheme 14, Eq. 5). Acenaphthylene (**2**), is oxidized to 1,8-naphthalene dicarboxylic acid which is used as an intermediate in the production of perylene tetracarboxylic acid derivatives that are fluorescent dyes[196] (Scheme 14, Eq. 6). Pyrene (**8**), is oxidized to 1,4,5,8-naphthalenetetracarboxylic acid, which is another pigment precursor (Scheme 14, Eq. 7). Benzanthrone (**70**) and KOH yield the dimer violanthrone (**73**).[4]

Chemists of the pioneering period published both on coal tar characterization and dye synthesis. It is interesting to read in the original papers, especially in *Chemische Berichte* and *Liebigs Annalen der Chemie* how the compounds were isolated, named and how their reactivity was systematically explored, merely to assign their correct structures. For example, in 1869 Graebe and Liebermann and independently Perkin, discovered alizarin after oxidation of anthracene to anthraquinone, while exploring the sulphonation of the latter.[197] This synthesis replaced the production of this dye from madder. They[197] also isolated pyrene (**8**), chrysene (**10**) and its homologue picene (**22**).[198] It took sometimes several decades before the correct structures were finally established.[199] To find leads for their thermal origin, purified PAH compounds were dropped into 'red hot' tubes, e.g. picene (**22**) was obtained from 1,2-di(α-naphthyl)ethene.[200] This was inspired by the early systematic approach of Berthelot, who showed that thermolysis of methane and subsequently acetylene,[201,202] led to build up of more complicated molecules like butadiene, benzene, styrene, naphthalene and anthracene.[7,203]

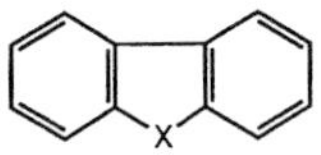

68 X = NH, O, S

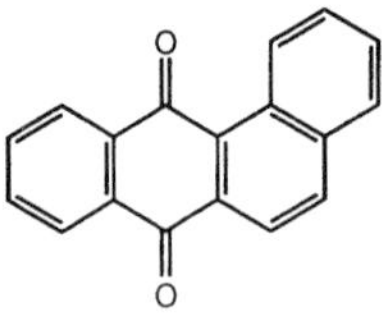

69

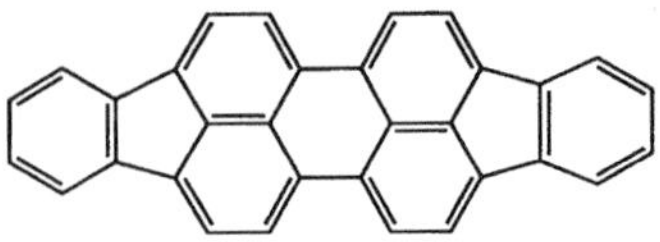

71

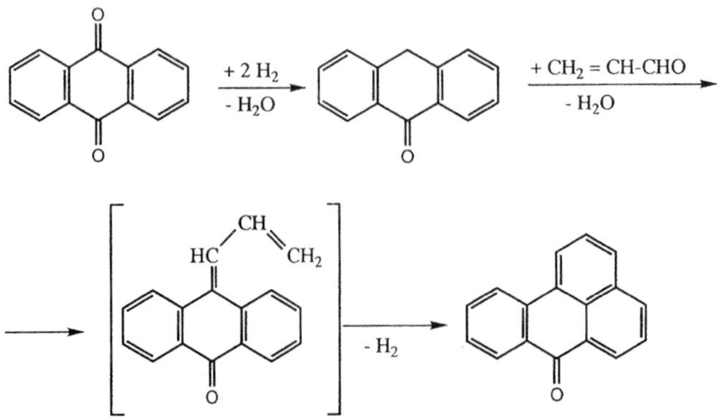

$$+ 2 H_2 \quad / \quad - H_2O$$

$$+ CH_2 = CH\text{-}CHO \quad / \quad - H_2O$$

(eq. 4)

$$- H_2$$

70

Scheme 14

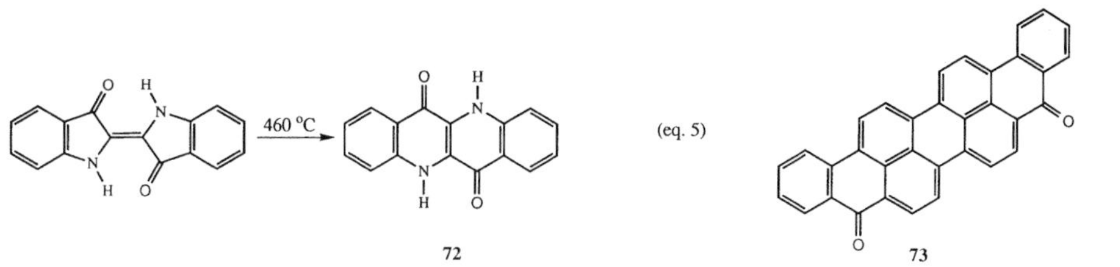

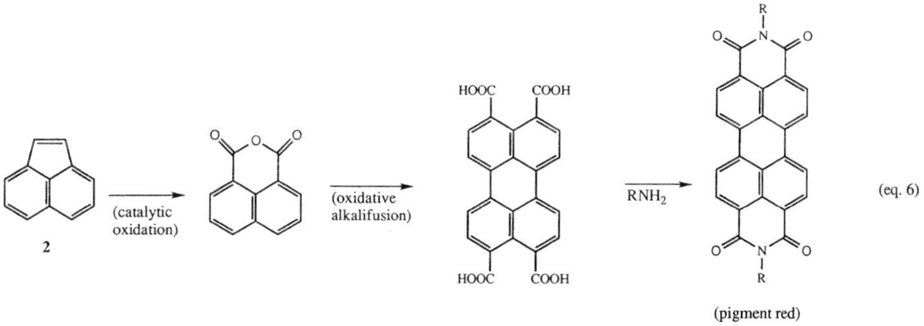

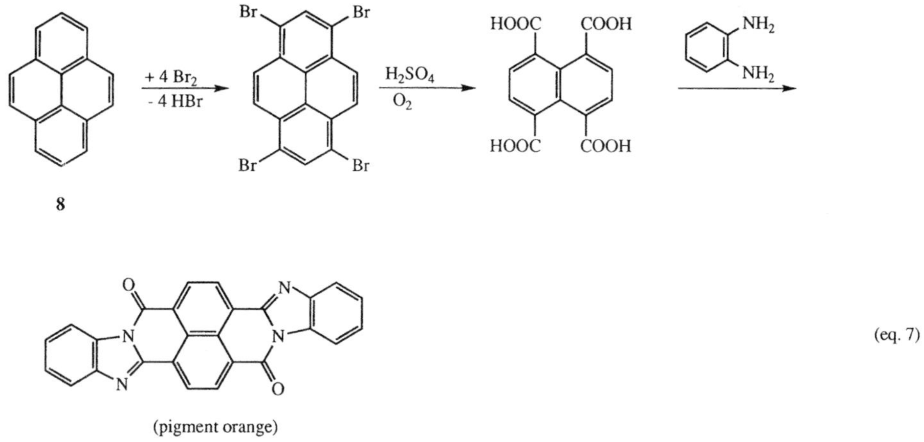

(eq. 5)

72   73

(eq. 6)

(pigment red)

2

8

(eq. 7)

(pigment orange)

Scheme 14   (*Continued*)

Berthelot thermal dimerisation of benzene to diphenyl, in 1867, is a synthetic classic, still applied in industry.[204] The thermal formation of fluorene from diphenyl-methane, respectively carbazole (**68**, X=NH) from diphenylamine by Graebe, are among the oldest examples of dehydrocyclization, summarized in a review.[59] It took about 100 years to find that these cyclizations, respectively, involve diphenyl carbene and diphenyl nitrene.[16,17] Indene (**35**) dimerizes surprisingly selectively at high temperature to chrysene (**10**). The latter is also produced by co-pyrolysis of benzofuran and naphthalene (Scheme 15, Eq. 8).[205] The mechanisms of these pyrolysis reactions are still speculative.[13,60,206]

The scientific exploration of coal tar progressed in the twentieth century. Separation methods, by (alkaline) extraction of coal tar distillation fractions were developed.[184] In addition, picrate formation and chemical modification of certain fractions by reacting them with maleic anhydride, became routine to isolate specific PAH. Benz[a]anthracene (**9**) was isolated in fair amounts from the chrysene coal tar fraction *via* its maleic anhydride adduct, followed by thermal decarboxylation.[207] Maleic anhydride treatment is industrially applied to purify phenanthrene from anthracene, by removing the Diels Alder adduct of the latter. Benzo[ghi]perylene (**74**) can be prepared[208] from perylene (**23**) *via* Diels Alder type addition of maleic anhydride in the presence of chloranil as oxidant and subsequent decarboxylation (Scheme 15, Eq. 9). In turn **74** and maleic anhydride yielded coronene (**25**) in a similar sequence.[208,209]

In 1955 all PAH and hetero PAH with established structures, isolated from coal tar at the time, *ca*. 300 constituents from over 10,000 estimated components, were listed with their first discoverers.[210] This list is probably the shortest, comprehensive documentation of the classical coal tar period. A more detailed extension of this review was published somewhat later.[50] At present some 600 PAH structures, including hetero-PAH, phenolics, amino- and methyl derivatives, are identified in coal tar.[185] Some curiosities like benzo[c]naphto[2,3-d]furan (brasan **17**, X=O) that were once isolated from the so called chrysene fraction in substantial quantities[211] and even chrysene (**10**) itself, have become increasingly less available from industrial coal tar operations.

When around 1930 high boiling PAH coal tar as recognized as carcinogenic, a new impulse was given to PAH coal tar research in order to isolate the active substances. Kilogram scales, sometimes tons, were required to isolate small amounts of individual PAH. From two tons of coal tar only a few grams of benzo[a]pyrene (**13**) and benzo[e]pyrene (**21**), were isolated.[212] A certified coal tar standard reference material gives an idea of the complexity of such separations.[19] With the discovery of carcinogenic activity, strategic synthesis developed as the preferred way to obtain pure PAH, over the tedious isolation procedures from coal tar. Multistep synthetic methodology was developed, often involving decisive pyrolytic steps, e.g. dehydrocyclization[55,203] and the Elbs reaction.[57] In addition to the maleic anhydride method, these reactions are probably the best known pyrolytic routes to larger PAH. The Elbs reaction involves cyclization of o-methyl diarylketones, that are rapidly accessible by Friedel-Crafts acylation. An example is the double Elbs reaction that yields dinaphtho[1,2-b:3',4'-k]chrysene (**75**, Scheme 15, Eq. 10).[213] The Elbs reaction may beneficially be run under FVT conditions (see ref. 16, page 3). The dehydrocyclization reaction is based on the simple old principle of heating up (hetero) PAH.[59] Various reagents are at hand to facilitate aromatization of partly saturated polycyclic systems.[18] A recent report[214] describes the coupling of stilbene and homologues, followed by oxidative dehydrocyclizations at room temperature, e.g. to tribenzo[b,n,pqr]perylene (**76**) and further to the extent of large graphite model structures as **77**. Photocyclization of stilbenes and biaryls to PAH, in the presence of oxidants has been widely explored.[215,216]

The classic approach of heating individual PAH for obtaining tars that are less complex than coal tar, was reexplored after world war II by distilling or

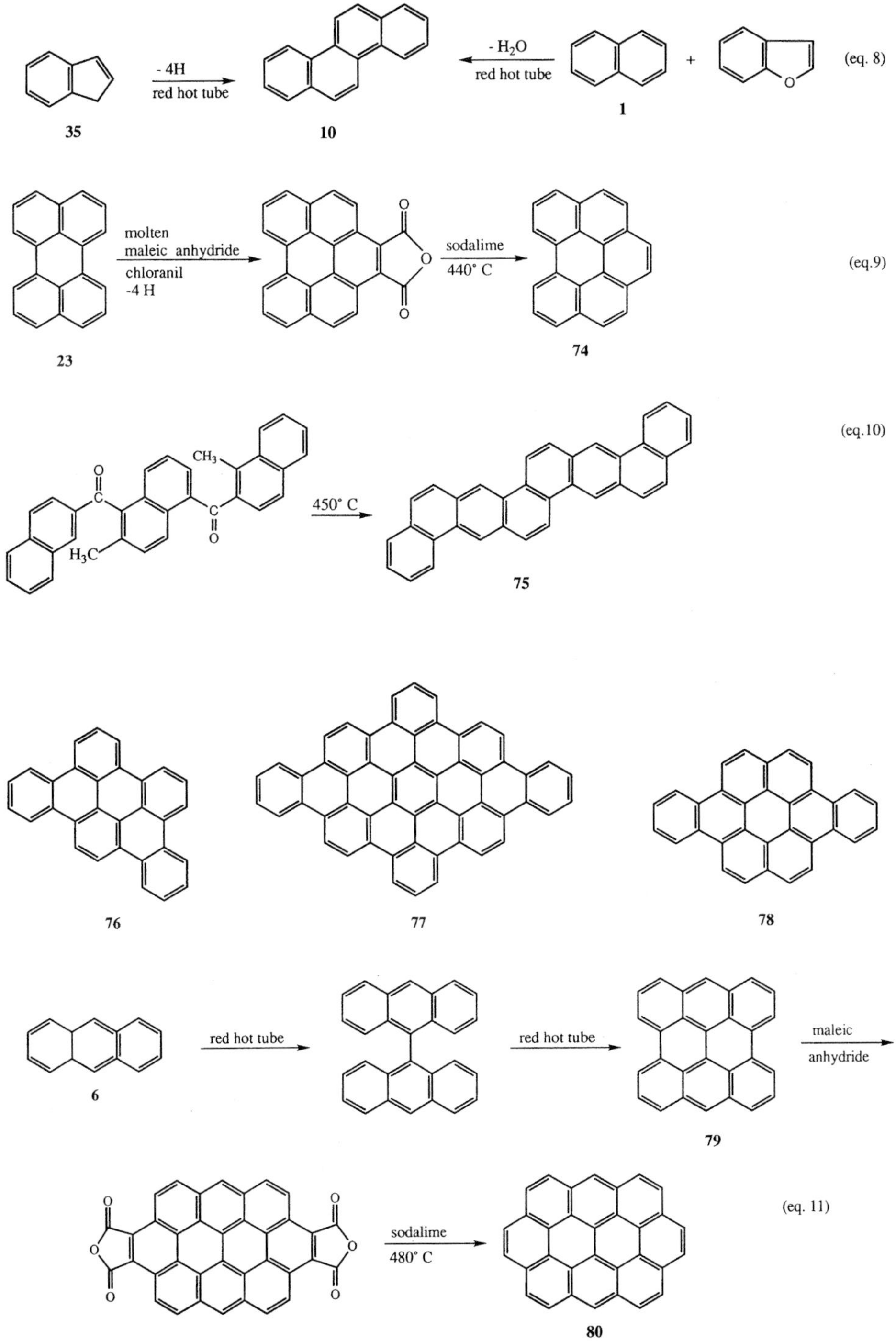

- 4H
red hot tube
- H$_2$O
red hot tube
+
O
(eq. 8)
35
10
1
molten
maleic anhydride
chloranil
-4 H
O
O
O
sodalime
440° C
(eq.9)
23
74
(eq.10)
CH$_3$
O
C
C
O
H$_3$C
450° C
75
76
77
78
red hot tube
red hot tube
maleic
anhydride
6
79
O
O
O
O
O
O
sodalime
480° C
(eq. 11)
80

Scheme 15

co-distilling them through 'red hot' tubes, e.g. the EPA compounds **1–8** . For example kilogram scale runs were made of phenanthrene[217] and of anthracene[218] and the product balances were accurately described. Maleic anhydride treatment of the obtained dehydrodimerisation mixtures gave substantial quantities of much larger PAH, for example dibenzo[aj]coronene (**78**) from the phenanthrene reaction and bis-anthene (**79**) and subsequently ovalene (**80**), from the anthracene condensates (Scheme 15, Eq. 11). More examples were reviewed and it was attempted to formulate a mechanism for the typical pyrolytic build up reaction patterns.[203]

6. COAL TAR AND PITCH VERSUS SOOT: MOLTEN PHASE AND GAS PHASE REACTIONS OF PAH

It is not always taken into account that some PAH, despite their formation under high temperature during combustion, are at ambient temperature prone to photo-oxidation,[219] or to condensation in the solid phase on melting.[220] CP-PAH are very typical for this behaviour. For example pyracylene (**81**), is generated in dilute gas phase at 1000°C from pyrenedione[221] or from 1,2,5,8-tetra(chloromethyl)naph-thalene,[222] although it rapidly deteriorates on standing at room temperature (Scheme 16, Eq. 12). The latter property may explain that pyracylene was sought without success in soot collected from wood.[223] The shortest preparation of **81** is by FVT at 1000–1100°C of 5-(1-chloroethenyl)acenaphthylene by *in situ* cycliz-ation of 5-ethynylacenaphthylene Scheme 16.[40]

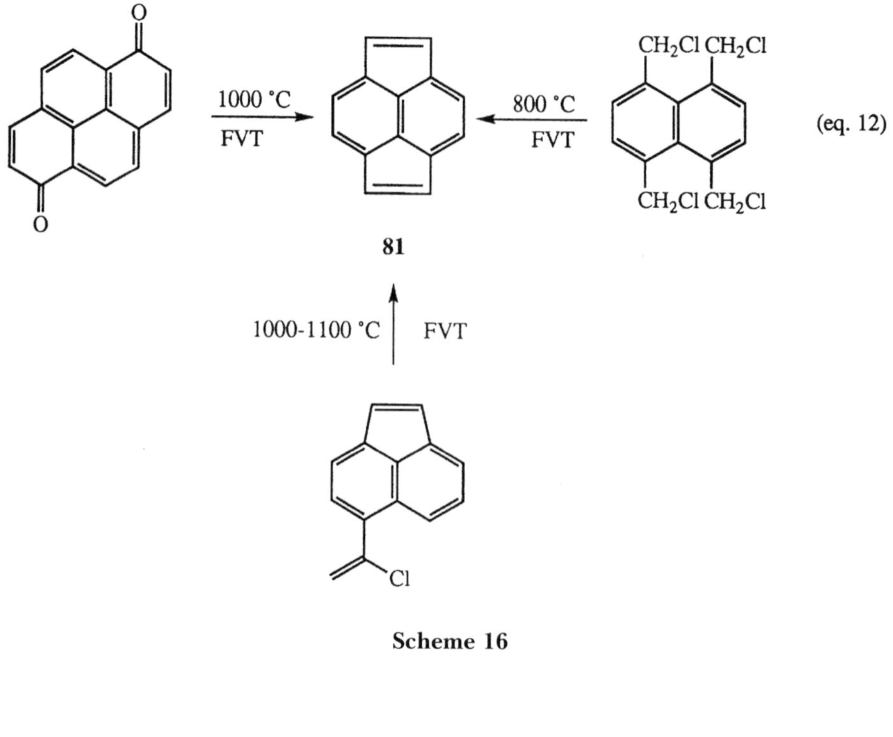

Scheme 16

Another example of a high temperature gas phase CP-PAH generation is cyclopent[hi]acephenanthrylene (CPAP, **42**), that was obtained by FVT at 1000°C from triphenylene.[43] It was also observed that cyclopenta[hi]aceanthrylene (**41**) rearranges at high temperature on the $C_{18}H_{10}$ energy surface to more stable CPAP (**42**, Eq. 13).[42] Despite its generation at very high temperature **42** rapidly polymerizes at its melting point of 170°C.[43] In contrast, triphenylene (**20**), is in its molten phase stable up to 600°C and is one of the most stable PAH.[220,224] As a rule many PAH give however pyrocondensations in the molten phase before volatilisation occurs.[224] The molten phase thermal condensation property has been utilized by functionalizing prepolymers with PAH as end caps, to make them crosslinkable on melting.[225] Under rigorous conditions, by melting PAH in the presence of metals like sodium, magnesium, zinc, or their salts, e.g. zinc chloride or aluminum chloride, even more possibilities arise to synthesize beautiful larger PAH structures.[1] When **20** is molten down at 140°C with a mixture of sodium chloride and aluminum choride, tetrabenzo[a,c,hi,qr]pentacene (**82**), a phosphorescent compound at low temperature, is formed (Scheme 17, Eq. 13).[226]

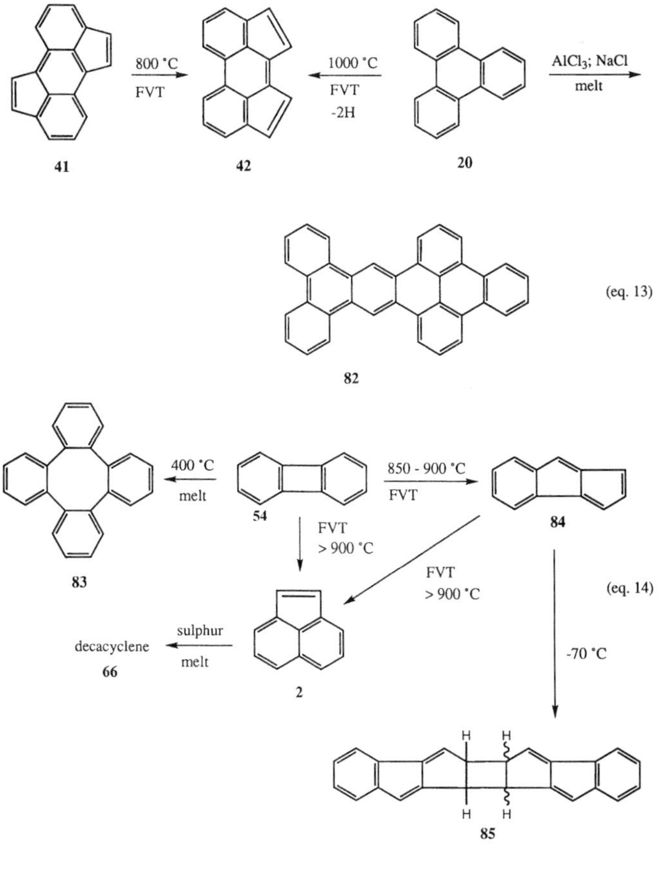

Scheme 17

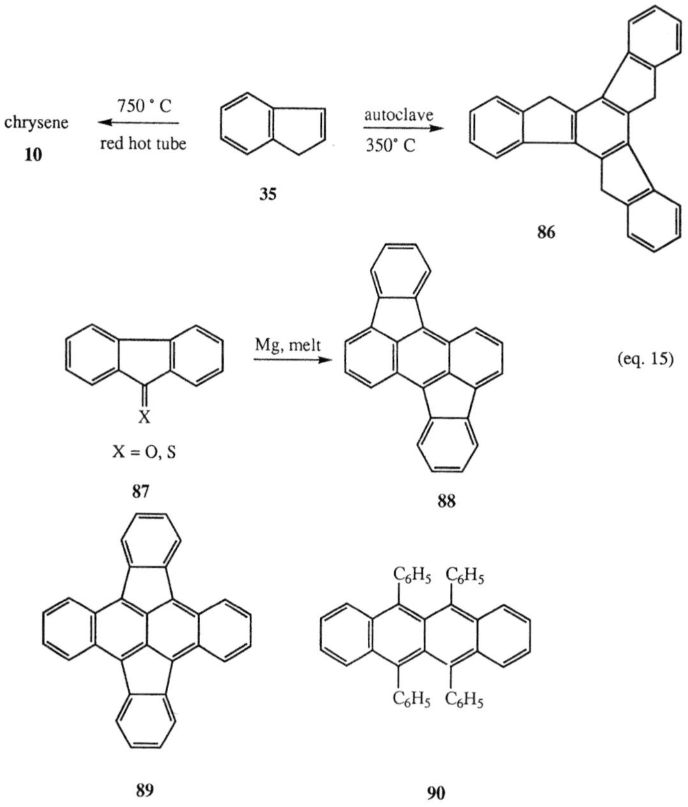

Scheme 17   *(Continued)*

A highly selective thermal reaction in the molten phase without a reagent is that of biphenylene (**54**). It dimerises quantitatively to tetraphenylene (**83**).[227,228] Biphenylene units incorporated into polymers are effective crosslinkers.[225,229] In the hot gas phase around 900°C, **54** rearranges however to cyclopent[a]indene (**84**) and in turn at somewhat higher temperature **84** rearranges to acenaphthylene (**2**). Compound **84** deposits in the form of its [2+2] dimer **85** after leaving hot zone (Scheme 17, Eq. 14).[230,231] When trapped from the hot gas phase on a liquid nitrogen temperature surface, **84** already starts to dimerize when the cold trap warms up above −70°C.[95,96] The melt condensation of **2** to larger PAH, e.g. decacyclene (**66**),[232] is long known.[233] On further heating, **2** carbonizes to a pitch with optically anisotropic mesophases.[189,234] Despite of this **2** is commonly observed in combustion.

Based on their condensation behaviour, the formation of the CP-PAH species **2**, **41**, **42**, **81**, **84**, as well as biphenylene (**54**), in the hot gas phase seems unlikely. However they are observed in soots, or can be synthesized under high temperature FVT conditions. The CP-PAH **2**, **36**, **37** and **40** were monitored in soot collected from a methane-dichloromethane flame.[235] Both ethylene flames[236] and brown coal combustion[11] give rise to significant amounts of CP-PAH.

The smallest CP-PAH compound, indene, (**35**), shows two different thermal reaction patterns, which are both of preparative interest. Condensation of **35** in the liquid phase yields isotruxene (**86**) and related structures[237] while upon gas phase pyrolysis chrysene (**10**), becomes the major product (Scheme 17; see also Scheme 15, Eq. 8).[60] Topological reactivity parameters of PAH in the liquid phase,[238] also in biochemical oxidation, are based on the form of the reactive edges of the flat PAH sheets. The terms bay-region, fjord-region, K-region, L-region, and M-region are very usefull and commonly applied.[2] Another classification[1,2] made in connection with the UV-Vis absorbtion characteristics, i.e. colour, is *kata*-annealated and *peri*-condensed. To estimate (aromatic) stability and reactivity towards intermolecular molten phase condensations, the empirical aromatic sextet rules for six membered ring PAH, introduced by Clar,[1] are worth noting.[239] The more a PAH can be drawn with fully aromatic rings (three conjugated double bonds per ring) and the less quinonoid moieties are necessary to get the π-system conjugated throughout the skeleton, the more stable it will be. For example, triphenylene (**20**), with all rings aromatic is a most stable PAH. Its isomer tetracene **19** or pentacene (**26**) can only be drawn with one genuine aromatic ring, with all other rings necessarily quinonoid. They are rather reactive and rapidly deteriorate on melting.[238] Precise PPP molecular orbital calculations, quantifying exitation energies and mobilities of the π-aromatic systems are in line with Clar's rules.[163] Despite their reactivity at ambient temperatures **19** and **26** are formed at high temperature and found in automobile exhaust and in coal tar. When kept in solution at room temperature, these compounds sensitize oxygen to its singlet state and then form endoperoxides.[1,2,240] Like anthracene **19** and **26**, are reactive dienes in the Diels-Alder reaction and undergo melt condensation and cyclodehydrogenation. Melting of fluorenone (**87**, X=O) with magnesium or alternatively heating thiofluorenone (**87**, X=S), yields red crystals of diin-deno[1,2,3-de:1',2',3'-jk] anthracene (rubicene **88**, Scheme 17, Eq. 15).[241] Many more interesting melt condensation reactions can be found in Clar's monograph.[1] For example diindeno[1,2,3-fg:1',2',3'-op]tetracene (**89**), a tetrabenzopyracylene, is a blue photooxidizable compound, that can be prepared by melt thermolysis of 5,11-diphenyl-6,12-dichloronaphthacene.[242] The popular photosensitizer 5,6,11, 12-tetraphenylnaphthacene (rubrene, **90**) has been reported from melt condensation of 3-chloro-1,3,3-triphenylpropyne.[243] The latter type of PAH are singlet oxygen scavengers, fluorescents and feature other light induced proper-ties.[219] As extended conjugated π-systems, many of these PAH, are of interest as laser dye and for solar energy collectors.[244] Fluorescence is also an analytical tool for determination of PAH.[245] Fluorescent PAH are utilized as probes for studying pollutant transport, for example in groundwater.[246]

7.   PYROLYSIS VERSUS COMBUSTION. PAH FROM COMBUSTION AND FLAMES

Pyrolysis and combustion are scarcely treated as subjects in organic chemistry textbooks.[247] On attempted identification of the core research activities in these very large subjects, no matter if it concerns alkanes or other hydrocarbons like olefins and aromatics (Part 4), the two subjects are seemingly non-related. Pyrolysis is typically an organic subject including some preparations. Due to the low selectivity and the large bodies of tar formed in various pyrolytic studies, complex work up and identification of the individual products is necessary to rationalize mechanistic patterns. It was concluded that "heating up organic molecules is one of the oldest, yet least understood ways to bring about chemical reactions".[248] Combustion research is approached with physical chemistry methodology and techniques, e.g. measuring energetics and kinetics in (cold) flames.[249,250] Partial combustion is also a subject of catalysis and technological research, since numerous selective oxidations, often utilizing metal catalysts, were developed as chemical processes.[251]

There is ample evidence that hydroxyl radicals are chain carriers in combustion. The latter abstract hydrogens, to end up as water.[97] When oxygen is taken up by a species it can be lost again as carbon monoxyde or carbon dioxyde, so that hydrocarbon species remain predominant in the hot combustion zone. The carbon–carbon bonds in the smaller non-ionic reactive molecules,[252] do not rapidly homolyse in either a hot pyrolysis tube or in a combustion chamber (See also Part 4). Neither do these molecules have labile hydrogen bonds, prone to rapid oxidative attack by hydroxyl radicals. Entropic factors disfavour condensation of these species in the high temperature zones and therefore they have a certain survival time. Their fate is then governed by the chance of escape to the colder edges of flames, or into the chimney of a combustion instalment. Here they will undergo sequential dimerisation and condensation reactions. This is what one observes while looking at a sooting flame or a smoking fireplace. Tracing of PAH in flames and soot is a longstanding topic.[253] Soot, deposited from a flame on a cold surface, goes through an entirely different formation process than coal tar, that is generated in the molten phase and worked up *via* high temperature distillation schemes (Part 5). Soot may still contain components that are sensitive to melt condensation and therefore are not found in coal tar.

It is already documented (Part 2) that burning of fossil fuels in power plants, aircraft and automobile engines,[20,254] burning of wood,[25] biomass[255] and coal in industry and household,[28] waste incineration[29] etc., but also forest fires and vulcanic activity leave residues and airborne particulate matter with an abundance of PAH.[2,256] Strictly spoken, burning of organic material, giving rise to PAH emission, is a naturally occurring chemical process, just like PAH formation in the geological process of coal deposition.[257,258] Our excessive use of energy, by combustion of organic fuel, makes that the accessory pollution residues are not in equilibrium with other natural processes, such as degradation of PAH by bacteria.[259,260] As a consequence PAH have become ubiquitous in the environment. The PAH concentration in urban areas is sometimes factors of ten, even hundred fold higher,[2] compared with remote places. Nevertheless, benzopyrenes have been found in rural areas and are probably indigenous to soil.[261] PAH occur even in the interstellar space,[262–264] although the correct identification remains questionable.[265] With the ever increasing global energy demand, either for gas, oil and ultimately for coal, governmental regulations for more efficient combustion are needed[29,266] to control PAH emissions towards safer limits.[267]

The detection of PAH in flames began with the isolation of benzene, from the condensate of the effluent gases of the sooting flame of burning fish oil, in 1825 by Faraday. He also associated soot formation from a burning candle with condensation of volatile gaseous species.[268,269] Thousands of reports exist on PAH formation in flames, from cigarette smoking,[33] automobile driving,[270] chemical kinetics of methane, ethane, ethene or acetylene flames[249] and kinetics and mutagenicity in an ethylene feeded reactor,[271] to dioxin formation in the Beilstein copper wire halogen test,[139] to mention a few. The detection of contaminating PAH is mainly focussed on the parent PAH systems, but dioxins, benzofurans, carbazoles, phenols, methyl-, amino- and nitro-PAH derivatives, are also frequently reported. In an extensive analytical study of soot collected from burning brown coal in a standard stove, over 170 different components, including many hetero-PAH, were identified.[28] Among the dominant products in this list are the EPA compounds phenanthrene (**5**), fluoranthene (**7**), pyrene (**8**), benz[a]anthracene (**9**), chrysene (**10**), and benzo[b]fluoranthene (benz[e] acephenanthrylene, **11**). Other major constituents are acephenanthrylene (**37**), benzo[a]chrysene (picene, **22**), benzo[ghi]fluoranthene (**32**), methyl-PAH and hetero-PAH, e.g. the benzonaphthopyrroles, -furans and -thiophenes, **17** and **18**. This list[28] may serve as a guide to look for contaminants in soot samples. The PAH were catagorized in types and molecular weight, including nitrogen, oxygen- and sulphur PAH. The product mix is much alike that of coal tar.[19,210] It should be emphasized that molecular weights above 300 become problematic for specified structure detection, as indicated in another paper.[23] The benzonaphthofurans **17** and **18**, X=O, were recently certified as reference compounds.[20] In the PAH monographs[1,2] hetero-PAH systems are not treated although they are prevalent in combustion. When formed in combustion, pyrolytic synthesis may give access to these heterocycles as well. Out of many possibilities, FVT preparations for **17**, X=O,[17] **18**, X=O[272] and of 7H-benzo[c]carbazole (**91**),[231] are shown in Scheme 18, Eqs. 16–18.

A marked difference between coal tar and combustion soot is that nitro-PAH are commonly present in the latter. Nitro-PAH were recently certified as reference compounds because they are found in ambient air, especially from diesel exhaust.[20] Notable examples are nitronaphthalenes, 1-nitropyrene (**92**) and 3-nitrofluoranthene (**93**). It was very recently reported[41] that **92**, **93**, and 6-nitrochrysene (**94**), when submitted to FVT between 750 and 800°C, equilibrate to the nitrite esters and then loose NO, with formation of the aryloxy radicals. The latter pick up hydrogen, because the corresponding hydroxy-PAH **95**, **96** and **97** can be isolated in fair yields (Scheme 19). This pathway confirms the classic observation that pyrolysis of nitrobenzene yields phenol.[248] Nitro-nitrite rearrangement has also been taken into account in kinetic studies of the thermal decomposition of nitroaromatic explosives of the trinitrotoluene (TNT) variety.[273] In the case of 6-chrysyloxy radical, further loss of CO to form 2,3-benzofluorene, (**17**, X=CH$_2$) occurred above 800 °C as a selective sequential pathway (Scheme 19, Eq. 19).[274] Hydroxy-PAH, much studied as metabolites,[5] are rarely observed in combustion.[20] The increase of urinary 1-hydroxypyrene (**95**), is often used as a biomarker to estimate total PAH exposure, for example of foundry or coke oven workers.[275]

(eq. 16)

(isobenzofuran)

17

(minor)

(eq. 17)

18

(eq. 18)

91

Scheme 18

Aryl-NO$_2$ $\xrightarrow[\text{FVT}]{750\,°C}$ Aryl-O-NO $\xrightarrow{- NO^•}$ Aryl-O$^•$ $\xrightarrow{RH}$ Aryl-OH

92	Aryl = 1-pyrenyl	**95**
93	Aryl = 3-fluoranthyl	**96**
94	Aryl = 6-chrysenyl	**97**

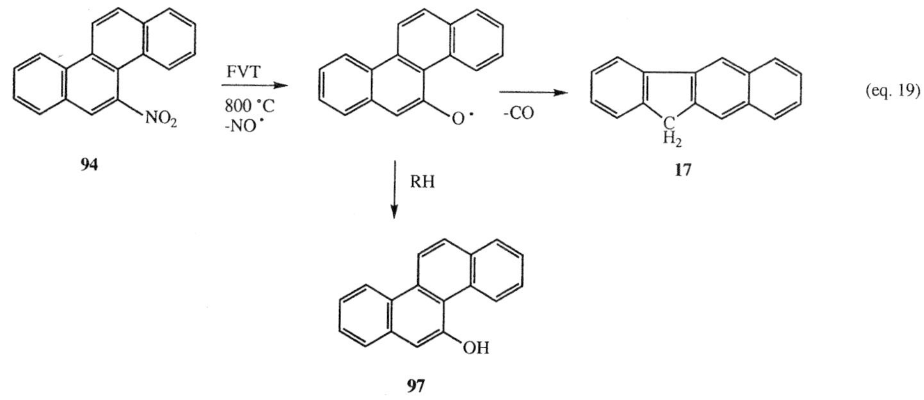

(eq. 19)

Scheme 19

Chemical analysis of soot materials is done by extraction at ambient temperature, in combination with chromatographic and spectroscopic techniques, without isolation of the individual components. Therefore, the more sensitive PAH and CP-PAH are more likely to be found in soot deposits than in coal tar. Thus soot is a correct substrate for tracing pyrolytic PAH formation patterns. The chimney part of a stove or incinerator is in fact a reactor where condensation reactions of intermediates, escaped from the flame part of the combustion instalment, can take place. At still lower temperatures PAH compounds condense on fly ash and nucleate to small soot conglomerates that adhere to dust and remain in the air as respirable particles. It is claimed that the kind of fuel, wood, diesel or gasoline, that has been burned and the type of engine, fire or factory, can be traced from soot nucleation patterns.[276] Particle growth is not much dependent on the type of fuel, although carbons from aromatic molecules present in the feed contribute somewhat more to the soot yield.[277] Diesel engines have high working temperatures so that nitrogen oxydes start to form. Nitro-PAH probably arise from combination of NO$_2$ and PAH radicals, that escape into the much cooler muffler part of the engine. The combustion temperature depends on the fuel and the construction of the burning chamber. A flame can be premixed with oxygen or is diffusion controlled.[278] A flame in the open air is almost suddenly quenched to room temperature and reactive intermediates escape from the hot area to form PAH and soot condensates. A different PAH mix was found in smoke from straw burned in the open air to remove it from the land, in comparison with soot from biomass used as fuel in an installation for energy supply.[279] Partial combustion is applied for making carbon black on large technical scale by blowing gasses or spraying fuel into flames.[280] Carbon black, diesel soot and titanium dioxide particles cause lung tumors in rodents.[281]

The formation of benzene in Faraday's fish oil flame[268] could mean that the six benzene carbons originate from specific cracking of the fuel to C_6 fragments. Not much later than Faraday, Berthelot observed that benzene also arises from methane burning. Around 1850 he started the first systematic study of pyrolysis mechanisms. Berthelots work lasted over a period of almost 70 years.[7,201,202] As early as 1866, based on methane and ethane and acetylene pyrolysis experiments, Berthelot concluded: in addition to pyrolytic reactions of decomposition, there are also reactions of synthesis. According to this work, acetylene and butadiene became the central intermediates in the build up of aromatics. Later benzyne and aryne intermediates were introduced in acetylene pyrolysis.[282] The intermediacy of CH, CH_2 and CH_3 carbene- and radical species in build up processes was postulated in an early stage.[283] Convincing prove for the intrinsic formation of larger PAH and CP-PAH in combustion from C_1 and C_2 species comes from burning studies of methane or ethane.[236,249] Oxidative coupling of methane to ethane, ethene and acetylene on technical scale is a research topic[129,284] like straight pyrolysis of methane above 1100°C for gasoline production with PAH tar formation.[127,128] The BASF acetylene production process from methane involves a premixed flame of the hydrocarbon and oxygen. Many different feeds are studied under controlled flame conditions and their effluents have been analysed, e.g. thirty-eight PAH were identified in fuel oil flames.[285]

We found it helpful to look into the Chemical Abstracts Collective Indexes under elementary species like C_2, C_2H_2 (acetylene), C_3, C_3H_4 (cyclopropene, allene, propyne), C_4, C_4H_2 (butadiyne), C_4H_4 (vinylacetylene, cyclobutadiene, cumulene), C_6H_2 (hexatriyne), C_4H_6 (butadiene), C_6H_4 (benzyne) etc. They play a major role in the build up of aromatics. Large series of references apply to each of these basic reactive species, which often come up in calculations or in plasma and flame type studies. Even the champion of antiaromaticity,[286] cyclobutadiene and many other transients can exist and be monitored in high temperature vacuum.[252] Shock tube techniques are applied to study kinetics of such small species.[250] C_2 and C_n carbon species have been reviewed.[112] The larger a species becomes, the more intermediates of the same molecular formula can exist on its energy surface. Some examples are C_6H_6[287] e.g., benzene and fulvene, C_7H_6 e.g. fulvenallene, ethynylcyclopentadiene, benzocyclopropene (Part 4), C_8H_6 e.g. ethynylbenzene, C_8H_8 e.g. cyclooctatetraene, styrene, o- and p-quinodimethane, benzocyclobutene.[288]

The discovery of [60]fullerene (C_{60}, **57**) and homologues in the carbon arc as well as in flames has refocussed the attention of chemists intensely on the mechanistic aspects of its build up from small species.[119,177,289] C_{60} and C_{70} ions were observed in a benzene flame directly coupled with mass spectrometry.[290] Fullerenes were also recovered from soot of benzene flames.[271,291] Series of PAH observed or proposed (e.g. from mass numbers) in flames and more or less associated with fullerene formation, include the following molecular formula, with some examples given:

- $C_{10}H_8$, naphthalene (**1**), azulene;
- $C_{12}H_8$, biphenylene (**54**), acenaphthylene (**2**); ethynylnaphthalene;
- $C_{12}H_{10}$, biphenyl, acenaphthene (**13**);
- $C_{14}H_8$, pyracylene (**81**);
- $C_{14}H_{10}$, anthracene (**6**), phenanthrene (**5**);
- $C_{16}H_{10}$, acephenanthrylene (**37**), fluoranthene (**7**), pyrene (**8**);
- $C_{18}H_{10}$, cyclopenta[cd]pyrene (**40**), benzo[ghi]fluoranthene (**32**), cyclopenta[cd]fluoranthene (**43**), cyclopent[hi]aceanthrylene (**41**), cyclopent[hi]acephenanthrylene (**42**);
- $C_{18}H_{12}$, triphenylene (**20**), chrysene (**10**); benz[a]anthracene (**9**), benzo[c]phenan-threne (**28**);
- $C_{20}H_{10}$, dicyclopentacepyrenes (**44**), corannulene (**47**);
- $C_{20}H_{12}$, perylene (**23**), benzofluoranthenes, benzopyrenes;
- $C_{22}H_{12}$, benzo[ghi]perylene (**14**), indeno[1,2,3-cd]pyrene (**15**), anthanthrene;
- $C_{22}H_{14}$, pentacene (**26**), picene (**22**), dibenz[ah]anthracene(**16**),
- $C_{24}H_{12}$, dicyclopenta[1,2,3-cd:1',2',3"-lm]perylene (**38**), coronene (**25**).

The individual PAH are mentioned in Part 2 and FVT reactions to prepare the CP-PAH are given in Part 3. Throughout the above series of C_nH_m derivatives, the sequential $C_{n+2}H_m$ ethynyl-PAH derivatives formed by addition of C_2, always apply as possible combustion intermediates. A series including ethynyl-PAH was discussed[292] as a sequence of the most thermodynamically stable PAH, called "stabilomers". In a generalized representation starting with the smaller species increments of C_2, respectively C_2H_2 (acetylene), are typical of combustion. Mixing acetylene into the benzene flame furthers the appearence of the above sequence. In fact a flame of any organic material can produce soot with compounds from the above sequence.

A recent observation is that PAH can loss two hydrogen atoms above 1000°C, e.g. $C_{18}H_{12}$ to $C_{18}H_{10}$, triphenylene (**20**) to cyclopent[hi]acephenanthrylene (**42**),[43] respectively, of benzo[c]phenanthrene (**28**) to benzo[ghi]fluoranthene (**32**) and cyclopenta[cd] pyrene.[94] Triphenylene (**20**), cyclopenta[cd]pyrene (**40**) and cyclo-pent[hi]acephenanthrylene (**42**) were recently monitored in high temperature pyro-lysis of pyrene (**8**) and explained *via* addition of C_2 fragments, thought to be forced out of the substrate during pyrolysis.[293]

Fullerene inspired studies are at the same time rationalizing formation of smaller PAH from flames and combustion, because their build up involves similar mechanistic principles. When one reaches $C_{30}H_{12}$, e.g. semibuckminsterfullerene (**49**)[100,101] the hydrogen percentage has leveled down considerably and the molecule contains already half the bowl of C_{60}. Then, to double the C_{30} half bowls to C_{60} in one step is speculative.[177,290,294] Growth *via* smaller reactive intermediates, i.e. C_2 or C_n, followed by Stone-Wales and Bergmann type rearrangements is another option.[177]

Several of the PAH and CP-PAH summarized above were observed after electric discharges between graphite electrodes in liquid toluene[295] and from carbon arc discharges in hydrogen donating atmosphere.[119,296] Formation of acetylene under these conditions was already known in the nineteenth century. With all this detailed work available, the general conclusion from the enormous numbers of articles spotting PAH, is that combustion, no matter what type of organic fuel is burned, invariably produces PAH and soot from the genotoxic variety.

8.   LARGE PAH STRUCTURES AND FULLERENES

Until the fullerenes were discovered large PAH were commonly thought to be graphite type flat structures, although this dogma was earlier refuted.[165] Flexibility of larger carbon clusters[297] is now intensely studied in connection with formation mechanisms of fullerenes[115,298] and buckytubes.[299] In pure form larger PAH are often insoluble in organic solvents, although they do not crystallize from tar mixtures. Compounds **64** and **65** were crystallized from molten pyrene.[172] Soot from the carbon arc containing C_{60} (**57**) dissolves much better than the pure C_{60}. For separation of **57** chromatographic procedures are reported.[300] The fullerenes are found in soot forced out of graphite electrodes in the high temperature condition of the carbon arc. This experiment was also run in a chlorine atmosphere. In this way perchloro derivatives of prefullerenes, e.g. perchloroacenapthylene and perchlorocorannulene (**98**) were obtained.[296] This resembles the formation in the carbon arc of similar parent PAH in hydrogen atmosphere[119] or in the presence of volatile organics like methanol.[296] Several PAH and CP-PAH were observed after electric discharges between graphite electrodes in liquid toluene.[295] Thermolysis of tetrachlorophthalic anhydride[301] affords a mixture of several C_6Cl_4 oligomers from which perchlorotriphenylene (**99**) can be isolated (Scheme 20, Eq. 20). Perchlorocoronene (**100**) is another recent chloro-PAH that can be converted to dodecakisphenyl coronenes (**101**), an interesting host molecule.[302]

Although [60]fullerene, (C_{60}, **57**) has been around (like many PAH) in sediments for millions of years[303,304] it was only recently discovered. Structure elucidation is unconceivable without spectroscopy, especially 1H- and 13C NMR in combination with mass spectrometry. For chemists C_{60} is more than a carbon allotrope. As a genuine organic compound its functionalization is intensively explored and its chemistry is well on its way.[305–307] Other interesting carbon configurations have been discussed.[179,308] Recent strategic synthesis of larger prefullerenes are the preparation of semibuckminsterfullerene $C_{30}H_{12}$, **49** and some C_{28} analogs (Part 3).[100] Similar attempts to bowl shaped systems were reported by pyrolysis of 4,4'-bi-4H-cyclopenta[def]phenanthrylidene (**102**) to tetrabenzo[de,hi,-mn,qr] naphthacene (**103**) and diindeno[4,3,2,1-opqr;4',3',2',1'-avut]picene, (**104**, Scheme 21, Eq. 21).[309]

Gel permeation chromatography is a method suited for separation of large PAH in petroleum residues.[310] Fetzer has isolated large PAH out of pitch residues and mixtures that he obtained from melt pyrolysis of smaller PAH and benzanthrones.[157] Several large structures were identified, e.g. **105** and **106** (Scheme 22). Fluorescence band measurements were reported as environmental references.[156]

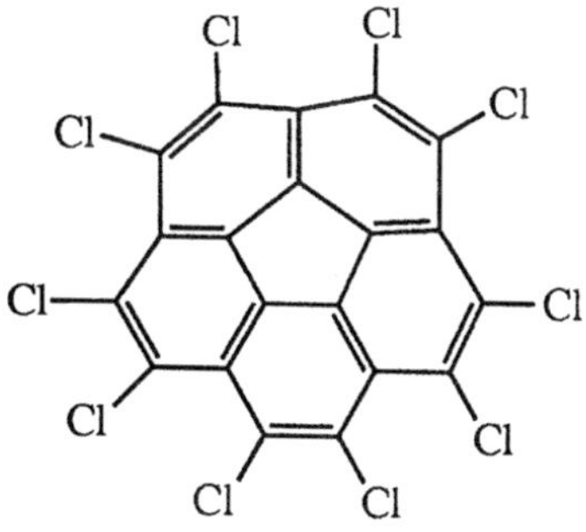

98

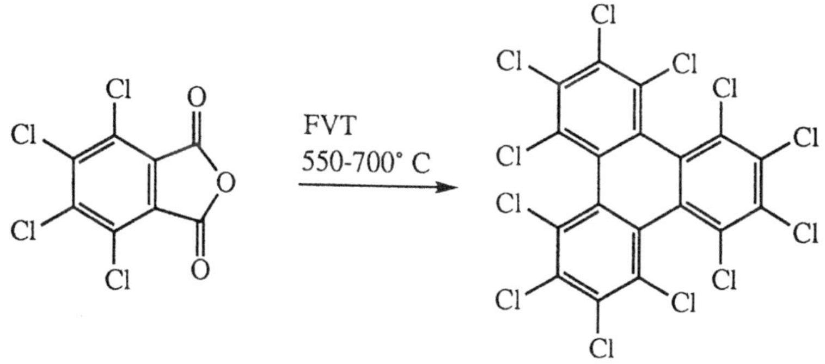

FVT
550-700° C

(eq. 20)

99

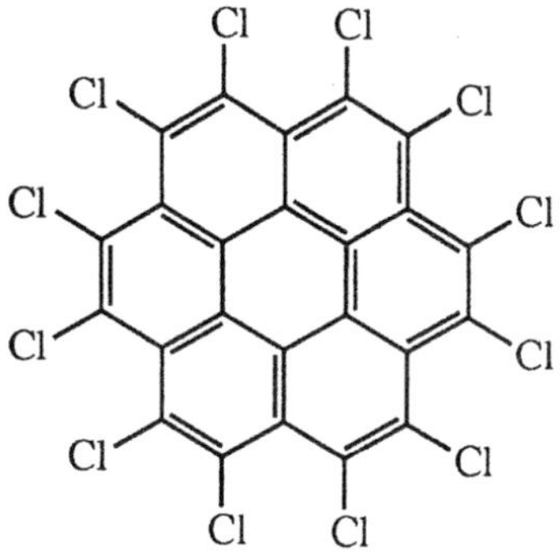

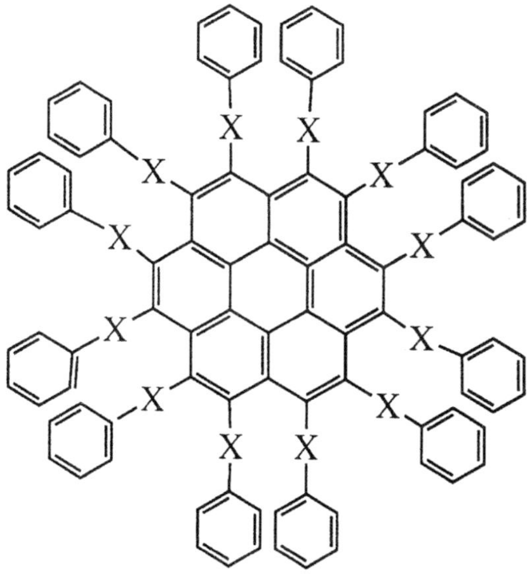

100

X = O, S

101

Scheme 20

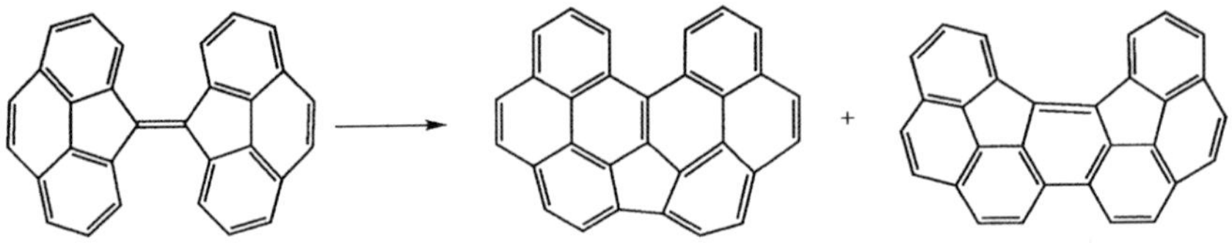

(eq. 21)

102   103   104

Scheme 21

The synthesis of PAH with only six membered rings includes dibenzo-[hi,qr]perylene (**107**) that is generated by FVT of the tosylhydrazone of 1-acenaph-thylaldehyde (Scheme 22, Eq. 22).[311] Synthesis of coronene homologues and larger PAH systems, was already shown in Part 5 with compounds **77** and **80**. The synthesis[240] of circumanthracene (**108**, Scheme 22, Eq. 23) is another example and preparation of the still larger circum[34]terphenyl (**109**) was recently achieved following the older strategy of maleic anhydride addition to dicoronylene (**65**).[51] Cyclopenta[1,2-a:3,4,5-b'c']dicoronene (**110**) was recently isolated as a second product from aluminumchloride catalyzed dimerisation of **25** (Scheme 22, Eq. 24).[312]

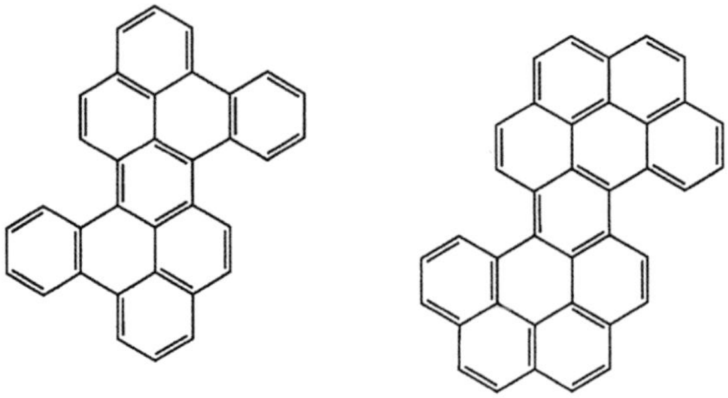

105   106

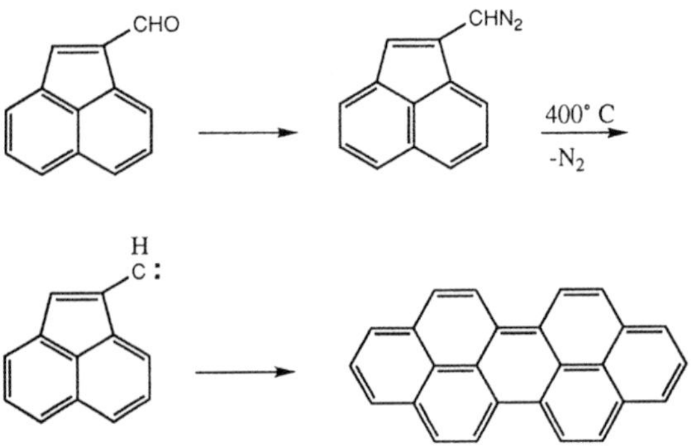

$\xrightarrow[-N_2]{400° C}$ (eq. 22)

107

Scheme 22

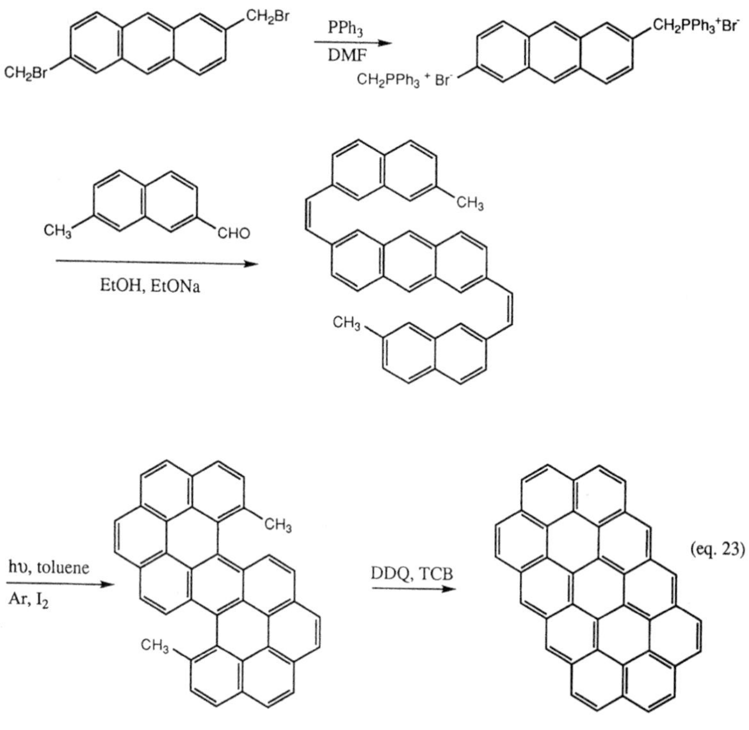

(eq. 23)

108

(eq. 24)

25

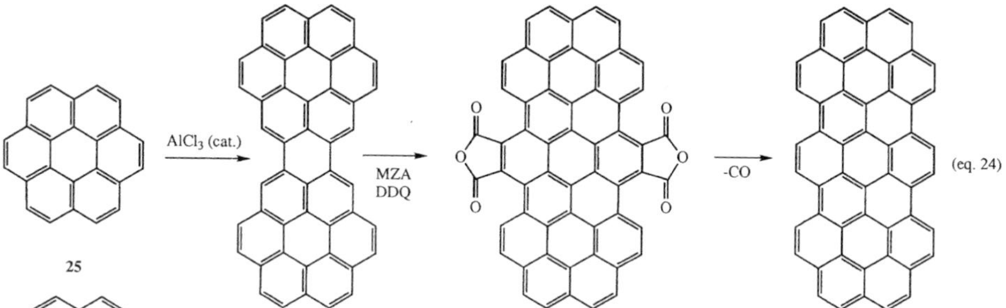

65

109

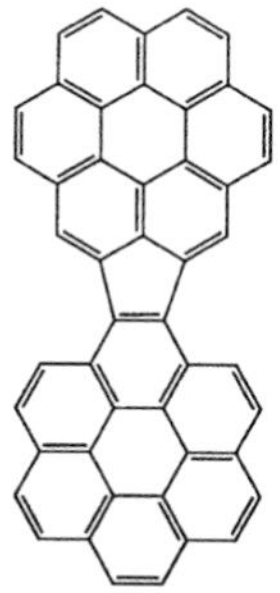

110

Scheme 22 *(Continued)*

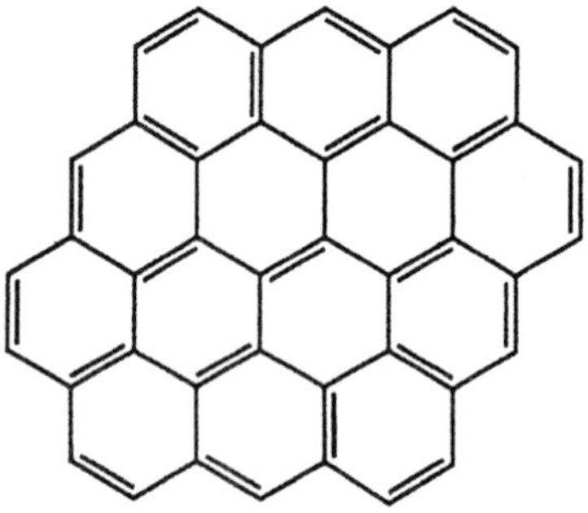

111 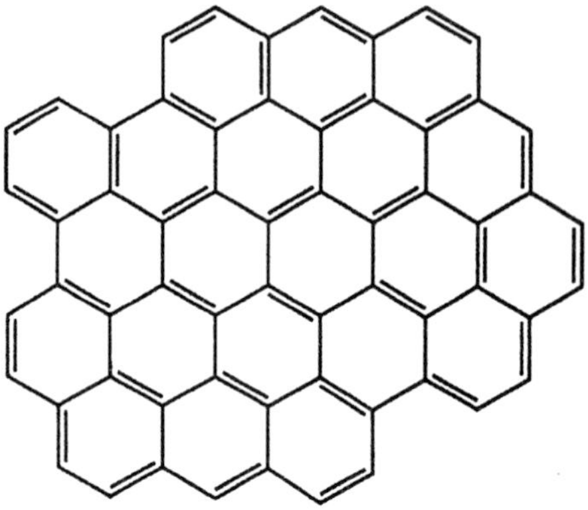   **112**

Scheme 23

Some large structures are associated with emissions observed from interstellar space,[262,313] e.g. **111** and in theoretical treatments, e.g. **112** (Scheme 23).[314] Calculations of topological variations in structure show that especially the larger PAH are capable to adapt themselves to many nonplanar, chiral and nonchiral molecular geometries.[297,315] Large PAH molecules have excited states close to the ground state, but they still have high resonance energies in their excited states so that they can survive in a hostile environment of the interstellar space.[313] [60]Fullerene (C_{60}, **57**) is an entirely sp^2 hybridized carbon system, with no hydrogens attached, so that it can not undergo substitution. It has however pronounced reactivity as dienophile and gives rapid addition reactions with nucleophiles. The orbital situation in C_{60} was related to the situation in PAH substructures, that repeat themselves on its surface, like pyracylene (**81**) and coranulene (**47**).[316,317]

9. AMBIGUITIES OF THE ENVIRONMENTALLY UBIQUITOUS PAH

Presently aromatics and their derivatives comprise about one third of the organic bulk chemicals.[4] The production of BTX aromatics is coupled with the bulk production of olefins by the petroleum industry. FCC gasoline fractions contain about 30% aromatics. Pyrolysis of crude oils and steam cracking of naphtha fractions between 800–900°C , takes by far the largest amount of organic material that is industrially converted. The residual pitches and soot particles contain PAH and therefore are potential carcinogens. Traces of PAH have been monitored in the fumes emerging from hot asphalt during road paving.[318] Soot particles from all sources including traffic exhaust, remain in the air and are threatening because they will be inhaled. Detrimental effects of inhaling are particularly evident from statistics of tobacco smoking.[319] These statistics are a sound basis for the obligatory warning on tobacco products.[320] Air cleaning programs are effective and quantify the amounts of organics involved.[321] The burning of candles for creating a romantic atmosphere, while at the same time liberating noxious aromatics, is generally not experienced as threatening. A dutch stamp designed to emphasize the uglier side of chemistry,[322] features PAH as air pollutants. On the other hand: do the ubiquitous PAH because of their mutagenic properties, play a role in the evolutionary processes on earth? Is C_{60} an exiting carbon allotrope, superstar molecule, a curiosity or a pollutant?

Awareness of the carcinogenic effects of soot dates back more than 200 years ago, when increase in scrotal cancer was observed among London chimney sweepers.[5] At the same time chimney soot was used as disinfectant for garden mould and for treating gangrenous ulcers. The disinfection property of soot is also evident from the traditional conservation of meats and fish by drying it on smouldering wood. Thermal appetizing of food products by pyrolytic treatment is often the finishing touch,[323] just like roasting of coffee.[324] Coal tar formulations are still popular as a medicine for treatment of skin diseases.[325] An application of coal tar in shampoos for prohibiting dendruf gave ten fold concentration increase in 1-hydroxypyrene (**95**) excretion in urine of the test persons.[326] The successful use of coal tar in medicine at first sight is a contradiction because of the capacity of carcinogenesis.

In 1947, 100 mg (!) of pure chrysene (**10**), was isolated from about 4 kg of swiss garden mould, by Soxhlet extraction.[327] Nobody at that time would probably be concerned with this observation but what impact would this have now. It was common use to deposit ash and soot from wood stoves in the garden both to fertilize and to desinfect. But why chrysene (**10**) as the dominant compound and no others? Chrysene is a major soot component and crystallizes very well from organic solvents. It is probably more persistent than other PAH that are more sensitive to oxygen, light or bacteria. More harmful residues from the past, left with waste deposits, usually in unlined pits, on the grounds of coal gasification plants, that were widely operated until world war II, are now cleaned up.[27] What to say about the PAH in the earth geological reserves of coal, oil and tar sands? Another remark can be made on carbon black that has very high PAH content[281] and makes up 30% of a rubber tire. What is the fate of the PAH molecules while the tires wear out during abrasion?

PAH have probably got so much attention in cancer research because the first carcinogenics were found in this class. This has analogy with the situation with steroids that were more explored than any other pharmaceutical class. The intrinsic stability makes PAH attractive for analytical research. There are numerous studies about the exposure of professionals like cooks, road and foundry workers to PAH of combustion origin. It is not generally known that combinations of PAH, for example dioxine and benzo[a]pyrene (**13**) give very different responses in physiological tests than the pure substances.[5] A new development is the use of PAH-carboxaldehydes as carrier group in anticancer drugs.[328]

Here we come to question the rational for budgetting the great effort of analytical tracing of PAH. Partial chemical problems are easily defined and, as a result, much detail has been reported on PAH contamination. Does this solve or create an environmental problem? It is unfair to simply blame the chemical and petroleum industry for causing environmental pollution. The problems, connected with our increasing use of energy, are forced upon us by an economic system, that only tolerates minor adaptations.

10.   CONCLUSIONS

The generic names smoke, soot, tar, pitch, asphalt, char, charcoal, coal, coke, graphite, diamond, all describe products resulting from the thermal forcing of organic material. PAH, or their smaller precursor fragments, are the nominators for understanding the chemistry that constitute all these phases. Once coal tar and PAH became on disposal, their potential uses became the incentive for further research that initiated rapid economic development. Firstly the growing steel industry required large scale coal pyrolysis for cokes supply, which became the source of the coal tar aromatic chemicals. The potential of these PAH chemicals, initially as precursors for synthetic dyes, stimulated the scientific endeavour far beyond the direct economic needs for production of the PAH with commercial value. It made organic chemistry a discipline in itself.

This review hopefully serves the interrelation of the many diverse topics in modern PAH chemistry, including the abundant combustion of fuels with its environmental impact. The discovery of the fullerenes is a driving force for visionary schemes with many applications of carbon. This will certainly radiate to other PAH systems. FVT can help to prepare PAH and to characterize their intermediates as shown for cyclopenta-PAH. Our own research is in essence the study of product mixtures obtained by FVT of individual PAH or of PAH precursors, that mimics flame pyro-synthesis and helps to describe the mechanisms of PAH build up processes in combustion. Recent FVT preparations of PAH do compare favourably with liquid phase synthesis. Professor Hurd, sixty years after he published his monograph,[7] sent one of us[13] the following comment on cigarette smoking: When we were heating a variety of compounds at 800–850°C, (*J. Am. Chem. Soc.*, 84, 4509 (1962)), which is in the temperature range of a burning cigarette, any application to smoking never entered our minds. Although the burning tip of a cigarette represents oxidation, the adjacent inner hot zone must involve pyrolysis. At this temperature we found that toluene and a wide array of other compounds gave rise to naphthalenes and tarry residues of the carcinogenic variety, whereas at 700°C these products were not formed. From this I have extrapolated the conclusion that burning cigarettes made from anything, corn, silk or shredded paper even, would produce these harmful tars. In other words 'less tar' in cigarette advertisements has little to do with the cancer problem. Instead, the 800°C temperature is the major cause of the trouble. This quotation illustrates the purpose of this PAH review: on coal tar, fullerenes, formation in pyrolysis, the environmental issues, but in particular on the prominent position of PAH in organic chemistry, in a nutshell.

REFERENCES

1. E. Clar, *Polycyclic Hydrocarbons*, Vols. I and II, Academic Press, London, UK (1964).
2. R.G. Harvey, *Polycyclic Aromatic Hydrocarbons: Chemistry and Carcinogenity*, Cambridge University Press, Cambridge, UK (1992).
3. The International Union of Pure and Applied Chemistry (IUPAC) sponsors PAH symposia, merely on fundamental issues. *8th International Symposium on Novel Aromatic Compounds*, (ISNA-8), Braunschweig, Germany. Plenary lectures published in *Pure and Appl. Chem.* **66**, 209 (1996). *7th International Symposium on Novel Aromatic Compounds*, (ISNA-7), Victoria, B.C., Canada. Plenary lectures published in *Pure and Appl. Chem.* **65**, 1 (1995).
4. H.G. Franck and J.W. Stadelhofer, *Industrielle Aromatenchemie. Rohstoffe, Verfahren, Producte*, Springer Verlag, Berlin, Germany (1987).
5. M.R. Osborne and N.T. Crosby, *Benzopyrenes*, Cambridge University Press, Cambridge, UK (1987).
6. D.W. van Krevelen, *Coal: Typology-Physics-Chemistry-Constitution*, Elsevier, Amsterdam, The Netherlands (1993).
7. C.D. Hurd, *The Pyrolysis of Carbon Compounds*, American Chemical Society Monograph Series, The Chemical Catalog Company, New York, USA (1929).
8. G. Grimmer, *The PAH story of Gernot Grimmer – 30 Years of PAH research at BIU*. Supplement to *Polycyclic Aromatic Compounds*, P. Garrigues and M. Lamotte, eds., Gordon and Breach Science Publishers, OPA, Amsterdam, The Netherlands, **3**, 1 (1993). This article includes a list with 216 of Grimmer's papers.
9. Biannual symposia concerning these topics, are held under the auspices of the International Society for Polycyclic Aromatic Compounds (ISPAC). ISPAC, c/o C.M. White, Pittsburg Energy Technology Center, P.O. Box 10940, Pittsburg, PA 15236, USA. *15th International Symposium on Polyaromatic Compounds*. September 19–22, Belgirate, Italy, Abstracts. Proceedings to be published in *Polycyclic Aromatic Compounds*, W. Karcher, ed., Gordon and Breach Science Publishers, OPA, Amsterdam, The Netherlands.
10. W. Karcher, J.J. Belliardo and J. Jacob, in *Polycyclic aromatic compounds*, P. Garigues and M. Lamotte, eds., Gordon and Breach Science Publishers, OPA, Amsterdam, The Netherlands, Supplement to Vol. 3, 43 (1993).
11. G. Grimmer and J. Misfeld, in *Environmental Carcinogens, PAH*, G. Grimmer, ed., CRC, Boca Raton FL., USA, 1 (1983).
12. R.G. Harvey and N.E. Geacintov, *Acc. Chem. Res.* **21**, 66 (1988).
13. U.E. Wiersum, *Jansen Chimica Acta* **10**(3), 3 (1992).
14. D. Koruga, S. Hameroff, J. Withers, R. Loufty and M. Sundareshan, *Fullerene C*$_{60}$, Elsevier, Amsterdam, The Netherlands (1993).
15. The terms 'Pyrolysis' and 'Thermolysis', used throughout this chapter, both signify transformation caused by heat. See ref.[7,16] FVP and FVT are essentially similar operations.
16. R.F.C. Brown, *Pyrolytic Methods in Organic Chemistry*, Organic Chemistry Monographs, Vol. 41, Academic Press, New York, USA (1980).
17. U.E. Wiersum, *Rec. Trav. Chim. Pays Bas* **101**, 317 and 365 (1982).
18. P.P. Fu and R.G. Harvey, *Chem. Rev.* **78**, 317 (1978).
19. S.A. Wise, B.A. Bruce, G.D. Byrd, S.N. Chesler, R.E. Rebbert and M.M. Schantz, *Anal. Chem.* **60**, 887 (1988).
20. J. Jacob, W. Karcher, J.J. Belliardo, R. Dumler and A. Boenke, *Fresenius J. Anal. Chem.* **340**, 755 (1991).
21. Y. Sakamoto and Y. Sakamoto, *Bull. Chem. Soc. Jpn.* **62**, 330 (1989).
22. Many PAH and CP-PAH systems, known and unknown, can be searched for in the Ring Index of Chemical Abstracts, where the nomenclature and numbering conventions can be found.
23. S.A. Wise, B.A. Benner, H. Liu and G.D. Byrd, *Anal. Chem.* **60**, 630 (1988).
24. S.A. Wise, L.C. Sander and W.E. May, *J. Chromatogr.* **642**, 329 (1993).
25. D.J. Freeman and F.C.R. Cattell, *Environ. Sci. Technol.* **24**, 1581 (1990).
26. J.E. Baker and S.J. Eisenreich, *Environ. Sci. Technol.* **24**, 342 (1990).
27. X. Yu, W. Xiaoping, R. Bartha and J.D. Rosen, *Environ. Sci. Technol.* **24**, 1732 (1990).
28. G. Grimmer, J. Jacob, K.W. Naujac and G. Dettbarn, *Anal. Chem.* **55**, 892 (1983).
29. P.H. Taylor, B. Dellinger and C.C. Lee, *Environ. Sci. Technol.* **24**, 316 (1990).

30. R.A. Hites, *Acc. Chem. Res.* **23**, 194 (1990).
31. P. Cikryt, *Nachr. Chem. Tech. Lab.* **39**, 648 (1991).
32. U.L. Nilson and C.E. Östman, *Environ. Sci. Technol.* **27**, 1826 (1993).
33. M.L. Lee, M. Novotny and K.D. Bartle, *Anal. Chem.* **47**, 405 (1976).
34. H. Wortham, G. Mouvier and P. Masclet, *Pollut. Atmos.* **135**, 28 (1993).
35. L.T. Scott and Roelofs, *J. Am. Chem. Soc.* **109**, 5461 (1987).
36. W. Jaworek and F. Vögtle, *Chem. Ber.* **124**, 347 (1991).
37. R.H. Mitchell, M. Chaudhary, R. Vaughan Williams, R. Fyles, J. Gibson, M.J. Ashwood-Smith and A.J. Fry, *Can. J. Chem.* **70**, 1015 (1992).
38. S. Nesnow, J. Lasley, S. Curti, J. Ross, G. Nelson, R. Sangaiah and A. Gold, in *Polycyclic aromatic compounds*, P. Garrigues and M. Lamotte, eds., Gordon and Breach Science Publishers, OPA, Amsterdam, Supplement to Vol. 3, 817 (1993).
39. M. Sarobe, J.W. Zwikker, J.D. Snoeijer, U.E. Wiersum and L.W. Jenneskens, *J. Chem. Soc., Chem. Commun.* 89 (1994).
40. L.W. Jenneskens, M. Sarobe and J.W. Zwikker, *Pure and Appl. Chem.* **68**, 2219 (1996).
41. U.E. Wiersum, *Polycyclic Aromatic Compounds* **11**, 291 (1996). See ref.[9]
42. M. Sarobe, J.D. Snoeijer, L.W. Jenneskens, J.W. Zwikker and J. Wesseling, *Tetrahedron Lett.* **36**, 9565 (1995).
43. R.H.G. Neilen and U.E. Wiersum, *J. Chem. Soc., Chem. Commun.* 149 (1996).
44. M. Sarobe, J.D. Snoeijer, L.W. Jenneskens, M.Q. Slagt and J.W. Zwikker, *Tetrahedron Lett.* **36**, 8489 (1995).
45. M. Sarobe, S. Flink, L.W. Jenneskens, B.L.A. van Poecke and J.W. Zwikker, *J. Chem. Soc., Chem. Commun.* 2415 (1995).
46. L.T. Scott and A. Necula, *J. Org. Chem.* **61**, 386 (1996).
47. A. Sygula, A.H. Abdourazak and P.W. Rabideau, *J. Am. Chem. Soc.* **118**, 339 (1996).
48. E. Clar, J.M. Robertson, R. Schlögl and W. Schmidt, *J. Am. Chem. Soc.* **103**, 1320 (1981).
49. K.F. Lang and M. Zander, *Chem. Ber.* **98**, 579 (1965).
50. K.F. Lang and I. Eigen, *Fortschr. Chem. Forsch.* **8**, 91 (1967).
51. M. Zander and W. Friederichsen, *Chemiker-Zeitung* **115**, 360 (1991).
52. R.G. Harvey, J. Pataki, C. Cortez, P. Di Raddo and C. Yang, *J. Org. Chem.* **56**, 1210 (1991).
53. U.E. Wiersum, *Aldrichimica Acta* **17**(2), 31 (1984).
54. G. Schaden, *J. Anal. Appl. Pyrol.* **7**, 135 (1985).
55. J.C. Fetzer, *Org. Prep. Proced. Int.* **21**, 47 (1989).
56. R.F.C. Brown, *Pure and Appl. Chem.* **62**, 1981 (1990).
57. L.F. Fieser, *Org. Reactions* **1**, 129 (1942).
58. J.C. Jutz, *Top. Curr. Chem.* **73**, 125 (1978).
59. C. Hansch, *Chem. Rev.* **53**, 353 (1953).
60. G.M. Badger and R.W.L. Kimber, *J. Chem. Soc.* 2746 (1960).
61. C. Wentrup, *Adv. Heterocycl. Chem.* **28**, 231 (1981).
62. J.I.G. Cadogen, C.L. Hickson and H. McNab, *Tetrahedron* **42**, 2135 (1986).
63. U.E. Wiersum, *Aldrichimica Acta* **14**(3), 53 (1981).
64. M.C. Lasne and J.L. Ripoll, *Synthesis* **121** (1985).
65. J.L. Ripoll and Y. Vallée, *Synthesis* 659 (1993).
66. V. Boekelheide, *Top. Curr. Chem.* **113**, 87 (1983).
67. S. Seel and F. Vögtle, *Angew. Chem. Int. Ed. Engl.* **31**, 528 (1992).
68. R. Gleiter and Kratz, *Acc. Chem. Res.* **26**, 311 (1993).
69. F. Toda and P. Garratt, *Chem. Rev.* **92**, 1685 (1992).
70. T. Yamato, A. Miyazawa and M. Tashiro, *Chem. Ber.* **126**, 2505 (1993).
71. H. Hopf and G. Maas, *Angew. Chem. Int. Ed. Engl.* **31**, 931 (1992).
72. V.E. Platonov and G.G. Yakobson, *Chem. Rev.* **5**, 297 (1984).
73. M. Look, in *Selections from the Aldrichimica Acta*, 1968–1982, Aldrich Chemical Company, Milwaukee, Wisconsin, USA, 40 (1984).
74. M. Ballester, *Adv. Phys. Org. Chem.* **25**, 267 (1989).
75. M.C. Campbell, R.E. Humphries and N.M. Munn, *J. Org. Chem.* **57**, 641 (1992).
76. A. Maquestiau, A. Tommassetti, C. Pedregal-Freire, J. Elguero, R. Flamang, U.E. Wiersum, H. Bender and C. Wentrup, *J. Org. Chem.* **51**, 306 (1986).

77. S. Iwatsuki, *Adv. Polym. Sci.* **58**, 93 (1984).

78. E. Schmohel, F. Ott., J. Breutenbach, M. Nieger and F. Vögtle, *Chem. Ber.* **126**, 2477 (1993).

79. J.L. Charlton and M.M. Alauddin, *Tetrahedron* **47**, 2873 (1987).

80. F. Vögtle and L. Rossa, *Angew. Chem. Int. Ed. Engl.* **18**, 514 (1979).

81. A.P.A. Crew, G. Jenkins, R. Storr and M. Yelland, *Tetrahedron Lett.* 1491 (1990).

82. R.F.C. Brown, *Rec. Trav. Chim. Pays Bas* **107**, 655 (1988).

83. R.F.C. Brown and F.W. Eastwood, *Synlett* 9 (1993).

84. (1-Chloroethenyl)-PAH (masked ethynyl-PAH) are superior for subliming into the quartz tube, compared to ethynyl-PAH, as the latter may suffer melt condensation before volatilisation. Ethynyl-PAH intermediates can also arise from C_2 insertion, (Part 7).

85. R.F.C. Brown, N. Browne, K. Coulston, F.W. Eastwood, M.J. Irvine, D.E. Pullin and U.E. Wiersum, *Aust. J. Chem.* **42**, 1321 (1989).

86. G. Mayer, J. Schrot, H.P. Reisenauer and R. Janoschek, *Chem. Ber.* **124**, 2617 (1991).

87. R. Schulz and A. Schweig, in *Structure and reactivity*, J.L. Liebman and Greenberg A., eds., VCH Publishers, Inc., Weinheim, Germany, 289 (1988).

88. T. Mosandi, O.C. Kappe, R. Flamang and C. Wentrup, *J. Chem. Soc., Chem. Commun.* 1571 (1992).

89. L.W. Jenneskens, U.E. Wiersum and J.L. Ripoll, *Tetrahedron Lett.* **29**, 6489 (1988).

90. B.B. Broere, P.P.J. Mulder, J. Cornelisse and J. Lugtenburg, *Rec. Trav. Chim., Pays Bas* **109**, 463 (1990).

91. P.P.J. Mulder, B.B. Boere, J. Cornelisse and J. Lugtenburg, *Rec. Trav. Chim., Pays Bas* **112**, 255 (1993).

92. K.K. Wang, *Chem. Rev.* **96**, 207 (1996).

93. L. T. Scott, M.M. Hashemi, T.H. Schulz and M.B. Wallace, *J. Am. Chem. Soc.* **113**, 9692 (1991).

94. M. Sarobe, L.W. Jenneskens and U.E. Wiersum, *Tetrahedron Lett.* **37**, 1121 (1996).

95. R.F.C. Brown, N. Choi, K. Coulston, F.W. Eastwood, U.E. Wiersum and L.W. Jenneskens, *Tetrahedron Lett.* **35**, 4405 (1994).

96. R.F.C. Brown, N. Choi and F.W. Eastwood, *Aust. J. Chem.* **48**, 185 (1995).

97. R.D. VanDell and L.A. Shadoff, *Chemosphere* **13**, 1177 (1984).

98. S.E. Stein and R.L. Brown, *J. Am. Chem. Soc.* **113**, 787 (1991).

99. W.E. Barth and R.G. Lawton, *J. Am. Chem. Soc.* **93**, 1730 (1971).

100. Z. Marcinov, F.R. Fronczek, L. Yen-Hsiang and P.W. Rabideau, *J. Org. Chem.* **60**, 7015 (1995).

101. P.W. Rabideau, A.H. Abdourazak, H.E. Folsom, Z. Marcinow, A. Sygula and R. Sygula, *J. Am. Chem. Soc.* **116**, 7891 (1994).

102. R. Faust, *Angew. Chem., Int. Ed. Engl.* **34**, 1429 (1995).

103. A.H. Abdourazak, Z. Marcinow, A. Sygula, R. Sygula and P.W. Rabideau, *J. Am. Chem. Soc.* **117**, 6410 (1995).

104. A. Corma and B.M. Wojciechowski, *Catal. Rev.* **27**, 29 (1985).

105. N.Y. Chen and T.F. Degnan, *Chem. Eng. Progr.* **84**, 32 (1988).

106. H. van Bekkum and H.W. Kouwenhoven, *Rec. Trav. Chim. Pays Bas* **108**, 283 (1989).

107. M.A. Davis, *Acc. Chem. Res.* **26**, 111 (1993).

108. P.F. Greenwood, M. G. Strachan, G.D. Willett and M.A. Wilson, *Org. Mass Spectrom.* **25**, 353 (1990).

109. L.W. Jenneskens, A.H.P. Krul, P.A. Kraakman, W.H. de Wolf and F. Bickelhaupt, *J. Org. Chem.* **51**, 2162 (1986).

110. U.E. Wiersum and L.W. Jenneskens, *Rec. Trav. Chim. Pays Bas* **110**, 129 (1991).

111. The latter type of transient cumulenylidene carbenes are interestingly also monitored in the interstellar space. P. Thaddeus, C.A. Gottlieb, R. Mollaaghababa and J.M. Vrtilek, *J. Chem. Soc., Faraday Trans.* **89**, 2125 (1993). See also ref.[118]

112. W. Weltner and R.J. VanZee, *Chem. Rev.* **89**, 1713 (1989).

113. P.K. Bachmann and R. Messier, *Chem. Eng. News* 24 (May 15, 1989).

114. F.C. Celii and J.E. Butler, *Annu. Rev. Phys. Chem.* **42**, 643 (1991).

115. M.A. Wilson, L.S.K. Pang, G.D. Willett, K.J. Fisher and I.G. Dance, *Carbon* **30**, 675 (1992).

116. P.S. Skell, J.J. Havel and M.M. McGlinchey, *Acc. Chem. Res.* **6**, 97 (1973).

117. G.H. Jeong, K. Klabunde, O. Pan, G.C. Paul and P.B. Shevlin, *J. Am. Chem. Soc.* **111**, 8784 (1989).

118. H.W. Kroto, A.W. Allaf and S.P. Balm, *Chem. Rev.* **91**, 1213 (1991).

119. T.M. Chang, A. Naim, S.N. Ahmed, G. Goodloe and P.B. Shevlin, *J. Am. Chem. Soc.* **114**, 7603 (1992).

120. G.G. Tibbetts, D.W. Gorkiewicz and R.L. Alig, *Carbon* **31**, 809 (1993).
121. F.D. Kopinke, W. Zychlinski, G. Böttcher and D. Rodewald, *Chem. Techn.* **42**, 325 (1990). F.D. Kopinke, G. Zimmermann and S. Nowak, *Carbon* **26**, 117 (1988).
122. P.F. Britt and A.C. Buchanan, III, *J. Org. Chem.* **56**, 6132 (1991) and references cited therein.
123. A.G. Oblad, H.G. Davis and R. Tracy Eddinger, eds., *Thermal Hydrocarbon Chemistry*, Advances in chemistry series 183, American Chemical Society, Washington D.C., USA (1979).
124. L.F. Albright, B.L. Crynes and W.H. Corcoran, eds., *Pyrolysis. Theory and industrial practice*, Academic Press, New York, USA (1983).
125. J.A. Fairburn, L.A. Behie and Svrcek, *Fuel* **69**, 1537 (1990).
126. F.P. Larkins and A.Z. Khan, *Aust. J. Chem.* **42**, 1655 (1989).
127. H.D. Gesser and L.A. Morton, *Catalysis Lett.* **11**, 357 (1991).
128. E.A. Blekan, R. Myrstad, O. Olsvik and O.A. Rokstad, *Carbon* **30**, 665 (1992).
129. Y. Amenomiya, V.I. Birss, M. Goledzinowski, J. Galuszka and A.R. Sanger, *Catal. Rev. – Sci. Eng.* **32**, 163 (1990).
130. G. Hutchings, J.S. Hargreaves, R.W. Joyner and C.J. Kiely, *Chem. Tech.* (November 25, 1994).
131. W. Kaminsky, *Macromol. Chem. Symp.* **48/49**, 381 (1991), 325 (1990).
132. R.A. Hawley-Fedder, M.L. Parsons and F.W. Karasek, *J. Chromatogr.* **315**, 201 (1984).
133. M.D. Banciu, M.D. Stanescu, C. Florea, A. Petride, C. Draghici and E. Cioranescu, *Bull. Soc. Chim. Fr.* **128**, 919 (1991).
134. R. Gleiter and U. Steurle, *Chem. Ber.* **122**, 2193 (1989).
135. Under suitable conditions PAH may reverse to polycyclic alkanes as well. On the walls of piping associated with the effluent cooling system of a petroleum hydrocracking plant, over 100 kg of almost pure perhydrocoronene precipitated. J.C. Fetzer, W.R. Biggs and M. Zander, *Z. Naturforsch.* **46a**, 291 (1991).
136. W.R. Roth, O. Adamczak, R. Breuckmann, H.W. Lennartz and R. Boese, *Chem. Ber.* **124**, 2499 (1991).
137. O.S.L. Bruinsma, P.J.J. Tromp, H.J.J. de Sauvage Nolting and J.A. Moulijn, *Fuel* **67**, 334 (1988).
138. D.C. DeJongh and D.A. Brent, *J. Org. Chem.* **35**, 4204 (1970).
139. B.M. Scholz-Böttcher, M. Bahadir and H. Hopf, *Angew. Chem. Int. Ed. Engl.* **31**, 443 (1992).
140. M.M. Suryan, S.A. Kafafi and S.E. Stein, *J. Am. Chem. Soc.* **111**, 1423 (1989).
141. J.D. Mold, R.E. Means and A.G. Kallianos, *Tobacco Sci.* **4**, 130 (1960).
142. R.H. Shen, S.A. Kafafi and S.E. Stein, *J. Am. Chem. Soc.* **111**, 1418 (1989).
143. M.L. Poutsma and C.W. Dyer, *J. Org. Chem.* **47**, 3367 (1982).
144. T. Yokono, T. Obara, Y. Sanada and K. Miyazawa, *Carbon* **22**, 169 (1984).
145. R.H. Schlosberg and A. Kurs, *J. Org. Chem.* **49**, 3032 (1985).
146. H-G. Schäfer, *Chemiker-Zeitung* **115**, 310 (1991).
147. L.M. Stock, *Acc. Chem. Res.* **22**, 427 (1989).
148. U.E. Wiersum, *Euchem conference on organic free radicals*, September 22–26, Assisi-Perugia, Italy, symposium lecture (1986).
149. Y. Uemichi, Y. Kashiwaya, A. Ayame and H. Kanoh, *Chem. Lett.* 41 (1984).
150. J. Zawadzki, *Infrared spectroscopy in surface chemistry of carbons, Chemistry and physics of carbon*, Vol. 21, P.A. Thrower, ed., Marcell Dekker, New York, USA (1990).
151. L.B. Ebert, J.C. Scanlon, A.R. Garcia, C.F. Pictroski and L.A. Gebhard, *Carbon* **28**, 912 (1990).
152. A. Alexiou and A. Williams, *Fuel* **73**, 1280 (1994).
153. E.K. Fields and S. Meyerson, *Chemtech.* 104 and 241 (1974).
154. I.S. Kozin, J.W. Hofstraat and U.E. Wiersum, unpublished results.
155. H.S. Gill, P.L. Kole, J.C. Wiley, K.M. Li and S. Higginbotham, *Carcinogenesis* **15**, 2455 (1994).
156. J.C. Fetzer and W.R. Biggs, *J. Chromatogr.* **642**, 319 (1993).
157. J.C. Fetzer and W.R. Biggs, *Polycyclic Arom. Compds.* **4**, 3 (1994).
158. R. Huisgen, *Angew. Chem. Int. Ed. Engl.* **25**, 297 (1986).
159. W.J. Wiswesser, *Aldrichimica Acta* **22**, 17 (1989).
160. Y. Tobe, *Top. Curr. Chem.* **172**, 1 (1994).
161. Z. Zhou and R.G. Parr, *J. Am. Chem. Soc.* **111**, 7371 (1989).
162. J. Ahira, *Bull. Chem. Soc. Jpn.* **66**, 57 (1993).
163. K. Hiruta, S. Tokita and K. Nishimoto, *J. Chem. Soc., Perkin Trans.* **2**, 1443 (1995).
164. H.W. Kroto, *Angew. Chem. Int. Ed. Engl.* **31**, 111 (1992).

165. H. Wynberg, W.C. Nieuwpoort and H.T. Jonkman, *Tetrahedron Lett.* 4623 (1973).
166. H.W. Kroto, J.P. Hare, A. Sarkar, K. Hsu, M. Terrones and J.R. Abeysinghe, *MRS Bulletin* **24** (11), 51 (1994).
167. L. T. Scott, M.M. Hashemi and M.S. Bratcher, *J. Am. Chem. Soc.* **114**, 1920 (1992).
168. A. Borchard, A. Fuchicello, K.V. Kilway, K.K. Baldridge and J.S. Siegel, *J. Am. Chem. Soc.* **114**, 1921 (1992).
169. K. Yamamoto, *Pure Appl. Chem.* **65**, 157 (1993).
170. K. Yamamoto, H. Sonobe, H. Matsubara, M. Sato, S. Okamoto and K. Kitaura, *Angew. Chem. Int. Ed. Engl.* **35**, 69 (1996).
171. I.C. Lewis, *Carbon* **20**, 519 (1982).
172. R. Goddard, M.W. Haenel, W.C. Herndon, C. Krüger and M. Zander, *J. Am. Chem. Soc.* **117**, 30 (1995).
173. R. Taylor, G.J. Langley, H.W. Kroto and D.R.M. Walton, *Nature* **366**, 728 (1993).
174. C. Crowley, H.W. Kroto, R. Taylor, D.R.M. Walton, M.S. Bratcher, P.C. Cheng and L.T. Scott, *Tetrahedron Lett.* **36**, 9215 (1995).
175. R. Taylor, H.W. Kroto, D.R.M. Walton and G.J. Langley, *Mol. Mat.* **4**, 7 (1994).
176. D.H.M. Ho and R.A. Pascal Jr., *Chem. Mater.* **5**, 1358 (1993).
177. N.S. Goroff, *Acc. Chem. Res.* **29**, 77 (1996).
178. R. Hoffmann, *New Scientist.* Nov 17, 44 (1990).
179. F. Diederich and Y. Rubin, *Angew. Chem. Int. Ed. Engl.* **31**, 1101 (1992).
180. J.F. van Oss, *Petroleum and organic chemicals, Materials and Technology*, Vol. IV, J.H. Longman, deBussy, Amsterdam, The Netherlands (1972).
181. Ullmann's Encyclopedia of Industrial Chemistry, fourth and fifth edn., VCH Verlag, Weinheim, Germany.
182. Kirk-Othmer, *Encyclopedia of Chemical Technology*, third and fourth edn., Wiley, New York, USA.
183. The coal tar history is a good example to substantiate the present belief that above all things market driven incentives for research generate commercial results. Evenso, curiosity was the major drive for the scientific exploration of coal tar, with an unforeseen but decisive impact on our economy.
184. K.F. Lang, *Angew. Chem.* **63**, 345 (1951).
185. G. Collin and H. Höke, *Tar and Pitch*, in Ullmann's Encyclopedia of Industrial chemistry, fifth edn., Vol. A 26, 91 (1995); Cf. G. Collin and M. Zander, Teer und Pech, Ibid., fourth edn., Bd. **22**, 411 (1982), with a list of individual PAH percentages.
186. A.A. Bright and L.S. Singer, *Carbon* **17**, 59 (1979).
187. M.A. May, M.E. Huston, M.R. Callstrom and A.G. Marshall, *Chem. Mater.* **5**, 648 (1993).
188. M. Zander and G. Collin, *Fuel* **72**, 1281 (1993).
189. N.A. Bacon, W.A. Barton, L.J. Lynch and D.S. Webster, *Carbon* **25**, 669 (1987).
190. K.R. Payne, ed., *Chemicals from coal: new processes*, Critical reports on Applied chemistry, Vol. 14, Wiley, New York, USA, 1 (1987). Cf. K.R. Payne, ed., Ibid., Vol. 9, 1 (1985).
191. The non-distillable tar residues containing large PAH, that are leaving the refineries, have become indispensable as building material for asphalt roads. In a way this can be seen as an unimposed form of waste management.
192. L. Wheatley, Y.A. Levendis and P. Vouros, *Environ. Sci. Technol.* **27**, 2885 (1993).
193. G. Wegner and K. Wagemann, *Adv. Materials* **6**, 629 (1994).
194. W. Kaminsky and H. Rössler, *Chemtech.* **108** (1992).
195. G. Haucke and B. Granefl, *Angew. Chem. Int. Ed. Engl.* **34**, 67 (1995).
196. H. Kaiser, J. Lindner and H. Langhals, *Chem. Ber.* **124**, 529 (1991).
197. C. Graebe and C. Liebermann, *Chem. Ber.* **3**, 742 (1870); C. Graebe and C. Liebermann, *Chem. Ber.* **3**, 636 (1870); Cf. Beilstein, *Handbuch der organischen Chemie*, Springer Verlag, Berlin, Germany, Hauptwerk, Bd VIII, 437 (1925).
198. C. Graebe and J. Walter, *Chem. Ber.* **14**, 175 (1881).
199. Beilstein. *Handbuch der organischen Chemie*, Springer Verlag, Berlin, Germany, Hauptwerk, Bd V (1922), with historical notes and references of individual coal tar constituents.
200. T. Hirn, *Chem. Ber.* **32**, 3341 (1899).
201. R. Meyer, *Chem. Ber.* **45**, 1609 (1912).
202. R. Meyer and K. Taeger, *Chem. Ber.* 1261 (1920).

203. G.M. Badger, *Progr. Phys. Org. Chem.* **3**, 1 (1965).
204. R.L. Jenkins, R. McCullough and C.F. Booth, *Ind. Eng. Chem.* **22**, 31 (1930).
205. A. Spilker, *Chem. Ber.* **26**, 1538 (1893).
206. R.F.C. Brown, F.W. Eastwood, K.J. Harrington and G.L. McMullen, *Aust. J. Chem.* **27**, 2393 (1974).
207. O. Kruber, *Chem. Ber.* **74**, 1688 (1941).
208. E. Clar and M. Zander, *J. Chem. Soc.* 4616 (1957).
209. A.G. Holba and E.J. Eisenbraun, *Org. Prep. Proced. Int.* **13**, 149 (1981).
210. O. Kruber, A. Raeithel and G. Grigoleit, *Erdöl und Kohle* **8**, 637 (1955).
211. O. Kruber and L. Rappen, *Chem. Ber.* **70**, 1556 (1937).
212. J.W. Cook, C.L. Hewitt and L. Hieger, *J. Chem. Soc.* 395 (1933).
213. L.F. Fieser and E.M. Dietz, *Chem. Ber.* **62**, 1827 (1929).
214. M. Müller, H. Mauermann-Düll, M. Wagner, V. Enkelmann and K. Müllen, *Angew. Chem. Int. Ed. Engl.* **34**, 1583 (1995).
215. F.B. Mallory and C.W. Mallory, *Org. Reactions* **30**, 1 (1984).
216. H-D. Becker, *Chem. Rev.* **93**, 145 (1993).
217. K.F. Lang, H. Buffeb and J. Kalowy, *Chem. Ber.* **93**, 303 (1960).
218. K.F. Lang and H. Buffeb, *Chem. Ber.* **94**, 1075 (1961).
219. J.A. Dodge, J.D. Bain and A.R. Chamberlain, *J. Org. Chem.* **55**, 4190 (1990) and references cited.
220. M. Zander, *Fuel* **65**, 1019 (1986).
221. G. Schaden, *J. Org. Chem.* **48**, 5358 (1983).
222. E.V. Dehmlov and M. Fründ, *Z. Naturforsch.* **47b**, 1647 (1992).
223. B. Freiermuth, S. Gerber, A. Riesen, J. Wirz and M. Zehnder, *J. Am. Chem. Soc.* **112**, 738 (1990).
224. M. Zander, *Erdöl und Kohle* **45**, 435 (1992).
225. T.A. Upshaw, J.K. Stille and J.P. Droske, *Macromolecules* **24**, 2143 (1991).
226. M. Zander, *Chem. Ber.* **92**, 2744 (1959).
227. L. Friedman and D.F. Lindow, *J. Am. Chem. Soc.* **90**, 2324 (1968).
228. T.C.W. Mak and H.C. Wong, *Top. Curr. Chem.* **140**, 141 (1987).
229. A. Sutter, P. Schmutz and C.S. Marvel, *J. Polym. Sci. Polym. Chem. Ed.* **20**, 609 (1982).
230. U.E. Wiersum and L.W. Jenneskens, *Tetrahedron Lett.* **34**, 6615 (1993).
231. R.F.C. Brown, K.J. Coulston and F.W. Eastwood, *Aust. J. Chem.* **47**, 411 (1994).
232. Y. Yamada and S. Toyoda, *Bull. Chem. Soc. Jap.* **46**, 3571 (1973).
233. K.F. Lang and M. Zander, *Chem. Ber.* **94**, 1871 (1961).
234. T. Imamura and M. Nakamizo, *Carbon* **17**, 507 (1979).
235. S. Krishnan and R.A. Hites, *Chemosphere* **9**, 679 (1980).
236. A.L. Lafleur, J.P. Longwell, L. Shirnamé-Moré, P.A. Monchamp, W.A. Peters and E.F. Plummer, *Energy Fuels* **4**, 307 (1990).
237. M. Zander and W.A. Franke, *Chem. Ber.* **101**, 212 (1968).
238. M. Zander, *Top. Curr. Chem.* **153**, 101 M. (1990).
239. H. Hosoya, *Top. Curr. Chem.* **153**, 255 (1990).
240. R.D. Broene and F. Diederich, *Tetrahedron Lett.* **32**, 5227 (1991).
241. V. Sachweh and H. Langhals, *Chem. Ber.* **123**, 1981 (1990).
242. C. Dufraisse, R. Buret and R. Girard, *Bull. Soc. Chim. Fr.* **53**, 782 (1932).
243. G. Wittig and D. Waldi, *J. Prakt. Chem.* **160**, 242 (1942).
244. H. Langhals, *Nachr. Chem. Tech.* **28**, 716 (1980).
245. U. Mazzucato and F. Momicchioli, *Chem. Rev.* **91**, 1679 (1991).
246. D.A. Backhus and P.M. Gschwend, *Environ. Sci. Technol.* **24**, 1214 (1990).
247. D. Barton and W.D. Ollis, *Comprehensive Organic Chemistry*, Pergamon Press, Oxford, UK, Vol. 1, 66 and Vol. 6, 974 (1979).
248. E.K. Fields and S. Meyerson, *J. Org. Chem.* **33**, 2315 (1968).
249. J.A. Miller, R.J. Kee and Westbrook, *Annu. Rev. Phys. Chem.* **41**, 345 (1990).
250. J.V. Michael and K.P. Lim, *Annu. Rev. Phys. Chem.* **44**, 429 (1990).
251. D.J. Hucknall, *Selective oxidation of hydrocarbons*, Academic Press, London, UK (1974).
252. C. Wentrup, *Reactive molecules*, John Wiley, New York, USA (1984).
253. J.B. Howard and J.P. Longwell, in *Polynuclear Aromatic Compounds, formation, metabolism and measurement*, M. Cooke and A.J. Dennis, eds., Batelle Press, Columbus, Ohio, USA,27 (1983).
254. H. Takada, T. Onda and N. Ogura, *Environ. Sci. Technol.* **24**, 1179 (1990).

255. D.J. Desilets, P.T. Kissinger, F.E. Lytie, M.A. Horne, M.S. Ludwiczak and R.B. Jacko, *Environ. Sci. Technol.* **18**, 386 (1984).

256. B.A. Benner, Jr., N.P. Bryner, S.A. Wise, G.W. Mulholland, R.C. Lao and M.F. Fingas, *Environ. Sci. Technol.* **24**, 1418 (1990).

257. W.A. Hartgers, J.S. Sinninghe Damste, J.W. de Leeuw, Y. Ling and G.N. Dyrkacz, *Energy Fuels* **8**, 1055 (1994).

258. J.S. Sinninghe Damste, J. Köster, M. Baas, M.P. Koopmans, H.M.E. van Kaam-Peters, J.A.J. Geenevasen and C. Kruk, *J. Chem. Soc. Chem. Commun.* 187 (1995).

259. M. Alexander, *Science* **211**, 132 (1981).

260. D. Bryniok, *Bioscope* **2**, 42 (1994).

261. M. Blumer, *Science* **134**, 474 (1962).

262. L.J. Allamandola, *Top. Curr. Chem.* **153**, 1 (1990).

263. F. Salama and L.J. Alamandola, *J. Chem. Soc. Faraday Trans.* **89**, 2177 (1993).

264. E. Herbst, *Annu. Rev. Phys. Chem.* **46**, 27 (1995).

265. S. Schlemmer, D.J. Cook, J.A. Harrison, B. Wurfel, W. Chapman and R.J. Saykally, *Science* **265**, 1686 (1994).

266. R. Sharifi, S.V. Pisupati and S. Scaroni, in Kirk-Othmer, *Encyclopedia of Chemical Technology*, fourth edn., Wiley, New York, USA, Vol. 6., 1049 (1993).

267. Evenso, the global demand for energy by a growing world population will continue to increase until fossil fuels run out. From this point of view PAH formation and control of carbon dioxide emissions, is only a derived problem from a potential catastrophe. Whatever the situation exactly is, the chemistry of PAH build up in combustion remains an intriguing subject.

268. J.M. Thomas, *J. Chem. Soc. Faraday Trans.* **87**, 2865 (1991).

269. H.W. Kroto, *J. Chem. Soc. Faraday Trans.* **87**, 2870 (1991).

270. R.N. Westerholm, T.E. Alsberg, Å.B. Frommelin and M.E. Strandell, *Environ. Sci. Technol.* **22**, 925 (1988).

271. A.L. Lafleur, J.B. Howard, J.A. Marr and T. Yadav, *J. Phys. Chem.* **97**, 13539 (1993).

272. M. Black, J.I.G. Cadogan and H. McNab, *J. Chem. Soc. Chem Commun.* 395 (1990).

273. T.B. Brill and K.J. James, *Chem. Rev.* **93**, 2667 (1993).

274. U.E. Wiersum, unpublished results (1995).

275. F. Jongeneelen, *Int. Arch. Occup. Environ. Health* **63**, 511 (1992).

276. A.F. Sarofim and J. VanderSande, *New Scientist* 25, (October 22, 1994).

277. H.S. Homan and W.K. Robbins, *Combustion and Flame* **63**, 177 (1986).

278. J.A. Barnard and J.N. Bradley, *Flame and Combustion*, Chapman and Hall, New York, USA (1985).

279. T.J. Mast, D.P.H. Hsieh and J.N. Seiber, *Environ. Sci. Technol.* **18**, 338 (1984).

280. J. Lahaye and G. Prado, in *Chemistry and Physics of Carbon: A series of advances* **14**, 167 (1977).

281. J. Gallagher, U. Heinrich, M. George, L. Hendee, D.H. Phillips and J. Lewtas, *Carcinogenesis* **15**, 1291 (1994).

282. E.K. Fields and S. Meyerson, *Tetrahedron Lett.* 571 (1967).

283. C.D. Hurd, A.R. Macon, J.I. Simon and R.V. Levetan, *J. Am. Chem. Soc.* **84**, 4509 (1962).

284. G.J. Hutchings, M.S. Scurrell and J.R. Woodhouse, *Chem. Soc. Rev.* **18**, 251 (1989).

285. J. Kozinski, *Combust. Flame* **96**, 249 (1994).

286. J. Ahira, *Sci. Am.* 44 (March, 1992).

287. For the C_6H_6 ensemble, 217 possible topological isomers were indicated. H. Bock, *Angew. Chem. Int. Ed. Engl.* **28**, 1627 (1989).

288. O.L. Chapman, J.W. Johnson, R.J. McMahon and P.R. West, *J. Am. Chem. Soc.* **110**, 501 (1988).

289. J. Osterodt, A. Zett and F. Vögtle, *Tetrahedron* **52**, 4949 (1996).

290. T. Baum, S. Löffler, P. Weilmünster and K.H. Homann, *Ber. Bunsenges. Phys. Chem.* **96**, 841 (1992).

291. C.J. Pope, J.A. Marr and J.B. Howard, *J. Phys. Chem.* **97**, 11001 (1993).

292. S.E. Stein, *Acc. Chem. Res.* **24**, 350 (1991).

293. J. Mukherjee, A.F. Sarofim and J.P. Longwell, *Combustion and flame* **93**, 191 (1994).

294. J.B. Howard, *Twenty-fourth Symposium on Combustion*, The Combustion Institute, Pittsburg, 933 (1992).

295. M.T. Beck, Z. Dinya, S. Kéki and L. Papp, *Tetrahedron* **49**, 285 (1993).

296. H. Schwarz, *Angew. Chem. Int. Ed. Engl.* **32**, 1412 (1993).

297. D. Bakowies and W. Thiel, *J. Am. Chem. Soc.* **113**, 3704 (1991).

298. T. Belz, H. Werner, F. Zemlin, U. Klengler, M. Wesemann, B. Tesche, E. Zeitler, A. Reller and R. Scöhlgel, *Angew. Chem. Int. Ed. Engl.* **33**, 1866 (1994).

299. X.K. Wang, X.W. Lin, S.N. Song, V.P. Dravid, J.B. Ketterson and R.P.H. Chang, *Carbon* **33**, 949 (1995).

300. A.D. Darwish, H. Kroto, R. Taylor and D.R.M. Walton, *J. Chem. Soc. Chem. Commun.* 15 (1994).

301. K. Shibata, A.A. Kulkarni, D.M. Ho and R.A. Pascal Jr., *J. Am. Chem. Soc.* **116**, 5983 (1994).

302. G.A. Downing, C.S. Frampton, D.D. MacNico and P.R. Mallinson, *Angew. Chem. Int. Ed. Engl.* **33**, 1587 (1994).

303. L. Becker, J.L. Bada, R.E. Winams, J.E. Hunt, T.E. Bunch and B.M. French, *Science* **265**, 642 (1994).

304. D. Heymann, L.P. Felipe Chibante, R.R. Brooks, W.S. Wolbach and R.E. Smalley, *Science* **265**, 645 (1994).

305. R. Taylor and R.M. Walton, *Nature* **363**, 685 (1993).

306. A.J. Janssen, J.C. Hummelen and F. Wudl, *J. Am. Chem. Soc.* **117**, 544 (1995).

307. A.B. Smith III, ed., *Fullerene Chemistry, Tetrahedron Symposia-in-print* **52**, 4925 (1996).

308. R.L. Johnston and R. Hoffmann, *J. Am. Chem. Soc.* **111**, 810 (1989).

309. S. Hagen, U. Nuechter, M. Nuechter and G. Zimmermann, *Tetrahedron Lett.* **35**, 7013 (1994).

310. J.F. McKay and D.R. Latham, *Anal. Chem.* **45**, 1050 (1973).

311. T.T. Coburn and W.M. Jones, *Tetrahedron Lett.* 3903 (1973).

312. M. Zander and W. Friedrichsen, *Z. Naturforsch.* **47b**, 1314 (1992).

313. J. Ahira, *Bull. Chem. Soc. Jpn.* **65**, 597 (1992).

314. J. Ahira, *Bull. Chem. Soc. Jpn.* **60**, 3581 (1987).

315. W.C. Herndon, P.C. Nowak, D.A. Connor and P. Lin, *J. Am. Chem. Soc.* **114**, 41 (1992).

316. F. Wudl, *Acc. Chem. Res.* **25**, 157 (1992).

317. R.C. Haddon, *Science* **261**, 1545 (1993).

318. H.C.A. Brandt, *Polycyclic Aromat. Compd.* **5**, 115 (1994).

319. T.D. McKenzie, C.E. Bartecchi and R.W. Schrier, *N. Engl. J. Med.* **330**, 907 and 975 (1994).

320. People may ask if working with PAH feels threatening. Since smoking has to be considered much worse than working with (non volatile) PAH, the statistics for smokers can also be seen as reassuring. Careful handling of PAH is important because crystals, e.g. of chrysene (**10**), may accumulate electrostatic charge, giving them a tendency to dust.

321. J.M. Lents and W.J. Kelly, *Sci. Am.* 18 (October 1993).

322. Z. Rappoport, *Acc. Chem. Res.* **25**, 24 (1992).

323. R.J. McGorrin and C.T. Ho, *Thermal generation of aromas*, ACS symposium series 409, T.H. Parliment, ed., American Chemical Society, Washington, USA (1989).

324. T.H. Parliment and H.D. Stahl, *Chemtech*, 38 (August 1995).

325. Martindale, in *The extra pharmacopoeia*, 29th edn., J.E.F. Reynolds, ed., The pharmaceutical Press, London, UK, 933 (1989).

326. F.J. van Schooten, E.J.C. Moonen, E. Rhijnsburger, B. van Agen, H.H.W. Thijsen and J.C.S. Kleinjans, *The Lancet* **344**, 1505 (1994).

327. W. Kern, *Helv. Chim. Acta* **30**, 1595 (1947).

328. K.W. Bair, C. Webster Andrews, R.L. Tuttle, V.C. Knick, M. Cory and D.D. McKee, *J. Med. Chem.* **34**, 1983 (1991).

4. VACUUM GAS–SOLID REACTIONS (VGSR): APPLICATION TO THE SYNTHESIS OF UNSTABLE SPECIES

JEAN-MARC DENIS and ANNIE-CLAUDE GAUMONT

*Synthèse et Electrosynthèse Organiques (UMR CNRS 6510),
Université de Rennes 1, 35042, Rennes France*

INTRODUCTION

Considerable interest has been devoted during the last two decades to the synthesis and the characterization of theoretically significant short-lived species, the multiple bonding involving third and higher row of the main group elements fulfilling the major place. To acceed to these molecules, two different techniques have been mainly used.[1] The first one, by far the most popular, is the unimolecular vacuum gas-phase thermal fragmentation of suitable precursors under dynamic conditions, usually called flash vacuum thermolysis (FVT). Documentation about this technique can be found in a comprehensive treatise[2] and several reviews.[3–7] The other technique, usually called vacuum gas–solid reaction VGSR has known a weaker development, although its contribution to the formation of many reactive intermediates like highly strained molecules, was in numerous cases the only conclusive one. The two procedures, very different from a technical point of view, are complementary and their synthetic potentialities are in numerous cases comparable. A review focused on the application of the VGSR procedure to the synthesis of highly strained alkenes has been recently presented.[8] The preparation of short-lived molecules by thermal decomposition of their precursors on heterogeneous catalysts and their real time detection by photoelectron spectroscopy (PES) have been elsewhere described.[9,10] The purpose of the present review is to describe the VGSR technique and its synthetic potential, mainly related to the synthesis of short-lived species. After a general survey dealing with gas–solid reactions, the first part of this chapter will be devoted to the description of the VGSR technique. Analogies and differences with FVT will then be precised. The second part will be devoted to the application of VGSR to the synthesis of short-lived species and to the determination of their chemical and physical properties.

1. GAS–SOLID REACTIONS, GENERAL SURVEY

Reactions between a gas and a solid play a major role in most chemical processing industries such as extraction of metal from their ores, combustion of solid fuels and pollution control.[11] Many important chemicals are manufactured by processes in which gases react on the surface of solid catalysts[12] or reagents[13] (for a recent application see for example reference 14). Another methodology involves the chemical transformation of a solid by reaction with a gaseous substrate. This process is known for more than one century; the main aspects and developments of this chemistry has been recently reviewed.[15]

In laboratory, gas–solid reactions *under atmospheric pressure* are of importance in several chemical processes. Absence of solvent and simplicity of the procedure constitute the major advantages. Thus, the 2- and 7-bromobicyclo heptenes, which were prepared in solution with poor yield by dehydrobromination of their corresponding precursors, are obtained more efficiently by a direct gas-phase HBr-elimination on a solid hydroxylic base (NaOMe or tBuOK).[16] Other cycloalkenes bearing five or six carbon atoms have also been efficiently synthesized.[17]

This simple technique was also used for the preparation of reactive alkenes unavailable in pure form by classical synthetic approaches in solution. Thus, the trimethylenecyclopropane ([3]-radialene), a poly cross-conjugated alkene was for the first time isolated by HBr-elimination of suitable precursors on a solid base[18]; the dibromide precursor was added dropwise *under atmospheric pressure* through a U-tube filled with KOH and heated at 150°C. The displacement of the gaseous flow was induced by a stream of helium. The product was condensed on a cold trap. It is stable for few days at -78°C but polymerizes upon warming at room temperature. This procedure was then extented to the preparation of the tetramethylenecyclopropane ([4]-radialene); the IR, Raman[19] and photoelectron spectra[20] were reported. Development of this chemistry in VGSR conditions will be described in Section 4.4.1.

Limitations of the gas–solid reactions under atmospheric pressure were however observed. Thus, attempts to generate the highly strained cyclopropene by dehydrohalogenation of the corresponding precursor over solid tBuOK under one atmosphere failed; isomeric 1-propyne was observed as major product[21]. This ring opening probably resulted from a too long transit time of the thermodynamically unstable species inside the reactor. This problem was solved by using dynamic gas–solid reactions under vacuum (see below). Principle of this technique will be presented, analysed, and then compared with the one of FVT. The main applications will then be developed in the last part of this chapter.

2. VACUUM GAS–SOLID REACTIONS

The dynamic gas–solid reactions under vacuum were found to be an efficient method for the preparation of short-lived species. We have called this technique "Vacuum Gas–Solid Reactions" (VGSR).[22] This term will be used in the following, even for the description of earlier experiments.

2.1. First Experiments

Interesting advances have been presented by Schenk *et al.* in 1933.[23,24] Thus to reduce the transit time of the short-lived species on the solid reagent, reactions were carried out under vacuum and the gaseous flow was condensed on a cold trap directly fitted on the reactor. Under these conditions, the formation of sulfur monoxide by dehalogenation of thionyl chloride or bromide on silver or antimony metal at 100–400°C under 1 mm Hg pressure was first claimed. The gas was in fact composed of an equimolecular mixture of SO_2 and S_2O coming from the disproportionation of sulfur monoxide. This transformation proceeds even at low temperature since the life-time of SO is about 2 ms (for a review on the lower oxides of sulfur, see reference 25). The efficiency of this technique was however demonstrated beyond this mistake: the labile disulfur monoxide has been cleanly formed by reaction under vacuum of thionyl chloride on solid silver sulfide[26] and more recently unambiguously characterized in real time by photoelectron spectroscopy[27] (see development and Scheme 4 in Section 4.2.).

2.2. Principle of the VGSR Technique

The formation of reactive or transient species by vacuum gas solid reactions (VGSR) and their direct characterization by low temperature NMR, solid film, matrix IR or by gas-phase spectroscopies has undergone a large development during the last two decades. A description of this technique in relation with the corresponding chemical processes and with experimental conditions is presented in the following.

For catalytic VGSR experiments, a tubular reactor (quartz or pyrex) is packed with catalyst pellets or covered by layers of catalyst or metal. For non-catalytic reactions, the reactor is in most cases a simple tubular pyrex tube since reactions do not usually occur at very high temperature. The tube can be either fully packed with the solid reagent (*crossed-bed reactor*) or only packed in half part of the section (*lecked-bed reactor*) (Figure 1).

Precursors are passed in the gas-phase under high vacuum through the tube containing the catalyst or the solid reagent. The reactions occur as the stream flows down the reactor. Gaseous products emerging at the exit are condensed on a trap cooled at 77 K (liquid nitrogen). The displacement of the gaseous flow is induced by the gradient of pressure; the lower pressure at the outlet being is due to the condensation of the products on the cold trap. Elemental chemical steps of this process involve diffusion of the substrate into the porous solid, chemisorption followed by surface reaction and finally desorption and diffusion of the product out of the solid.

The rate of the reaction is depending on the temperature, the concentration of the solid reagent and the pressure near the solid boundary layer. The

crossed-bed reactor   lecked-bed reactor

Figure 1 Main types of reactors for VGSR experiments.

temperature can be considered as constant during the reaction. The determination of the two other parameters is much more complex. In the *crossed-bed* reactors, the pressure decreases strongly along the tube to become very weak at the exit; the reagent near the entry is first consumed. In the *lecked-bed* reactors (half-filled reactors), the solid reagent localized at or near the surface is first consumed. The concentration of the reagent are changing and the pressure at this level are mainly function of the effective diffusivity of the gas into the solid. These two parameters in the different parts of the section (on going from the surface to the bottom, the pressure and the concentration increase) and along the length of the reactor (lower pressure and higher concentration are expected on going from the inlet to the outlet). Consequently, concentration and pressure are changing with the elapse of time at any part of the solid. These gradients are incompatible with simple mathematical modelings. This explains why for particular applications the processes have mainly evolved according to operating experience rather than as a result of an analytic design.

2.3. Research of the Best Experimental Conditions

Most of the VGSR experiments involve reactions between a gas and a solid reagent with formation of reactive species. In the non-catalytic reactions, the solid reagent is consumed. And a new solid is recovered at the end of the reaction (stoechiometric transformation).

$$A(g) + B(s) \rightarrow C(g) + D(s)$$

Different factors are to be optimized to assure a complete and univocal transformation of the starting material and to reduce the decomposition of the reactive species. The most significant are presented in the following. They are mainly related to the nature of the solid reagent, the transit time and the design of the reactor. A rapid comparison beween VGSR and FVT techniques will then be presented.

2.3.1. Solid reagent

The solid reagent and the new solid resulting from the chemical transformation must be thermally stable at the temperature of the reaction and unreactive toward the products which are formed. Under these conditions, higher is the reactivity of the solid reagent, lower will be the temperature of the reaction and consequently the decomposition of the product.

The parameters are chosen to assure a complete transformation of the substrate along the reactor. The amount of reagent required for a complete transformation will be mainly function of the solid porosity towards the precursor. In the case of highly porous reagents, a small excess is generally sufficient. As an example, only 1.2 equivalents of N-chlorosuccinimide was needed to assure a total transformation of amines into N-chloramines[28] even for molar scale. However, to secure the total transformation, at least a fourfold excess was generally used.[29] Inorganic compounds present a weaker porosity than the organic ones and consequently must be used in a very large excess.

Supported reagents were used[30,31] in the first VGSR experiments. We have then observed that using the solid reagent in pure form gave generally better results. The rate of product formation and the porosity of the solid, both higher in the later experiments, explain mainly this difference.

2.3.2. Transit time

To lower the decomposition of reactive species, short transit time of products into the reactor are required. Best results are expected if the experiments are carried out under high vacuum with a weak gradient in pressure. These conditions are fulfilled with a *lecked-bed* reactor in which the gradient is very weak on going from the bottom of the tube to the surface as well as between the two ends. The experiments fully agree with this assumption. As an example, we have nearly avoided the decomposition of the spiropentene starting from the bromo-spiropentane (Section 4.4., Table 2) by using a *lecked-bed* reactor instead of the *crossed bed* one precedently used in our first experiments.[32] In these conditions, the average transit time measured by coupling the VGSR reactor to a mass spectrometer (real time analysis) is close to $2\,\text{s}$.[21]

Short transit times are expected with weakly porous reagents such as inorganic bases. Their use allowed to prepared compounds which are considered as transient species. One of the best examples is the formation of phosphaalkenes by HCl-elimination of the precursor on solid potassium carbonate (Section 4.4.).[33]

2.3.3. VGSR apparatus

The VGSR apparatus illustrated in Figure 2 takes into account the different data precised above. The solid reagent is distributed between two pads of glass wool (half-filled reactor). The tubular reactor is then introduced into a horizontal oven and then fitted onto the vacuum line. To maintain a constant temperature inside the reactor, it is important to note that the heating zone must overflowed the belt occupied by solid and pads (for a first description of the apparatus, see 34). The volume, the diameter of the tubular reactor and the swept volume of the vacuum pump are to be adapted to the scale of the chemical transformations. As an example, for reactions going up to gram scale, a simple pyrex tube (ϕ: 3 cm; l: 30 cm) coupled to a vacuum pump (swept volume, $4\,\text{m}^3/\,\text{h}$; pressure, 10^{-1} to $10^{-2}\,\text{hPa}$) can be adopted. All the factors which increase the velocity of the gaseous flow are to be taken into account. Thus, the displacement of the flow being induced by condensation of the gas on the cold trap, it is of importance to maintain along the vacuum line large diameters glass tubes (ϕ: 3 cm) and to fit the trap as close as possible to the oven.

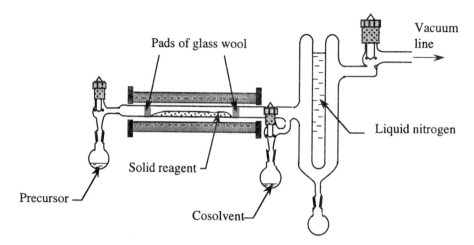

Figure 2   VGSR apparatus.

2.3.4.   *Spectroscopic analysis of the reactive species*

The short-lived species thus generated are characterized in solution (low temperature NMR), in the gas-phase (mass spectrometry (MS), photoelectron (PES) or millimeter-wave spectroscopies (MWS)) or in the solid phase (matrix IR spectroscopy), depending on their structure and their stability.

   – *Liquid-phase spectroscopies.* Most of the species have been characterized by low temperature NMR. Descriptions concerning the vacuum transfer of the products from the cold trap into the NMR tube can be found in the literature.[3–8,34]

   – *Gas-phase spectroscopies.* To avoid the polymerisation of the short-lived species, the tubular reactor is directly fitted onto the spectrometer (MS, MW or PE spectrometers) and the resulting products flowing into the cell are analyzed in real time. This technique, similar to the one applied for the detection of reactive species produced by FVT, has been largely used.[2–9,35]

   – *Solid film or matrix IR or UV spectroscopies.* Transient species can be condensed on a KBr window (solid film at the liquid nitrogen temperature) or co-condensed with argon into a matrix surface at 12 K and then analysed by IR or FT-IR spectroscopies. These two techniques already applied for FVT are well documented (see for example references 8,36–38). The apparatus allowing the preparation of benzocyclobutadiene by reduction of the diiodobenzocyclobutane precursor[38] and its characterisation by IR and UV on argon matrix is described as an example in Figure 3 (for the description of this reaction, see Section 4.2).

2.4.   Comparison between VGSR and FVT Techniques

It seems useful to compare the parameters of VGSR and FVT techniques and to precise the chemical transformations involved for each of them (Table 1). Both

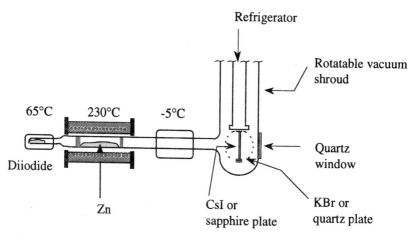

Figure 3 Apparatus used for the generation and the IR and UV analysis of benzocyclobutadiene (according to reference 38).

processes use a heated tubular reactor fitted on a vacuum line. Similar procedure are elsewhere employed for isolation and spectroscopic characterization of reactive species. These external aspects probably explain why the fundamental differences between these two techniques appeared sometime not very well arrested. The FVT involves monomolecular thermal fragmentations (mainly pericyclic reactions, retro-Diels-Alder, Retro-ene... and rearrangement reactions). "The VGSR experiments involve a reaction between an adsorbed gas and a solid reagent". The classical reactions of the organic synthesis can be considered as operating for the formation of reactive species under these conditions. However, as it will be mentioned further, only few of them have been developed so far. It was elsewhere taken advantage of the lower temperature in VGSR experiments for the preparation of thermodynamically unstable species like cyclopropenes. Although the FVT technique appears *a priori* more extendable, the two processes have their own sphere of efficiency and can be considered as complementary for the synthesis of unstabilized species. The main parameters of each process are collected in Table 1.

2.5. Multistep Gas-Phase Sequence

In numerous cases, several different reactions have been performed along a vacuum line (multistep gas-phase sequence). This mainly concerned VGSR/VGSR, VGSR/FVT, or FVT/VGSR coupling reactions, but other various "one line reactions" have been also described. Advantages of these processes will be in each case precised. We present as an example (Figure 4) the preparation of a pure sample of aldimine $H_2C{=}NH$ by a one line sequence involving N-chlorination of methylamine on solid NCS and dehydrochlorination of the short-lived N-chloramine intermediate on solid tBuOK.[28] If necessary, another reactor containing adipic

Table 1 Comparison between VGSR and FVT techniques [a]

	VGSR	FVT
Type of reactions	– Reactions are induced by a solid reagent – The classical reactions of organic synthesis can be involved In some cases, multistep gas-phase sequence are operating	– Monomolecular thermal fragmentations – Mainly pericyclic reactions (retro-Diels-Alder, retro-ene…), and rearrangements
Gaseous flow	– Gaseous flow of volatile precursors on solid reagent	– Gaseous flow inside an empty quartz tube
Temperature	– Reactions at moderate temperature (20–200°C)	– Fragmentation at very high temperature ($>600°C$)
Pressure	– Low pressure (10^{-1}–10^{-2} hPa)	– Very low pressure (10^{-4}–10^{-6} hPa)
Transit time	– Around 1 to 2 s for lecked beds (longer transit time for crossed beds)	– Very short transit time in the hot zone (*ca.* 10^{-2} s)
Detection	– Low NMR temperature – IR/UV on solid film or matrix – Real time analysis of the gaseous flow by mass, photo-electron or microwave spectroscopies	– Same experiments
Limitations	– Precursors and products must be volatile and solid reagent stable at the temperature of the experiment	– Products must be thermodynamically stable

[a] Conditions for non-catalytic reactions

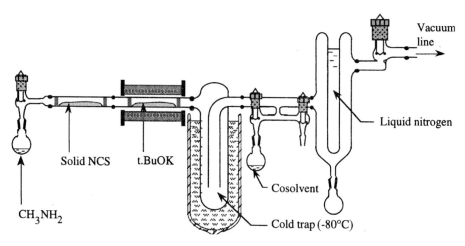

Solid NCS   t.BuOK

CH$_3$NH$_2$

Vacuum line

Liquid nitrogen

Cosolvent

Cold trap (-80°C)

Figure 4   Example of a multistep gas-phase sequence: preparation of the methanimine.[28]

acid can be added at the final step to remove the unreacted amine. The non-volatile impurities are eliminated in a first trap cooled at $-80°C$ and the final product are condensed on a liquid nitrogen cold trap (for description of this reaction, see Section 4.1.).

3.   CATALYTIC VACUUM GAS–SOLID REACTIONS

We have filled in this heading the reactions which do not involve a stoechiometric consumption of the solid phase by the gaseous substrate. The use of catalytic VGSR for the preparation of reactive species has been fruitful. In these reactions, experimental conditions, in particular reaction temperature, are close in general from those observed under flash vacuum thermolysis conditions. The first studies started in the 70's have revealed the great potential of this technique when it is associated with real time photoelectron spectroscopy analysis of the gaseous flow (VGSR/PES coupling).

3.1.   Catalytic VGSR/PES Coupling

The following experiments involve thermal fragmentations of various precursors on elemental silicon, cristalline boron or activated charcoal and concern the preparation of dihalogenosilylenes, HBS and F$_3$CBr respectively.
    Dihalogenosilylenes SiX$_2$ are coordinatively unsaturated species which can be inserted into many bonds. They were formed by heterogeneous catalytic fragmentations under VGSR conditions and identified by PES. Silicon difluoride was obtained by passing BaSiF$_6$ over silica heated to *ca* 1150°C.[39] Syntheses of silicon

dichloride,[40] silicon dibromide, and silicon diiodide[41] were achieved upon heating the corresponding tetrahalosilane derivatives over silica. The conditions were optimized by PES in a *one line* gas analysis at short time intervals. Assignments of the reactive species were correlated with those of related series of molecules.

The thioborine, HBS, was generated by passing H_2S through a horizontal tube packed with cristalline boron. The effluent from the reactor was directly analysed by MW,[42] high resolution IR,[43] and PE spectroscopies.[44]

Considerable interest was attached to a facile preparation of trifluorobromomethane, a highly effective fire-extinguishing agent. The possible passway involving the gas-phase bromination of trifluoromethane was optimized by photoelectron gas analysis. Conversion and yields were optimized for each catalyst.[45] The best results were observed using a mixture of $FeCl_3/KBr$ on activated charcoal (Scheme 1).

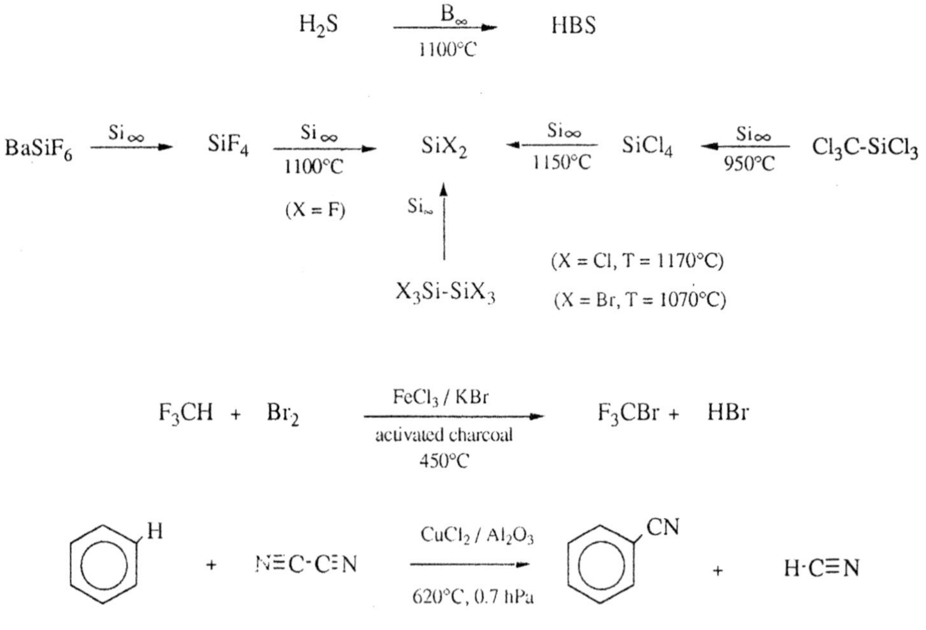

$$H_2S \xrightarrow[1100°C]{B_\infty} HBS$$

$$BaSiF_6 \xrightarrow{Si_\infty} SiF_4 \xrightarrow[1100°C]{Si_\infty} SiX_2 \xleftarrow[1150°C]{Si_\infty} SiCl_4 \xleftarrow[950°C]{Si_\infty} Cl_3C\text{-}SiCl_3$$

$(X = F)$ Si_∞

$X_3Si\text{-}SiX_3$ $(X = Cl, T = 1170°C)$

$(X = Br, T = 1070°C)$

$$F_3CH + Br_2 \xrightarrow[\substack{\text{activated charcoal} \\ 450°C}]{FeCl_3 / KBr} F_3CBr + HBr$$

[benzene ring]—H + N≡C-C≡N $\xrightarrow[620°C, 0.7 \text{ hPa}]{CuCl_2 / Al_2O_3}$ [benzene ring]—CN + H-C≡N

Scheme 1

Saturated and unsaturated hydrocarbons were cyanated in the gaseous phase with cyanogen at high temperature. The cyanation of benzene on various catalysts in VGSR conditions was optimized on the base of the characteristic PE spectra of cyanogen and HCN and the parameters (pressure and temperature) were precised for each of them.[46] The best results were obtained with $CuCl_2$ and Al_2O_3 in the presence of hydrogen at 620°C and at a pressure of 0.7 hPa (Scheme 1).

Other studies allowed to precise the mechanism of the thermal and hetero-geneous catalysed decomposition of azo compouds[47] and the dimerisation of dimethylcarbene on Raney nickel surface.[48]

3.2. Catalytic Vacuum Gas–Solid Dehydration

Siloles and germoles without substitution at the ring carbon tend to dimerise and were for a long time only characterized by chemical trapping.[49] The VGSR process allowed for the first time to independently isolate C-unsubstituted dimethylsiloles by catalytic dehydration of the corresponding 3-hydroxy-metallacyclopentene over alumina at 290°C[50,51] or by HBr-elimination from the 3-bromodimethylsilylcyclopentene over solid tBuOK.[21] The catalytic dehydration was then extended to the preparation of 3,4-C-dimethyl-silole and-germole.[52] However, presence in the last example of isomers bearing a methylene group in exo position cannot be avoided[52,53] (Scheme 2).

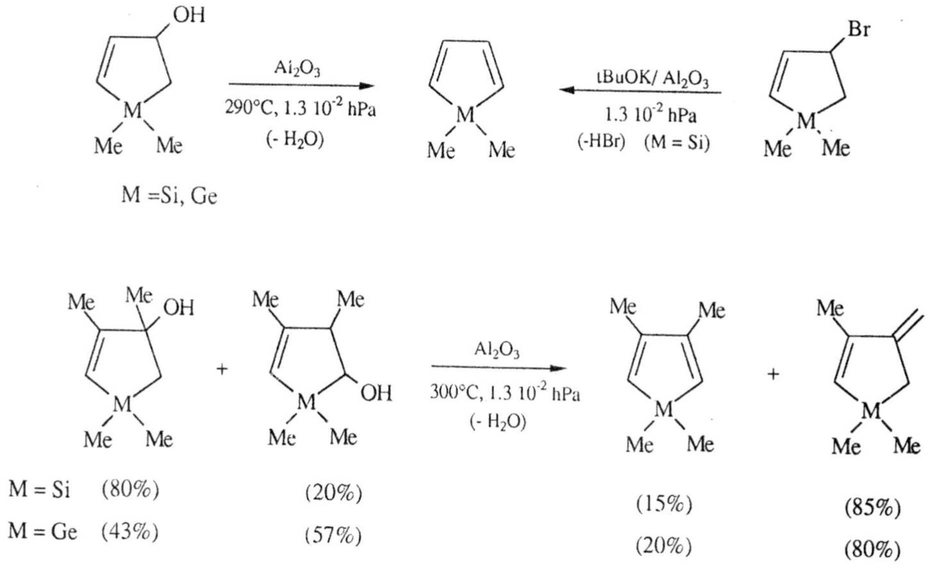

Scheme 2

4. NON-CATALYTIC VACUUM GAS–SOLID REACTIONS

The non catalytic VGSR processes involve N-chorination, dehalogenation, dehydrocyanation, dechlorosilation, dehydrohalogenation reactions and base-induced rearrangement of suitable precursors. Heteroalkenes and heteroalkynes constituted the main targets of these reactions, but various other structures have been also investigated. The VGSR applications in this area will be presented in the following according to the type of reaction.

4.1. N-chlorination of Amines

Primary N-chloramines **A** are important starting materials in organic synthesis. They are for example used as aminating agents of alkenes, aldehydes, and Grignard reagents,[54] or more recently as precursors of aldimines.[28] Because of their poor stability due to the N—H activation by the presence of halogen in an intrinsically basic media and their tendency to disproportionate easily even in the gas-phase,[55] they have been for a long time poorly studied. They are usually prepared in water with chlorinating agents like chlorine or sodium hypochlorite, or in an organic solvent with N-chlorosuccinimide (NCS). However, these approaches cannot be used for the preparation of volatile species since most of them present an intrinsical instability and are potentially blasting agents in the pure state. The volatile N-chloramines, including N-primary amines **A**, secondary N-chloramines **B** and N,N-dichloramines C, were conveniently prepared in VGSR conditions on solid NCS.[28,56-59] The main results are collected in Scheme 3.

$$R^1-N\overset{\displaystyle Cl}{\underset{\displaystyle H}{\Big\langle}} \qquad R^1-N\overset{\displaystyle Cl}{\underset{\displaystyle R^2}{\Big\langle}} \qquad R^1-N\overset{\displaystyle Cl}{\underset{\displaystyle Cl}{\Big\langle}}$$

A   **B**   **C**

Primary N-chloramines **A** were obtained by slowly passing the precursors on solid NCS (*lecked-bed reactor*) at room temperature. Products were directly collected on a cold trap (77 K) and then analysed by NMR at low temperature (−50°C) to avoid their disproportionation. This approach is compatible with various substituents;[28] the yields are excellent (between 85 to 88%). The unreacted primary amine (< 15%) can be eliminated on solid adipic acid in a multistep sequence.[57]

A number of secondary dialkyl- or cyclic N-chloramines **B** were prepared in a nearly quantitative yield by N-chlorination of secondary amines over solid NCS (*lecked-bed*)[29,57-60] and were stored at low temperature (−30°C). The reaction can be extended to molar scale[21] (amine/NCS, ratio 1:1.2). This heterogeneous approach allowed to prepare very efficiently and under safety conditions various N-chloramines which were for most of them either unknow or only accessible in solution without purification.

No general access to N,N-dichloramines **C** was reported in the literature before 1985. Photoelectron spectra of methyldichloramine were first obtained by observation of the products formed by gas-phase reaction of CH_3NH_2 with chlorine.[55] First attempts to prepare the pure products by using a large excess (fivefold excess) of NCS in VGSR conditions led to a mixture of mono and dichloramines in variable ratio. The complete bis-N-chlorination of amines was finally secured in excellent yield using NCS supported on silica at room temperature[57] (*lecked-bed*).

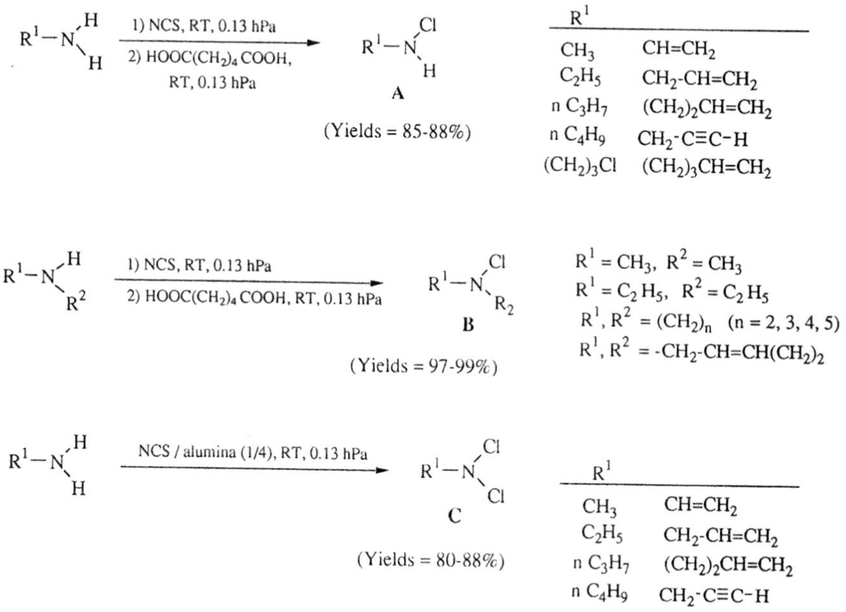

$$R^1-N\overset{H}{\underset{H}{}} \xrightarrow[\text{RT, 0.13 hPa}]{\text{1) NCS, RT, 0.13 hPa} \atop \text{2) HOOC(CH}_2)_4\text{COOH,}} R^1-N\overset{Cl}{\underset{H}{}}$$

A

(Yields = 85-88%)

R^1	
CH_3	$CH=CH_2$
C_2H_5	$CH_2\text{-}CH=CH_2$
$n\,C_3H_7$	$(CH_2)_2CH=CH_2$
$n\,C_4H_9$	$CH_2\text{-}C\equiv C\text{-}H$
$(CH_2)_3Cl$	$(CH_2)_3CH=CH_2$

$$R^1-N\overset{H}{\underset{R^2}{}} \xrightarrow[\text{2) HOOC(CH}_2)_4\text{COOH, RT, 0.13 hPa}]{\text{1) NCS, RT, 0.13 hPa}} R^1-N\overset{Cl}{\underset{R_2}{}}$$

B

(Yields = 97-99%)

$R^1 = CH_3,\ R^2 = CH_3$
$R^1 = C_2H_5,\ R^2 = C_2H_5$
$R^1, R^2 = (CH_2)_n\ (n = 2, 3, 4, 5)$
$R^1, R^2 = -CH_2\text{-}CH=CH(CH_2)_2$

$$R^1-N\overset{H}{\underset{H}{}} \xrightarrow{\text{NCS / alumina (1/4), RT, 0.13 hPa}} R^1-N\overset{Cl}{\underset{Cl}{}}$$

C

(Yields = 80-88%)

R^1	
CH_3	$CH=CH_2$
C_2H_5	$CH_2\text{-}CH=CH_2$
$n\,C_3H_7$	$(CH_2)_2CH=CH_2$
$n\,C_4H_9$	$CH_2\text{-}C\equiv C\text{-}H$

Scheme 3

4.2. Dehalogenation

The dehalogenation of dihalo precursors on solid reagent mainly concerned inorganic compounds and was often considered as a heterogeneous catalysed reaction.[10] One of the first experiments was the preparation of disulfur monoxide. The best approach involved the reaction of thionyl chloride over silver sulfide at 160°C under VGSR conditions[25,26] (Scheme 4). The species is stable in the condensed phase at low temperature but polymerizes on warming or in the gas phase under a partial pressure ($> 1\,hPa$). The unambiguous characterization of this labile species was obtained by PE real time analysis (VGSR/PES coupling). The first six bands were identified[27,61] and compared with those previously reported for S_2O.

A number of short-lived molecules containing low coordinated main group elements bonded to phosphorus and sulfur were accessible by heterogeneous gas-phase α-dehalogenation of suitable precursors over heated metal. Thus, $O=P-Cl$ and $S=P-Cl$ were prepared in the gas-phase by dechlorination at 900°C of the corresponding trichlorides on silver turnings and identified by PES real time analysis.[10,62] Products were also detected by analysis of the gaseous flow by mass spectrometry,[63–65] and by IR and Raman on argon matrix at 15 K.[64,66] In another experiment, $F-P=S$ was formed by debromination of Br_2FPS on silver wool at 500°C and then identified by PES.[67]

Conditions for dechlorination of the bis-dichlorophosphine acetylene on solid catalyst $[MgCl_2/MgO/SiO_2]$ were monitored by VGSR/PES and VGSR/MS coupling. Trichlorophosphine and P4 were predominantly produced accompanied by

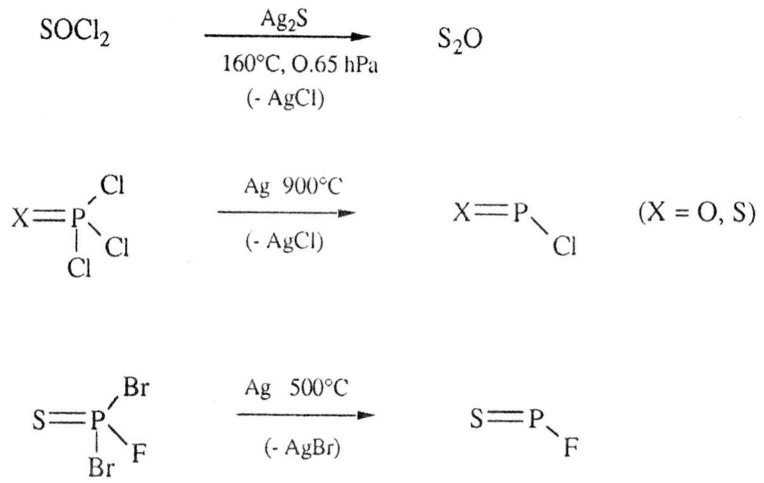

$$SOCl_2 \quad \xrightarrow[\substack{160°C,\ 0.65\ hPa \\ (-\ AgCl)}]{Ag_2S} \quad S_2O$$

$$X{=}P\underset{\underset{Cl}{|}}{\overset{\diagup Cl}{\diagdown}}{}_{Cl} \quad \xrightarrow[\substack{(-\ AgCl)}]{Ag\ \ 900°C} \quad X{=}P\diagdown_{Cl} \qquad (X = O, S)$$

$$S{=}P\underset{\underset{Br}{|}}{\overset{\diagup Br}{\diagdown}}{}_{F} \quad \xrightarrow[\substack{(-\ AgBr)}]{Ag\ \ 500°C} \quad S{=}P\diagdown_{F}$$

Scheme 4

minor amounts of 1,4-diphosphaacetylene.[10,65] The P-chlorophosphaalcyne was also produced by reduction of Cl_3CPCl_2 over granular zinc at 530°C and analyzed by microwave[67,68] and photoelectron spectroscopies[69] (Scheme 5). As predicted, the dipole moment was found to be very small (0.056). Some very weak microwave transitions were detected and the Cl—C and C≡P bond lengths determinated. The assignments of the four PE peaks were supported by *ab initio* SFC calculations.

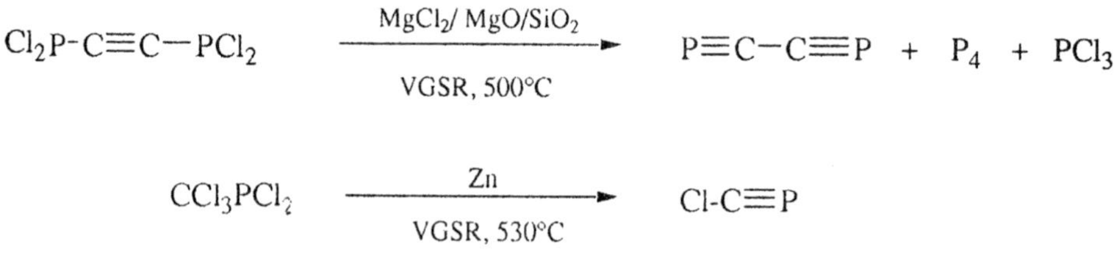

$$Cl_2P{-}C{\equiv}C{-}PCl_2 \quad \xrightarrow[\substack{VGSR,\ 500°C}]{MgCl_2/\ MgO/SiO_2} \quad P{\equiv}C{-}C{\equiv}P \ + \ P_4 \ + \ PCl_3$$

$$CCl_3PCl_2 \quad \xrightarrow[\substack{VGSR,\ 530°C}]{Zn} \quad Cl{-}C{\equiv}P$$

Scheme 5

The VGSR technique associated to photoelectron spectroscopy gas analysis allowed also to precise the mechanism of chemical reactions. Thus, the formation of a phosphinidene intermediate [RP→Mg] by heterogeneous dechlorination of R—PCl_2 on Mg was evidenced.[70]

In few examples, the short-lived species proceeding from a first reaction react in real time on a new reagent in a same vacuum line according to a gas-phase multistep sequence to lead to another transient intermediate. Thus, O=P—Cl previously formed by dechlorination of the trichlorophosphine oxide precursor on silver metal at about 1000°C[71] was directly oxidized without con-densation[72] and the resulting PO$_2$Cl characterized by MS (see description of the apparatus Figure 5). Heat of formation was derivated from mass spectrometric equilibrium measurements. In another experiment, the methyl(thioxo)phos-phine, previously formed by reaction of magnesium with the dichloro precursor, was chemically trapped by Mn$_2$(CO)$_{10}$ (Scheme 6). The stable isolated complex was fully characterized.[73] Similar multistep sequences allowed to prepare and to determine heats of atomisation and ionisation potentials of AsOCl and SbOCl.[74]

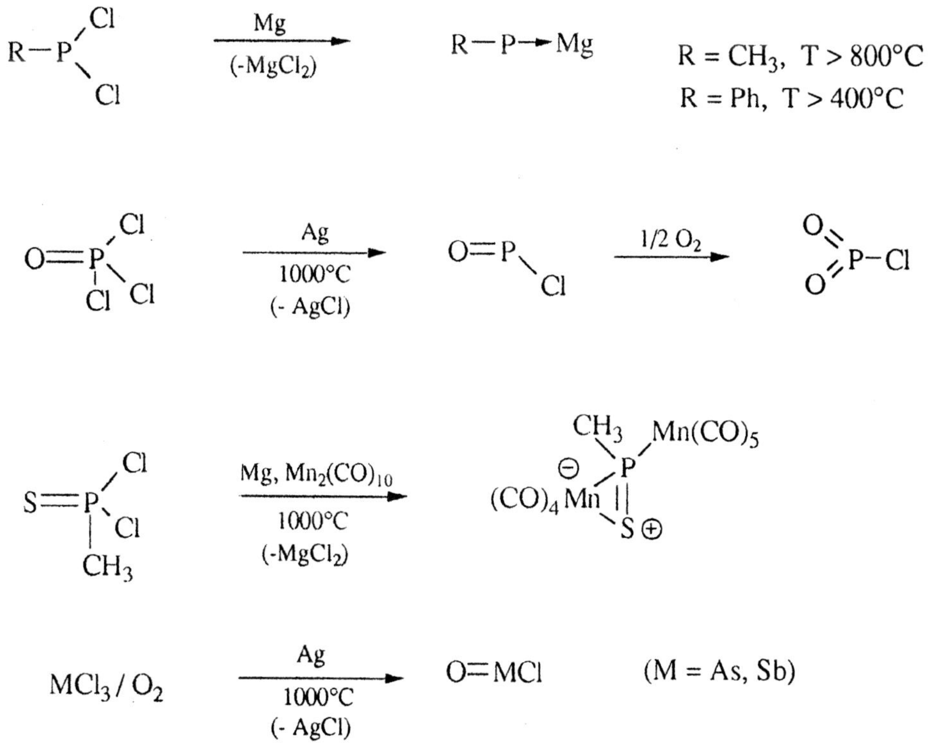

$$R-P\begin{smallmatrix}Cl\\ \\Cl\end{smallmatrix} \xrightarrow[(-MgCl_2)]{Mg} R-P\rightarrow Mg$$

R = CH$_3$, T > 800°C

R = Ph, T > 400°C

$$O=P\begin{smallmatrix}Cl\\ |\\Cl\end{smallmatrix}Cl \xrightarrow[\substack{1000°C\\(-AgCl)}]{Ag} O=P\diagdown Cl \xrightarrow{1/2\,O_2} \begin{smallmatrix}O\\ \diagdown\\O\end{smallmatrix}P-Cl$$

$$S=P\begin{smallmatrix}Cl\\ |\\ \ \ Cl\\CH_3\end{smallmatrix} \xrightarrow[\substack{1000°C\\(-MgCl_2)}]{Mg,\ Mn_2(CO)_{10}} \begin{smallmatrix}CH_3 \quad Mn(CO)_5\\ \diagdown \ \diagup\\ \ominus \quad P\\(CO)_4\,Mn\ ||\\ \diagdown S\ \oplus\end{smallmatrix}$$

$$MCl_3\,/\,O_2 \xrightarrow[\substack{1000°C\\(-AgCl)}]{Ag} O=MCl \qquad (M = As,\ Sb)$$

Scheme 6

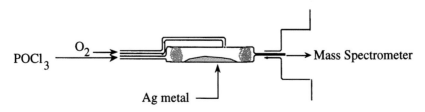

POCl$_3$ $\xrightarrow{O_2}$ Mass Spectrometer

Ag metal

Figure 5   VGSR/MS coupling: formation and characterization of PO$_2$Cl.[72]

Only a few short-lived organic compounds have been prepared by dehalogena-tion of their polyhalo precursors in VGSR conditions. The PE spectra of the dichloroketene (Cl_2C=C=O) which is an intermediate proceeding from de-halogenation of its corresponding precursors on Zn metal at 380–660°C, was assigned by radical cation state comparison based on MNDO calculations.[75]

A special mention should be devoted to the clean generation and matrix isolation in argon at 8 K of benzocyclobutadiene, an organic molecule of theo-retical interest. This short-lived species was formed by passing under high vacuum ($1.3 \cdot 10^{-6}$ hPa) the *cis* 1,2-diiodobenzocyclobutene, previously sublimated in a first oven, over zinc powder heated to 230°C (Scheme 7). After deposition, the vacuum shroud was rotated 90° for IR (CsI plate) or UV (sapphire plate) spectroscopies. The transient intermediate then undergoes dimerisation upon warming the plates above 75 K.[38] This easy 1,2-deiodation should be very useful for the preparation of highly strained molecule, but to our knowledge, this reaction has not been further extended. The apparatus (Figure 3, Section 2.3.) can be considered as having the typical design of those currently used for the characterisation of transient species in the VGSR/IR coupling experiments.

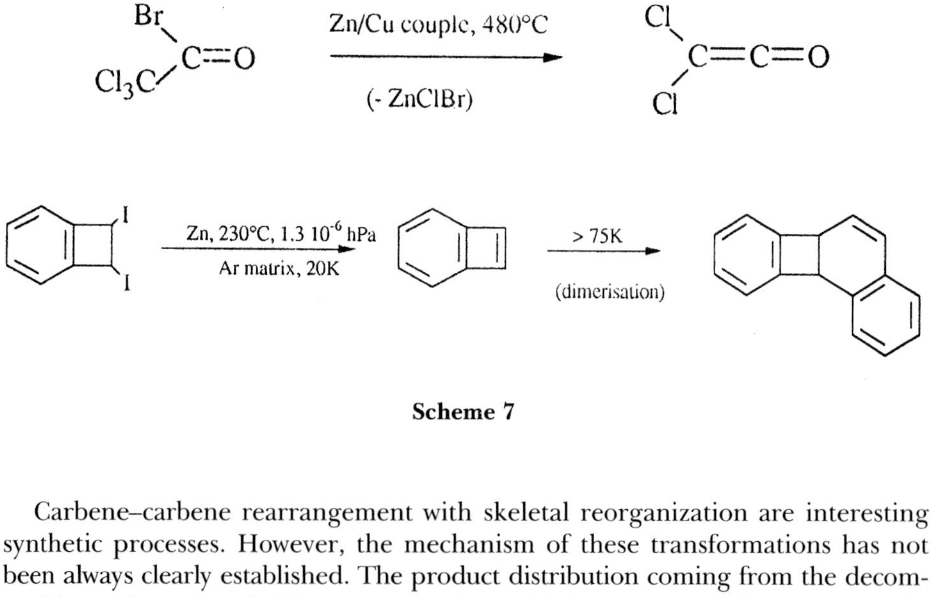

Scheme 7

Carbene–carbene rearrangement with skeletal reorganization are interesting synthetic processes. However, the mechanism of these transformations has not been always clearly established. The product distribution coming from the decom-position of norcarnelylidene and norbornelylidene, intermediates previously formed by α-debromination of their precursors, was found to be dramatically different when the reaction was induced by a solution of MeLi in ether or by MeLi in the solid state (VGSR conditions, *crossed-bed reactor* filled with glass turning coated with MeLi). Thus, reaction of the dibromide precursors **A** or **B** with MeLi in ether gave nearly exclusively the bromomethylnorbornene **E** while the two highly-strained alkenes **C** and **D** became the major products with MeLi in the solid state.[76] In another experiment, debromination of vinyldibromocyclo-propane **F** on solid methyl lithium allowed to isolate the vinylallene intermediate **G** together with cyclopentadiene **H** *via* the carbene intermediates **I** and **J**, the oligomerisation of **G** being mainly observed in solution. In these two examples, the decomposition of the short-lived species was avoided or strongly reduced by the rapid evacuation of products under vacuum (VGSR process). The main results are summarized in Scheme 8.

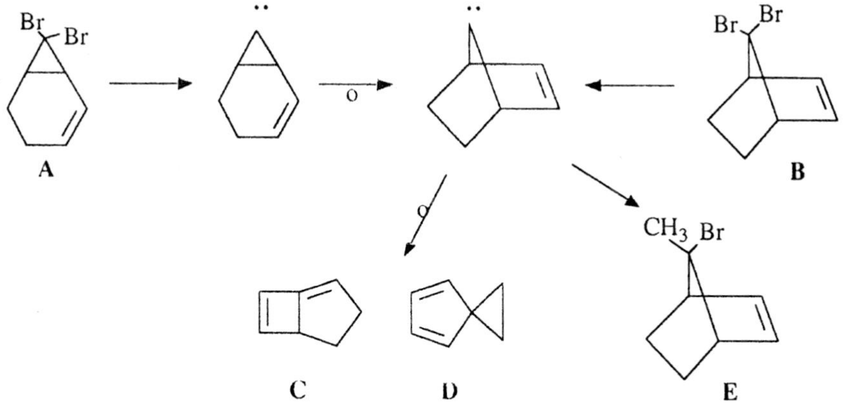

A or B		C + D	E
(reaction conditions)		(yield (%))	(yield (%))
MeLi in ether		2	98
MeLi on glass turnings		30 (starting from A)	70
MeLi on glass turnings		60 (starting from B)	40

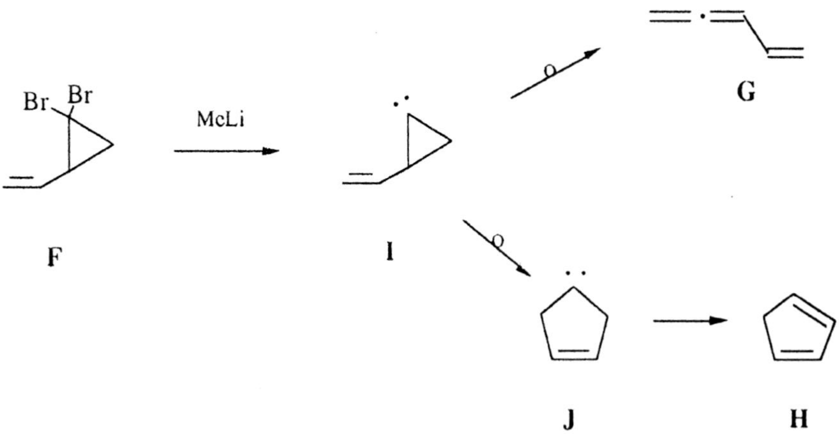

Scheme 8

4.3. α-Elimination of HCl

Whereas α-HX-elimination reactions in solution have been largely used for the construction of cyclopropane rings (intermolecular [1 + 2] cycloadditions), allenes formation, and skeletal reorganizations,[77] only a few compounds have been synthesized in VGSR conditions. By passing the 9-bromobicyclo [6.1.0] nonane over supported tBuOK in VGSR conditions, we have observed as main product the cyclic allene formed by a sequence involving a 1,1-HBr-elimination followed by a cyclopropylidene/allene rearrangement (86%). The two bicyclic [6.1.0] nonene isomers were also observed as minor products. They are probably formed by a 1,2-dehydrobromination followed by a subsequent base-induced rearrangement of the 1,3-bridged cyclopropene intermediate[21] (for isolation of products having such a structure, see reference 8) (Scheme 9). In solution and with the same base, only the elimination/rearrangement products were observed. This different behaviour was explained in terms of steric hindrance of the solid reagent.

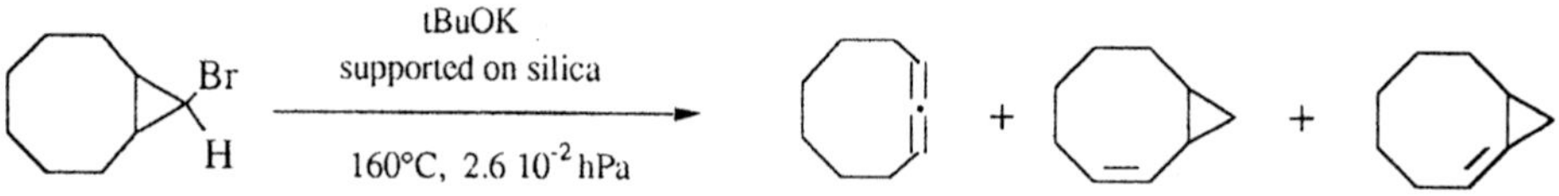

Scheme 9

4.4. β-Elimination of XY (X = H, SiMe$_3$; Y = Cl, Br, CN)

Most of the short-lived species were synthesized by a VGSR methodology involving a 1,2-elimination of XY from suitable precursors (X = H, SiMe$_3$; Y = Cl, Br, CN). This process allowed to prepare highly strained alkenes and heteroalkenes, alkynes, and cumulenes with a nitrogen, sulfur, phosphorus or arsenic atom.

4.4.1. High strained alkenes

(a) β-dehydrohalogenation Radialenes possess an uninterrupted cyclic arrangement of cross-conjugated π systems. Interest in this class of compounds has significantly increased since they are potential candidates for the construction of organic conductors and ferromagnets.[78] Applications of gas–solid reactions constitute an interesting method for the preparation of such structures. Thus, the [3][18] and [4][79] radialenes have been efficiently prepared by bis- or tetrakis-dehydrobromination of their corresponding precursors on dried KOH. The [3] radialene was also prepared by dehydroiodination of the triiodotrimethylcyclopropane[80] and the [4] radialene by reduction of the dichloromethylcyclobutene over zinc or dehydrobromination of the tetrabromomethylcyclobutane over solid KOH[79,81,82] (Scheme 10).

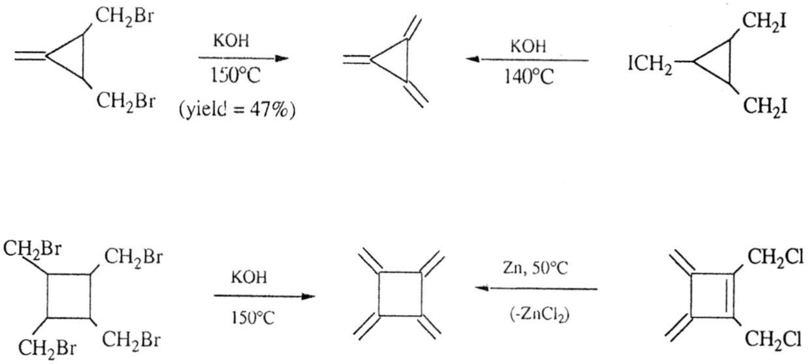

Scheme 10

Our involvement in gas–solid reactions started with the chemistry of cyclopropene rings and with other small ring structures. Attempts to prepare cyclopropenes by FVT usually failed. Thus, a mixture of acetylenes was the only isolated compounds in the thermolysis of methylenecyclopropene anhydrides.[83] Numerous other examples can be found in the literature.[3,5] Application of the VGSR methodology to the preparation of these thermodynamically unstable species appeared useful. Optimization of the parameters was first established by selecting the cyclopropene parent compound as a synthetic target. Its thermal unstability was first checked: the ring opening leading to propyne was observed by FVT around 300°C.[86] In contrast, the cyclopropene ring was in part preserved in VGSR conditions with a column fully packed with tBuOK supported on Chromosorb[30] (crossed-bed reactor). However, the reaction in these conditions appeared to be only of analytical significance (Table 2). Other solid bases like Li$_2$CO$_3$/LiBr, CaO, Na$_2$O led only to a complete opening of the cycle.

Our second target was the spiropentene, a very high strained molecule (strained energy ≈ 90 kcal mol^{-1} [85]). If the opening of this compound was in part avoided when the HBr-elimination of the bromocyclopentane was performed in solution (DMSO) and under vacuum (80 hPa), only 5% of spiropentene was observed in VGSR conditions using a crossed-bed reactor fully filled with tBuOK supported on silica, the main product being the cyclopropyl acetylene.[32]

We have two years later revisited this work[86] and observed in both examples a dramatic change when using a lecked-bed reactor half-filled with solid tBuOK: the ring opening of cyclopropene was nearly avoided and the desired product was isolated with a good yield on gram scale. Furthermore, the spiropentene became the major product (≈ 65%) in the HBr-elimination of the corresponding precursor (Scheme 11). We assumed that the very weak pressure maintained along the tube with the lecked-bed reactor assured a short transit time of the strained molecules in the oven. The ring opening of the thermaly unstable molecules can be thus avoided or minimized (see above for discussions). These results are summarized in Table 2.

Table 2   Formation of cyclopropenes under various experimental conditions

Alkene	Reactor	Base (T°C)	Yield (%)	Comments
Cyclopropane	crossed-bed (VGSR)	tBuOK supported on chromosorb W (160)	75	analytical significance, (presence of propyne)
Cyclopropene	crossed-bed (VGSR)	tBuOK supported (160)	14	poor yield in gram-scale experiments [presence of the precursor (62%) and propyne)]
Cyclopropane	lecked-bed (VGSR)	solid tBuOK (160)	62	gram-scale experiments. Yields are strongly improved
Spiropentene	DMSO (solution)	tBuOK, solution (90)	60–70	presence of 2–4% of ethynyl-cyclopropane
Spiropentene	Crossed-bed (VGSR)	tBuOK supported on chromosorb W (180)	2–4	mainly ethynylcyclopropane
Spiropentene	lecked-bed (VGSR)	solid tBuOK (160)	65–75	presence of 5% of ethynyl-cyclopropane

High strain of spiropentene was revealed by its unusual chemical reactivity. Addition of tBuOH at 20°C was observed in the presence of base. Reaction with cyclopentadiene and furan at 0°C gave nearly exclusively the endo adduct and an endo–exo mixture respectively (1:3)[32] (Scheme 12).

Methylenecyclopropene is the simplest member of cross-conjugated hydrocarbons. It has been the subject of numerous theoretical calculations.[8] Its generation was for a long time attempted in solution under a variety of conditions. Indirect evidence was provided, but its strong reactivity towards nucleophiles prevented its isolation. The first synthesis was independently achieved by Billups *et al.*[31] and Staley and Norden[87] in 1984 by the use of the VGSR technique. The procedure involved the dehydrohalogenation of chloro- or bromomethylenecyclopropane on tBuOK supported on Chromosorb W. The same compound was recently prepared by bis-dehydrohalogenation˙ of the bromocyclopropane precursor over

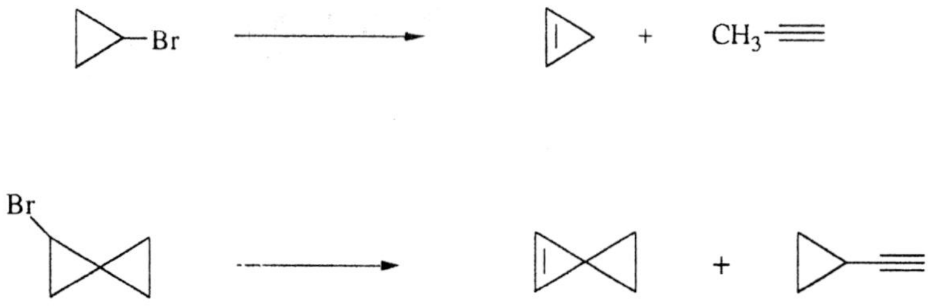

Scheme 11

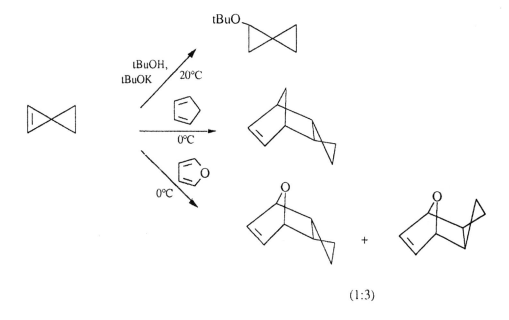

Scheme 12

solid tBuOK (*lecked-bed reactor*) of 1-bromo-2-chloromethylcyclopropane, a more easily available precursor[88] (Scheme 13).

The structure of methylenecyclopropene was fully established by NMR,[31,87] IR,[31] microwave,[89] photoelectron,[90] and millimeter and submillimeter-wave spectroscopies.[88] A few other cyclopropene derivatives have also been prepared. Attempts to synthesize by the same methodology the spiropentadiene, a highly strained molecule (heat of formation of 157.4 kcal from calculation[91]), by

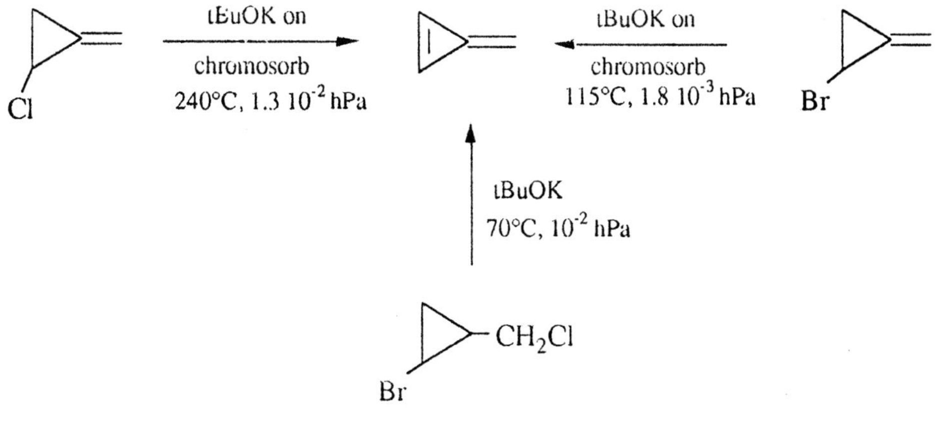

Scheme 13

bis-dehydrochlorination of the dichlorospiropentane precursor were however unsuccessful. The vinylcyclopropene isomer which was first isolated as the main product at low temperature[92] dimerises upon heating ($> -60°C$) into the divinyl-tricyclohexane which is also rather unstable. The *ortho* -divinylbenzene was the first isolated product stable at room temperature. In this experiment, the elimination temperature (tBuOK supported on Chromosorb at 320°C) was probably too high to avoid the ring opening of the cyclopropene ring intermediate (Scheme 14). As it will be seen further, better results were obtained by dehalosilation promoted by solid fluoride salts.

(b) β-halosilane elimination   The conversion in solution of β-halosilanes into alkenes using fluoride salts was first reported by Chan and Massuda[93]. The feasibility of extending this reaction in the gas-phase was further demonstrated by Billups and Lin.[8,92] The following experiments show interest and powerfulness of this reagent for the synthesis of highly strained molecules. Thus, the synthesis of a cyclopropene ring was achieved by room temperature β-chlorosilane elimination of a halocyclopropane precursor with Bu$_4$NF coated on glass helices in VGSR conditions (*crossed-bed reactor*) (Scheme 15). This solid base also allowed to prepare at room temperature the spiropentadiene among a complex mixture of by-products by a bis ClSiMe$_3$-elimination from the corresponding precursor.[94] The product is highly unstable (^{1}H NMR, $-105°C$) and polymerized upon warming but was however trapped at low temperature ($-78°C$) with cyclopentadiene (Diels-Alder adduct).

   The 3,3'-cyclopropenylcyclopropene, a valence isomer of benzene,[95] was prepared by a similar approach. The X-ray cristal structure (103 K) revealed an *anti* conformation in the solid state while the photoelectron[96] spectrum was rationalized in terms of a mixture of the *anti* and *syn* conformers in the gas phase. Two other bicyclopropenyl isomers were also prepared and chemically trapped by Diels-Alder cycloaddition with cyclopentadiene.[95]

   In an other experiment, the first 1,3-bridged cyclopropene was synthesized by passing the precursor over nBu$_4$NF deposed on glass helices.[97] Dimerisation (ene-reaction) occured readily below $-90°C$. The chemical properties of this dimer have been studied in detail. In another study, a cyclopropene-vinyl carbene rearrangement has been furthermore evidenced by the formation of a chlorocycloheptadiene derivative starting from a bicyclic silylated dichlorocyclopropane (Scheme 16).

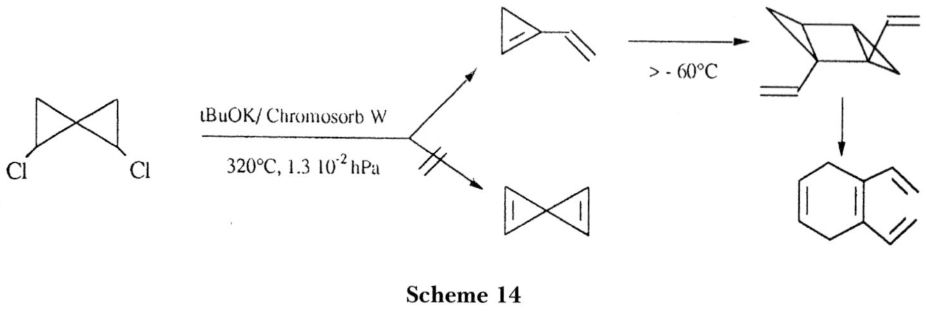

Scheme 14

The dehalosilation promoted by fluoride anion has been recently used to pre-pare a variety of monosubstituted cyclopropenes containing additional unsatu-rations.[98] The β-chlorotrimethylsilylcyclopropane precursors were prepared by classical procedure. Dehalosilation with Bu_4NF on glass helices occured in mild conditions (25°C, $4 \cdot 10^{-2}$ hPa). Cyclopropenes bearing a vinyl or ethynyl group in 1 or 3 position have been prepared (Scheme 17). The stability of the former has been found to be very weak.

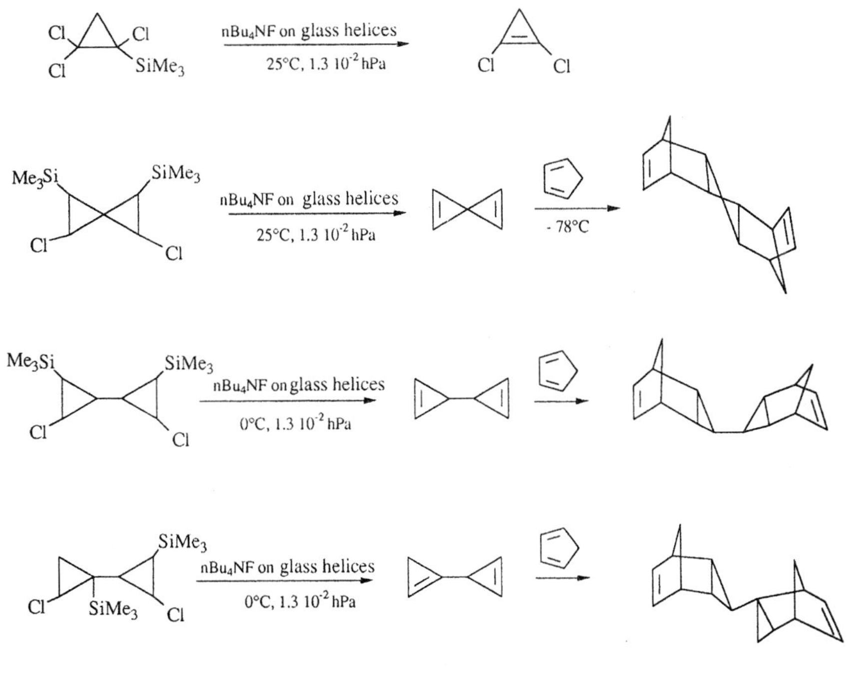

Scheme 15

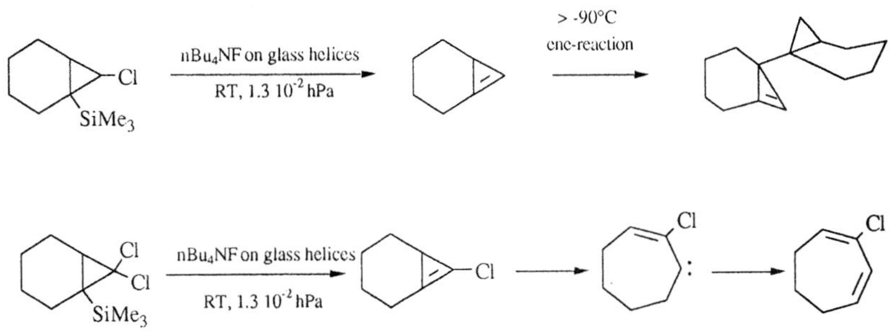

Scheme 16

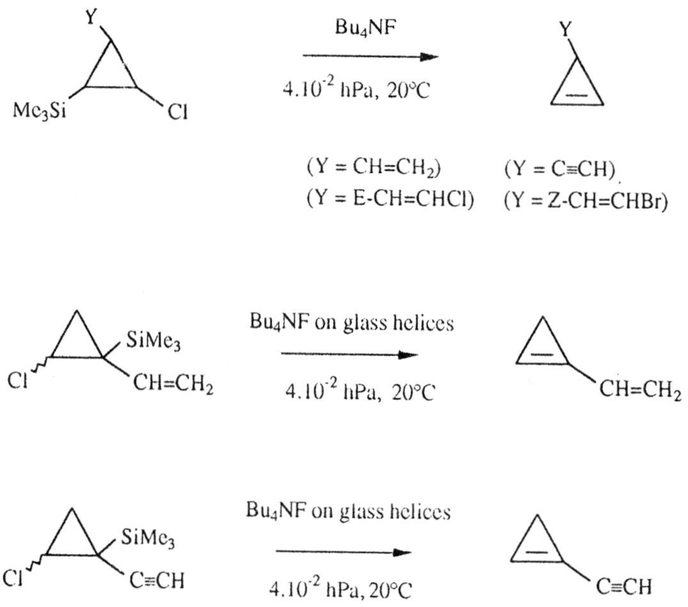

$$(Y = CH=CH_2) \qquad (Y = C\equiv CH)$$
$$(Y = E\text{-}CH=CHCl) \qquad (Y = Z\text{-}CH=CHBr)$$

Scheme 17

4.4.2. Short-lived imines by HCl-elimination

Imines constitute one the most important functions of organic chemistry. If the N-trisubstituted and N-disubstituted derivatives bearing alkyl or aryl groups at carbon and nitrogen are usually stable, N-unsubstituted aldimines (R(H)C=NH), N-unsubstituted cyclic imines, and methylenimines (H_2C=NR) present an intrinsical reactivity which prevented for a long time their isolation. Several of them were for the first time detected mainly by PES in the pyrolysis products of cyclic s-triazines, cyclic alkylamines, or alkylazides (see 99 and reference cited therein). The VGSR technique has brought during the last two decades a decisive contribution to the synthesis of the "missing links" and has allowed their isolation and their spectroscopic characterization (NMR, IR, PES, and MWS) by two different routes: dehydrochlorination of N-chloramines and dehydrocyanation of α-acetonitriles. Short-lived aldimines were obtained by the former (present Section) and methylenimines mainly by the later (Section 4.4.3.).

(a) N-unsubstituted (E)- and (Z)- aldimines   N-unsubstituted imines are short-lived intermediates which are involved in numerous reactions such as chemical or enzymatic oxidations of primary amines, or dehydration of "aldehyde ammonia" (see 28 and reference cited therein). Some of them are considered to be of prebiotic importance.[100] Methanimine, the simplest derivative, has been detected in the interstellar space.[101] We have prepared in good yields these short-lived species by dehydrochlorination of secondary N-chloramines on solid tBuOK (*lecked-bed reactor*) in VGSR conditions.[28] The chloramine precursors were first synthesized by N-chlorination of the corresponding amines on solid NCS in VGSR conditions and then slowly revaporized on tBuOK after condensation on a cold trap (see Section 4.1). Since much of them decompose upon warming, we solved this problem by performing both N-chlorination of amines and HCl-elimination of the resulting N-chloramines in the same vacuum line without condensation of the intermediates (Scheme 18 and Figure 4, Section 2.5.). Amine impurities were removed, if necessary, by adding a solid adipic acid bed (three steps VGSR sequence). This procedure allowed to prepare gram scale amounts of various unsaturated aldimines in avoiding hazardous manipulations of the unstable N-chloramines intermediates.

Unsubstituted (E)- and (Z)-aldimines were characterized by low temperature IR (77 K), ^{1}H, and ^{13}C NMR (CFCl$_3$ + CD$_2$Cl$_2$, $-100°$C) spectroscopies.[29,59,60] The stereochemistry of the (E)- and (Z)-isomers was established by comparison of the $^3J_{HH}$ coupling constants of aldimines with those of H$_2$C=NH.[56] Presence of the two diastereoisomers implies a slow inversion of the N—H bond on the NMR scale. Aldimines can be kept in solution at low temperature ($< -80°$C); a trimerisation was observed above $-70°$C as it was shown for the ethyl, propyl, and butylimine derivatives (formation of s-triazines). In absence of solvent, oligomerization occured above $-120°$C[28,102] leading in the case of methanimine to the formation of azaadamantane by an aldol-type reaction, among other more complex structures (see 56,59 and reference cited therein).

The dehydrochlorination of N-chloroallylamine over tBuOK led to a mixture of the two 1-azadiene isomers (E/Z ratio; 70:30, 30% yield), and to 1-azabicyclobutane (50%).[28] The formation of the later product can be formally explained by a mechanism involving an α-dehydrochlorination followed by a (1 + 2) cycloaddition of the discret nitrene on the double bond. The following results agree with

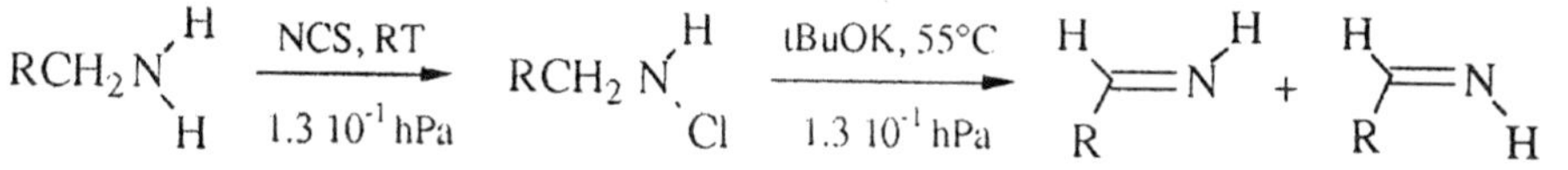

R	Yield (%)	cis / trans
CH$_3$	89	2.3 / 1
C$_2$H$_5$	85	2.6 / 1
nC$_3$H$_7$	55	3.0 / 1
CH=CH$_2$	30	2.7 / 1
CH$_2$-CH=CH$_2$	53	2.3 / 1
(CH$_2$)$_2$CH=CH$_2$	52	2.3 / 1

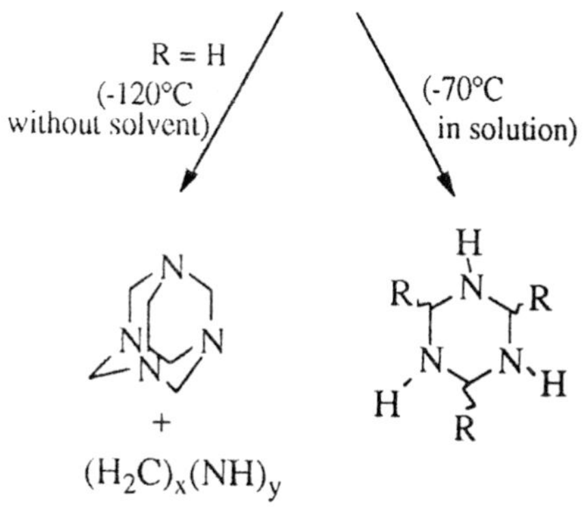

Scheme 18

this mechanism.[29] They involve DCl-elimination from the N-deuterated chloro-propylamine with formation of the two N—H imine isomers (1,2-hydrogen shift) and skeletal reorganizations (presence of N-methylpropylimine and Δ^1-pyrroline in the HCl-elimination of isopropylchloramine and cyclobutylchloramine respectively[29,59]) (Scheme 19). However, the two mechanisms, α- and β-eliminations, are probably occuring since azetine issued from a ring enlargement, and cyclo-propylimine, the expected β-elimination product, were independently detected respectively by low temperature NMR[29] and by photoelectron spectroscopy[59,99] in the HCl-elimination of cyclopropylchloramine.

(b) N-*unsubstituted cyclic imines*   2H-azirine was formed by thermal decomposition of vinyl azide and identified by microwave,[103] photoelectron,[104] and NMR spectroscopy;[105] the mechanism of the reaction was also precised.[99,104] Rota-tional spectrum was measured in the frequencies range of 50–300 GHz; quadru-pole and spin-rotation coupling constants were also given.[106] Its stability was found to be higher than previously reported; oligomerization was only observed

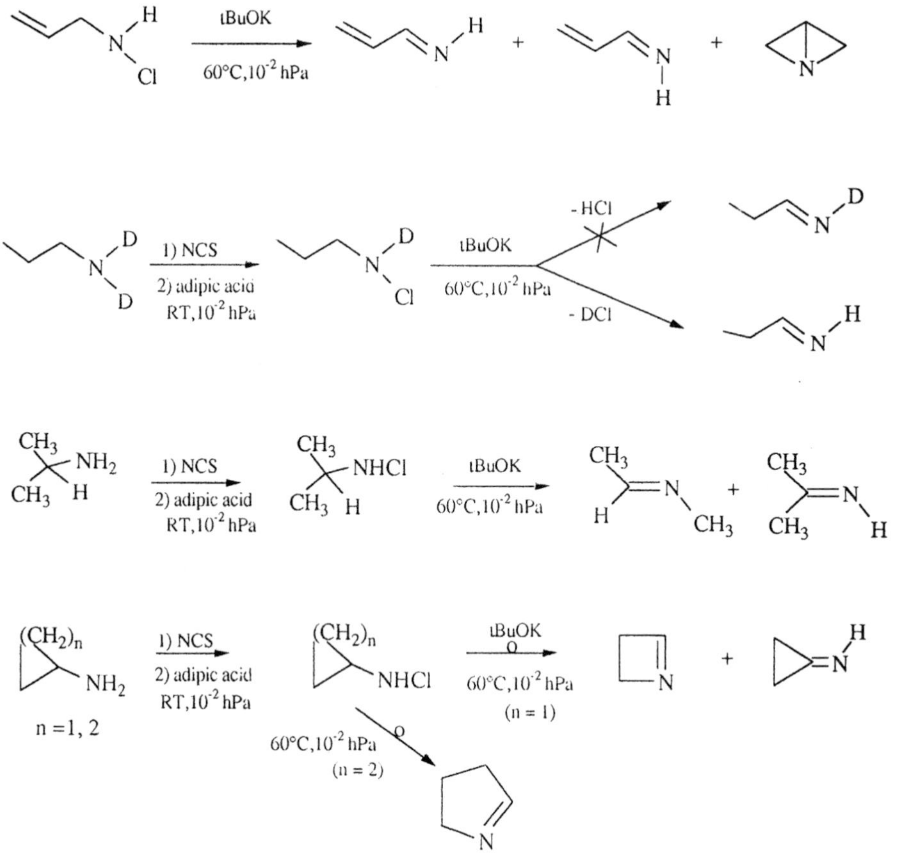

Scheme 19

in solution above 0°C.[105] Attempts to prepare azirine by HCl-elimination from N-chloroaziridine were unsuccessful, the only isolated products being acetonitrile, isonitrile, and aziridine[60] (Scheme 20). Use of the vinyl azide as an isolated starting material for a preparative scale synthesis of 2H-azirine was excluded since this compound is a highly sensitive material which may detonate on mechanical shocks, on heating (80°C),[107] or under vacuum even at low temperature.[108] We have observed the formation of this intermediate by HCl-elimination of chloroethylazide on a solid base.[105] Gram-scale amount of 2H-azirine[60,99] were prepared in safety conditions using a one line multistep sequence (VGSR/FVT

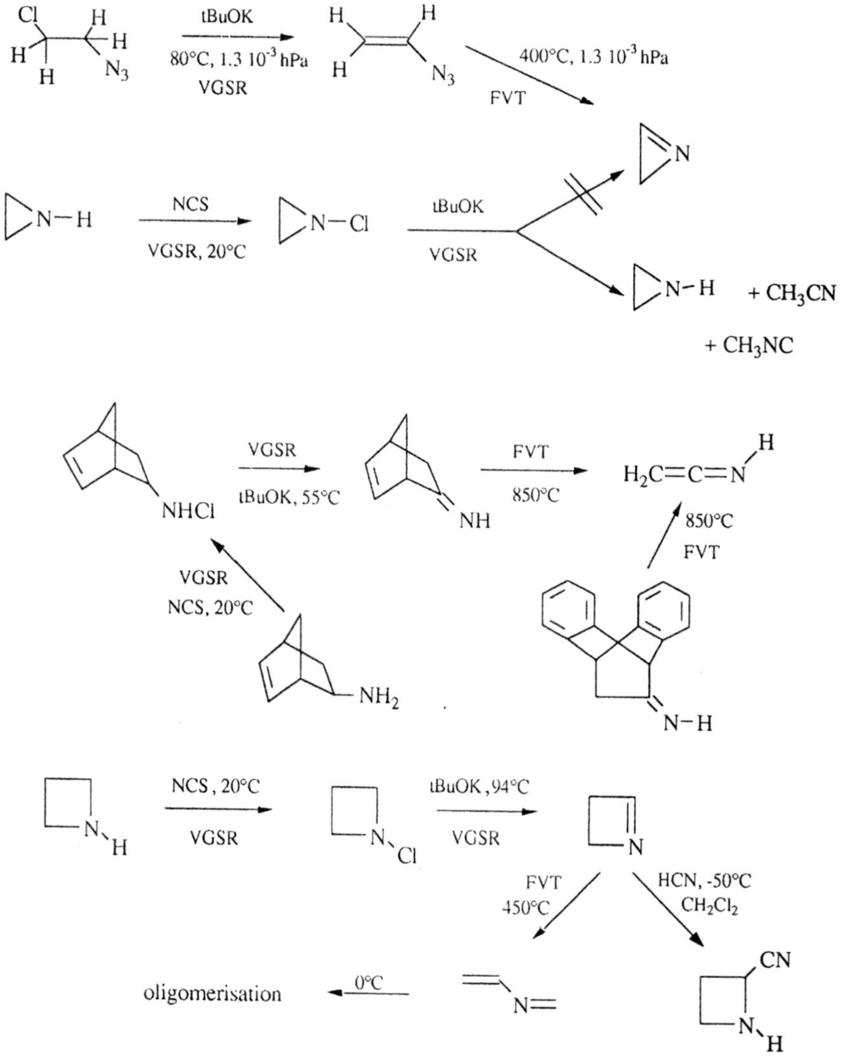

Scheme 20

coupling) involving HCl-elimination of chloroethylazide, condensation of the resulting tBuOH in a dry-ice trap, and subsequent thermolysis of the volatile vinyl azide under high dilution conditions.

Vinylideneamine, a metastable isomer of methylcyanide, is a molecule of astrophysical interest.[109] It was synthesized by FVT (retro Diels-Alder from the anthracene adduct) or by a multistep sequence involving N-chlorination reaction of aminonorbornene on solid NCS (VGSR conditions), HCl-elimination from the resulting chloramine on solid tBuOK (VGSR conditions), and retro-Diels-Alder (FVT) of the imino adduct[105] (Scheme 20). The IR spectrum (77 K) was in good agreement with the calculated one,[110] but the high reactivity of the species precluded NMR measurements and chemical trapping.

1-Azetine is too unstable to be isolated by HCl-elimination from N-chloro-azetidine in solution. This short-lived species was obtained in a nearly quantitative yield by dehydrohalogenation of the precursor over silica supported tBuOK in VGSR conditions.[111] The use of a "one line" multistep sequence involving N-chlorination of azetidine on solid NCS followed by HCl-dehydrohalogenation of the N-chloramine intermediate on pure tBuOK avoided the manipulation of this intermediate. Azetine was fully characterized (^{1}H, ^{13}C NMR, HRMS, IR and UV at 77 K.).[111,59,60] The half-life in solution in a sealed degassed NMR tube (CFCl$_3$, 36°C) is about 90 min; no trimeric product was detected. Addition of HCN at -50°C led to the corresponding nitrile derivative (Scheme 20). Upon heating azetine under FVT conditions (oven temperature 450°C, pressure $1.3 \cdot 10^{-4}$ hPa), the 2-azabutadiene was formed; oligomerisation in solution occured around 0°C.

Application of the one line N-chlorination/elimination sequence involving solid NCS as chlorinating agent and potassium alcoholate or KOH as solid base also allowed the preparation of Δ^1-pyrroline, and Δ^1-piperideine (Scheme 21). This procedure avoids isolation of the N-chloropyrrolidine intermediates which is known to explode in the condensed phase.[57] As expected,[112] the trimerisation was observed upon warming (-20°C).[102,58,60]

Dihydropyridines are of importance in biological systems and have been postulated as intermediates in the biosynthesis of alkaloids. Calculations predict that 5,6-dihydropyridine is the most unstable of the five isomers.[113] This later compound was synthesized by retro Diels-Alder (FVT, 520°C) or by a one line VGSR sequence involving the N-chlorination of tetrahydropyridine followed by HCl-elimination of the corresponding N-chloramine (solid K$_2$CO$_3$, 300°C). A more direct route involves HCl-elimination and retro Diels-Alder on K$_2$CO$_3$

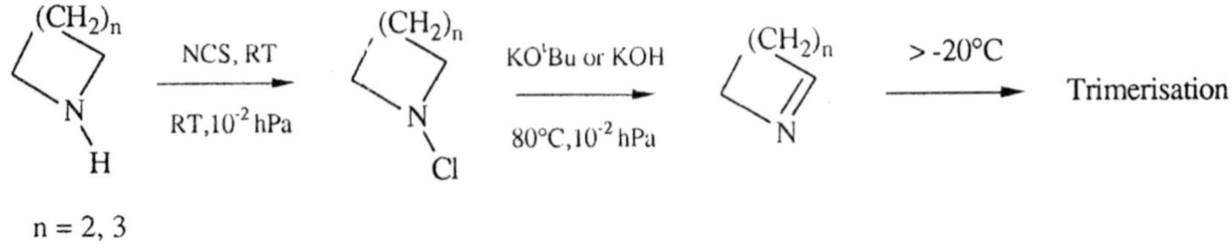

Scheme 21

heated at 450°C. The product has been characterized by low temperature NMR spectroscopy. Oligomerisation occured above $-80°C$[114] (Scheme 22).

Photoelectron and microwave spectroscopies Bock and Dammel have extensively studied short-lived aldimines in gas-phase by means of real time analysis PE-spectroscopy. They have optimized the parameters of the reaction by multistep VGSR procedure as described above. Details can be found in reviews[99,115] and papers.[116,58,59] Comparison beween the electronic structure of $H_2C=NCl$ and $H_2C=PCl$ has been elsewhere achieved by PES.[117] The microwave spectra of 1-azetine and 2-azabutadiene have been assigned. They allowed to determinate the dipole moments and nuclear quadrupole coupling constants.[118]

(*c*) *Methanimines by HCl-elimination of N-chloramines* Methanimines were prepared by HCl-elimination from their N-dimethylchloramine precursors in a one line N-chlorination/elimination sequence. N-methylmethanimine and the all deuterated derivative $D_2C=N—CD_3$ were thus obtained in a nearly quantitative yield by elimination of chloramines on tBuOK at 90°C (Scheme 23). Products were characterized by ^{13}C NMR, and by their millimeter-wave spectra (VGSR/MWS coupling). The rotational constants, centrifugal distortion constants and internal rotation barrier were determined. The quadrupole coupling constants were also measured[119] for the perdeuterated species.

In solution, the thermodynamic internal imines were the only products to be isolated in the base-promoted HCl-elimination of N-chloro N-methylalkylamines; only traces of the external methanimine isomer were detected as an intermediate by chemical trapping.[120] The VGSR technique (solid tBuOK) allowed to accede for the first time and in good yield to these short-lived intermediates. The elimination under these heterogeneous conditions was found to be regioselective, the kinetic product being mainly observed (*ca* 70% *vs* 30% for internal imines).[29]

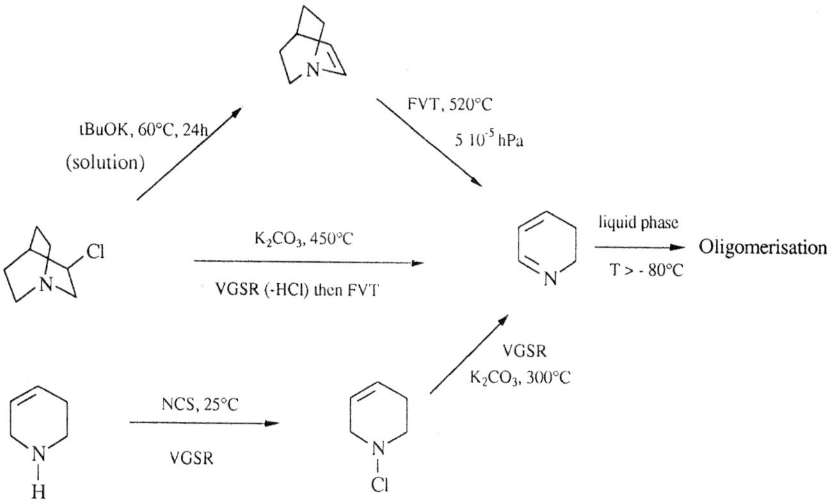

Scheme 22

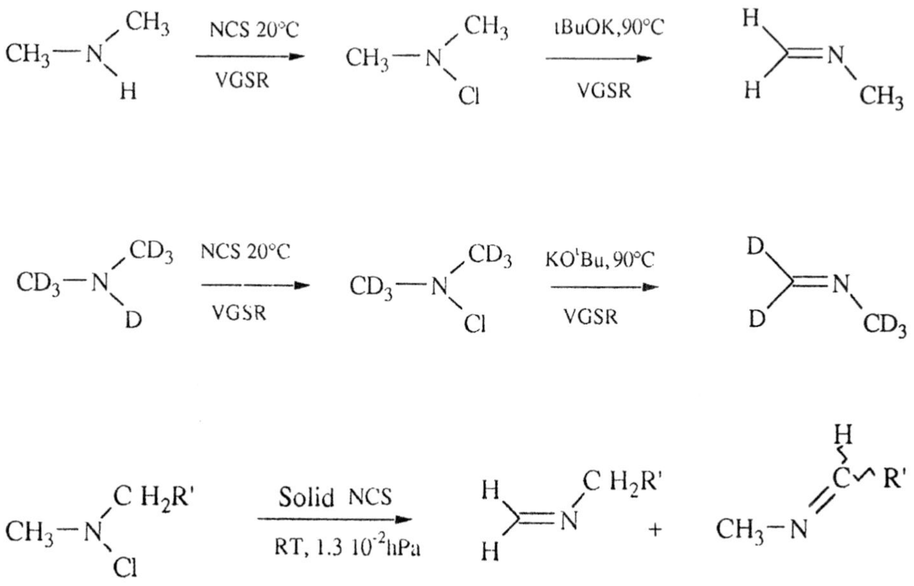

R'= H, alkyl

Scheme 23

This orientation, very different from the one observed in solution, can be attributed to steric interactions of associated bases,[121,122] which are higher with bases in the solid state (the crystalyzed tBuOK is under tetrameric aggregates[123]) than in solution. The synthetic interest of this reaction is however limited since the two isomers cannot be separated.

4.4.3. *Methanimines by* HCN-*elimination of α-aminonitriles*

α-Aminonitriles are recognized as precursors of stable imines and more recently as N-methyleneamine equivalents.[124] The methanimine parent compound **1** has been detected in the gas-phase decomposition products of α-aminoacetonitrile.[125] Methanimine **1** and hydrogen cyanide **2** were observed in the microwave spectrum of α-aminonitrile. Since recombination of **1** and **2** occured in the condensed phase at very low temperature ($< 77\,K$), an important role of **2** in prebiotic and interstellar chemistry has been postulated[126] (Scheme 24). This experiment indicated furthermore that HCN should be removed to avoid recombination into α-aminonitrile.

N-chloroaldimines were prepared in a *one line* sequence involving dehydrocyanation of α-aminonitriles and N-chlorination of the resulting imines. The electronic structure was established by PES and compared with that of the P-chlorophosphaethene analogue. Incidence of the chlorine substitution on the electronic structure of heteroatomic double bond was also precised[117] (Scheme 24).

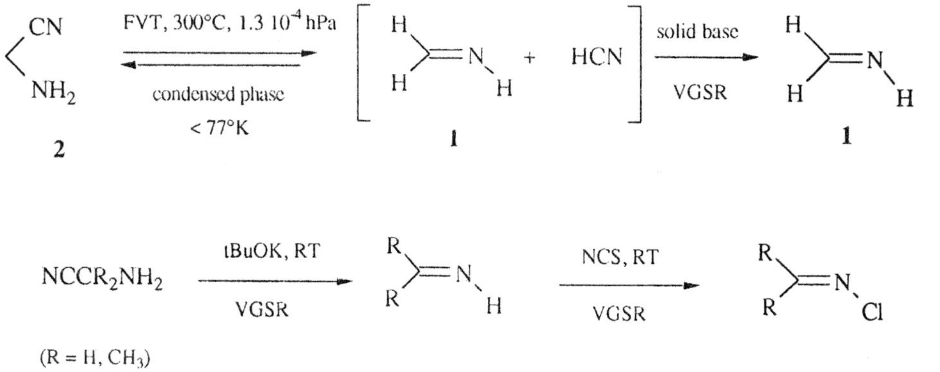

Scheme 24

N-Trimethylsilylimines are important intermediates in organic synthesis. Their ability to undergo nucleophilic addition reactions is well established and was used in the synthesis of nitrogen containing molecules such as β-lactams.[127] They were formed in solution by condensation of lithium hexamethyldisilylamide on the corresponding aldehyde and trapped *in situ* with lithium enolate of esters.[128] The reactive silylimines were for the first time isolated in stable conditions in gram scale by dehydrocyanation of the corresponding N-silylated α-aminonitriles on solid KOH at 60°C (VGSR conditions) and were fully characterized.[129] The alkyl group at the silicon atom can be modulated. Oligomerisation in solution occured from -40 to -10°C, depending on the substituents (Scheme 25).

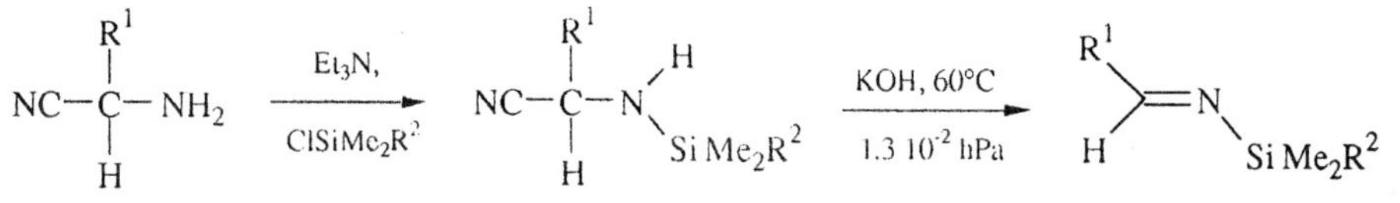

R^1	H	Me	Et	Pr	H	Me	Et	H
R^2	Me	Me	Me	Me	tBu	tBu	tBu	thexyl

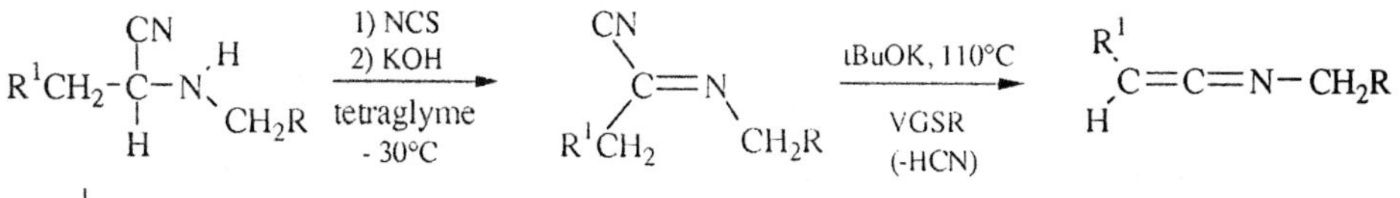

R	H	Me	H	Me
R^1	H	H	Me	Me

Scheme 25

   Stable substituted ketenimines are accessible by base-promoted dehydrocyanation of α-cyano-enamines, but the reported reaction conditions are not
compatible with less substituted alkyl species.[130] A convenient synthesis of
simple ketenimines involves dehydrocyanation of imidoyl cyanides using the
VGSR procedure.[22] The cyanide precursors were obtained in a one pot procedure by N-chlorination of aminonitriles and subsequent HCl-elimination
of the resulting N-chloramines in solution at low temperature ($-30°C$). The
dehydrocyanation occured on solid tBuOK at $110°C$. Ketenimines decompose
slowly in absence of solvent. A small amount of the azadiene isomer was also
observed.
   Short-lived aldimines were isolated as single products by dehydrocyanation
of N-alkyl-substituted α-aminonitriles on solid tBuOK (VGSR technique).
However, the presence of by-products in the elimination of functionalized aminonitriles limits interest of this approach. They were more cleanly prepared
by a *one line* sequence involving flash vacuum thermolysis of α-aminonitriles and gas-phase HCN removing on ROK[29,126,131] (R = H, tBu) (FVT/VGSR
coupling). In the absence of solid base, recombination and formation of the
starting material was observed below $-50°C$. A number of functionalized
methanimines were thus efficiently prepared. By changing the nature of the
base, either 2-azabutadiene (tBuOK) or chloroethylimine (KOH) were obtained
starting from the chloroethylaminonitrile precursor. The thermolysis of the
propargylaminonitrile at $240°C$ gave a mixture of the two imine isomers.
However, a whole isomerisation of propargylmethanimine into allenylmetha-
nimine was observed by increasing the thermolysis temperature ($300°C$). All the
products were characterized by low temperature ^1H and ^{13}C NMR ($-90°C$).
Their oligomerisation was observed upon warming the solution above $-60°C$[131]
(Scheme 26).

4.4.3. Short-lived thioketones

Most of aliphatic and aromatic thioaldehydes are too reactive species to be
isolated in stable conditions, even at low temperature. Some of them were
characterized by chemical trapping[132] or by real time gaseous flow PES analysis[133]
starting from various precursors. Formation of methanethial ($H_2C=S$) and
ethanethial ($CH_3(H)C=S$) by dehydrocyanation of the corresponding thiocyan-
hydrins on solid base (K_2CO_3 or CaO at $200°C$) was evidenced by VGSR/HRMS
coupling, including CAD MIKE experiments, and by low temperature IR spectroscopy[134] (condensation of the gaseous flow on a KBr window at 77K). In the
same way, the VGSR (K_2CO_3 or CaO, $200°C$, $1.3 \cdot 10^{-3}$ hPa) of different mono-
and dialkylthiocyanhydrins gave the expected thioketones accompanied by their
enethiol tautomers.[135]
   In other experiments, the elusive monomeric thioformyl cyanide $NC(H)C=S$,
a possible cosmic species, has been generated in the gas-phase, either by FVT
(retroene reaction of allylcyanomethyl sulfide) or by VGSR (dehydrochlorination
of cyanomethylsulfenyl chloride over potassium carbonate).[136] Analysis of the
gaseous flow in real time by millimeter wave spectroscopy allowed the determination of the molecular constants (Scheme 27).

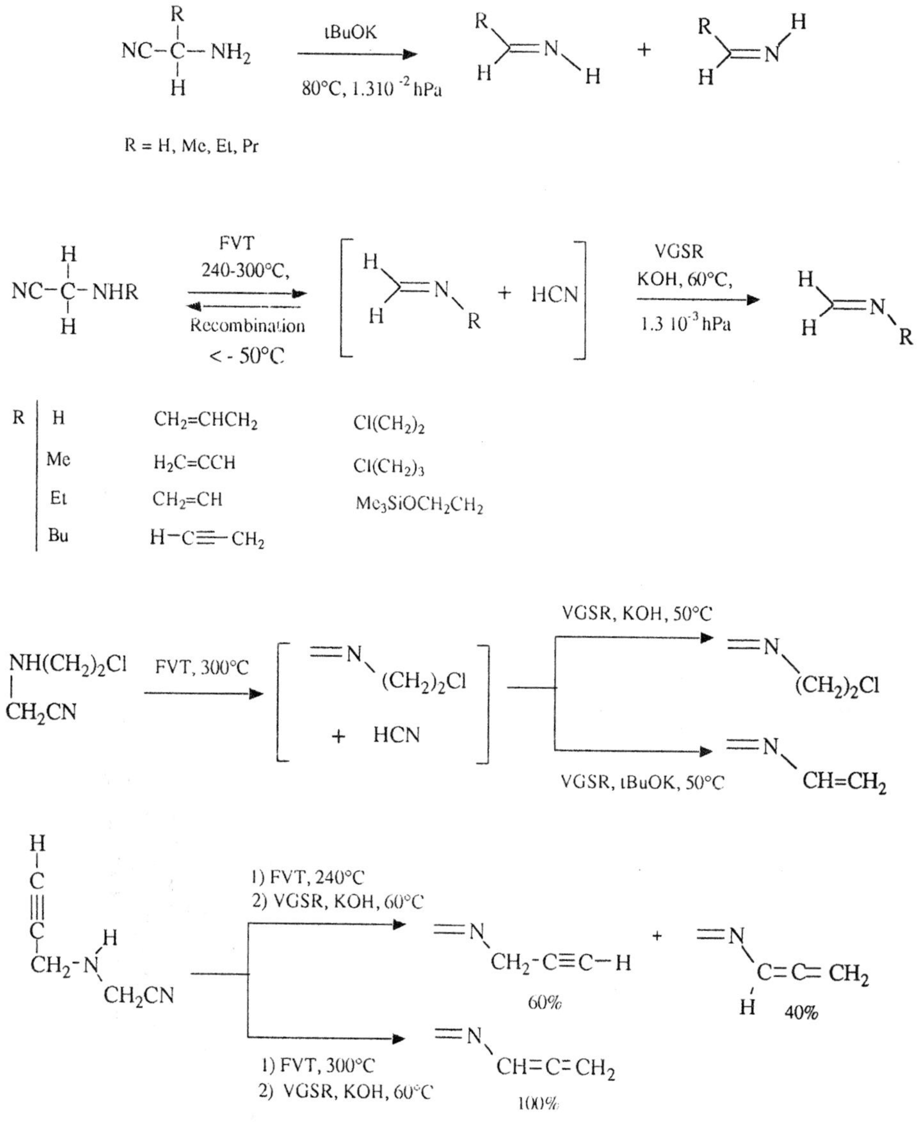

$$
\underset{\underset{H}{\overset{R}{|}}}{NC-C-NH_2} \xrightarrow[80°C, 1.3\,10^{-2}\,hPa]{tBuOK} \quad \overset{R}{\underset{H}{>}}=N\overset{\diagup H}{\diagdown_H} \quad + \quad \overset{R}{\underset{H}{>}}=N\overset{\diagup H}{\diagdown_H}
$$

R = H, Me, Et, Pr

$$
\underset{\underset{H}{\overset{H}{|}}}{NC-C-NHR} \underset{\underset{<-50°C}{Recombination}}{\overset{FVT\ 240-300°C,}{\rightleftharpoons}} \left[\overset{H}{\underset{H}{>}}=N_R \quad + \quad HCN \right] \xrightarrow[1.3\,10^{-3}\,hPa]{VGSR\ KOH,\ 60°C,} \overset{H}{\underset{H}{>}}=N_R
$$

R			
H	CH₂=CHCH₂	Cl(CH₂)₂	
Me	H₂C=CCH	Cl(CH₂)₃	
Et	CH₂=CH	Me₃SiOCH₂CH₂	
Bu	H−C≡−CH₂		

$$
\underset{\underset{CH_2CN}{\overset{NH(CH_2)_2Cl}{|}}}{} \xrightarrow{FVT,\ 300°C} \left[=N_{\diagdown (CH_2)_2Cl} \quad + \quad HCN \right]
$$

VGSR, KOH, 50°C → =N⟍(CH₂)₂Cl

VGSR, tBuOK, 50°C → =N⟍CH=CH₂

1) FVT, 240°C
2) VGSR, KOH, 60°C → =N⟍CH₂-C≡C−H 60% + =N⟍C=C=CH₂ (H) 40%

1) FVT, 300°C
2) VGSR, KOH, 60°C → =N⟍CH=C=CH₂ 100%

Scheme 26

4.4.4. Short-lived phosphaalkenes, phosphaalkynes, and phosphallenes

The ability of phosphorus to form pπ–pπ bonds with carbon atoms was demonstrated for the first time in 1961 with the formation of H—C≡P by passage of a stream of PH₃ between an electric arc of graphite electrodes.[137] This unstable compound remained a simple chemical curiosity during near two decades. A renewed interest started with the first synthesis in solution of P–C multiple bond

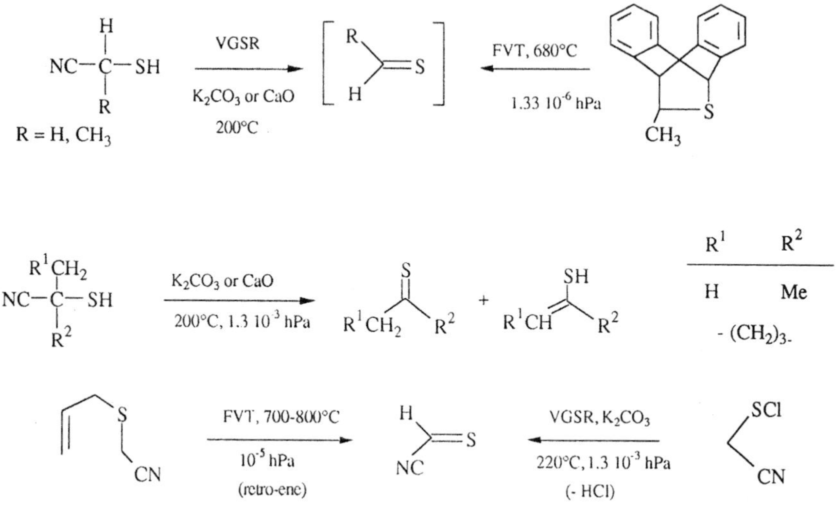

Scheme 27

derivatives stabilized by bulky substituents and by the generation and spectro-scopic characterization in the gas-phase of the simplest of them. This chemistry has known during the last 15 years an impressive development. A comprehensive treatise[138] and several reviews[139] dealing with their preparation, characterization, reactivity, and structural properties have been provided.

(a) *Simple phosphaalkenes* The simplest phosphorus–carbon multiple bond de-rivatives were first generated in the gas-phase.[139] Thus, the formation of $H_2C{=}P{-}H$, $H_2C{=}P{-}Cl$, $F_2C{=}P{-}H$ and $H_3C{-}C{\equiv}P$ was achieved by thermal HX-elimination from suitable P-haloprecursors under FVT conditions. Their structural data were obtained by gaseous flow analysis (FVT/MWS and FVT/PES coupling).[35] However, in these conditions, the reverse reaction due to the presence of HX prevented their isolation in solution.

The P-chlorophosphaethene $H_2C{=}PCl$ was detected by MWS in the thermal decomposition products of CH_3PCl_2 under FVT conditions[140]. The P-chlorophos-phaalkene $H_2C{=}P{-}Cl$ and phosphaalkyne $H{-}C{\equiv}P$ were formed in a one line sequence by thermal dehydrochlorination of the dichlorophosphine in FVT conditions and HCl-removing on solid KOH at $-78°C$[141,142] or tricyclohexyl-s-triazine at room temperature.[143] The P-chlorophosphapropyne were formed by a similar approach but the selectivity of the thermal elimination reaction was poor (Scheme 28). Attempts to form these different structures by HCl-elimination on solid base (VGSR) were unsuccessful.[21]

It was shown that HX-elimination is strongly favored when the leaving group (X = F, Cl) is bonded to the carbon in α-position to the phosphorus atom. Thus, $F_2C{=}P{-}H$ and $F{-}C{\equiv}P$ were formed by passing at room temperature the trifluoromethyl precursor through a U-tube containing KOH pellets[144]

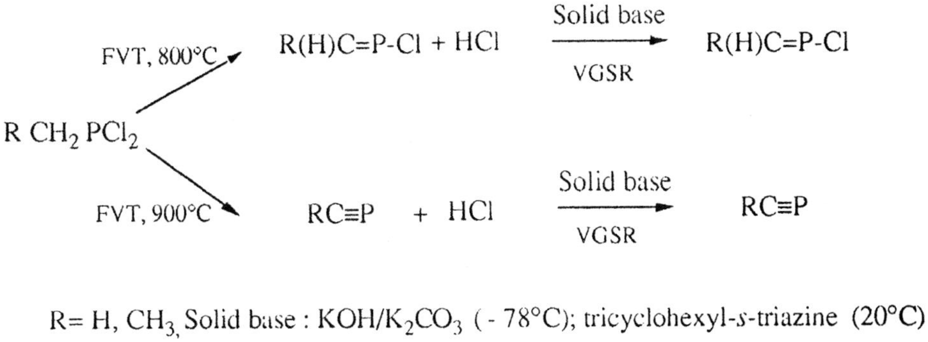

$$\text{R CH}_2\text{ PCl}_2 \begin{cases} \xrightarrow{\text{FVT, 800°C}} \text{R(H)C=P-Cl + HCl} \xrightarrow[\text{VGSR}]{\text{Solid base}} \text{R(H)C=P-Cl} \\ \\ \xrightarrow{\text{FVT, 900°C}} \text{RC}{\equiv}\text{P} \;+\; \text{HCl} \xrightarrow[\text{VGSR}]{\text{Solid base}} \text{RC}{\equiv}\text{P} \end{cases}$$

R= H, CH$_3$, Solid base : KOH/K$_2$CO$_3$ (- 78°C); tricyclohexyl-s-triazine (20°C)

Scheme 28

or through a spiral glass tube filled with broken KOH (*crossed-bed reactors*). Products were characterized by microwave,[145,146] photoelectron[146] and IR spectroscopies.[146–148] This procedure has been extended to the synthesis of CF$_3$P=CF$_2$[149–152] (Scheme 29). Absence of HX allowed to isolate and characterize these products in solution by low temperature NMR.[144,149] Thus, the VGSR process appeared here to be cleaner than FVT.

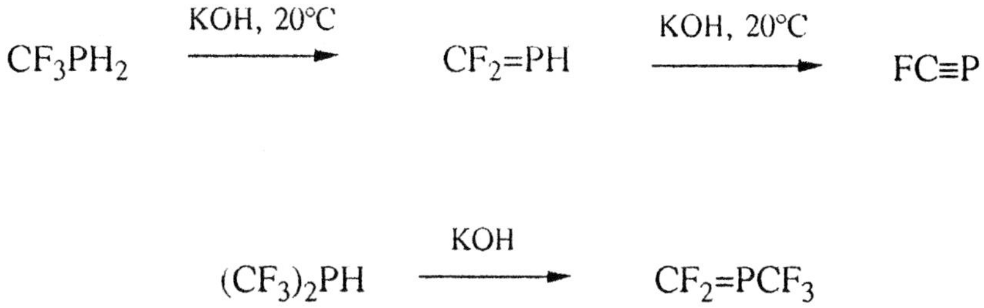

$$\text{CF}_3\text{PH}_2 \xrightarrow{\text{KOH, 20°C}} \text{CF}_2{=}\text{PH} \xrightarrow{\text{KOH, 20°C}} \text{FC}{\equiv}\text{P}$$

$$(\text{CF}_3)_2\text{PH} \xrightarrow{\text{KOH}} \text{CF}_2{=}\text{PCF}_3$$

Scheme 29

The transient methylidenephosphine H$_2$C=P—H was formed by passing ClCH$_2$PH$_2$ in VGSR conditions over K$_2$CO$_3$ heated at 150–200°C and was characterized in the gas-phase by HRMS, PES, and in solid state by IR after condensation of the gaseous flow on a KBr window cooled at 77 K.[153,154] Its stability was evaluated by the rapid decrease of the v_{PH} and $v_{P=C}$ stretching frequencies on the KBr window ($\tau_{1/2} \approx 10$ min). Dehydrochlorination of the P- and C-substituted α-chlorophosphines led to the corresponding phosphaalkenes (characterization by VGSR/IR, MS,[155,156] and PES coupling[154]) (Scheme 30). Attempts to characterize the parent compound by NMR failed, oligomerization occuring during the transfer from the cold trap to the NMR tube.[153] The characterization and the chemical trapping were however accomplished when the elimination was performed in solution in the presence of the required Lewis base

polymerization
- 20°C
in solution
IR 77K
VGSR
K2CO3
1.3 10-3 hPa
ClCH2PH2
DABCO
THF, - 50°C
High Resolution
Mass Spectroscopy
Chemical trapping (solution)
RCH(Cl)P(H)R'
VGSR
K2CO3, 100-250°C
(E)
(Z)
R = CH3, R' = H
R = H, R' = CH3

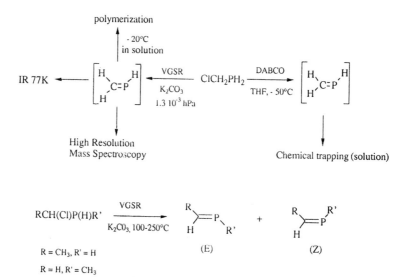

Scheme 30

and trapping agent.[153,155,156] Thus, dehydrochlorination in VGSR conditions and in solution appeared to be complementary.

(*b*) *Simple phosphabutadienes* Unsubstituted 1- and 2-phosphabutadienes were formed by 1,4- and 1,2-HCl elimination from their chloroalkyl precursors (Scheme 31). The 2-phosphabutadiene is sufficiently stable to be analyzed by ^1H and ^{13}C NMR. The structure of the two products was precised by real time gas analysis (VGSR/MS and VGSR/PES)[157]. Although the ring closure of the two structures was predicted to be thermoneutral from theoretical calculations,[158,159] the cyclic dihydrophosphete isomers were not detected upon warming the solution at room temperature; a decomposition was only observed.

(*c*) *Phosphaalkynes* Phosphaalkynes are useful building blocks in synthesis. They react for example with dienes, dipoles, and nucleophiles to lead to the corresponding cycloadducts or adducts. Complexation with transition metal opened a

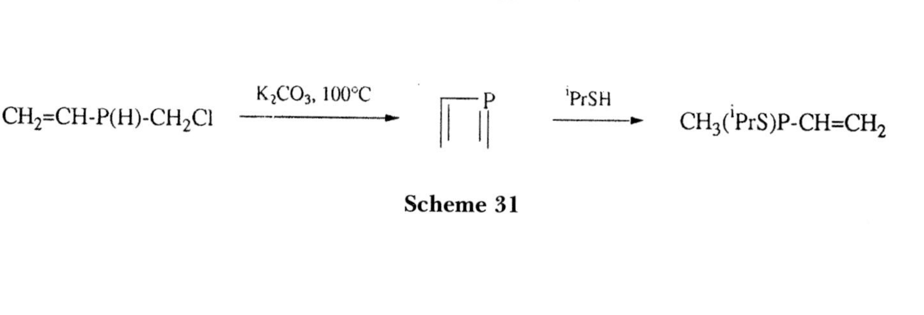

ClCH2-CH=CH-PH2
K2CO3, 100°C
iPrSH
CH2=CH-CH2P(H)SiPr
CH2=CH-P(H)-CH2Cl
K2CO3, 100°C
iPrSH
CH3(iPrS)P-CH=CH2

Scheme 31

new field in organometallic chemistry.[160] They were mainly prepared by three efficient routes which involved respectively elimination of ClSiMe$_3$ or HCl on solid KOH from their corresponding precursors. The third approach, which involves the base-promoted rearrangement of primary ethynylphosphines will be presented in Section 4.5.

Dechlorosilation of suitable precursors occured under vacuum by slow addition of the P-silylated phosphaalkene precursors on solid NaOH heated to 110–160°C. The phosphaalkynes were evacuated under vacuum as they were formed and condensed on a cold trap.[161–163] This procedure can be considered as a vacuum gas–solid reaction since the vaporization of the substrate occured on the contact of the solid base. This approach constitutes the best procedure for the preparation of volatile phosphaalkynes bearing tertiary and secondary carbon atom. However, the required stability of the precursors and the products limited the extend of this method.

The second approach involved the bis-dehydrochlorination of a-dichlorophosphines on solid K$_2$CO$_3$[164] in VGSR conditions. This method allowed to synthesize phosphaalkynes bearing a primary alkyl group or a heteroelement (Cl, SiMe$_3$), compounds which are not accessible by the dechlorosilation route (Scheme 32); the C-chlorophosphaalkene intermediates were not observed. Alkylphosphaalkynes can be kept few days in solution at room temperature but rapidly oligomerize in the condensed phase. The half-live of H—C≡P is about 6 h in solution at room temperature. The photoelectron spectrum of ClC≡P has been also reported.[165]

4.5. Base-Promoted Rearrangements

The base-promoted rearrangement on solid base constitutes an useful procedure when the desired target is thermodynamically more stable than other isomers. This process allowed to synthesize simple phosphaalkynes and more recently the arsapropyne.

4.5.1. Phosphaalkynes

It was predicted from theoretical calculations that the stability of the C$_2$H$_3$P isomers increased on going from ethynylphosphine to phosphaallene and finally to phophaalkyne.[166] The experiments were in good agreement with these calculations: the phosphaalkynes (R = H, Me or phenyl groups) were the sole observed products by passing the ethynylphosphine isomers on solid K$_2$CO$_3$ at room temperature in VGSR conditions (Scheme 32). The corresponding P-unsubstituted phosphaallene intermediates were never observed (low temperature ^{31}P NMR). The formation of these short-lived species was finally proved by performing these rearrangements in solution in the presence of a Lewis base (Et$_3$N, 10°C or DBU, − 90°C) and a thiol. The expected adducts were isolated.[167]

A mixture of P-methylethynylphosphine and P-methylphosphaallene was observed in the same molar ratio [8 : 92] respectively by gas-phase HCl-elimination

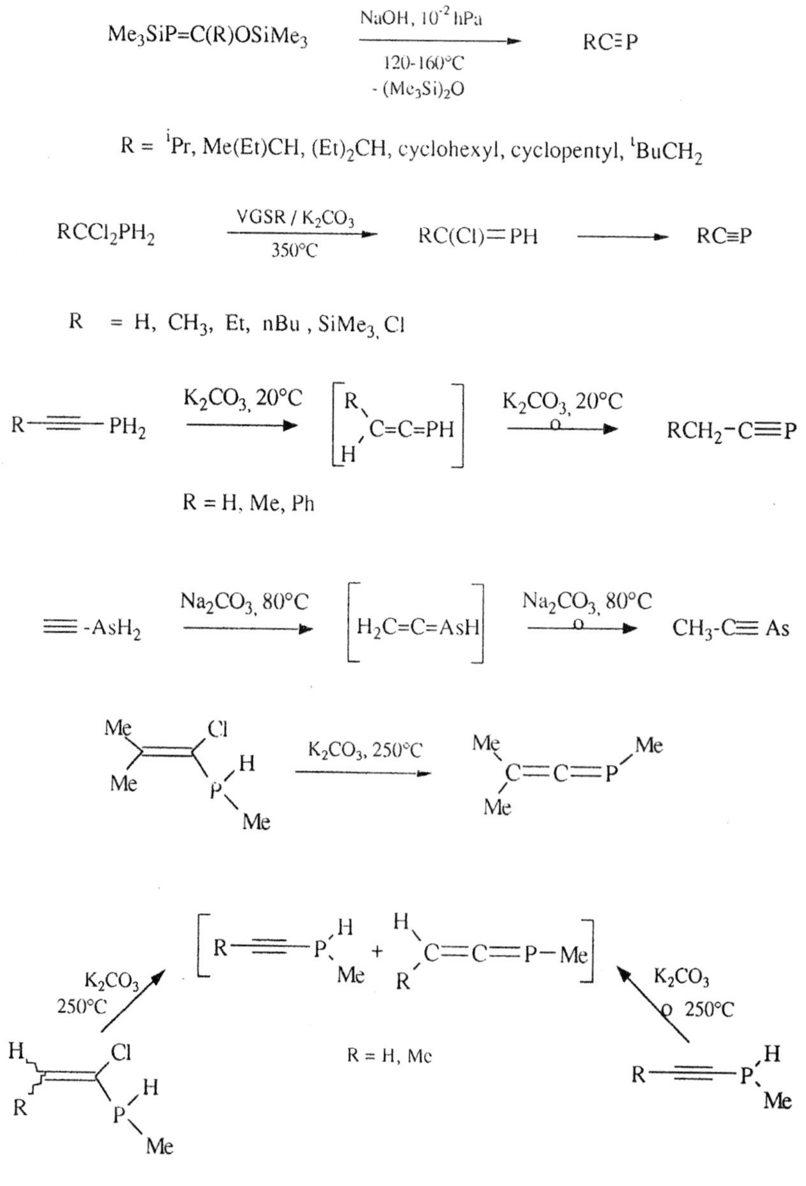

$$Me_3SiP{=}C(R)OSiMe_3 \quad \xrightarrow[\substack{120\text{-}160°C \\ -(Me_3Si)_2O}]{NaOH,\ 10^{-2}\,hPa} \quad RC{\equiv}P$$

R = ^{i}Pr, Me(Et)CH, (Et)$_2$CH, cyclohexyl, cyclopentyl, ^{t}BuCH$_2$

$$RCCl_2PH_2 \quad \xrightarrow[350°C]{VGSR\,/\,K_2CO_3} \quad RC(Cl){=}PH \quad \longrightarrow \quad RC{\equiv}P$$

R = H, CH$_3$, Et, nBu , SiMe$_3$, Cl

$$R{\equiv\!\equiv}PH_2 \quad \xrightarrow{K_2CO_3,\ 20°C} \quad \left[\begin{array}{c} R \\ \\ H \end{array}\!\!C{=}C{=}PH \right] \quad \xrightarrow{K_2CO_3,\ 20°C} \quad RCH_2{-}C{\equiv}P$$

R = H, Me, Ph

$${\equiv\!\equiv}{-}AsH_2 \quad \xrightarrow{Na_2CO_3,\ 80°C} \quad \left[H_2C{=}C{=}AsH \right] \quad \xrightarrow{Na_2CO_3,\ 80°C} \quad CH_3{-}C{\equiv}As$$

Scheme 32

of α-chlorovinylphosphines or by vaporisation of secondary ethynylphosphines on solid K$_2$CO$_3$. Analogous structures were observed with the C-methyl derivative, the molar ratio beween the two isomers being however different [55 : 45]. These results were rationalized in terms of a tautomeric equilibrium between ethynyl-phosphines and phosphaallenes. The P-substituted phosphaallenes generated by VGSR were collected by classical procedures and characterized by low tempera-ture NMR and chemical trapping.[168]

4.5.2. *Arsapropyne*

Compounds containing —As=C < and As≡C— bonds are less stable than their phosphorus counterparts. Only the kinetically stabilized derivatives have been isol ated so far. A simple arsapropyne was recently prepared by base-induced rearrangement of the ethynylarsine isomer on solid sodium carbonate. The arsa-allene which is probably involved as intermediate has not been detected. Arsap-ropyne which has been fully characterized by NMR and microwave spectroscopies exhibits as expected a low stability ($\tau_{1/2}$ *ca* 1 h in the gas-phase)[169] (Scheme 32).

ACKNOWLEDGEMENTS

The authors would like to express their thanks to their own co-workers whose work has been reviewed here. They would like to mention among them Rosalie Niamayoua which started the work in 1979 and has isolated the first short-lived species by using the VGSR technique and J.C. Guillemin whose persevering dynamism has greatly contributed to assure the development of this work.

REFERENCES

1. Photochemical reactions have been not considered in this review.
2. R.F.C. Brown, *Pyrolytic Methods in Organic Chemistry*, R.H. Wasserman, Ed., Academic Press: New York (1980).
3. Review: G. Seybold, *Angew. Chem. Int. Ed. Engl.*, **16**, 365 (1977).
4. Review: U.E. Wirsum, *Recl. Trav. Chim. Pays-Bas*, **101**, 317 (1982).
5. Review: M. Karpf, *Angew. Chem. Int. Ed. Engl.*, **25**, 414 (1986).
6. Review: M.C. Lasne and J.L. Ripoll, *Synthesis*, 121 (1985).
7. Review: J.L. Ripoll and Y. Valle, *Synthesis*, 659 (1993).
8. Review: W. E. Billups and D.J. McCord, *Angew. Chem. Int. Ed. Engl.*, **33**, 1332 (1994).
9. Review: H. Bock and B. Solouki, *Angew. Chem. Int. Ed. Engl.*, **20**, 427 (1981).
10. (a) H. Bock, *Phosphorus, Sulfur, Silicon and Relat. Elem.*, **49**, 3 (1990); (b) H. Bock, B. Solouki, S. Aygen, M. Bankmann, O. Breuer, R. Dammel, J. Dörr, M. Haun, T. Hirabayashi, D. Jaculi, J. Mintzer, S. Mohmand, H. Müller, P. Rosmus, B. Roth, J. Wittmann and H.P. Wolf, *J. Mol. Struct.*, **173**, 31 (1988).
11. See for example A.N. Gokarn and A.R. Pande, *Ind. Chem. Ing.*, **33**, 52 (1991).
12. For a review, see for example J. Szekely, *Chem. React. Eng. Proc. Int. Sym.*, 517 (1976).
13. J.A. Barnard, W.J. Thomas and S.P.S. Andrew, *Compr. Chem. Kinet.*, C.H. Bamford, C.F.H. Tipper and R.G. Compton, Eds., from kinetics and chemical technology, Elsevier, Amsterdam (1985).
14. G. Kaupp and D. Matthies, *Chem. Ber.*, **119**, 2387 (1986).
15. R. Lamartine, *Mol. Cryst. Liq. Cryst.*, **186**, 53 (1990).
16. M.J. Tremeling, S.P. Hopper and J.A. Evans, *Tetrahedron Lett.*, 1153 (1976).
17. R. Niamayoua, Dissertation N° 771, Université de Lille (1979).
18. E.A. Dorko, *J. Am. Chem. Soc.*, **87**, 5518 (1965).
19. F.A. Miller, F.R. Brown and K.H. Rhee, *Spectrochim. Acta*, **28**, 1467 (1972).
20. T. Bally, U. Buser and E. Haselbach, *Helv. Chim. Acta*, **61**, 38 (1978).
21. J.M. Denis, Unpublished results.
22. B. De Corte, J.M. Denis and N. De Kimpe, *J. Org. Chem.*, **52**, 1147 (1987).
23. P.W. Schenk and H. Platz, *Z. Anorg. Allg. Chem.*, **215**, 112 (1933).
24. P.W. Schenk and H. Triebel, *Z. Anorg. Allg. Chem.*, **229**, 305 (1936).

25. P.W. Schenk and R. Steudel, *Angew. Chem. Int. Ed. Engl.*, **4**, 402 (1965).
26. P.W. Schenk and R. Steudel, *Z. Anorg. Allg. Chem.*, **342**, 253 (1966).
27. H. Bock, B. Soluki, P. Rosmus and R. Steudel, *Angew. Chem. Int. Ed. Engl.*, **12**, 933 (1973).
28. J.C. Guillemin and J.M. Denis, *Angew. Chem. Int. Ed. Engl.*, **21**, 690 (1982) and *Angew. Chem. Suppl.*, **21**, 1515 (1982).
29. J.C. Guillemin and J.M. Denis, *Tetrahedron*, **44**, 4431 (1988).
30. J.M. Denis, R. Niamayoua and A. Lablache-Combier, *Tetrahedron Lett.*, **21**, 515 (1980).
31. W.E. Billups, L.J. Lin and E.W. Casserly, *J. Am. Chem. Soc.*, **106**, 3698 (1984).
32. R. Bloch and J.M. Denis, *Angew. Chem. Int. Ed. Engl.*, **19**, 928 (1980).
33. B. Pellerin, P. Guenot and J.M. Denis, *Tetrahedron Lett.*, **28**, 5811 (1987).
34. J.C. Guillemin and J.M. Denis, *Tetrahedron*, **44**, 4431 (1988).
35. H.W. Kroto, *Chem. Soc. Rev.*, **435** (1982).
36. Y. Malecot, J.L. Ripoll and A. Thuillier, *J. Chem. Res.*, S 86; M 959 (1983).
37. K. Ohno, E. Kurita, M. Kawamura and H. Matsuura, *J. Am. Chem. Soc.*, **109**, 5614 (1987).
38. O.L. Chapman, C.C. Chang and R.N. Rosenquist, *J. Am. Chem. Soc.*, **98**, 261 (1976).
39. N.P.C. Westwood, *Chem. Phys. Lett.*, **25**, 558 (1974).
40. H. Bock, B. Solouki and G. Maier, *Angew. Chem. Int. Ed. Engl.*, **24**, 205 (1985).
41. H. Bock, M. Kremer, M. Dolg and H.W. Preuß, *Angew. Chem. Int. Ed. Engl.*, **30**, 1186 (1991).
42. E.F. Pearson and R.V. McCormick, *J. Chem. Phys.*, **58**, 1619 (1973).
43. R. Sams and A.G. Maki, 28th Colombus Symposium Abstract (1973).
44. K.W. Kroto, R.J. Suffolk and N.P.C. Westwood, *J. Phys. Lett.*, **22**, 495 (1973).
45. (a) H. Bock, J. Minszer, J. Wittmann and J. Russow, *Angew. Chem. Int. Ed. Engl.*, **19**, 147 (1980);
    (b) H. Bock, J. Wittmann, J. Mintzer and J. Russow, *Chem. Ber.*, **115**, 2346 (1982).
46. H. Bock, J. Wittmann and H.J. Arpe, *Chem. Ber.*, **115**, 2327 (1982).
47. H. Bock and B. Berkner, *Z. Naturforsch.*, **47b**, 1551 (1992).
48. H. Bock, G. Tschmutowa and H.P. Wolf, *J. Chem. Soc., Chem. Commun.*, 1068 (1986).
49. G.T. Burns and T.J. Barton, *J. Organomet. Chem.*, **209**, C25 (1981).
50. A. Laporterie, J. Dubac, P. Mazerolles and H. Iloughmane, *J. Organomet. Chem.*, **216**, 321 (1981).
51. A. Laporterie, G. Manuel, J. Dubac, P. Mazerolles and H. Iloughmane, *J. Organomet. Chem.*, **210**, C33 (1981).
52. A. Laporterie, G. Manuel, J. Dubac and P. Mazerolles, *Nouv. J. Chem.*, **6**, 67 (1982).
53. A. Laporterie, G. Manuel, H. Iloughmane and J. Dubac, *Nouv. J. Chem.*, **8**, 437 (1984).
54. W.E. Bachmann, M.P. Cava and A.S. Dreiding, *J. Am. Chem. Soc.*, **76**, 5554 (1954).
55. E.N. Nagy-Felsobuki and J.B. Peel, *J. Chem. Soc., Faraday 2*, **74**, 1927 (1978).
56. B. Braillon, M.C. Lasne, J.L. Ripoll and J.M. Denis, *Nouv. J. Chem.*, **6**, 121 (1982).
57. J.C. Guillemin and J.M. Denis, *Synthesis*, **12**, 1131 (1985).
58. H. Bock and R. Dammel, *Chem. Ber.*, **120**, 1961 (1987).
59. H. Bock and R. Dammel, *Chem. Ber.*, **120**, 1971 (1987).
60. J.C. Guillemin and J.M. Denis, *Tetrahedron*, **44**, 4447 (1988).
61. D.C. Frost, S.T. Lee and C.A. McDowell, *Chem. Phys. Lett.*, **22**, 243 (1973).
62. M. Binnewies, B. Soluki, H. Bock, R. Becherer and R. Ahlrichs, *Angew. Chem. Int. Ed. Engl.*, **23**, 7315 (1984).
63. Von M. Binnewies, M. Lakenbrink and H. Schnöckel, *Z. Anorg. Allg. Chem.*, **497**, 7 (1983).
64. Von H. Schnöckel and M. Lakenbrink, *Z. Anorg. Allg. Chem.*, **507**, 70 (1983).
65. H. Bock and Bankmann, *Phophorus, Sulfur, Silicon and Relat. Elem.*, **53**, 167 (1990).
66. Von M. Binnewies, *Z. Anorg. Allg. Chem.*, **507**, 66 (1983).
67. T.J. Dennis, S. Firth, H.W. Kroto, D.R.M. Walton and C.Y. Mock, *J. Chem. Soc., Chem. Commun.*, 1430 (1990).
68. S. Firth, S. Khalaf and H.W. Kroto, *J. Chem. Soc., Faraday Trans.*, **88**, 3393 (1992).
69. T.J. Dennis, S. Firth, H.W. Kroto, G.Y. Matti, C.Y. Mock and R.J. Suffolk, *J. Chem. Soc., Faraday Trans.*, **87**, 917 (1991).
70. H. Bock and M. Bankmann, *J. Chem. Soc., Chem. Commun.*, 1131 (1989).
71. Von M. Binnewies, M. Lakenbring and H. Schnöckel, *Z. Anorg. Allg. Chem.*, **497**, 7 (1983).
72. Von M. Binnewies, *Z. Anorg. Allg. Chem.*, **507**, 77 (1983).
73. E. Lindner, K. Auch, W. Hiller and R. Fawzi, *Angew. Chem. Int. Ed. Engl.*, **23**, 320 (1984).
74. Von M. Binnewies, *Z. Anorg. Allg. Chem.*, **505**, 32 (1983).

75. H. Bock, T. Hirabayashi and S. Mohmand, *Chem. Ber.*, **114**, 2595 (1981).
76. U.H. Brinker and J. Ritzer, *J. Am. Chem. Soc.*, **103**, 2116 (1981).
77. Review: W. Kirmse, *"Carbene chemistry"*, Academic Press, New York (1971).
78. Review: H. Hopf and G. Maas, *Angew. Chem. Int. Ed. Engl.*, **31**, 931 (1992).
79. T. Bally, U. Buser and E. Haselbach, *Helv. Chim. Acta*, **38**, 61 (1978).
80. P.A. Waitkus, L.I. Peterson and G.W. Griffin, *J. Am. Chem. Soc.*, **88**, 181 (1966); P.A. Waitkus, E. Sanders, L.I. Peterson and G.W. Griffin, *J. Am. Chem. Soc.*, **89**, 6318 (1967).
81. F.A. Miller, F.R. Brown and K.H. Rhee, *Spectrochimica Acta Part A*, **28**, 1467 (1972).
82. L. Trabert and H. Hopf, *Liebigs Ann. Chem.*, 1786 (1980).
83. E. Hedaya, M.E. Kent, D.W. McNeil, F.P. Lossing and T. McAllister, *Tetrahedron Lett.*, 3415 (1968).
84. G.L. Closs and K.O. Krantz, *J. Org. Chem.*, **31**, 638 (1966).
85. J. Kao and J. Radom, *J. Am. Chem. Soc.*, **100**, 760 (1978).
86. J.M. Denis, unpublished results; these modifications have been at first applied for the imine synthesis in VGSR conditions (Reference. 28).
87. S.W. Staley and T.D. Norden, *J. Am. Chem. Soc.*, **106**, 3699 (1984).
88. M. Bogey, M. Cordonnier, J.L. Destombes, J.M. Denis and J.C. Guillemin, *J. Mol. Spectrosc.*, **149**, 230 (1091).
89. T.D. Norden, S.W. Staley, W.H. Taylor and M.D. Harmony, *J. Am. Chem. Soc.*, **108**, 7912 (1986).
90. S.W. Staley and T.D. Norden, *J. Am. Chem. Soc.*, **111**, 445 (1989).
91. I. Shavitt, D.W. Ewing and J.E. Del Bene, *J. Am. Chem. Soc.*, **113**, 9389 (1991).
92. W.E. Billups and L.J. Lin, *Tetrahedron*, **42**, 1575 (1986).
93. T.H. Chan and D. Massuda, *Tetrahedron Lett.*, **39**, 3383 (1982).
94. W.E. Billups and M.M. Haley, *J. Am. Chem. Soc.*, **113**, 5084 (1991).
95. W.E. Billups and M.M. Haley, *Angew. Chem. Int. Ed. Engl.*, **28**, 1711 (1989).
96. R. Bose, D. Bläser, R. Gleiter, K.H. Pfeifer, W.E. Billups and M.M. Haley, *J. Am. Chem. Soc.*, **115**, 743 (1993).
97. W.E. Billups, G.A. Lee, B.E. Arney Jr. and K.H. Whitmire, *J. Am. Chem. Soc.*, **113**, 7980 (1991).
98. M.M. Haley, B. Biggs, W.A. Looney and R. D. Gilbertson, *Tetrahedron Lett.*, **36**, 3457 (1995).
99. Review: H. Bock and R. Dammel, *Angew. Chem. Int. Ed. Engl.*, **26**, 504 (1987).
100. S.L. Miller and L.E. Orgel, *The Origin of life on Earth*, Prentice-Hall, NY, USA (1974).
101. P.D. Godfrey, R.D. Brown, B.J. Robinson and M.W. Sinclair, *Astrophys. Lett.*, **13**, 119 (1973).
102. J.C. Guillemin, Dissertation, Université de Rennes, N°445, 1986.
103. R.G. Ford, *J. Am. Chem. Soc.*, **99**, 2389 (1977).
104. H. Bock, R. Dammel and S. Aygen, *J. Am. Chem. Soc.*, **105**, 7681 (1983).
105. J.C. Guillemin and J.M. Denis, *J. Chem. Soc., Chem. Commun.*, **238** (1983).
106. M. Bogey, J.L. Destombes, J.M. Denis and J.C. Guillemin, *J. Mol. Spectrosc.*, **115**, 1 (1986).
107. R.H. Wiley and J. Moffat, *J. Org. Chem.*, **22**, 190 (1957).
108. J.M. Denis and J.C. Guillemin, unpublished results.
109. S. Green and E. Herbst, *Astrophys. J.*, **229**, 121 (1979).
110. M.E. Jacox, *J. Am. Chem. Soc.*, **99**, 2389 (1977).
111. J.C. Guillemin, J.M. Denis and A. Lablache-Combier, *J. Am. Chem. Soc.*, **103**, 468 (1981).
112. B.K. Ogawa, Y. Numura, Y. Takeuchi and S. Tomada, *J. Chem. Soc. Perkin Trans.* **1**, 3031 (1982).
113. N. Bodor and R. Pearlman, *J. Am. Chem. Soc.*, **100**, 4946 (1978).
114. M.C. Lasne, J.L. Ripoll, J.C. Guillemin and J.M. Denis, *Tetrahedron Lett.*, **25**, 3847 (1984).
115. R. Dammel, Dissertation, Frankfurt am Main (1985).
116. R. Dammel, H. Bock and J.M. Denis, *Chem. Phys. Lett.*, **102**, 239 (1983).
117. S. Lacombe, B. Pellerin, J.C. Guillemin, J.M. Denis and G. Pfister-Guillouzo, *J. Org. Chem.*, **54**, 5958 (1989).
118. M. Sugie, H. Takeo and C. Matsumura, *J. Am. Chem. Soc.*, **111**, 906 (1989).
119. J. Demaison, J. Burie, J.M. Denis and B.P. Van Eijck, *J. Mol. Spectrosc.*, **107**, 250 (1984).
120. F.E. Scully, Jr. and R.C. Davis, *J. Org. Chem.*, **43**, 1467 (1978).
121. Review: R.A. Bartsch, *Acc. Chem. Res.*, **80**, 453 (1980).
122. R.A. Bartsch, G.J. Bracken and I. Yilmaz, *Tetrahedron Lett.*, **23**, 2109 (1979).
123. E. Weiss, H. Alsdorf and H. Kühr, *Angew. Chem. Int. Ed. Engl.*, **6**, 801 (1967).

124. L.E. Overman and R.M. Burk, *Tetrahedron Lett.*, **25**, 1635 (1984).

125. R.D. Brown, P.D. Godfrey and D.A. Winkler, *Aust. J. Chem.*, 951 (1985).

126. J.C. Guillemin, J.M. Denis, M. Bogey and J.L. Destombes, *Tetrahedron Lett.*, **27**, 1187 (1986).

127. D.C. Ha, D.J. Hart and T.K. Yang, *J. Am. Chem. Soc.*, **106**, 4816 (1984); R.W. Colvin and D.G. McGarry, *J. Chem. Soc., Chem. Commun.*, 1986 (1985).

128. G. Cainelli, D. Giacomini, M. Panunzio, G. Martelli and G. Spunta, *Tetrahedron Lett.*, **28**, 5369 (1987).

129. J.C. Guillemin, L. Ammi and J.M. Denis, *Tetrahedron Lett.*, **29**, 1287 (1988).

130. N. De Kimpe, R. Verhé, L. De Buyck, J. Chys and N. Schamp, *Synthesis*, **895** (1978).

131. J.C. Guillemin and J.M. Denis, *J. Chem. Soc., Chem. Commun.*, 951 (1985).

132. E. Vedejs, T.H. Eberlein, D.J. Mazur, C.K. McClure, D.A. Perry, R. Ruggeri, E. Schwartz, J.S. Stults, D.L. Varie, R.G. Wilde and S. Wittenberger, *J. Org. Chem.*, **51**, 1556 (1986).

133. H. Bock, T. Hirabayashi and S. Mohmand, *Chem. Ber.*, **115**, 492 (1982).

134. L. Waszneh, J.C. Guillemin, P. Guenot, Y. Valle and J.M. Denis, *Tetrahedron Lett.*, **29**, 5899 (1988).

135. A.C. Gaumont, L. Waszneh and J.M. Denis, *Tetrahedron*, **47**, 4927 (1991).

136. M. Bogey, C. Demuynck, J.L. Destombes, A.C. Gaumont, J.M. Denis, Y. Vallée and J.L. Ripoll, *J. Am. Chem. Soc.*, **111**, 7399 (1989).

137. T.E. Gier, *J. Am. Chem. Soc.*, **83**, 1764 (1961).

138. M. Regitz and O.J. Scherer, Eds., *Multiple Bonds and Low Coordination in Phosphorus Chemistry*, Thieme: Stuttgart, (1990).

139. See A.C. Gaumont and J.M. Denis, *Chem. Rev.*, **94**, 1413 (1994) and references therein.

140. M.J. Hopkinson, H.W. Kroto, J.F. Nixon and N.P.C. Simmons, *J. Chem. Soc., Chem. Commun.*, **513**, (1976).

141. T.A. Cooper, H.W. Kroto, J.F. Nixon and O. Ohashi, *J. Chem. Soc., Chem. Commun.*, 333 (1980).

142. E.P.O. Fuchs, M. Hermesdorf, W. Schnurr, W. Röesch, H. Heydt and M. Regitz, *J. Oganomet. Chem.*, **338**, 329 (1988).

143. B. Pellerin, J.M. Denis, J. Perrocheau and R. Carrié, *Tetrahedron Lett.*, **27**, 5723 (1986).

144. H.E. Eshtiagh-Hosseini, H.W. Kroto and J.F. Nixon, *J. Chem. Soc., Chem. Commun.*, **653** (1979).

145. H.W. Kroto, J.F. Nixon and N.P.C. Simmons, *J. Mol. Spectrosc.*, **82**, 185 (1980).

146. H.W. Kroto, J.F. Nixon, N.P.C. Simmons and N.P.C. Westwood, *J. Am. Chem. Soc.*, **100**, 446 (1978).

147. K. Ohno, H. Matsuura, H.W. Kroto and H. Murata, *Chem. Lett.*, 981 (1982).

148. K. Ohno, E. Kurita, M. Kawamura and H. Matsuura, *J. Am. Chem. Soc.*, **109**, 5614 (1987).

149. H.E. Eshtiagh-Hosseini, H.W. Kroto and J.F. Nixon, *J. Organomet. Chem.*, **181**, C1 (1979).

150. H.E. Eshtiagh-Hosseini, H.W. Kroto, J.F. Nixon and O. Ohashi, *J. Organomet. Chem.*, **296**, 351 (1985).

151. J. Grobe, *J. Organomet. Chem.*, **296**, 351 (1985).

152. A.B. Burg, *Inorg. Chem.*, **22**, 2573 (1983).

153. B. Pellerin, P. Guenot and J.M. Denis, *Tetrahedron Lett.*, **28**, 5811 (1987).

154. S. Lacombe, D. Gonbeau, J.L. Cabioch, B. Pellerin, J.M. Denis and G. Pfister-Guillouzo, *J. Am. Chem. Soc.*, **110**, 6964 (1988).

155. J.L. Cabioch, *Thése de l'Université de Rennes I*, order N° 364 (1989).

156. X. Morise, *Thése de l'Université de Rennes I*, order N° 625 (1991).

157. J.C. Guillemin, J.L. Cabioch, X. Morise, J.M. Denis, S. Lacombe, D. Gonbeau, G. Pfister-Guillouzo, P. Guenot and P. Savignac, *Inorg. Chem.*, **32**, 5021 (1993).

158. S.M. Bachrach and M. Liu, *J. Am. Chem. Soc.*, **113**, 7929 (1991).

159. S.M. Bachrach and M. Liu, *J. Org. Chem.*, **57**, 209 (1992).

160. Reviews: (a) J.F. Nixon, *Chem. Rev.*, **88**, 1327 (1988); (b) M. Regitz, *Angew. Chem. Int. Ed. Engl.*, **100**, 1541 (1988); (c) M. Regitz, *Chem. Rev.*, **90**, 191 (1990).

161. M. Regitz, *Chem. Rev.*, **90**, 191 (1990).

162. W. Rösch, U. Vogelbacher, T. Allspach and M. Regitz, *J. Organomet Chem.*, **306**, 39 (1986).

163. M. Regitz, *Bull. Soc. Chim. Belg.*, **101**, 359 (1992).

164. J.C. Guillemin, T. Janati, P. Guenot, P. Savignac and J.M. Denis, *Angew. Chem., Int. Ed. Engl.*, **30**, 196 (1991).

165. S. Lacombe, G. Pfister-Guillouzo, J.C. Guillemin and J.M. Denis, *J. Chem. Soc., Chem. Commun.*, **403** (1991).

166. M.T. Nguyen, L. Landuyt and L.G. Vanquickenborne, *J. Chem. Soc. Faraday Trans.*, **90**, 1771 (1994).
167. J.C. Guillemin, T. Janati and J.M. Denis, *J. Chem. Soc., Chem. Commun.*, 415 (1992).
168. J.C. Guillemin, T. Janati, J.M. Denis, P. Guenot and P. Savignac, *Tetrahedron Lett.*, **35**, 245 (1994).
169. J.C. Guillemin, L. Lassalle, P. Drean, G. Wlodarczak and J. Demaison, *J. Am. Chem. Soc.*, 116, 8930 (1994).

5. GENERATION BY FLASH VACUUM OR FLOW THERMOLYSIS OF DOUBLE-BONDED SILICON AND GERMANIUM SPECIES

JEAN-LOUIS RIPOLL

Laboratoire de Chimie Moléculaire et Thio-organique, UMR CNRS-ISMRA, Université de Caen, 6 Boulevard Maréchal Juin, 14050 Caen, France

INTRODUCTION

The chemistry of double-bonded silicon and germanium species has undergone an outstanding development during the two past decades and several reviews have been devoted to this subject.[1-8] In this chapter, the interest will be focused on the contribution brought in that area by gas-phase reactions and limited to those cases using thermal processes and dynamic conditions, namely flash vacuum and flow thermolysis. To our knowledge, no double-bonded tin derivatives have been generated until now under such conditions, likely owing to the easy loss of metallic tin at the temperature required for thermolysis.

In many reported examples, these species have been generated as transient intermediates, undirectly characterized after oligomerization or chemical trapping. Furthermore, the use of low temperature matrix or gas-phase spectrometries has allowed the direct observation of several of them. It should be noted that, if static conditions of heating are sometimes suitable in the former cases, the latter ones require always flash vacuum or flow thermolysis (photochemical methods not being considered here).

The first part of the chapter will be thus concerned with the generation of non-isolated intermediates and subdivided according to the unimolecular thermal process involved, such as fragmentations ([2+2] and [4+2] cycloreversions, retroene reactions), 1,1- 1,2- and 1,4-eliminations, silatropic rearrangements. In the second part, aimed at the direct spectrometric identification of double-bonded silicon or germanium compounds, the classification will be made depending on the method used: infrared or ultraviolet spectroscopy in matrix, and, in the gas phase, photoelectron, mass, and other spectrometries. The bibliography, covering the period from 1972 to 1992 (partly 1993), should not be regarded as fully exhaustive. The generation of silylenes ($R_2Si:$) has been recently surveyed[9,10] and will be considered here only in connection with that of silenes or disilenes.

Conditions of thermolysis (the word "thermolysis" is preferred to "pyrolysis" with the same meaning): The reader's attention should be drawn on the hereafter often used terms "flash vacuum thermolysis" (abbreviated FVT) and "flow thermolysis". True FVT conditions, in which high vacuum or dilution by an inert gas, as well as very short transit times in and after the hot zone, preclude mostly bimolecular and catalytic processes, remain as limit cases approached principally in direct spectrometric investigations (part II). More frequently, both terms refer with some overlapping to a continuum of conditions ranging from true FVT to atmospheric pressure flow thermolysis, and gas-phase bimolecular reactions are as a rule observed, owing to the high reactivity of the involved silicon or germanium unsaturated species. This is most evidently the case for cothermolyses or flow thermolyses in a non-inert gas acting as trapping agent. However, it is sometimes difficult in these thermolyses to be sure, whether a given reaction takes actually place in the gas phase, or only during condensation in the cold trap.

1.   GENERATION OF NON-ISOLATED INTERMEDIATES

By Thermal Fragmentations

[2+2] cycloreversions

Since the first report (1966) on the thermal cleavage, under flow thermolysis conditions, of 1,1-dimethylsilacyclobutane (**1**, R=H) into 1,1-dimethylsilene (**2**, R=H), characterized as its cyclo-dimer 1,1,3,3-tetramethyl-1,3-disilacyclobutane (**3**, R=H) (see ref. 11 and previous papers cited), the gas-phase thermal decomposition of 1,1-disubstituted silacyclobutanes has constituted the most often used access to the corresponding silenes.

The flow thermolysis (nitrogen, 610°C) of **1** (R=H), in the presence of a sixfold excess of acetonitrile, led in 34% yield, by addition of this latter on **2** (R=H), to trimethylsilylacetonitrile (**4**).[12] The successful use of silicon halides as trapping agents indicates clearly the strong polarization of the Si–C double bond; for example, compound **5** was obtained in 44% yield with silicon tetrafluoride and only traces of the cyclo-dimer **3** (R=H) were in that case detected.[13] The investigation of the FVT of **1** (R=H) by coupling with an infrared cryostat (77 K) showed also the cyclo-dimerization of 1,1-dimethylsilene in the gas phase, resulting in the trapping of **3** (R=H) on the plate, when the distance between oven and plate was sufficient (25 cm). On the other hand, a shorter distance (3 cm) resulted in the immediate polymerization of the monomer on the plate and no dimer was observed.[14]

A similar thermolysis of 2-substituted silacyclobutanes **1** (R=Me, Ph) showed that the preponderant pathway includes an initial C^2–C^3 bond cleavage giving first a biradical which is then cleaved into ethylene and the corresponding C-substituted silenes **2**. Compounds **2** were isolated as their products **6** and **7** of addition with phenol, and of insertion with hexamethylcyclotrisiloxane, respectively. As an example, phenoxysilanes **6** were obtained in *ca.* 60% yield and 12–30:1 R≠H/R=H ratio.[15] A similar conclusion concerning the site of initial ring cleavage in the flow thermolysis of **1** (R=Me) was reached in an independent investigation.[16]

The FVT of silacyclobutanes has been used as the method of generation of silenes in an extensive investigation concerning the reactivity of these

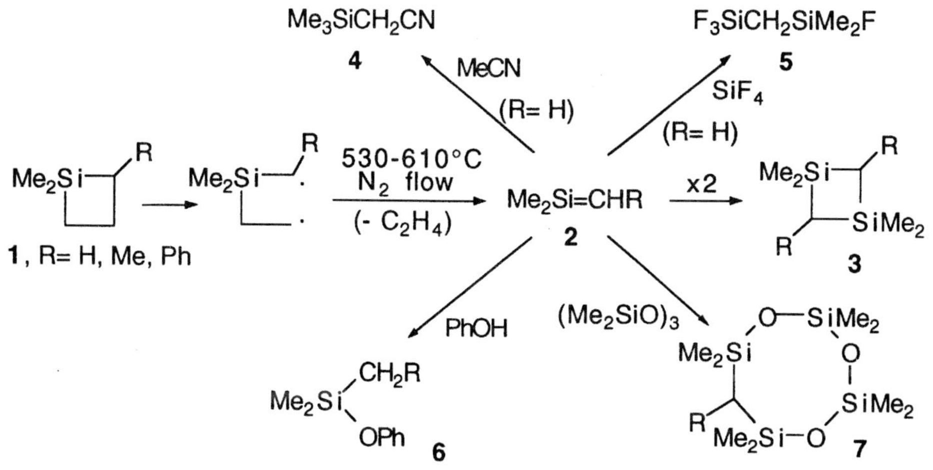

Me₃SiCH₂CN
4

F₃SiCH₂SiMe₂F
5

$$Me_3SiCH_2CN \quad F_3SiCH_2SiMe_2F$$

Scheme 1

compounds.[17-21] These thermolyses, performed on 3 g quantities of starting material, led in 10–100% yield to many alkyl-, vinyl-, aryl-, chloro-, or amino-1,3-disilacyclobutanes. The examples providing the best yields are given below.[17] On the other hand, the flow thermolysis of 1-chlorosilacyclobutanes appeared to be less selective, partial loss of hydrogen chloride taking always place and lowering the yields of 1,3-disilacyclobutanes under these conditions (Scheme 2).[22,23]

In a recent report,[24] the FVT of 1-methyl-1-vinylsilacyclobutane (**8**) has been shown to give, besides the cyclo-dimer of 1-methyl-1-vinylsilene, methylsilylene

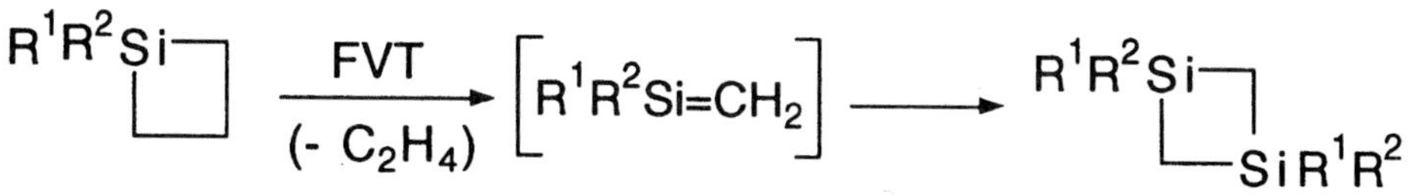

R^1	R^2	FVT temp. (°C)	yield (ref.[17]) mmol / %
Me	Me	625	15 / 100
Me	Vi	625	10 / 75 (Z+E)
Me	Me₂N	625	8.4 / 72 (Z+E)
Me₂N	Me₂N	625	6.7 / 70
Me	Cl	670	7.9 / 63 (Z+E)
Cl	Cl	715	4.6 / 43

Scheme 2

formed by loss of allene from silirane **9**. This silylene was characterized by its cycloadducts with 2,3-dimethyl-1,3-butadiene and isoprene. 1,1-Divinylsilacyclo-butane reacted in a similar fashion. The flow thermolysis of **8** in the presence of methanol or phenol showed in the last case, beside the usual addition product **10**, the formation of diphenoxysilane **11** resulting from the intermediacy of a silicenium ion **12**.[25] Other 1-alkenylsilacyclobutanes behave similarly.[26] Another interesting feature is the rearrangement observed in the FVT of silacyclobutane **13**. In place of the expected 1,2-elimination yielding 1-methylenesilacyclobutane, 1,3-disilacyclobutane **14** was obtained *via* loss of ethylene and rearrangement of silene **15**.[27]

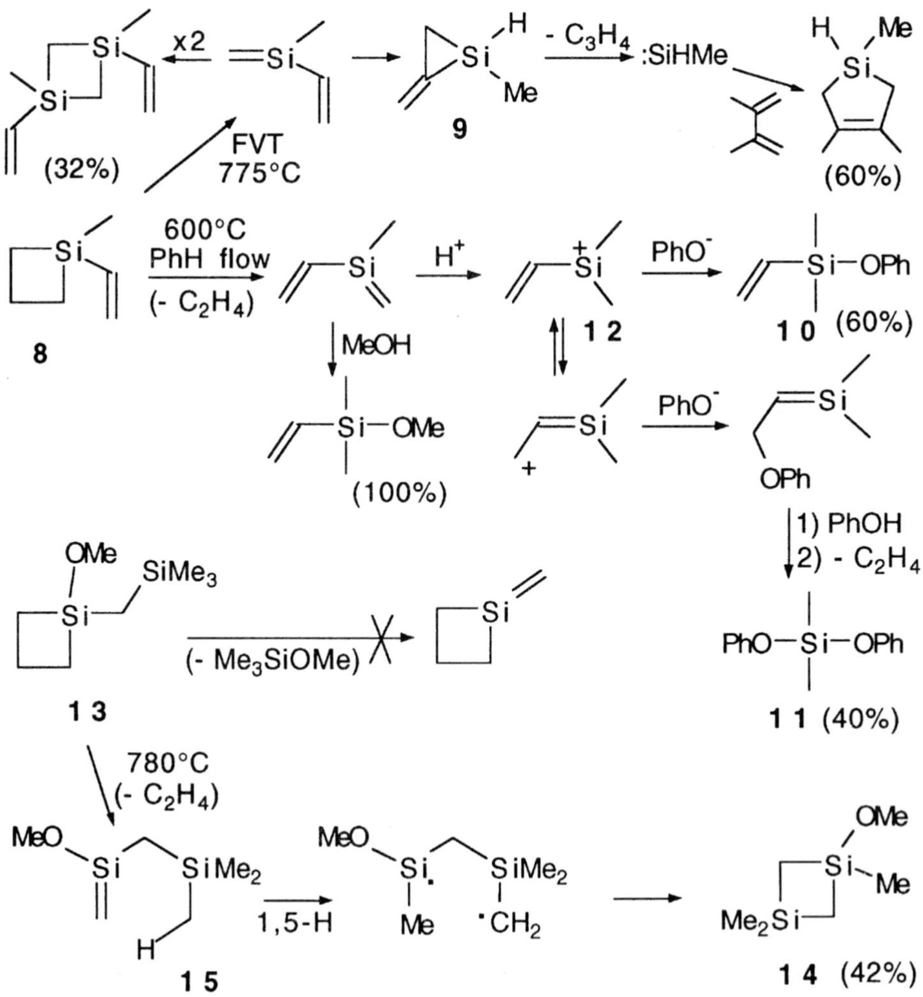

Scheme 3

The FVT of a solution in hexane of the photochemically obtained 3-vinyl-silacyclobutane **16** led, at the moderate temperature of 450°C, to bicyclodisilane **17** resulting from the isomerization of silene **18**. Silacyclohexene **19**, the [4+2] cyclo-adduct of **18** and the co-formed butadiene, was indeed also present as major product.[28]

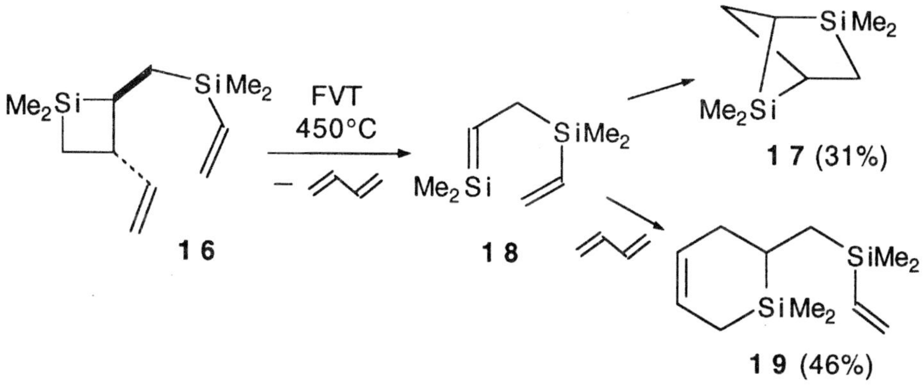

Scheme 4

In the case of 1-monosubstituted or non-substituted silacyclobutanes, the results of thermolysis are complicated by the possible isomerization of the first generated silenes into dimethyl- or methylsilylenes.[29]

The low pressure flow thermolysis of 1-methyl-1-silacyclobutane (**20**) led to 1,3-disilacyclobutanes **21** and **22**, reported to be the major evolution products of dimethylsilylene **23**, and to dimethylsilylsilacyclobutane **24** resulting from the insertion of **23** into the starting material **20**.[30] Trapping of the products by 1,3-butadiene or trimethylsilane gave also products such as **25** or **26**, involving the intermediacy of **23**.[31]

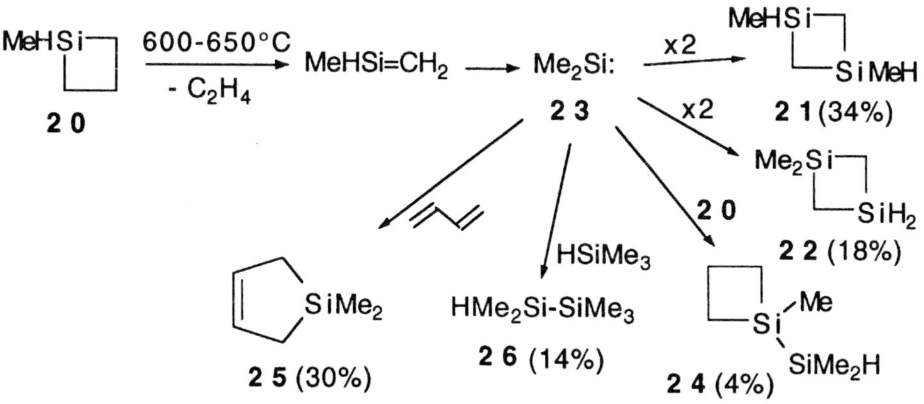

Scheme 5

A similar silene → silylene rearrangement, giving 1,3-disilacyclobutane **27** in nearly quantitative yield, has been observed in the FVT of trimethylsilylsilacyclobutane **28**.[32]

The thermolysis of unsubstituted silacyclobutane (**29**), conducted in a high vacuum flow system with a tenfold excess of 1,3-butadiene, led to products **30–32** resulting from trapping of silene, silylene, and predominantly methylsilylene, respectively[33] (see also refs. 34–36 for other kinetic, mechanistic and spectroscopic investigations concerning the low pressure thermolysis of **29** and related sila- and disilacyclobutanes).

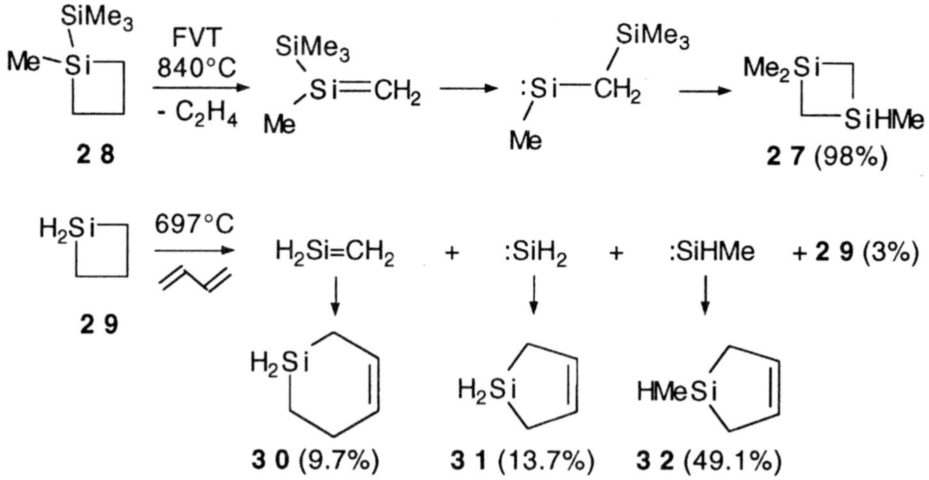

Scheme 6

The cleavage of a siloxetane into an alkene and an intermediate silanone giving cyclic oligomers has been recognized since a long time in the trapping of silenes by non-enolizable carbonyl compounds.[37–40] For example, the flow thermolysis of 1,1-dimethylsilacyclobutane in the presence of benzophenone led to the isolation of 1,1-diphenylethylene as well as of cyclotri- and tetrasiloxanes **33** and **34**, the tri- and tetramers of dimethylsilanone (**35**), in 73, 28, and 5.5% yield, respectively.[38] The use in this reaction of N-phenylbenzaldimine **36** as co-reactant, in addition to benzophenone, resulted in the obtention of benzophenone phenylimine **37**, demonstrating the intermediate formation of silanimine **38** (Scheme 7).[41]

The use of acrolein as carbonyl compound allows the possibility of a competitive [4+2] cycloaddition with the intermediate silene, and both types of products, namely oxasilacyclohexene **39**, and those derivating from siloxetane **40**, respectively, have been effectively obtained in that case with a [4+2]/[2+2] ratio of *ca.* 1:4 (Scheme 8).[42,43]

Other examples of siloxetane formation and cleavage are provided by the flow thermolysis of phenylsilylketones **41**, where siloxetanes **42**, formed by C—H insertion of siloxycarbenes **43**, underwent Si–C and Si–O bond cleavages in almost

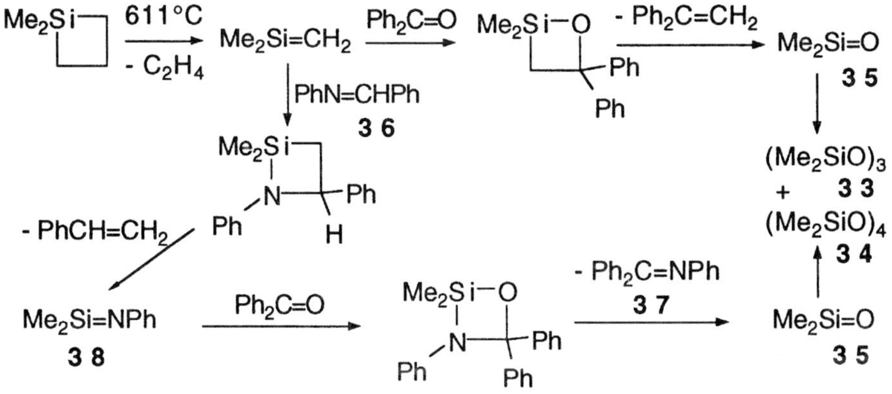

Scheme 7

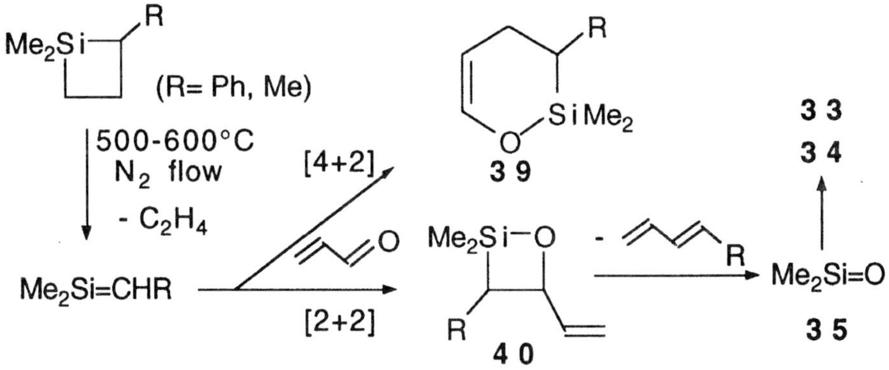

Scheme 8

comparable ratio,[44] and by the FVT of compound **44** which gave a 1,5-H migration and ring closure to **45** in place of the foreseen elimination of Me₃SiOMe (Scheme 9).[45]

6-oxa-3-sila-(or 3-germa-)bicyclo[3.1.0]hexanes **46**, when subjected to flow or flash thermolysis, are interesting precursors of 1,1-dimethylsilanones and -germanones, *via* the corresponding siloxetane or germoxetane **47**. The formation of 1,1-dimethylgermanone (**48**) has been confirmed by insertion in compounds such as **49** and the reaction applied, when starting from compound **50**, to the generation of monomeric SiO₂ (Scheme 10).[46-48]

Thiobenzophenone reacted with 1,1-dimethylsilene in the same way as benzophenone, providing the first generation of transient 1,1-dimethylsilanethione (**51**), isolated as its cyclo-dimer **52**. 1,1-diphenylsilanethione was similarly obtained, starting from 1,1-diphenylsilacyclobutane.[49] Several related thermolyses, involving **51** as an intermediate, have been since reported.[50-52] 1,1-Dimethylgermanethione (**53**) has been also synthesized as a transient species in the flow thermolysis of thiagermetane **54**, giving cyclodithiatrigermane **55** as final product (Scheme 11).[53]

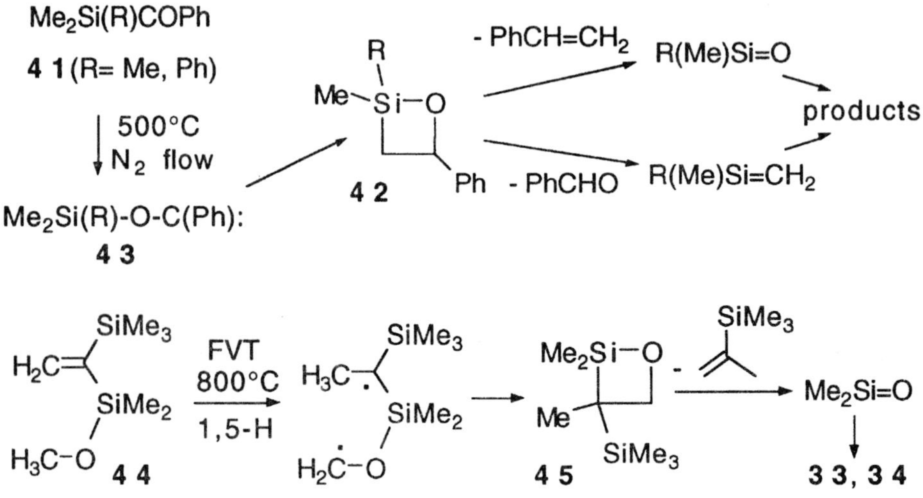
Me2Si(R)COPh
4 1 (R= Me, Ph)
500°C
N2 flow
Me2Si(R)-O-C(Ph):
4 3
R
Me
Si
O
Ph
4 2
- PhCH=CH2
R(Me)Si=O
- PhCHO
R(Me)Si=CH2
products
SiMe3
H2C
SiMe2
H3C-O
4 4
FVT
800°C
1,5-H
H3C
H2C-O
Me2Si-O
Me
SiMe3
4 5
Me2Si=O
3 3, 3 4

Scheme 9

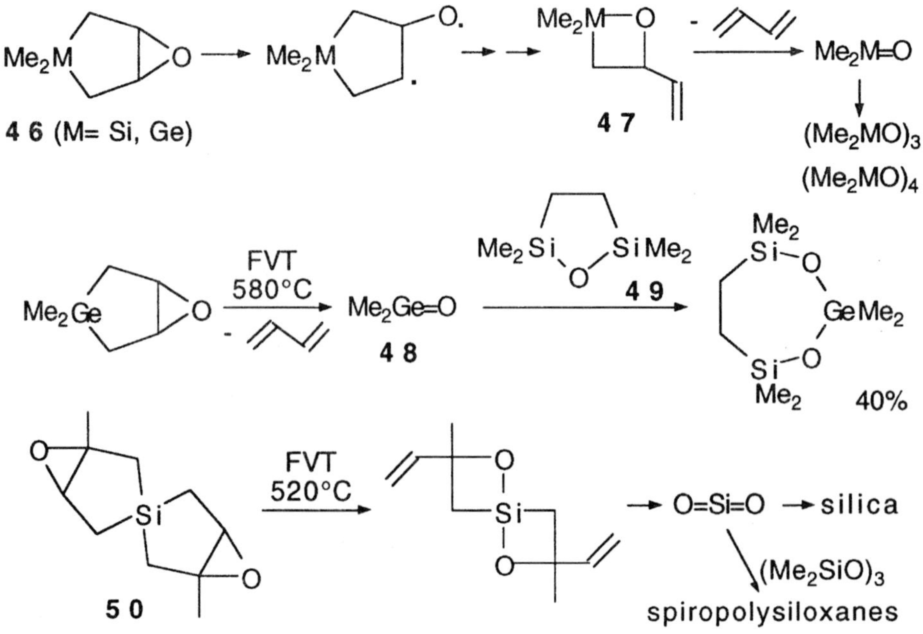
Me2M
O
4 6 (M= Si, Ge)
Me2M-O
4 7
Me2M=O
(Me2MO)3
(Me2MO)4
Me2Ge
FVT
580°C
Me2Ge=O
4 8
Me2Si
SiMe2
O
4 9
Me2
Si
GeMe2
Si
Me2
40%
5 0
FVT
520°C
O=Si=O
silica
(Me2SiO)3
spiropolysiloxanes

Scheme 10

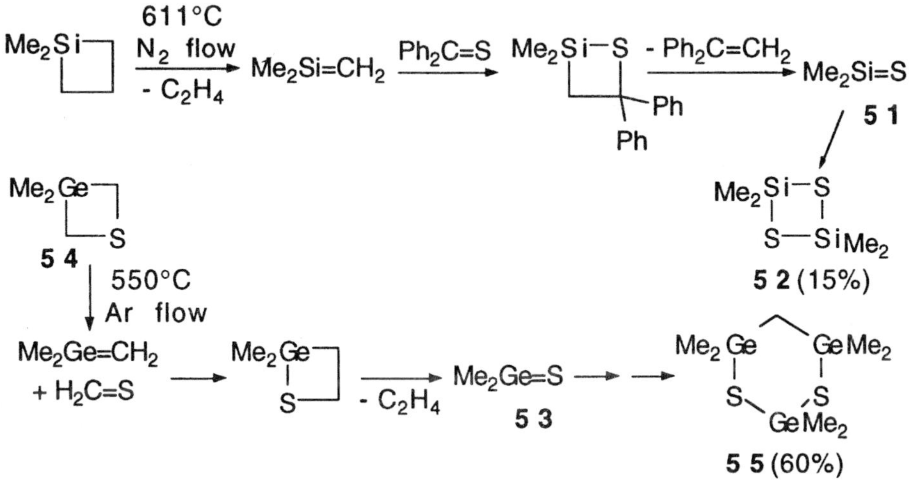

Scheme 11

[4+2] cycloreversions

Since the first reports on the generation of 1,1-dimethylsilene or 1,1-diethyl-germene from the bicyclic precursors **56** and **57**, respectively,[54,55] the flow or flash thermolysis of Diels-Alder adducts has constituted an alternative and efficient pathway to silicon and germanium unsaturated compounds. The Si-monomethyl-ated analogue of **56** has been also used as a precursor in the investigations concerning the methylsilene → dimethylsilylene rearrangement[31,32] (see above, [2+2] cycloreversions).

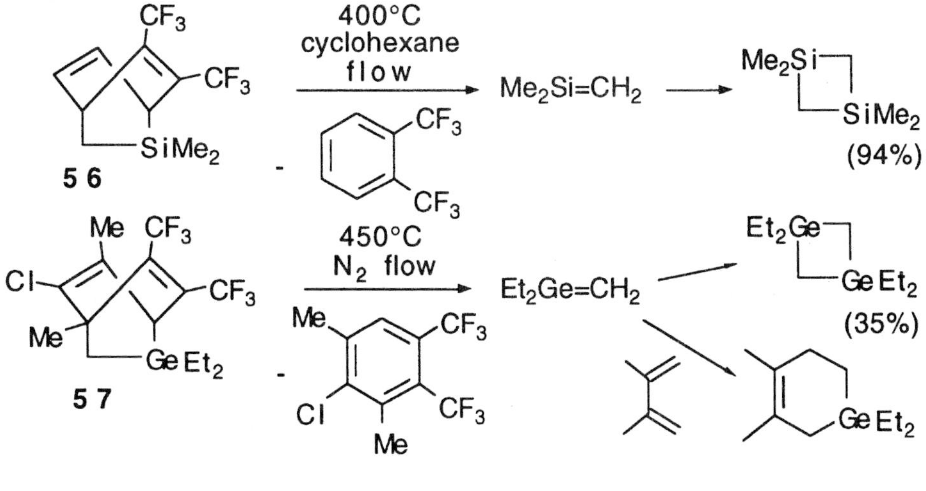

Scheme 12

Neopentylsilenes have been generated by retro-Diels-Alder reaction of cyclo-pentadiene and anthracene adducts **58** and **59**. In the FVT of **58**, silene **60** underwent a partial retro-ene cleavage giving isobutene and dimethylvinylsilane (**61**).[56] The FVT at 550°C of (Z) or (E)-**59** in the presence of trimethylmethoxy-silane led to the same mixture of diastereoisomers **62**, a stereospecific reaction being on the other hand achieved in sealed tube at 300°C.[57]

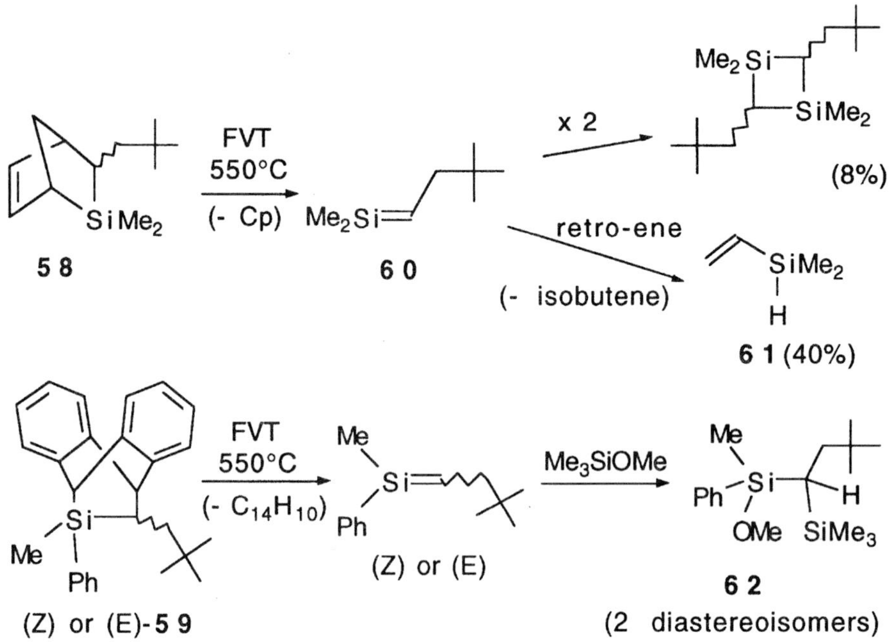

Scheme 13

The FVT of naphthalene and biphenyl adducts **63** and **64** gave the first evidence for the transient disilene **65**, yielding as main final products 1,3-disilacyclobutanes **21** and **22**.[58] Silylene **66** has been shown to be an intermediate in these gas-phase reactions of **65**.[59] When performed in the presence of benz-aldehyde, this thermolysis gave evidence for the transient presence of tetra-methyldisilene, 1,1-dimethyl-2-phenylsilene and dimethylsilanone, in the same reaction.[60,61] When silapyran **67** was thermolyzed in a hexafluorobut-2-yne flow in the presence of the cyclic disiloxane **49**, silanone **68**, formed by retro-Diels-Alder reaction of the transient adduct **69**, was trapped as compound **70**. A possible isomerization of **68** into silylene **71** was not observed (Scheme 14).[62]

An interesting example related to the above [4+2] cycloreversions is provided by the homo-retro-Diels-Alder reaction of the [4+4] adduct **72**, giving anthracene and 1,4-disilabenzene **73** upon gas-phase thermolysis. Compound **73** was either trapped to yield the corresponding adducts **74** and **75** by flow thermolysis with methanol and acetylene, respectively, or led to the bicyclic compound **76** by radical isomerization, upon FVT in the absence of any other reactant (Scheme 15).[63,64]

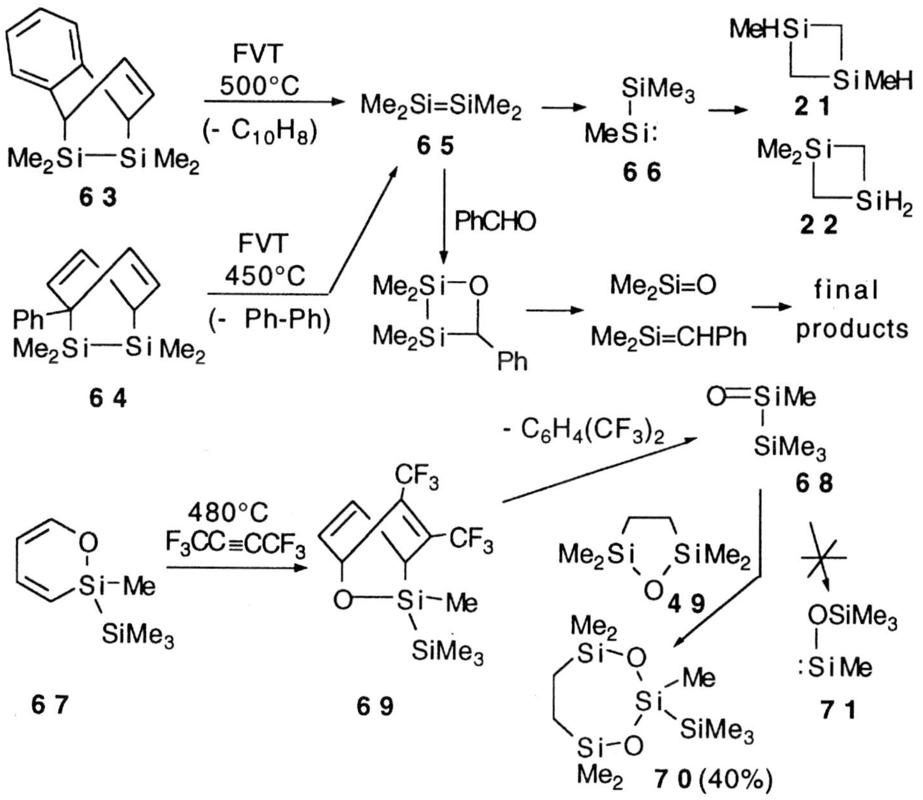

Scheme 14

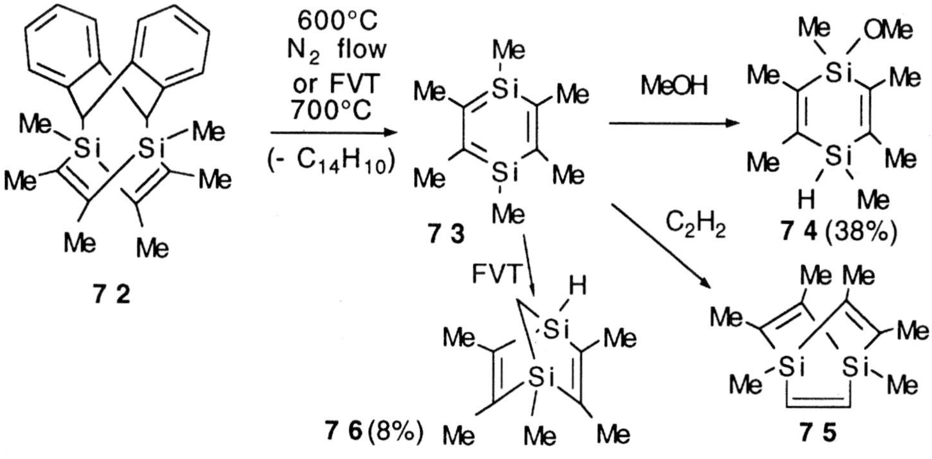

Scheme 15

Retro-ene reactions

Several unsaturated, silicon containing species have been synthesized by retro-
ene reactions, often of appropriate allylsilanes. However, the gas-phase thermolysis of allyltrimethylsilane itself (**77**) has been the subject of controversies
concerning the obtained products and involved mechanisms. Conclusive work
demonstrated the occurence in this case of a retro-ene pathway, leading to 1,1-di-
methylsilene, taking place at low pressures, while, at higher ones, a bimolecular
radical process yields trimethylvinyl- and tetramethylsilanes.[65]

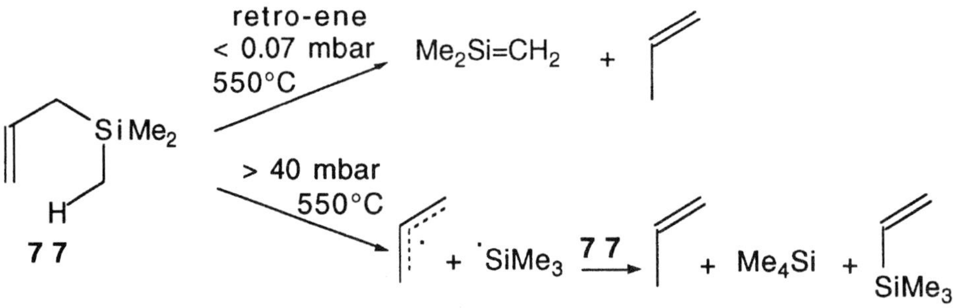

Scheme 16

For the ethenodihydroanthracenic adduct **79**, a competition was expected
between retro-Diels-Alder (path A) and retro-ene (path B) cleavages. It has been
possible in this case to drive the reaction specifically towards paths A or B by
changing the thermolysis conditions. The retro-ene reaction B, although break-
ing a strong Si–O bond, appeared here to be easier (likely by removal of etheno
bridge strain) when compared to vinyloxysilane **80** which was cleaved only
partially (40%) even at 1000°C (Scheme 17).[66]

A simple access to silacyclobutenes **81a–d** was reached, *via* the cyclisation of the
intermediate silabutadienes **82**, by FVT of diallylsilanes **83**.[67] The kinetics and
mechanism of this reaction have been investigated in the case of diallyldimethylsilane (**83a**).[68] 1-Silatoluene (**84**) has been obtained by retro-ene reaction of the
allylic silacyclohexadiene **85** and characterized by chemical trapping, in acetylene
and hexafluorobut-2-yne flows, as its adducts **86** and **87**, respectively (Scheme 18).[69]

The transient C-unsubstituted silole **88** (5-methyl-5-silacyclopentadiene) has
been obtained by retro-ene reaction of allylsilacyclopentene **89**, followed by 1,5-H
shift to silicon. Compound **88** finally dimerized or was trapped by other
dienophiles to give Diels-Alder adducts.[70] In the FVT of allylsilaindan **90**, the
intermediacy of the primary retro-ene product, silaindene **91**, and of isobenzo-
silole **92**, was demonstrated by trapping them with methanol and hexafluorobut-
2-yne, respectively (Scheme 19).[71]

Stable C-methylated siloles possessing a Si–H bond have been also synthesized:
the FVT of precursors **93** gave the corresponding siloles **94**, accompanied by

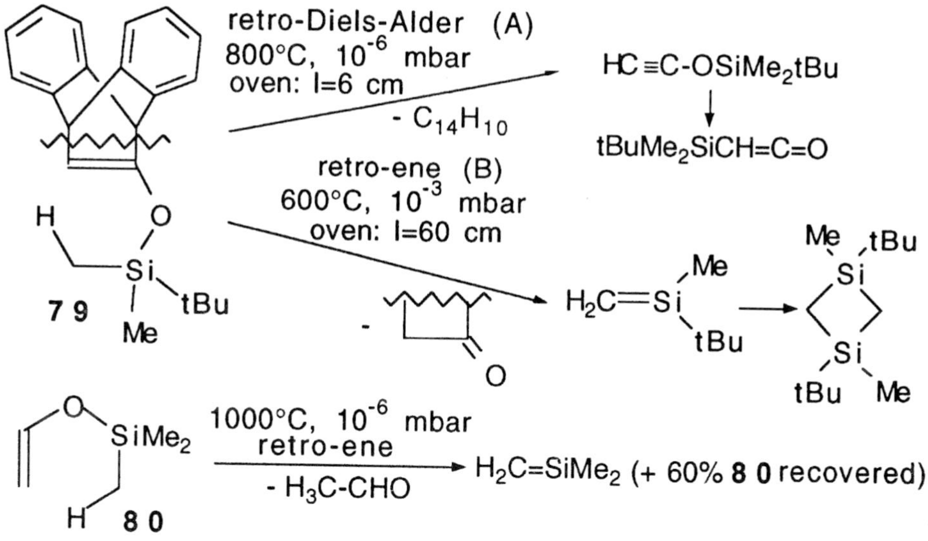

retro-Diels-Alder (A)
800°C, 10^{-6} mbar
oven: l=6 cm
- $C_{14}H_{10}$

$HC \equiv C\text{-}OSiMe_2tBu$

$\downarrow$

$tBuMe_2SiCH=C=O$

retro-ene (B)
600°C, 10^{-3} mbar
oven: l=60 cm

$H_2C=Si\begin{smallmatrix}Me\\\\tBu\end{smallmatrix}$ → (Si ring with Me, tBu, Si, tBu, Me)

7 9

H / Si / tBu / Me

1000°C, 10^{-6} mbar
retro-ene
- $H_3C\text{-}CHO$

$H_2C=SiMe_2$ (+ 60% **8 0** recovered)

8 0 O / SiMe₂ / H

Scheme 17

FVT
710°C
$-\bigtriangleup\text{-}R^3$

R^3
SiR^1R^2
H
R^3

8 3 a-d

$\overset{SiR^1R^2}{\underset{R^3}{\diagup}}$

8 2 a-d

$\square\text{-}SiR^1R^2$
R^3

8 1 a-d

	R^1	R^2	R^3	yield
a	Me	Me	H	25%
b	Me	Me	Me	36%
c	-(CH₂)₄-		H	9%
d	-(CH₂)₅-		H	18%

428°C
C_2H_2 flow
or 450°C
$F_3CC\equiv CCF_3$
$-\diagdown\diagup$

(Si–Me cyclohexadiene) **8 5**

(Si ring) **8 4**
Si
Me

→ (bicyclic Si-Me structure)
R   R
8 6 (R= H, 34%)
8 7 (R= CF₃, 20%)

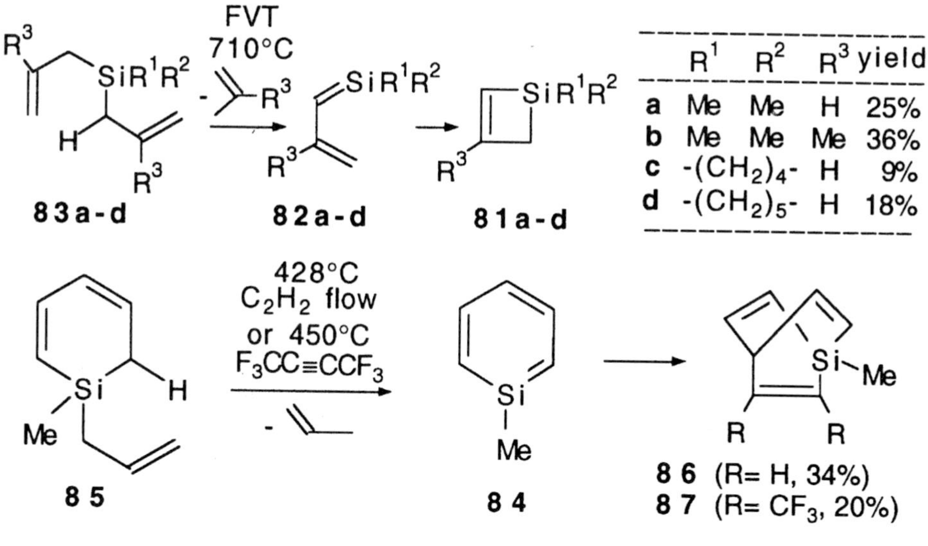

Scheme 18

minor amounts of their transoid isomers **95**. Compounds **94** were stable enough to be distilled under reduced pressure. On the other hand, upon FVT, the germa-analogue of **93** produced only 2,3-dimethylbuta-1,3-diene resulting from a cheletropic extrusion of germylene.[72] The FVT of allylsilane **96** led similarly to the generation of 6,6-dimethyl-6-silafulvene **97**, characterized by dimerization and trapping with dienophiles (Scheme 20).[73,74]

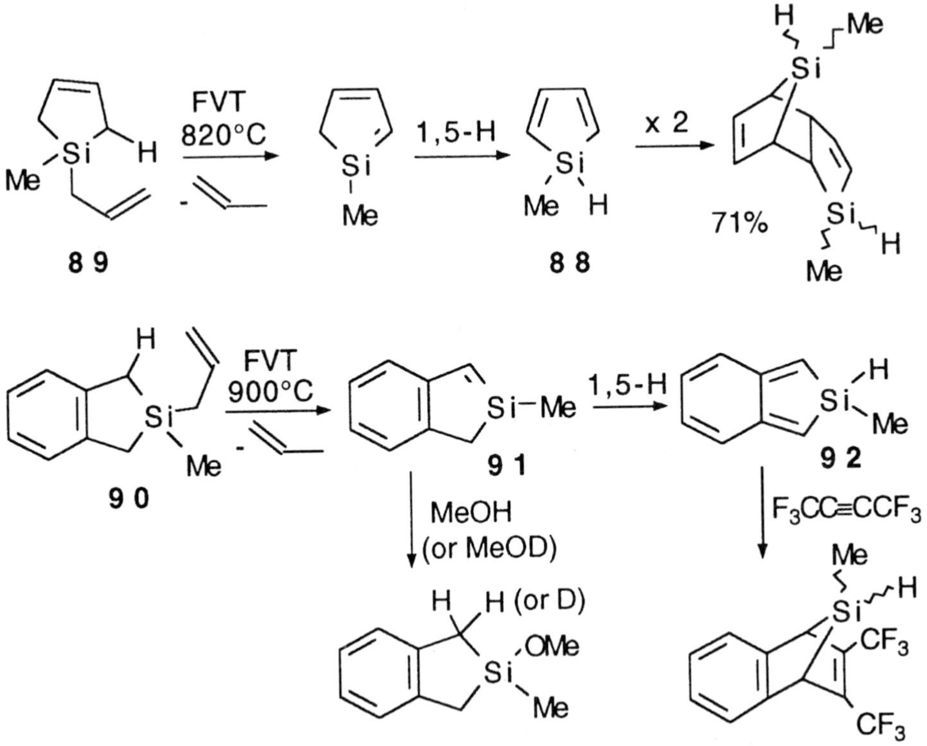

Scheme 19

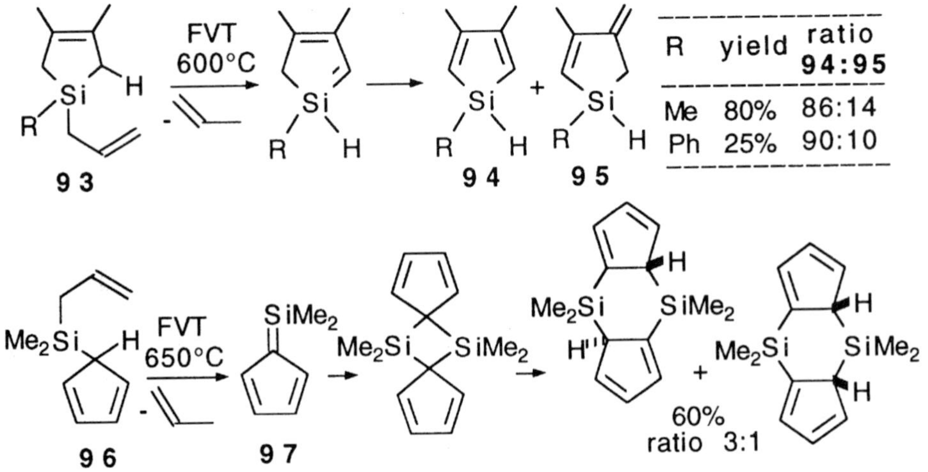

R	yield	ratio 94:95
Me	80%	86:14
Ph	25%	90:10

Scheme 20

Allyloxydimethylsilane (**98**) was cleanly converted, under FVT conditions, into dimethylsilanone by a concerted retro-ene reaction. However, in the case of compound **99**, radical stabilization induced a competition between this reaction and homolytic cleavage into isoprene and dimethylsilanol **100**.[75]

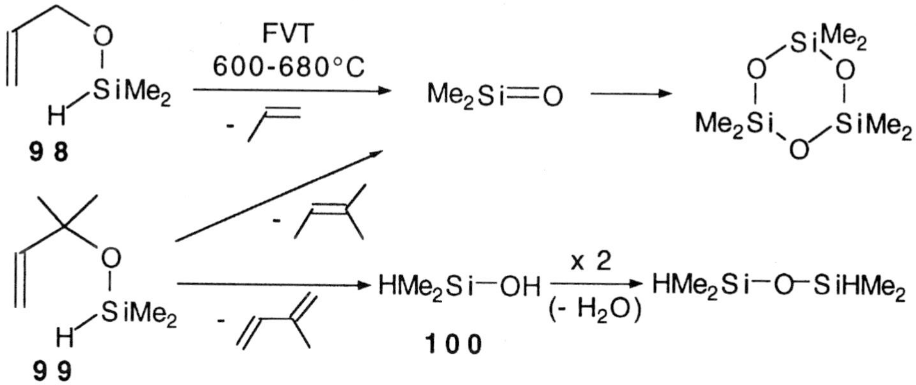

Scheme 21

An investigation, aimed at the direct characterization of monomeric 1,1-dimethylsilanimine (**101**) by FVT/MS coupling, remained unsuccessful. Its cyclo-dimer **102**, and adduct with *t*-butanol (**103**), were however identified, as well as silanimine **104** and the cyclo-dimer **105** of silanimine **106**. The here-involved thermal processes, starting from precursors **107** and **108**, are retro-ene reaction, [2+2] cycloreversion, and 1,2-elimination of isobutene or methylsilene.[76]

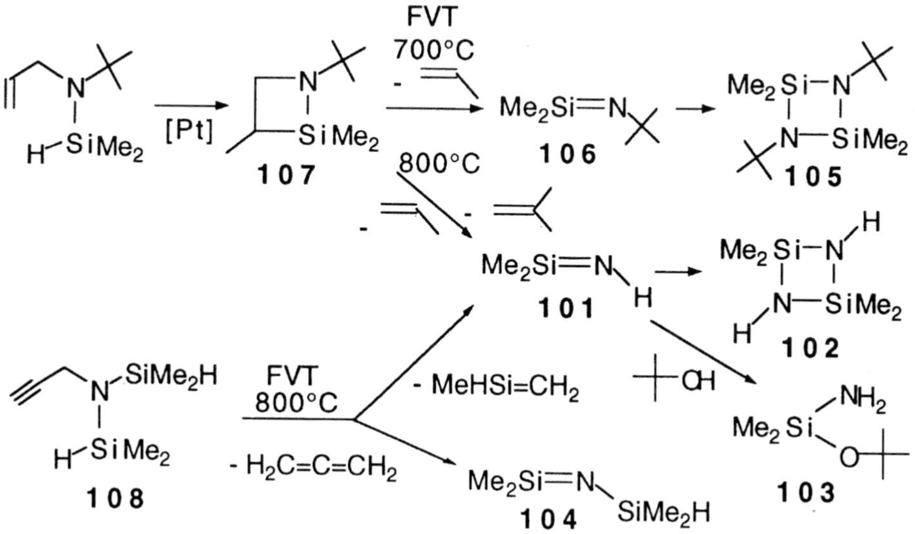

Scheme 22

The generation of unsubstituted silanone (**109**), silanethione (**110**), and silan-imine (**111**), molecules of possible cosmochemical importance, was also attempted in view to determine their millimeter-wave spectra. The FVT of retro-ene pre-cursors **112–115**, performed in coupling with a millimeter-wave spectrometer, did not allow the detection of these highly reactive species, but that of silicon mon-oxide (**116**), silicon monosulfide (**117**), and hydrogen silaisocyanide (**118**), result-ing from an easy thermal dehydrogenation of compounds **109–111**.[77]

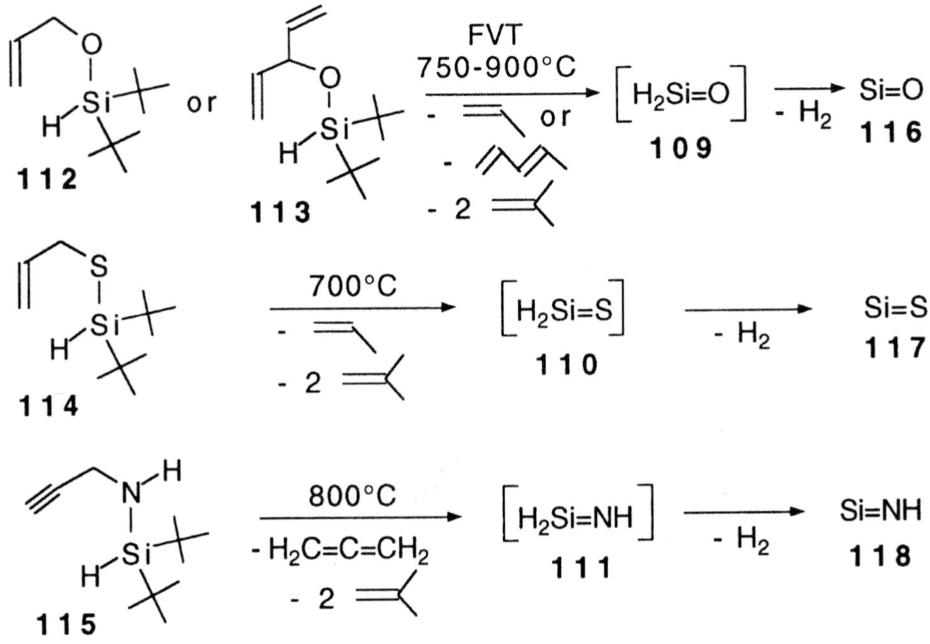

Scheme 23

By Eliminations

1,1-eliminations

The primary products of the here-relevant 1,1-eliminations initiated by flash or flow thermolysis of silicon or germanium organic compounds are divalent species such as silylcarbenes or -nitrenes, silylenes and germylenes. These transient inter-mediates then undergo isomerizations or dimerizations giving silicon or ger-manium double bonds.

The gas-phase thermolysis of trimethylsilyldiazomethane (**120**) yielded as major product 1,3-disilacyclobutane **121**. Compounds **122** or **123** were obtained when the thermolysis was performed in the presence of isopropanol or benzal-dehyde, respectively. All these results are coherent with the intermediacy of trimethylsilene **124**.[78] Similarly, whereas products **125** and **126** in the flow thermolysis of trimethylsilylphenyldiazomethane (127) resulted directly from silylcarbene 128, the obtention of compounds such as 129 or 130 showed the intermediacy of silene 131.[79,80]

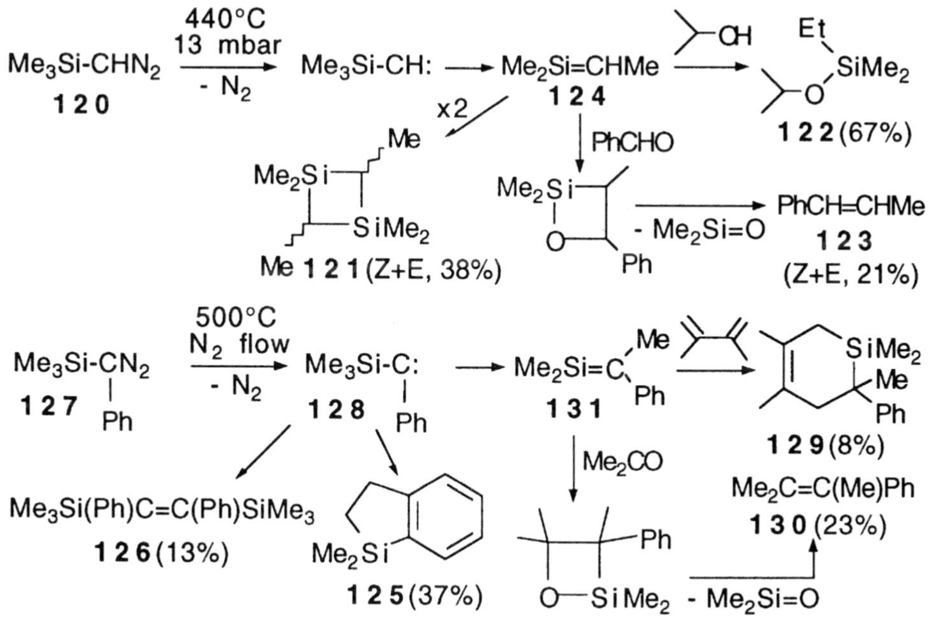

Scheme 24

An investigation of the flow or flash vacuum thermolysis of bis(trimethylsilyl)-, trimethylgermyltrimethylsilyl-, and bis(trimethylgermyl)diazomethanes (132–134), revealed in all cases a quantitative carbene isomerization giving the intermediate silenes and germene 135–137, characterized by their dimerization products and adducts with 1,3-dienes and benzaldehyde.[81]

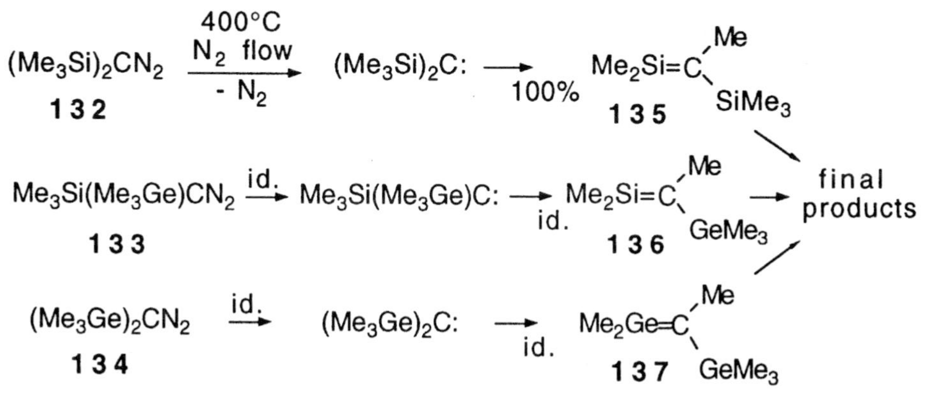

Scheme 25

The gas-phase thermolysis of silyl azides **138–140** led to silanimines **141–144**, resulting from the isomerization of the firstly formed nitrenes and trapped by hexamethylcyclotrisiloxane or hexamethylcyclodisilazane, as well as, in the case of **144**, by carbonyl compounds.[82]

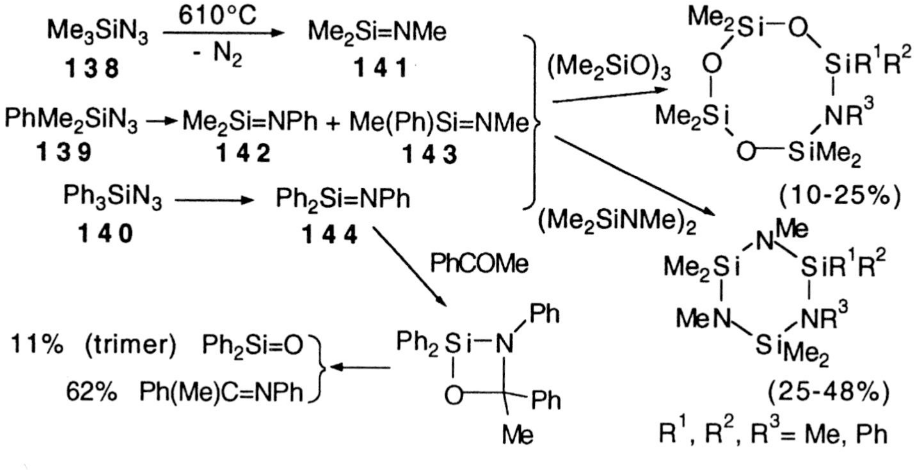

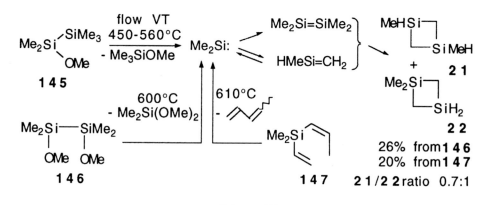

Scheme 26

After some controversies about the nature of the transient double-bonded silicon species in the 1,1-eliminations involving firstly a silylene, it has been established that Si=Si and Si=C intermediates can be formed and give as main products the same 1,3-disilacyclobutanes,[83,84] as observed in the flow thermolysis of disilanes **145**[83] and **146**,[85] and of propenylvinylsilane **147**.[86] Thus, the interpretation of thermal reactions involving silylenes should be made with caution.

Scheme 27

In a similar manner, silylene **148** can undergo, besides the insertion reaction giving **149**, a 1,2-trimethylsilyl shift to yield silene **150**, the formation of which was proved by cothermolysis of precursor **151** with alcohols and benzophenone, yielding compounds **152** and **153**, respectively.[87]

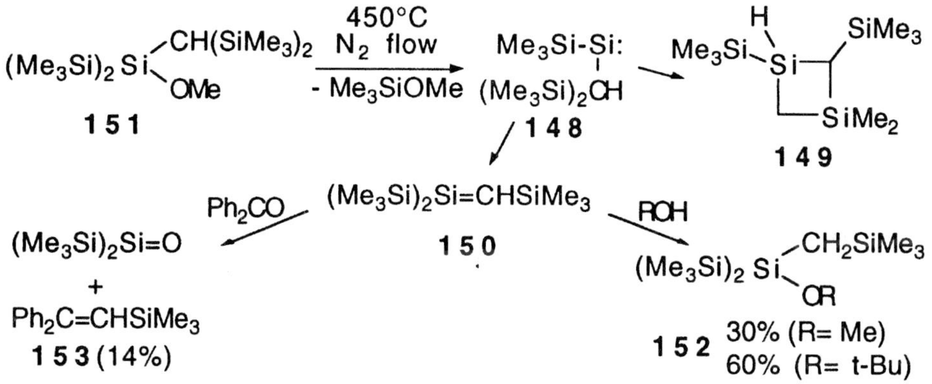

Scheme 28

Cyclopropylphenylsilylene, obtained by flow thermolysis of precursor **154**, underwent a rearrangement followed by ring opening to silene **155**, which was trapped by alcohols and benzophenone.[88] Silylene **156**, resulting from the FVT of **157**, behaved similarly, giving finally the dimer of vinylsilene **158**.[89]

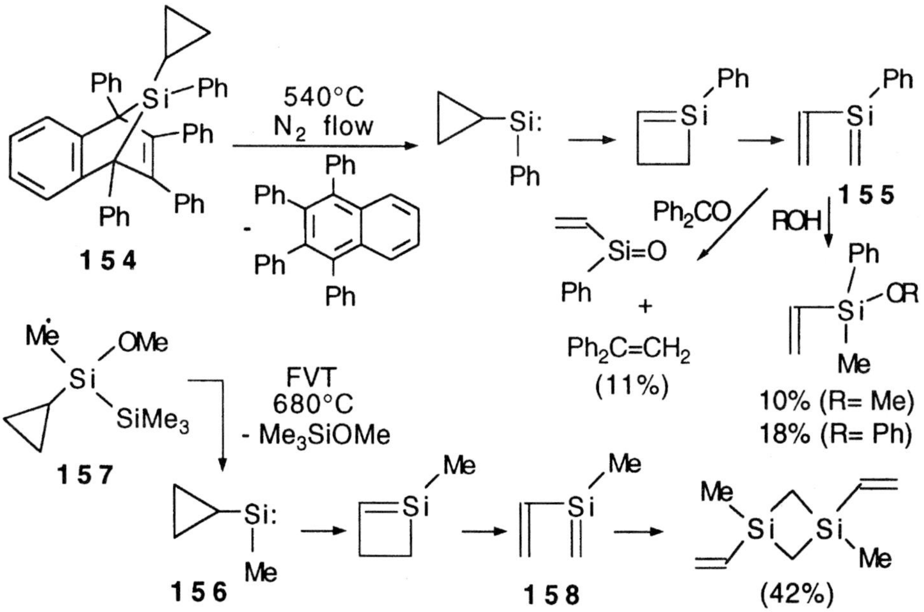

Scheme 29

The interconversion silylene **159** → silene **160** → silylene **161** has been pointed out, in the FVT of precursor **162**, by obtention of adducts **163** and **164** in the presence of 2,3-dimethyl-1,3-butadiene.[90] A similar reactivity was observed for silylene **165**, yielding silylene **166** *via* silene **167**.[91]

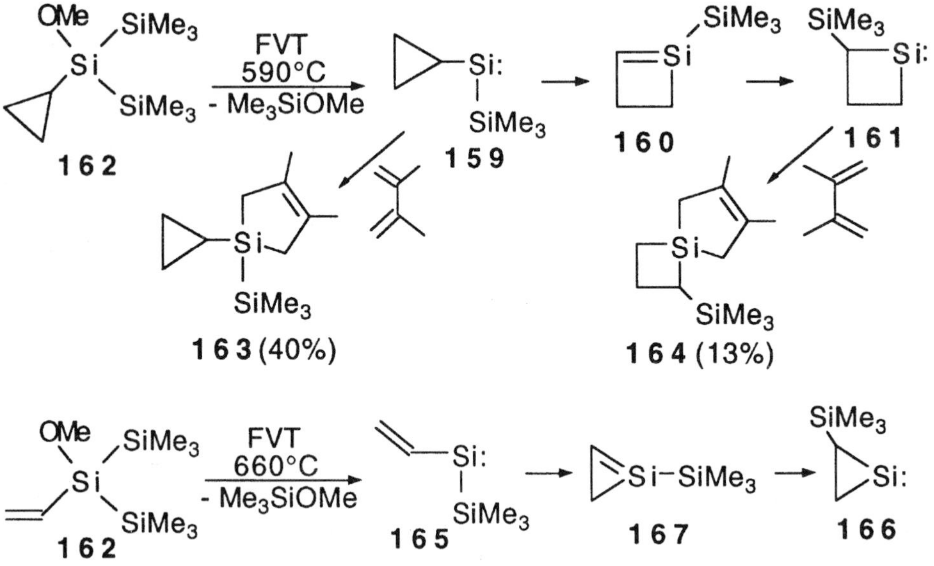

Scheme 30

A silylene → disilene → silylene rearrangement has been also observed in the flow thermolysis of trisilane **168** in the presence of trimethyl- and ethyl-dimethylsilanes. Among the obtained products, trisilanes **169** and **170**, respectively, are those resulting from insertion in the rearranged silylene **171**. The likely intermediacy of disilene **172** in this rearrangement has been confirmed by its direct trapping by 1,3-butadiene to give disilacyclohexene **173**.[92] The dimerization of dimethylgermylene into tetramethyldigermene **174** was proposed in the flow thermolysis, effected at 250°C in a gas chromatograph injector, of compound **175** with adamantanethione which gave the [2+2] adduct **176** in 50–60% yield.[93] However, in a paper reporting also the generation of **174** from **175** and its characterization in matrix by Raman spectroscopy, the authors privileged an insertion of dimethylgermylene into **175**, followed by a [4+2] cycloreversion, to give **174**[94] (cf. part II, IR and UV spectroscopy) (Scheme 31).

The easy thermal isomerization of silylene **177** to silanone **178**, obtained as its cyclo-trimer **179**, has been observed in the FVT of diallyloxydisilane **180** and confirmed by trapping of **177** and **178** by butadiene and dimethoxydimethylsilane, respectively (Scheme 32).[95]

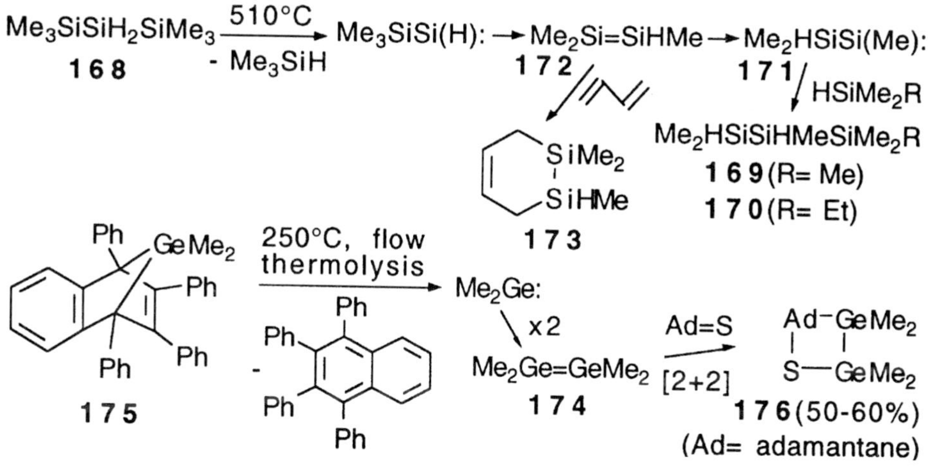

$$Me_3SiSiH_2SiMe_3 \xrightarrow[- Me_3SiH]{510°C} Me_3SiSi(H): \rightarrow Me_2Si=SiHMe \rightarrow Me_2HSiSi(Me):$$

168   **172**   **171**

HSiMe$_2$R

$$Me_2HSiSiHMeSiMe_2R$$

169 (R= Me)
170 (R= Et)

SiMe$_2$
SiHMe

173

175

$$\xrightarrow[thermolysis]{250°C,\ flow} Me_2Ge:$$

Ph, Ph, Ph, Ph (naphthalene structure)

$$\searrow \times 2$$

$$Me_2Ge=GeMe_2 \quad \underset{[2+2]}{\overset{Ad=S}{\longrightarrow}}$$

174

Ad−GeMe$_2$
| |
S−GeMe$_2$

176 (50-60%)

(Ad= adamantane)

Scheme 31

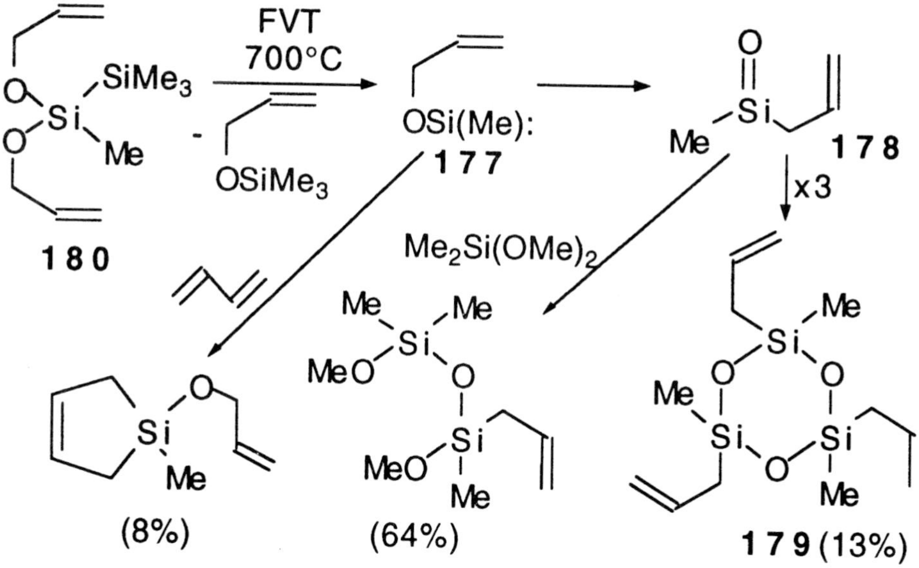

180

$$\xrightarrow[- \ OSiMe_3]{FVT\ 700°C}$$

OSi(Me):

177

Me$_2$Si(OMe)$_2$

178

$\downarrow \times 3$

(8%)

Me, Me — Si — MeO, O — Si, MeO, Me

(64%)

179 (13%)

Scheme 32

1,2- and 1,4-eliminations

Several silenes have been directly generated by 1,2-elimination of fluoro- or, most often, methoxytrimethylsilanes under flow or flash vacuum thermolysis conditions. Thus, the flow thermolysis of **181** led to three disilaindans resulting from successive isomerizations of the firstly obtained silene **182**. The expected adduct of **182**, compound **183**, was also obtained when **181** was cothermolyzed with methanol.[96]

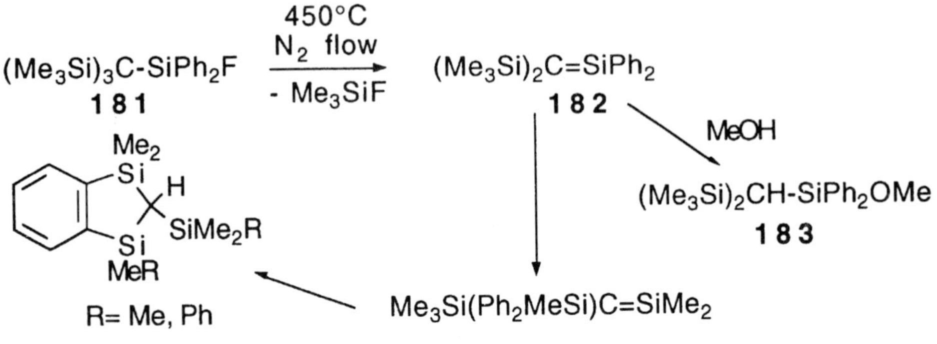

Scheme 33

Silafulvene **97**[74] (see also retro-ene reactions), silabutadiene **184**,[97] and sila-toluene **84**[98] have been conveniently generated by 1,2- or 1,4-elimination of methoxytrimethylsilane. This route has been also used as an alternative pathway to 1-methylsilene[31] (cf. [2+2] cycloreversions).

Several sila-aromatics have been also obtained in excellent conditions by 1,4-dehydrogenation of 1-sila-2,5-cyclohexadienes and, as a rule, directly character-ized by IR/UV spectroscopy in matrix.[64,99–102] Thus, these cases will be treated in the corresponding section (part II).

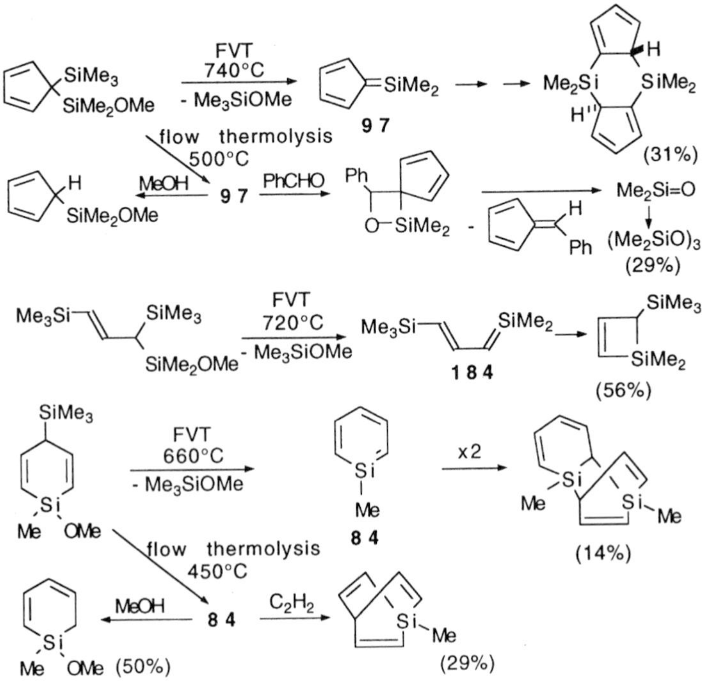

Scheme 34

The FVT of trisilazane **186** gave, upon loss of methoxytrimethylsilane and 1,3-rearrangements, several 1,3-cyclodisilazanes obtained in a total yield of 96% and isolated by GC. Silanimines **187–189**, involved in the generation of the main isomers of these cyclodisilazanes, have been also trapped with hexamethylcyclo-trisiloxane.[103,104]

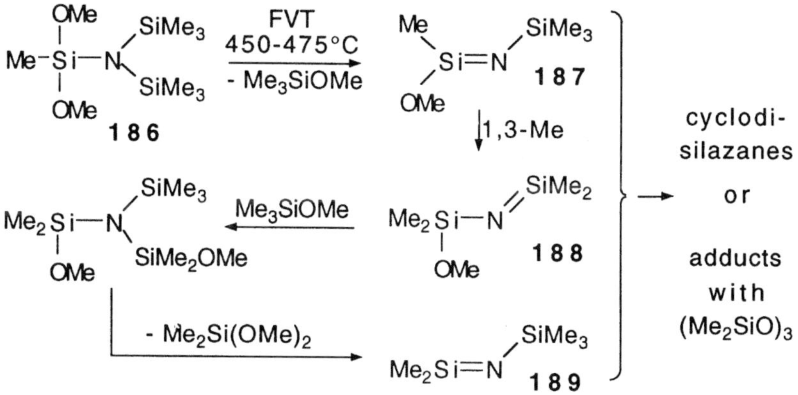

Me-Si—N with OMe, SiMe3, SiMe3 groups **186** $\xrightarrow[\text{- Me}_3\text{SiOMe}]{\text{FVT} \ 450\text{-}475°C}$ Me, OMe, SiMe3 Si=N **187**

187 $\downarrow$ 1,3-Me

Me2Si—N with SiMe3, OMe groups $\xleftarrow{\text{Me}_3\text{SiOMe}}$ Me2Si—N with SiMe2, OMe groups **188** (SiMe2OMe)

$\downarrow$ - Me2Si(OMe)2

Me2Si=N with SiMe3 **189**

cyclodi-silazanes

or

adducts with (Me2SiO)3

Scheme 35

By Rearrangements

The first evidence for a transient silene in a thermal rearrangement was provided by the (Z)→(E) isomerization of (Z)-propenylallylsilane **190**, taking place, *via* silene **191**, by a sila-Cope process.[105] However, the migration of silyl groups is a more general pathway in this sub-heading. For example, a 1,3-Si migration was preferred to account for the formation of 1,1-dimethylsilene, besides other products, in the thermolysis of divinyldisilane **192**.[106]

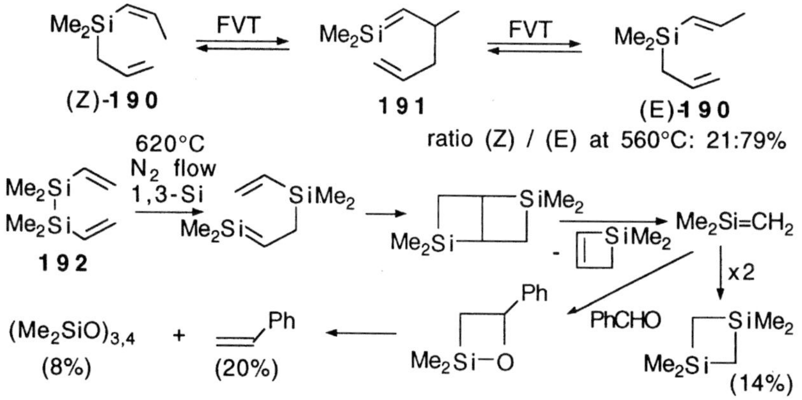

Me2Si (Z)-**190** $\underset{\xleftarrow{}}{\overset{\text{FVT}}{\xrightarrow{}}}$ Me2Si **191** $\underset{\xleftarrow{}}{\overset{\text{FVT}}{\xrightarrow{}}}$ Me2Si (E)-**190**

ratio (Z) / (E) at 560°C: 21:79%

Me2Si, Me2Si **192** $\xrightarrow[\text{1,3-Si}]{620°C \ \text{N}_2 \ \text{flow}}$ SiMe2, Me2Si $\rightarrow$ SiMe2, Me2Si $\xrightarrow[-\text{SiMe}_2]{}$ Me2Si=CH2

$\downarrow$ x2

(Me2SiO)3,4 (8%) + Ph (20%) $\leftarrow$ Me2Si—O Ph $\xleftarrow{\text{PhCHO}}$ Me2Si—SiMe2 (14%)

Scheme 36

Silapyrans **193** have been synthesized by FVT of compounds **194**, the first step in this reaction being a 1,5-Si shift to give a siladiene which then loses a tetramethylsilane molecule. These silapyrans are active dienes allowing an easy access to silanones by a sequence of [4+2] cycloaddition and cycloreversion[107,62] (see [4+2] cycloreversions). On the other hand, siladiene **195**, having no methoxy group, evoluted further by 1,5-H shifts to yield the silylated dienes **196–198**.[108]

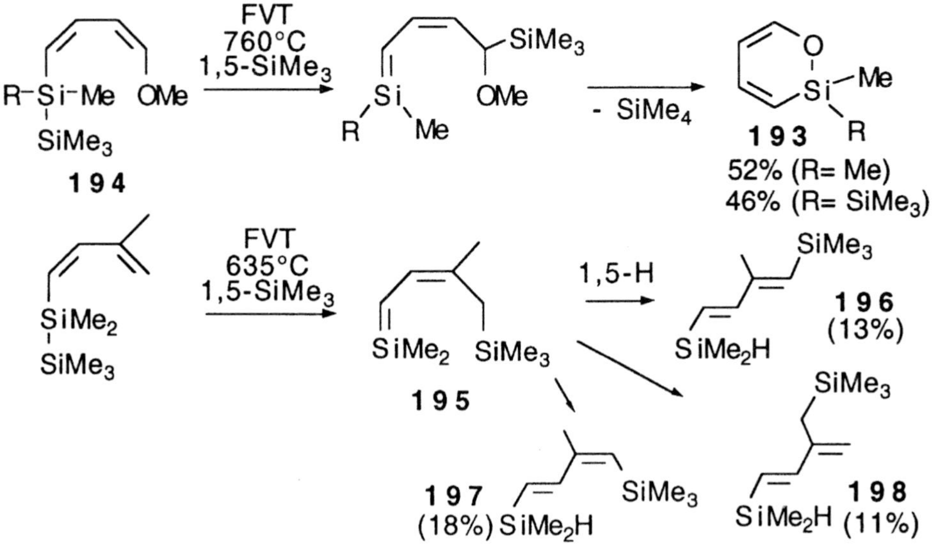

Scheme 37

Hydridosilylketene or -thioketene **199** underwent a very efficient extrusion of silanone or silanethione **200** upon FVT. The mechanism, privileged by the authors for this interesting reaction, is a retro-Wolff rearrangement involving as first step a 1,2-silyl shift.[109,110]

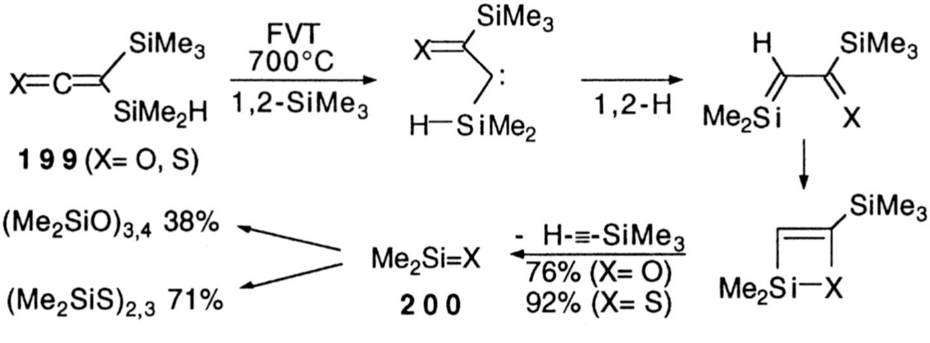

Scheme 38

2. DIRECT SPECTROMETRIC INVESTIGATION

Infrared and Ultraviolet Spectroscopy

Since the first attempts to observe monomeric 1,1-dimethylsilene, in the FVT of 1,1-dimethylsilacyclobutane, by infrared spectroscopy at 77 K,[111,14] the coupling of FVT and matrix isolation techniques allowed to report a number of absorption data concerning substituted and unsubstituted silenes (Table 1).

Technically, the products of thermolysis (either FVT or flow thermolysis in argon or sometimes nitrogen) are condensed with a large excess of the same inert gas (*ca.* one thousandfold) on an optical window refrigerated by liquid helium. After recording the spectrum of the obtained solid matrix at 10 K, the temperature is raised up to 35 K, causing the matrix to melt and the reactive silene to disappear, generally by cyclo-dimerization. The recording of a new spectrum at this temperature allows to assign by difference the bands belonging to the silene in the first-recorded spectrum.

The [2+2] cycloreversion of silacyclobutanes has been often used in this context. 1,1-Dimethylsilacyclobutane[112–115] and 1,1,3-trimethylsilacyclobutane[116,114] led thus to determine the infrared spectrum of 1,1-dimethylsilene, with a $\nu_{Si=C}$ assigned at 1003.5 cm^{-1}.[114,115,117–119]

The unsubstituted silene $H_2Si=CH_2$ (**201a**), generated by [4+2] cycloreversion of the Diels-Alder adduct **202** between 1-silacyclohexa-2,4-diene and hexafluorobut-2-yne, was chemically characterized as its insertion and cycloaddition derivatives **203** and **204** with hexamethyltrisiloxane and 2,3-dimethylbutadiene, respectively. The isolation of the FVT products in an Ar matrix at 10 K allowed to determine the IR and UV spectra of the monomeric silene **201a**, as well as those of silenes **201b–f**.[120,121] Furthermore, the isomerization of **201a** into silylene **205** was shown to occur reversibly by irradiation of the matrix at a suitable wavelength.[122,123]

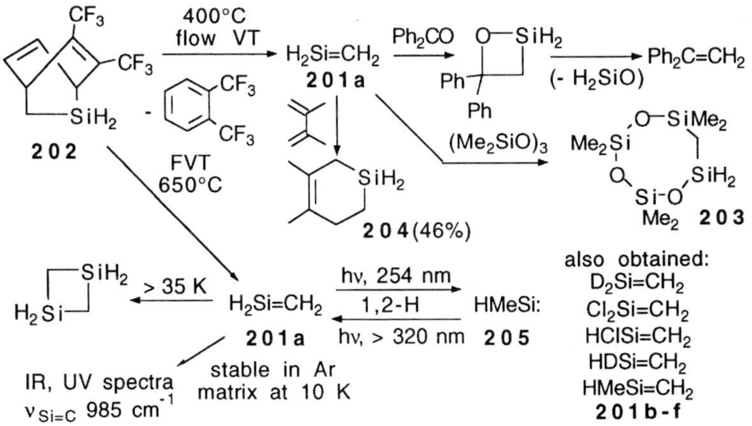

Scheme 39

In other experiments using the infrared spectroscopy in matrix as the detection method, silenes have been generated by retro-ene reaction or 1,1-elimination. Thus, the FVT of allyltrimethylsilane led to dimethylsilene, characterized as above by its spectrum in Ar matrix at 10 K, besides other products including those resulting from a competitive radical process.[124] A similar investigation concerning the FVT of diallyldimethylsilane gave only, *via* the non-identified dimethylsilabutadiene **82a**, dimethylsilacyclobutene **81a** and degradation products[125] (cf. part I, retro-ene reactions). 1,1,2-trimethylsilene (**124**) has been generated, *via* trimethylsilylcarbene, by FVT of trimethylsilyldiazomethane (**120**) (cf. part I, 1,1-eliminations) and some bands of its infrared spectrum, recorded in Ar matrix at 10 K, were assigned[126] (Table 1).

Tetramethyldisilene (**65**)[127] and tetramethyldigermene (**174**)[94] have been obtained by thermolysis of precursors **206** and **175**, respectively, and trapped as their products of [4+2] cycloaddition with dienes. Compounds **65** and **174** were also directly characterized by Raman and infrared spectroscopy in matrix (cf. also part I, [4+2] cycloreversions and 1,1-eliminations) (Scheme 40).

Several sila- and disila-aromatics, generated either by retro-ene reaction or dehydrogenation of 1-sila-2,5-cyclohexadienes, have been characterized by infrared and ultraviolet spectroscopy in Ar matrix. Representative spectral data are gathered together in Table 2.

The generation of 1-silatoluene (**84**) by retro-ene reaction allowed to record the IR and UV spectra of this compound at 23–26 K.[128] Silabenzene (**207**) has been obtained by the same reaction, starting from precursors **208** and **209**, and also by 1,4-dehydrogenation of silacycalohexadiene **210** (the disfavored

Table 1   IR and UV data for FVT generated silenes

Silene	IR (cm^{-1})	UV (nm)	Ref.
	Absorption bands (Ar matrix, 10 K)		
$Me_2Si{=}CH_2$	642.9, 817.5, 825.2, 1003.5, 1251.0, 1259.0		114
$(CD_3)_2Si{=}CH_2$	579.2, 687.5, 769.2, 1001.4, 1015.5		114
$Me(CD_3)Si{=}CH_2$	606.0, 755.2, 817.5, 975.5, 1016.5, 1025.0, 1259.5		117
$Me_2Si{=}CD_2$	508.0, 535.3, 617.1, 776.5, 819.8, 895.1, 1117.5, 1251.0, 1259.5		119
$(CD_3)_2Si{=}CD_2$	501.6, 651.2, 732.0, 866.5, 1002.5, 1005.0,1028.0, 1112.0		119
$H_2Si{=}CH_2$	741, 817, 927, 985, 1350, 2219, 2239	*258*	121
$D_2Si{=}CH_2$	396, 719, 759, 952, 1335, 1600, 1635	*258*	121
$Cl_2Si{=}CH_2$	593, 732, 1008	*246*	121
$HClSi{=}CH_2$	539, 544, 699, 843, 984, 2230	*255*	121
$HDSi{=}CH_2$	395, 425, 735, 738, 839, 869, 1338, 1620, 2231		121
$HMeSi{=}CH_2$	610, 711, 810, 899, 991, 1255, 1291, 2182	*260*	121
$Me_2Si{=}CHMe$	607.0, 643.9, 793.7		126

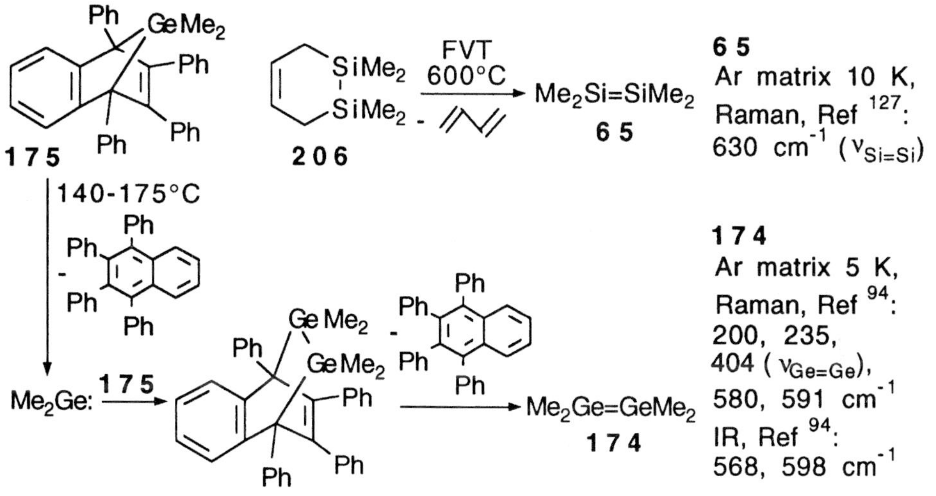

Scheme 40

1,2-dehydrogenation in the case of **211** required a too high temperature and **207** was not obtained in that case). Compound **207** was shown to be stable in matrix and its spectra were recorded. Upon irradiation of the matrix, a reversible isomerization of **207** into Dewar-silabenzene **212** was also observed.[129,99,100] A relative stabilization was obtained in the case of the crowded derivatives **213**, generated by 1,4-elimination of methoxytrimethylsilane, and their spectra were recorded, in the condensed phase, at temperatures $\leqslant 90\,\mathrm{K}$.[130]

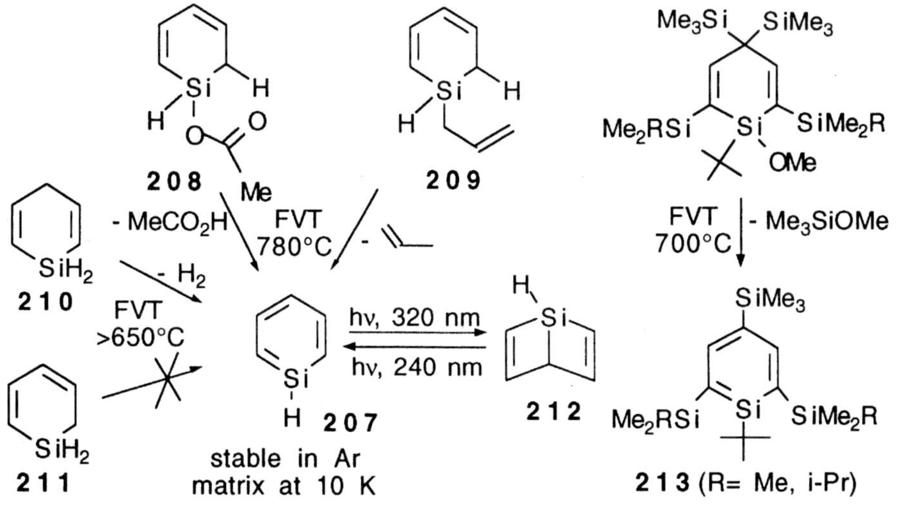

Scheme 41

The FVT of disilacyclohexadiene **214**, although more difficult than for **210**, led to 1,4-disilabenzene (**215**), identified by its IR and UV spectra in matrix.[101] On the other hand, similar spectra of hexamethyldisilabenzene **73**, obtained from disilacyclohexadiene **216** and indirectly well-characterized (see part I, [4+2] cycloreversions), could not be measured.[64] 9-Silaanthracene (**217**), as well as its 9-phenyl and 9,10-diphenyl derivatives, were also synthesized and characterized by their ultraviolet spectra at 15 K.[102]

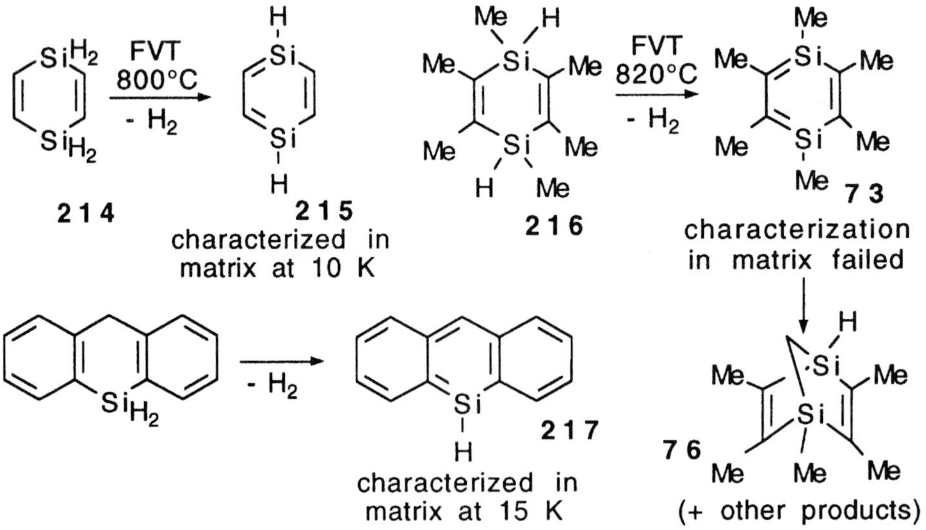

Scheme 42

Table 2   IR and UV data for FVT generated sila-aromatics

Sila-aromatic	Absorptions [Ar matrix, cm^{-1} (IR) or nm (UV)]	Ref.
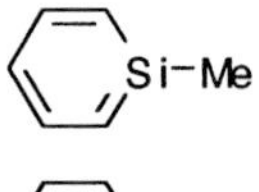 (structure) Si–Me	IR: 563, 655, 697, 770, 842, 883, 890, 900, 965, 980, 1268, 1360, 1410, 1500, 1530	
	UV: 300.5, 307.3, 313.7, 322.1	128
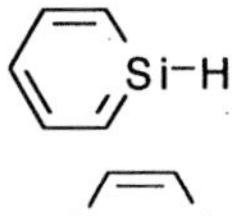 (structure) Si–H	IR: 419, 565, 568, 698, 700, 716, 720, 886, 1069, 1259, 1354, 1409, 1502, 1526, 2219, 2244, 3030–3060	
	UV: 212.0, 272.0, 298.0, 305.0, 313.4, 320.6	100
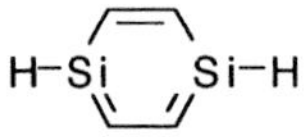 H–Si Si–H	IR: 1272.8 (only band assigned)	
	UV: 275, 340, 385, 396, 408	101
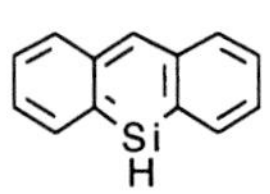 (structure) Si H	UV: 362, 380, 400, 420, 444, 468, 494	102

Dimethylsilanone (**35**) and its hexadeuterated analogue have been generated from precursors **46**, **218** and **98**[131-133] (cf. part I, thermal fragmentations). Methyl 1-sila-acetate (**219**), accompanied by dimethyldioxasilirane (**220**), resulted from the generation by FVT of dimethylsilylene from **146** (cf. part I, 1,1-eliminations) in the presence of small amounts of oxygen,[134,135] and dimethylgermanethione (**53**) was obtained by thermal monomerization of its trimer **221**.[136] All these compounds have been characterized by their infrared spectrum in argon matrix (Table 3). An attempt to observe dimethylsilanethione (**51**) in the FVT of precursors **52**, **222** and **223** remained on the other hand unsuccessful, the infrared spectrum of the obtained matrix showing in all cases the presence of **52** in place of the monomeric silanethione **51**.[137,138]

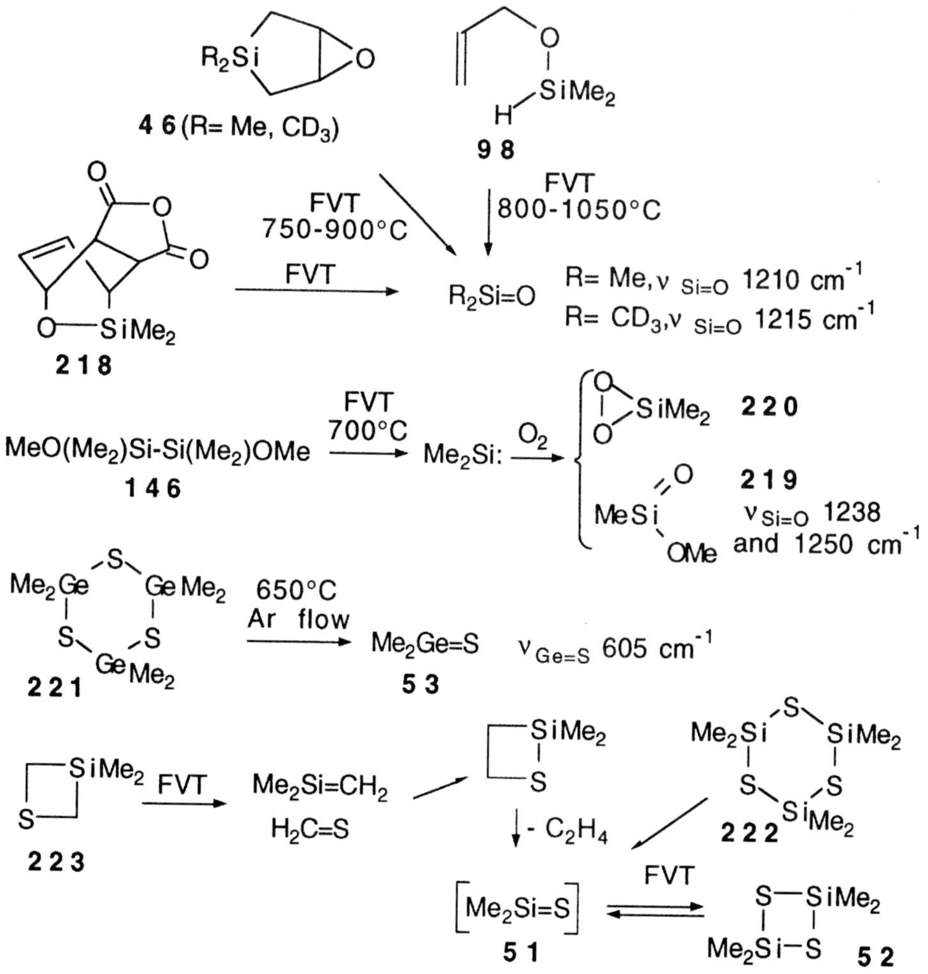

Scheme 43

Table 3 IR Data for FVT generated silanones, silaester and germanethione

Compound	IR absorption bands in Ar matrix	Ref.
$Me_2Si{=}O$	657, 770, 798, 822, 1210, 1240, 1244	132
$(CD_3)_2Si{=}O$	674, 685, 712, 995, 1007, 1032, 1215	132
$Me(OMe)Si{=}O$	1090.1, 1238.1, 1249.7	134
$Me_2Ge{=}S$	516, 605, 753, 761, 850, 1229, 1390	136

Photoelectron Spectrometry

The coupling of FVT with photoelectron spectrometry (PES) has been, until now, the most used method for the real time, gas-phase detection of double-bonded silicon and germanium reactive species. This spectrometry allows in fact an actual insight into the structure of these molecules, ensuing from the determination of their ionization potentials (IP's), collected in Table 4.

The first IP's of 1,1-dimethylsilene[139,140] and unsubstituted silene **201a**[141] have been thus obtained by FVT of 1,1-dimethylsilacyclobutane, and of silabicyclo-octadiene **202**, respectively. The retro-ene reaction of 1-allyl-1-silacyclohexa-2, 4-dienes **85** and **209** led to the obtention of 1-silatoluene (**84**)[142] and silabenzene (**207**).[143] Several 4-alkyl-1-sila-(or 1-germa)toluenes resulted in the same way from the FVT of the corresponding 1-allyl-1-sila-(or 1-germa)cyclohexa-1, 4-dienes.[144] Silabenzene has been also generated by dehydrogenation of 1-silacyclohexa-2,5-diene **210**.[145] All these sila- and germa-aromatics have been characterized in the gas phase by their photoelectron spectra.

The monomerization under FVT conditions of the cyclo-trimers **222** and **221** of 1,1-dimethylsilanethione (**51**) and 1,1-dimethylgermanethione (**53**) allowed also to record the PE spectra of these monomeric species.[146–148] The spectrum of 1,1-dimethylgermanone (**224**) could not be obtained from its cyclo-tetramer **225**,

Table 4 PE Data for some silicon and germanium double-bonded compounds

Compound	IP's (eV, He I photoionization)	Ref.
$Me_2Si{=}CH_2$	8.3	139
$Me_2Si{=}CH_2$	7.98	140
$H_2Si{=}CH_2$	8.97	141
1-silatoluene	7.7, 9.1	142
silabenzene	8.12, 9.46, 13.74	145
$Me_2Si{=}S$	9.15, 10.10, 11.05, 12.10	146
$Me_2Ge{=}S$	8.6, 9.55, 10.75, 11.47	146
$Me_2Ge{=}O$	9.7, 10.2, 11.0, 12.1	149

due to the too high temperature required. On the other hand, monomeric **224** was obtained and its PE spectrum determined, taking advantage of the low temperature (95°C) [2+2] cycloreversion of germa-heterocycle **226**.[148,149]

Trimethylsilanimine and -germanimine (**227**) were possible products in the FVT of trimethylsilyl and -germyl azides (**228**). These compounds were however not observed, but hydrogen sila- and germaisocyanides **229** were characterized by PES.[150] Also, the FVT/PES of triazido(phenyl)silane (**230**) gave the related phenyl silaisocyanide **231**.[151]

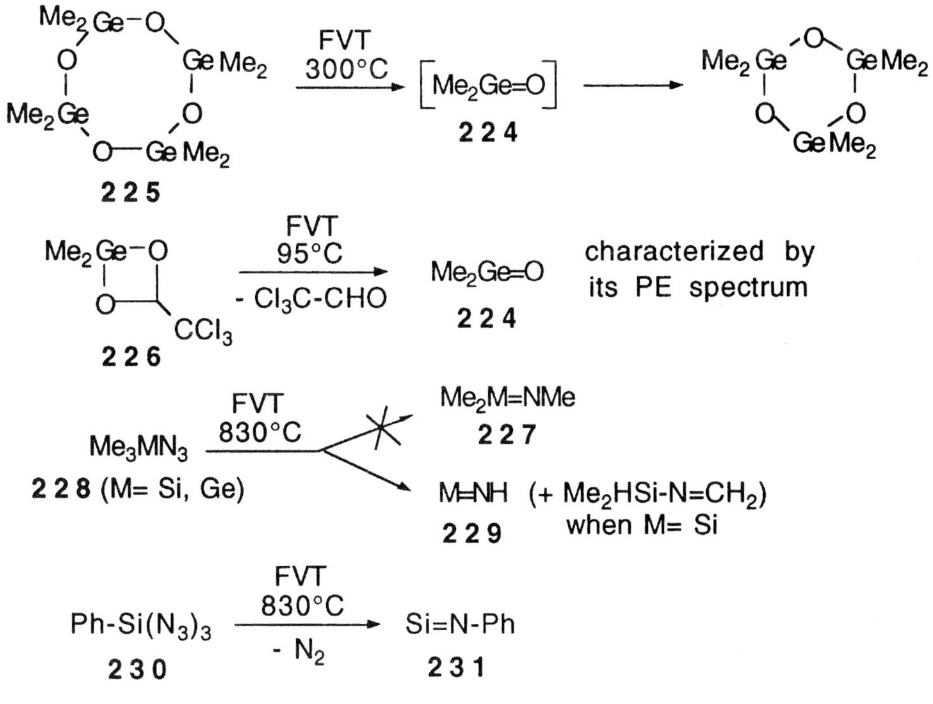

Scheme 44

Other Spectrometries

Several reactive molecules containing silicon or germanium double bonds have been characterized by coupling the FVT oven with a mass spectrometer. The FVT of 7-(dimethylgerma)norbornadiene **175** led to tetramethyldigermene (**174**), directly identified by IR in matrix (see part II, IR and UV spectroscopy), but also by its mass spectrum.[94] 1-Silatoluene (**84**) and silabenzene (**207**) have been observed in the gas phase by PES (see above) and by mass spectrometry.[142,152] This last method allowed also to characterize monomeric N-phenyl-1,1-dimethyl-silanimine (**38**), generated by retro-ene reaction of N-allyl-N-phenyl-1,1-dimethylsilanamine under FVT conditions.[153]

The well-controlled technique of very low pressure pyrolysis (VLPP)[154] has been used in connection with mass spectrometry in a physicochemical investigation of the thermal cleavage of 1,1-dimethylsilacyclobutane into 1,1-dimethylsilene.[155] This silene, produced by the same reaction in a high temperature supersonic nozzle system, has been also studied in the gas phase by electron diffraction[156] and microwave spectrometry,[157] providing thus precise informations concerning its structure.

CONCLUSION

This literature survey points out the contribution of FVT and flow thermolysis to the generation of silicon and germanium double-bonded species. Although the involved thermal reactions can give good results in condensed phase and static conditions, especially when they are effected at moderate temperature in solution with the formed transient species trapped *in situ* by an appropriate pre-added reagent, the here-considered gas-phase dynamic conditions allow to investigate a much broader scope of reactivity, as shown by the number and diversity of the examples reported in part I.

The most specific application of FVT and flow thermolysis in that field is indeed, as presented in part II, the direct spectrometric characterization of the highly reactive monomeric sila- and germa-heteroalkenes and a number of spectral data have been already collected, either in argon matrix or in the gas phase. However, the obtained informations remain presently incomplete and some classes such as silanimines, phosphasilenes, germenes, germa-aromatics, and others, are practically unknown in that respect. Much work is yet to be done before the comprehensive knowledge of these monomeric species.

REFERENCES

1. L.E. Gusel'nikov, N.S. Nametkin, and V.M. Vdovin, *Accounts Chem. Res.* **8**, 18 (1975).
2. P. Jutzi, *Angew. Chem. Int. Ed. Engl.* **14**, 232 (1975).
3. L.E. Gusel'nikov and N.S. Nametkin, *Chem. Rev.* **79**, 529 (1979).
4. G. Raabe and J. Michl, *Chem. Rev.* **85**, 419 (1985).
5. J. Dubac and A. Laporterie, *Chem. Rev.* **87**, 319 (1987).
6. G. Raabe and J. Michl, *The Chemistry of Organic Silicon Compounds* (S. Patai and Z. Rappoport, Eds.), Vol 2, J. Wiley, New York, 1989, p. 1015.
7. J. Barrau, J. Escudié, and J. Satgé, *Chem. Rev.* **90**, 283 (1990).
8. J. Satgé, *J. Organomet. Chem.* **400**, 121 (1990).
9. M.J. Almond, *Short-lived molecules* (Ellis Horwood, Chichester, 1990), 93.
10. E.A. Chernyshev and N.G. Komalenkova, *Russian Chem. Rev.* **59**, 531 (1990).
11. M.C. Flowers and L.E. Gusel'nikov, *J. Chem. Soc. (B)*, 419 (1968).
12. R.D. Bush, C.M. Golino, D.N. Roark, and L.H. Sommer, *J. Organomet. Chem.* **59**, C17 (1973).
13. R.D. Bush, C.M. Golino, and L.H. Sommer, *J. Am. Chem. Soc.* **96**, 7105 (1974).
14. N.S. Nametkin, L.E. Gusel'nikov, E.A. Volnina, E.N. Burdasov, and V.M. Vdovin, *Dokl. Akad. Nauk SSSR* **214**, 818 (1974).
15. C.M. Golino, R.D. Bush, P. On, and L.H. Sommer, *J. Am. Chem. Soc.* **97**, 1957 (1975).
16. T.J. Barton, G. Marquardt, and J.A. Kilgour, *J. Organomet. Chem.* **85**, 317 (1975).

17. N. Auner and J. Grobe, *J. Organomet. Chem.* **188**, 151 (1980).
18. N. Auner and J. Grobe, *J. Organomet. Chem.* **190**, 129 (1980).
19. N. Auner and J. Grobe, *J. Organomet. Chem.* **197**, 13 (1980).
20. N. Auner and J. Grobe, *J. Organomet. Chem.* **197**, 147 (1980).
21. N. Auner and J. Grobe, *J. Organomet. Chem.* **222**, 33 (1981).
22. P. Jutzi and P. Langer, *J. Organomet. Chem.* **202**, 401 (1980).
23. L.E. Gusel'nikov, V.M. Sokolova, E.A. Volnini, Z.A. Kerzina, N.S. Nametkin, N.G. Komalenkova, S.A. Bashkirova, and E.A. Chernyshev, *Dokl. Akad. Nauk SSSR* **260**, 348 (1981).
24. J. Grobe and H. Ziemer, *Z. Naturforsch.* **48b**, 1193 (1993).
25. G. Bertrand, G. Manuel, and P. Mazerolles, *Tetrahedron Lett.* 2149 (1978).
26. G. Bertrand, G. Manuel, and P. Mazerolles, *Tetrahedron* **37**, 2875 (1981).
27. T.J. Barton, G.T. Burns, and D. Gschneider, *Organometallics* **2**, 8 (1983).
28. R.T. Conlin and K.L. Bobbitt, *Organometallics* **6**, 1406 (1987).
29. R. Walsh, *J. Chem. Soc., Chem. Commun.* 1415 (1982).
30. R.T. Conlin and D.L. Wood, *J. Am. Chem. Soc.* **103**, 1843 (1981).
31. R.T. Conlin and Y.W. Kwak, *Organometallics* **3**, 918 (1984).
32. T.J. Barton, S.A. Burns, and G.T. Burns, *Organometallics* **1**, 210 (1982).
33. R.T. Conlin and R.S. Gill, *J. Am. Chem. Soc.* **105**, 618 (1983).
34. I.M.T. Davidson, A. Fenton, S. Ijadi-Maghsoodi, R.J. Scampton, N. Auner, J. Grobe, N. Tillman, and T.J. Barton, *Organometallics* **3**, 1593 (1984).
35. N. Auner, I.M.T. Davidson, S. Ijadi-Maghsoodi, and F.T. Lawrence, *Organometallics* **5**, 431 (1986).
36. N. Auner and J. Grobe, *Z. Anorg. Allg. Chem.* **459**, 15 (1979).
37. D.N. Roark and L.H. Sommer, *J. Chem. Soc., Chem. Commun.* 167 (1973).
38. C.M. Golino, R.D. Bush, D.N. Roark, and L.H. Sommer, *J. Organomet. Chem.* **66**, 29 (1974).
39. R.D. Bush, C.M. Golino, G.D. Homer, and L.H. Sommer, *J. Organomet. Chem.* **80**, 37 (1974).
40. C.M. Golino, R.D. Bush, and L.H. Sommer, *J. Am. Chem. Soc.* **97**, 7371 (1975).
41. C.M. Golino, R.D. Bush, and L.H. Sommer, *J. Am. Chem. Soc.* **96**, 614 (1974).
42. P.B. Valkovich and W.P. Weber, *J. Org. Chem.* **40**, 229 (1975).
43. P.B. Valkovich and W.P. Weber, *J. Organomet. Chem.* **99**, 231 (1975).
44. W. Ando, A. Sekiguchi, and T. Migita, *J. Am. Chem. Soc.* **97**, 7159 (1975).
45. S. Bain, S. Ijadi-Maghsoodi, and T.J. Barton, *Organometallics* **6**, 1136 (1987).
46. G. Manuel, G. Bertrand, W.P. Weber, and S.A. Kazoura, *Organometallics* **3**, 1340 (1984).
47. K.T. Kang, G. Manuel, and W.P. Weber, *Chem. Lett.* 1685 (1986).
48. G.K. Henry, D.R. Dowd, R. Bau, G. Manuel, and W.P. Weber, *Organometallics* **5**, 1818 (1986).
49. L.H. Sommer and J. McLick, *J. Organomet. Chem.* **101**, 171 (1975).
50. L.E. Gusel'nikov, V.M. Sokolova, E.A. Volnina, V.G. Zaikin, N.S. Nametkin, M.G. Voronkov, S.V. Kirpichenko, and V.V. Keiko, *J. Organomet. Chem.* **214**, 145 (1981).
51. L.E. Gusel'nikov, V.V. Volkova, E.A. Volnina, V.G. Avakian, and N.S. Nametkin, *Dokl. Akad. Nauk SSSR* **274**, 1106 (1984).
52. L.E. Gusel'nikov, V.V. Volkova, V.G. Avakian, E.A. Volnina, V.G. Zaikin, N.S. Nametkin, A.A. Polyakova, and M.I. Tokarev, *J. Organomet. Chem.* **271**, 191 (1984).
53. J. Barrau, G. Rima, M. El-Amine, and J. Satgé, *J. Organomet. Chem.* **345**, 39 (1988).
54. T.J. Barton and E.A. Kline, *J. Organomet. Chem.* **42**, C21 (1972).
55. T.J. Barton, E.A. Kline, and P.M. Garvey, *J. Am. Chem. Soc.* **95**, 3078 (1973).
56. R.T. Conlin, M.P. Bessellieu, P.R. Jones, and R.A. Pierce, *Organometallics* **1**, 396 (1982).
57. P.R. Jones and M.E. Lee, *J. Am. Chem. Soc.* **105**, 6725 (1983).
58. D.N. Roark and G.J.D. Peddle, *J. Am. Chem. Soc.* **94**, 5837 (1972).
59. W.D. Wulff, W.F. Goure, and T.J. Barton, *J. Am. Chem. Soc.* **100**, 6236 (1978).
60. T.J. Barton and J.A. Kilgour, *J. Am. Chem. Soc.* **96**, 2278 (1974).
61. T.J. Barton and J.A. Kilgour, *J. Am. Chem. Soc.* **98**, 7231 (1976).
62. G. Hussmann, W.D. Wulff, and T.J. Barton, *J. Am. Chem. Soc.* **105**, 1263 (1983).
63. J.D. Rich and R. West, *J. Am. Chem. Soc.* **104**, 6884 (1982).
64. K.M. Welsh, J.D. Rich, R. West, and J. Michl, *J. Organomet. Chem.* **325**, 105 (1987).
65. T.J. Barton, S.A. Burns, I.M.T. Davidson, S. Ijadi-Maghsoodi, and I.T. Wood, *J. Am. Chem. Soc.* **106**, 6367 (1984).
66. Z. Jabry, M.C. Lasne, and J.L. Ripoll, *J. Chem. Research* (S), 188 (1986).

67. E. Block and L.K. Revelle, *J. Am. Chem. Soc.* **100**, 1630 (1978).

68. N. Auner, I.M.T. Davidson, and S. Ijadi-Maghsoodi, *Organometallics* **4**, 2210 (1985).

69. T.J. Barton and G.T. Burns, *J. Am. Chem. Soc.* **100**, 5246 (1978).

70. T.J. Barton and G.T. Burns, *J. Organomet. Chem.* **179**, C17 (1979).

71. T.J. Barton and G.T. Burns, *Organometallics* **1**, 1455 (1982).

72. J.P. Béteille, G. Manuel, A. Laporterie, H. Iloughmane, and J. Dubac, *Organometallics* **5**, 1742 (1986).

73. Y. Nakadaira, H. Sakaba, and H. Sakurai, *Chem. Lett.* 1071 (1980).

74. T.J. Barton, G.T. Burns, E.V. Arnold, and J. Clardy, *Tetrahedron Lett.* **22**, 7 (1981).

75. T.J. Barton and S. Bain, *Organometallics* **7**, 528 (1988).

76. M. Letulle, A. Systermans, J.L. Ripoll, and P. Guenot, *J. Organomet. Chem.*, **484**, 89 (1994).

77. A. Chivé, V. Lefèvre, A. Systermans, J.L. Ripoll, M. Bogey, and A. Walters, *Phosphorus Sulfur Silicon*, **91**, 281 (1994).

78. R.L. Kreeger and H. Shechter, *Tetrahedron Lett.* 2061 (1975).

79. W. Ando, A. Sekiguchi, J. Ogiwara, and T. Migita, *J. Chem. Soc., Chem. Commun.* 145 (1975).

80. W. Ando, A. Sekiguchi, A.J. Rothschild, R.R. Gallucci, M. Jr. Jones, T.J. Barton, and J.A. Kilgour, *J. Am. Chem. Soc.* **99**, 6995 (1977).

81. T.J. Barton and S.K. Hoekman, *J. Am. Chem. Soc.* **102**, 1584 (1980).

82. D.R. Parker and L.H. Sommer, *J. Organomet. Chem.* **110**, C1 (1976).

83. I.M.T. Davidson, S. Ijadi-Maghsoodi, T.J. Barton, and N. Tillman, *J. Chem. Soc., Chem. Commun.* 478 (1984).

84. S.F. Rickborn, D.S. Rogers, M.A. Ring, and H.E. O'Neal, *J. Phys. Chem.* **90**, 408 (1986).

85. R.T. Conlin and P.P. Gaspar, *J. Am. Chem. Soc.* **98**, 868 (1976).

86. P.P. Gaspar and D. Lei, *Organometallics* **5**, 1276 (1986).

87. A. Sekiguchi and W. Ando, *Tetrahedron Lett.* **24**, 2791 (1983).

88. W. Ando, Y. Hamada, and A. Sekiguchi, *J. Chem. Soc., Chem. Commun.* 787 (1982).

89. T.J. Barton, G.T. Burns, W.F. Goure, and W.D. Wulff, *J. Am. Chem. Soc.* **104**, 1149 (1982).

90. S.A. Burns, G.T. Burns, and T.J. Barton, *J. Am. Chem. Soc.* **104**, 6140 (1982).

91. T.J. Barton and G.T. Burns, *Tetrahedron Lett.* **24**, 159 (1983).

92. B.H. Boo and P.P. Gaspar, *Organometallics* **5**, 698 (1986).

93. W. Ando, T. Tsumuraya, and A. Sekiguchi, *Tetrahedron Lett.* **26**, 4523 (1985).

94. P. Bleckmann, R. Minkwitz, W.P. Neumann, M. Schriewer, M. Thibud, and B. Watta, *Tetrahedron Lett.* **25**, 2467 (1984).

95. L. Linder, A. Revis, and T.J. Barton, *J. Am. Chem. Soc.* **108**, 2742 (1986).

96. C. Eaborn, D.A.R. Happer, P.B. Hitchcock, S.P. Hopper, K.D. Safa, S.S. Washburne, and D.R.M. Walton, *J. Organomet. Chem.* **186**, 309 (1980).

97. G.T. Burns and T.J. Barton, *J. Organomet. Chem.* **216**, C5 (1981).

98. T.J. Barton and M. Vuper, *J. Am. Chem. Soc.* **103**, 6788 (1981).

99. G. Maier, G. Mihm, and H.P. Reisenauer, *Chem. Ber.* **115**, 801 (1982).

100. G. Maier, G. Mihm, R.O.W. Baumgärtner, and H.P. Reisenauer, *Chem. Ber.* **117**, 2337 (1984).

101. G. Maier, K. Schöttler, and H.P. Reisenauer, *Tetrahedron Lett.* **26**, 4079 (1985).

102. Y. van den Winkel, B.L.M. van Baar, F. Bickelhaupt, W. Kulik, C. Sierakowski, and G. Maier, *Chem. Ber.* **124**, 185 (1991).

103. S.A. Kazoura and W.P. Weber, *J. Organomet. Chem.* **268**, 19 (1984).

104. S.A. Kazoura and W.P. Weber, *J. Organomet. Chem.* **271**, 47 (1984).

105. J. Slutsky and H. Kwart, *J. Org. Chem.* **38**, 3658 (1973).

106. T.J. Barton and W.D. Wulff, *J. Organomet. Chem.* **168**, 23 (1979).

107. T.J. Barton and W.D. Wulff, *J. Am. Chem. Soc.* **101**, 2735 (1979).

108. T.J. Barton, W.D. Wulff, and S.A. Burns, *Organometallics* **2**, 4 (1983).

109. T.J. Barton and B.L. Groh, *J. Am. Chem. Soc.* **107**, 7221 (1985).

110. T.J. Barton and G.C. Paul, *J. Am. Chem. Soc.* **109**, 5292 (1987).

111. T.J. Barton and C.L. McIntosh, *J. Chem. Soc., Chem. Commun.* 861 (1972).

112. A.K. Mal'tsev, V.N. Khabashesku, and O.M. Nefedov, *Izv. Akad. Nauk SSSR, Ser. Khim.* 1193 (1976).

113. A.K. Mal'tsev, V.N. Khabashesku, and O.M. Nefedov, *Dokl. Akad. Nauk SSSR* **233**, 421 (1977).

114. O.M. Nefedov, A.K. Mal'tsev, V.N. Khabashesku, and V.A. Korolev, *J. Organomet. Chem.* **201**, 123 (1980).
115. L.E. Gusel'nikov, V.V. Volkova, V.G. Avakian, and N.S. Nametkin, *J. Organomet. Chem.* **201**, 137 (1980).
116. A.K. Mal'tsev, V.N. Khabashesku, and O.M. Nefedov, *Izv. Akad. Nauk SSSR, Ser. Khim.* 2152 (1979).
117. A.K. Mal'tsev, V.N. Khabashesku, and O.M. Nefedov, *J. Organomet. Chem.* **226**, 11 (1982).
118. V.N. Khabashesku, E.G. Baskir, A.K. Mal'tsev, and O.M. Nefedov, *Izv. Akad. Nauk SSSR, Ser. Khim.* 238 (1983).
119. A.K. Mal'tsev, V.N. Khabashesku, and O.M. Nefedov, *J. Organomet. Chem.* **271**, 55 (1984).
120. G. Maier, G. Mihm, and H.P. Reisenauer, *Angew. Chem. Int. Ed. Engl.* **20**, 597 (1981).
121. G. Maier, G. Mihm, and H.P. Reisenauer, *Chem. Ber.* **117**, 2351 (1984).
122. H.P. Reisenauer, G. Mihm, and G. Maier, *Angew. Chem. Int. Ed. Engl.* **21**, 854 (1982).
123. G. Maier, G. Mihm, H.P. Reisenauer, and D. Littmann, *Chem. Ber.* **117**, 2369 (1984).
124. A.K. Mal'tsev, V.A. Korolev, N.D. Kagramanov, and O.M. Nefedov, *Izv. Akad. Nauk SSSR, Ser. Khim.* 1078 (1983).
125. V.A. Korolev, A.K. Mal'tsev, and O.M. Nefedov, *Izv. Akad. Nauk SSSR, Ser. Khim.* 711 (1985).
126. A.K. Mal'tsev, V.A. Korolev, V.N. Khabashesku, and O.M. Nefedov, *Dokl. Akad. Nauk SSSR* **251**, 1166 (1980).
127. A. Marchand, P. Gerval, F. Duboudin, M.H. Gaufryau, M. Joanny, and P. Mazerolles, *J. Organomet. Chem.* **267**, 93 (1984).
128. C.L. Kreil, O.L. Chapman, G.T. Burns, and T.J. Barton, *J. Am. Chem. Soc.* **102**, 841 (1980).
129. G. Maier, G. Mihm, and H.P. Reisenauer, *Angew. Chem. Int. Ed. Engl.* **19**, 52 (1980).
130. P. Jutzi, M. Meyer, H.P. Reisenauer, and G. Maier, *Chem. Ber.* **122**, 1227 (1989).
131. V.N. Khabashesku, Z.A. Kerzina, A.K. Mal'tsev, and O.M. Nefedov, *Izv. Akad. Nauk SSSR, Ser. Khim.* 1215 (1986).
132. V.N. Khabashesku, Z.A. Kerzina, E.G. Baskir, A.K. Mal'tsev, and O.M. Nefedov, *J. Organomet. Chem.* **347**, 277 (1988).
133. V.N. Khabashesku, Z.A. Kerzina, A.K. Mal'tsev, and O.M. Nefedov, *J. Organomet. Chem.* **364**, 301 (1989).
134. A. Patyk, W. Sander, J. Gauss, and D. Cremer, *Angew. Chem. Int. Ed. Engl.* **28**, 898 (1989).
135. M. Trommer, W. Sander, and A. Patyk, *J. Am. Chem. Soc.* **115**, 11775 (1993).
136. J. Barrau, V. Balaji, and J. Michl, *Organometallics* **8**, 2034 (1989).
137. L.E. Gusel'nikov, V.V. Volkova, V.G. Avakian, N.S. Nametkin, M.G. Voronkov, S.V. Kirpichenko, and E.N. Suslova, *Dokl. Akad. Nauk SSSR* **272**, 892 (1983).
138. L.E. Gusel'nikov, V.V. Volkova, V.G. Avakian, N.S. Nametkin, M.G. Voronkov, S.V. Kirpichenko, and E.N. Suslova, *J. Organomet. Chem.* **254**, 173 (1983).
139. T. Koenig and W. McKenna, *J. Am. Chem. Soc.* **103**, 1212 (1981).
140. J.M. Dyke, G.D. Josland, R.A. Lewis, and A. Morris, *J. Phys. Chem.* **86**, 2913 (1982).
141. P. Rosmus, H. Bock, B. Solouki, G. Maier, and G. Mihm, *Angew. Chem. Int. Ed. Engl.* **20**, 598 (1981).
142. H. Bock, R.A. Bowling, B. Solouki, T.J. Barton, and G.T. Burns, *J. Am. Chem. Soc.* **102**, 429 (1980).
143. B. Solouki, P. Rosmus, H. Bock, and G. Maier, *Angew. Chem. Int. Ed. Engl.* **19**, 51 (1980).
144. G. Märkl, D. Rudnick, R. Schulz, and A. Schweig, *Angew. Chem. Int. Ed. Engl.* **21**, 221 (1982).
145. H. Bock, P. Rosmus, B. Solouki, and G. Maier, *J. Organomet. Chem.* **271**, 145 (1984).
146. C. Guimon, G. Pfister-Guillouzo, H. Lavayssiere, G. Dousse, J. Barrau, and J. Satgé, *J. Organomet. Chem.* **249**, C17 (1983).
147. J.L. Garcia, D. Gonbeau, G. Pfister-Guillouzo, M. Roch, and J. Weber, *Can. J. Chem.* **63**, 1518 (1985).
148. G. Pfister-Guillouzo and C. Guimon, *Phosphorus Sulfur* **23**, 197 (1985).
149. C. Guimon, G. Pfister-Guillouzo, G. Rima, M. El Amine, and J. Barrau, *Spectrosc. Lett.* **18**, 7 (1985).
150. C. Guimon and G. Pfister-Guillouzo, *Organometallics* **6**, 1387 (1987).
151. H. Bock and R. Dammel, *Angew. Chem. Int. Ed. Engl.* **24**, 111 (1985).

152. A.K. Mal'tsev, N.D. Kagramanov, I.O. Bragilevskii *et al.*, *Izv. Akad. Nauk SSSR, Ser. Khim.* 1051 (1989).
153. J.M. Denis, P. Guenot, M. Letulle, B. Pellerin, and J.L. Ripoll, *Chem. Ber.* **125**, 1397 (1992).
154. D.M. Golden, G.N. Spokes, and S.W. Benson, *Angew. Chem. Int. Ed. Engl.* **12**, 534 (1973).
155. S. Basu, I.M.T. Davidson, R. Laupert, and P. Potzinger, *Ber. Bunsenges. Phys. Chem.* **83**, 1282 (1979).
156. P.G. Mahaffy, R. Gutowsky, and L.K. Montgomery, *J. Am. Chem. Soc.* **102**, 2854 (1980).
157. H.S. Gutowsky, J. Chen, P.J. Hajduk, J.D. Keen, and T. Emilsson, *J. Am. Chem. Soc.* **111**, 1901 (1989).

6. GAS PHASE SYNTHESIS OF THIOCARBONYL, SELENOCARBONYL AND TELLUROCARBONYL COMPOUNDS

NADIA PELLOUX-LÉON and YANNICK VALLÉE

*LEDSS, UMR CNRS-Université Joseph Fourier,
UJF, BP 53, 38041 Grenoble, France*

INTRODUCTION

In general, compounds bearing a carbon-sulfur double bond are much more stable than would believe the organic chemist unfamiliar with them. For instance, saturated dithioesters and thionoesters can be prepared by usual methods and will stay for weeks on the bench without appreciable change.[1] No special techniques are needed to synthesize them. On the over hand, most thioaldehydes and thioketenes, as well as their oxidized analogues, sulfines and sulfenes, are extremely reactive species which cannot be isolated under normal conditions (Scheme 1). The best way to study their reactivity is to generate them *in-situ* for trapping experiments.[2] However, this method, although useful for preparative purposes, does not allow the direct observation of the reactive intermediates, and without the help of FVT most thioaldehydes would have remained phantom molecules.

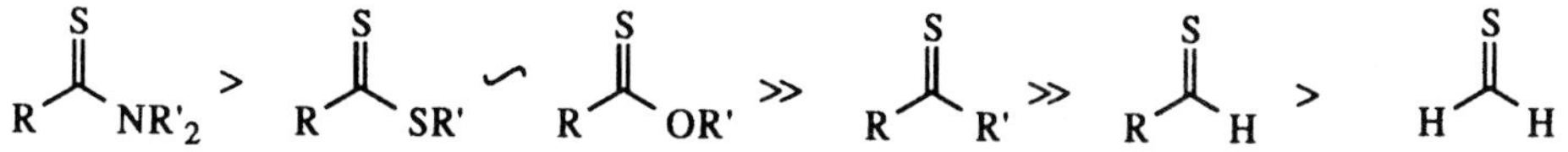

Scheme 1   Relative stability of thiocarbonyls.

In this chapter, we will review the uses of FVT for the synthesis of thioaldehydes, unstable thioketones, thioketenes, unsaturated thiocarboxylic acid derivatives, sulfines (thioaldehyde *S*-oxides) and sulfenes (thioaldehyde *S,S*-dioxides). One part will present the applications of FVT to the synthesis of selenocarbonyl compounds, mainly selenoaldehydes. Finally, a short section will describe the very few results concerning the thermolysis of tellurium compounds.

1.   THIOALDEHYDES

Among the reactive thioaldehydes[3] the most studied has been by far the parent compound, methanethial (thioformaldehyde). We will first discuss the results about this compound.

Methanethial

Because it is the most simple thioaldehyde and also perhaps because it is an interstellar molecule,[4] methanethial has been the subject of numerous reports. Even though, it was postulated as early as last century, it was not before the sixties that the first believable evidences of methanethial as a monomeric species appeared.[5,6] In 1970, Johnson et al. reported that methanethial was produced by thermolysis of dimethyldisulfide (Scheme 2). They were able to clearly identify it by microwave spectroscopy.[7] In 1971, Johns and Olsen studied its IR spectrum.[8] Also using dimethyldisulfide as precursor, Kroto and Suffolk were able to obtain its photoelectron spectrum for the first time.[9] The first ionisation potential was found to be 9.34 eV. Methanethiol, hydrogen sulfide and carbon disulfide were also detected.

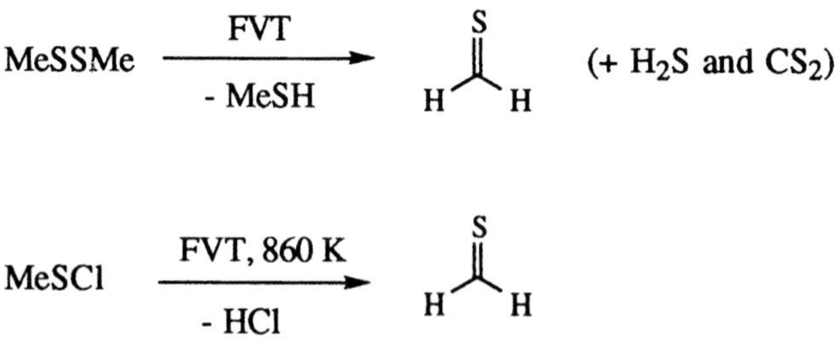

$$\text{MeSSMe} \quad \xrightarrow[\text{- MeSH}]{\text{FVT}} \quad \underset{H \quad H}{\overset{S}{\|}} \quad (\text{+ H}_2\text{S and CS}_2)$$

$$\text{MeSCl} \quad \xrightarrow[\text{- HCl}]{\text{FVT, 860 K}} \quad \underset{H \quad H}{\overset{S}{\|}}$$

Scheme 2

Methylsulfenyl chloride is another precursor for methanethial. Upon heating above 860 K, it is quantitatively cleaved into hydrogen chloride and the desired thioaldehyde. This method was used to record a clean PE spectrum of methanethial.[10,11] In this case, gaseous ammonia was carefully introduced in to a reaction chamber placed at the oven exit. Ammonium chloride was formed from the reaction of HCl with NH_3, thus allowing one to obtain pure monomeric methanethial. The recorded He(I) photoelectron spectrum showed five bands at 9.38, 11.76, 13.85, 15.20 and 19.9 eV.

Other precursors for methanethial are: allyl methyl sulfide[11] (retro-ene reaction), methyl thiocyanate,[11] trithiolane[12] and 1,3-dithietane[13] (Scheme 3). Upon FVT, 1,3-dithietane S-oxide 1 gives rise to two interesting small molecules: methanethial and its S-oxide.[13]

The retro-Diels-Alder reaction has also been used to obtain methanethial (Scheme 4).[14] Thiophosgene reacts with anthracene to give the adduct 2 which was reduced by lithium aluminium hydride.[15] When the obtained methanethial–anthracene adduct 3 was thermolysed at 650°C, anthracene was quantitatively recovered at the oven exit and H_2CS was characterized by PE spectroscopy.

Methanethial can be obtained by thermolysis of the phtalimide derivative 4. At 135–150°C, this compound is cleaved into phtalimide, methanal and methanethial which was chemically trapped.[16]

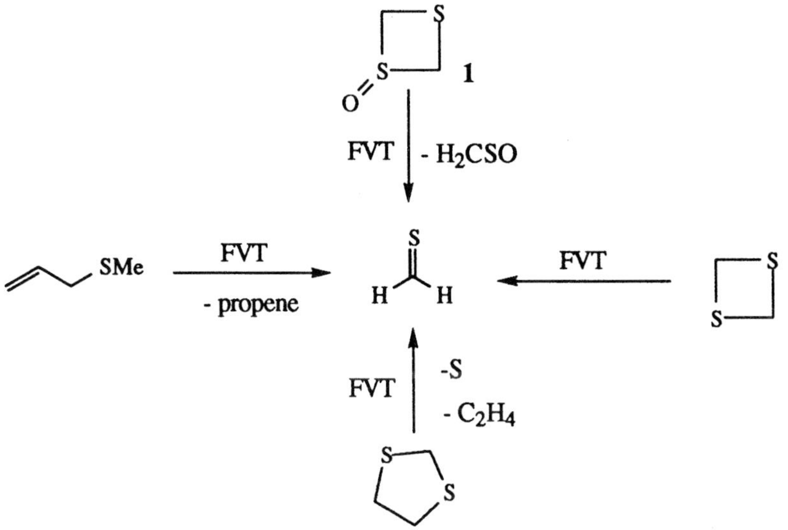

Scheme 3

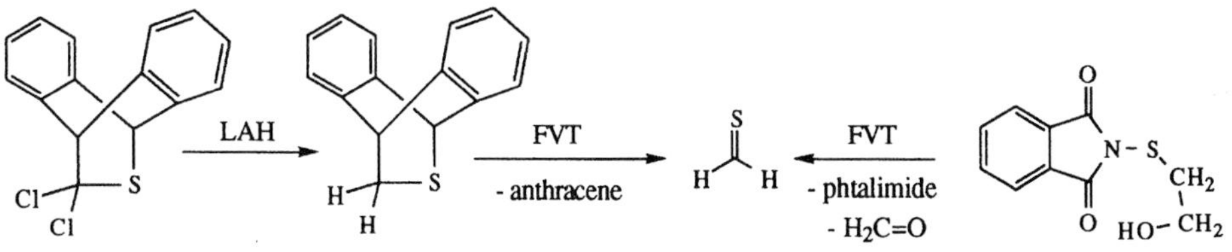

Scheme 4

Other Thioaldehydes

Some of the ways used to obtain methanethial can be extended to other thioaldehydes. This is the case of the retro-ene and the retro-Diels-Alder reactions.

De Mayo and coworkers[17,18] have synthesized ethanethial (thioacetaldehyde) by FVT of allyl ethyl sulfide (Scheme 5). This thioaldehyde was characterized by infrared spectroscopy.[19] The strong band at 1067 cm^{-1} which was also observed after thermolysis of 11-thia-12-methyl-9,10-dihydro-9,10-ethanoanthracene **5** can be attributed to the $v_{C=S}$ vibration. Another strong band at 1146 cm^{-1} is attributed to the v_{C-C} vibration. Ethenethiol **6**, which is the tautomeric form of ethanethial, was not detected during these thermolysis. This enethiol and two of its substituted analogues were synthesized later under FVT conditions by a retro-Diels-Alder reaction (Scheme 5).[20]

Trifluoroethanethial (Scheme 5) was obtained by Schuler and Sundermeyer by thermolysis of trifluoromethyldithiolan S,S-dioxide **7**.[21] Upon heating, this

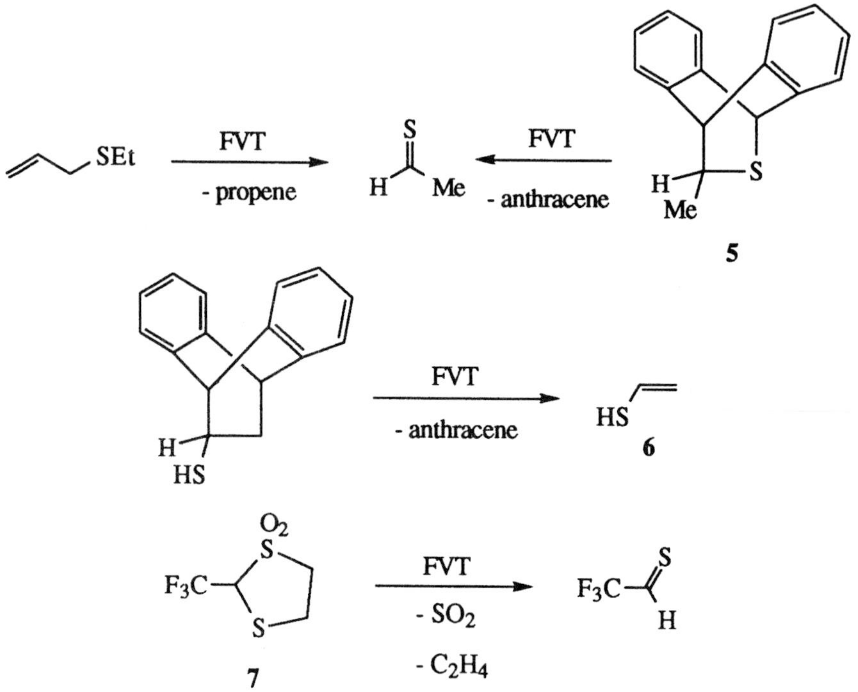

Scheme 5

heterocycle was cleaved into sulfur dioxide, ethylene and the expected halo-thioal-dehyde which was trapped by various dienes. The infrared spectrum of trifluoro-ethanethial was recorded and showed bands at 1144, 1155, 1290, 1292 and 1355 cm^{-1}. None of these bands can be clearly attributed to a pure $v_{C=S}$ vibration.

Thiobenzaldehyde (Scheme 6) has been obtained by various methods includ-ing the retro-ene[18] and rDA reactions.[22,23] Also, the thermolysis of diphenyl-1,2,4-trithiolan 8 has been used to generate this simple aromatic thioaldehyde.[11]

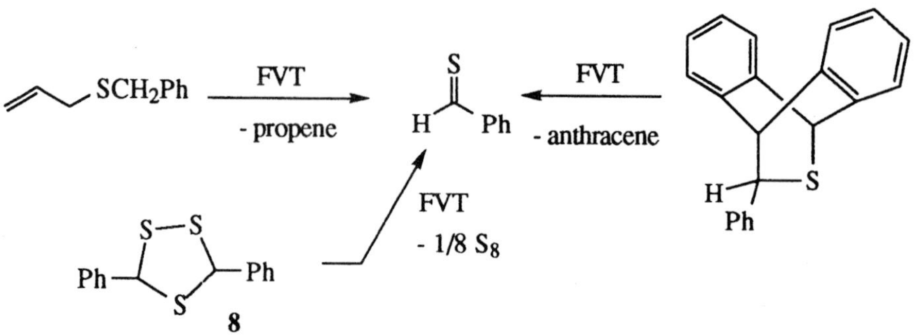

Scheme 6

The thermolysis of diallyl sulfide has been used to synthesize propenethial (thioacrolein). This α,β-unsaturated thioaldehyde is extremely unstable and dimerizes readily to a mixture of isomeric dithiins **9** and **10** (Scheme 7). When thermolyzed, these dithiins undergo a retro-Diels-Alder reaction and give back monomeric propenethial. This thioaldehyde has been characterized by infrared,[18] photoelectron[24,25] and microwave[26] spectroscopies. Propynethial, the triple bonded analogue of thioacrolein, was also synthesized by a retro-ene reaction.[27]

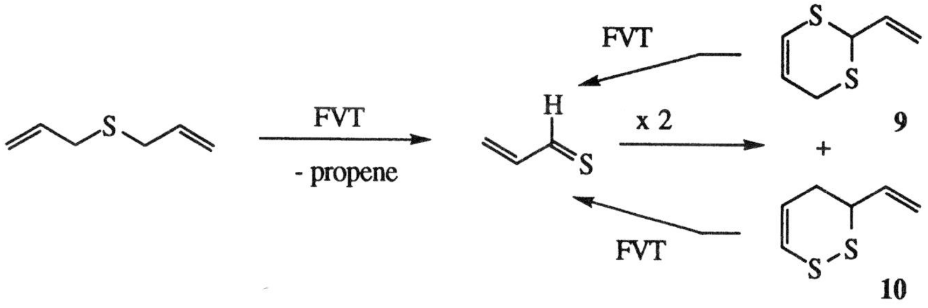

Scheme 7

The two parts of thioformyl cyanide (Scheme 8), the cyano and thioformyl groups, are often present in interstellar molecules and this α-functionalized thioaldehyde has been the subject of quite a lot of work. It was obtained for the first time under FVT conditions by the retro-ene reaction of allylthioacetonitrile and characterized by microwave spectroscopy.[28] The same precursor and the cyanothianorbornene **11** were used to obtain its photoelectron spectrum.[29] The visible spectrum of this blue thioaldehyde showed a maximum at 611 nm. This maximum was attributed to the n$\rightarrow\pi^*$ transition. Kappe *et al.* recorded its infrared spectrum and assigned bands at 2221 and 1103 cm^{-1} to the C$\equiv$N and C$=$S stretching vibrations.[30]

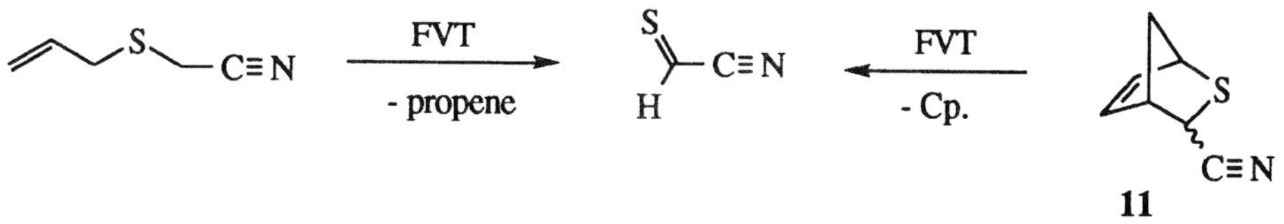

Scheme 8

Thioxoethanal (monothioglyoxal) is another simple α-functionalized thioaldehyde. It was obtained by Torres *et al.* by photolysis of vinylene thiocarbonate **12**.[31] The same precursor was used under FVT conditions by McNaughton and Bruget in a millimeter wave study (Scheme 9).[32] On the other hand, when 2,3-dihydro-1,4-oxathiin **13** was heated at 720–850°C, it underwent a retro-Diels-Alder reaction and also gave thioxoethanal. This method was used to record its

photoelectron spectrum[33] and to trap it using various diene. Trapping with cyclopentadiene gave the expected adduct **14** as a 4/1 mixture of stereoisomers in 53%
yield. The thermolysis of 1,4-oxathiins was later extended to the synthesis of
various α-oxothioaldehydes and thioketones.[34]

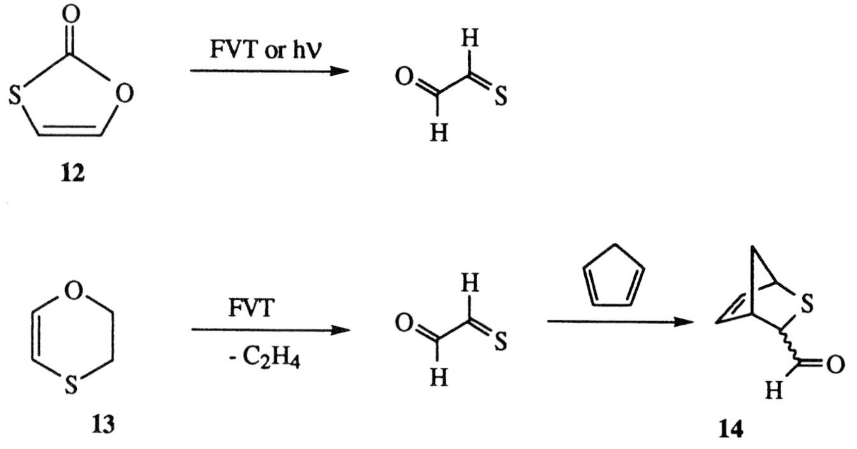

Scheme 9

In contrast with α-oxothiocarbonyl compounds which clearly exist in an open
O=C—C=S form,[33] α-dithiocarbonyl compounds are generally known in their
closed cyclic isomeric form, i.e. dithietes. As so, thioxoethanethial (dithioglyoxal)
cannot be regarded as a thioaldehyde. Selected papers dealing with 1,2-dithiete
are listed in Ref.[35]

The retro-Diels-Alder reaction has also been used to obtain the enol and
enethiol forms of the simplest β-oxo-thioaldehyde, 3-oxopropanethial.[36,37]

The synthesis of α-iminothioaldehydes was recently attempted (Scheme 10).[38,39]
Methylimino-ethanethial was found to be unstable and undergoes a cyclization to
2,3-dihydrothiazole **15**. This heterocycle, which is a secondary enamine, was
observed at −60°C. Upon heating to room temperature, it leads to its isomer,
2,5-dihydrothiazole **16**. The mechanism of this heterocyclization has been studied
by *ab initio* calculations.[40]

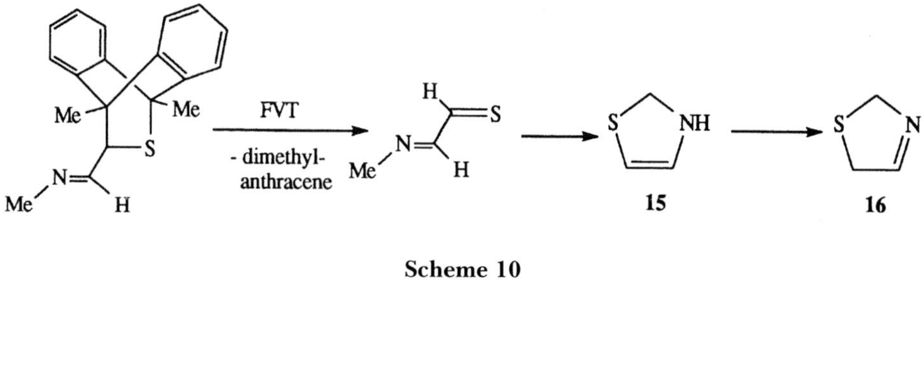

Scheme 10

2. THIOKETONES

Thioketones are much more stable than thioaldehydes. Many of them can be synthesized by thionatation of the corresponding ketones under usual conditions and tolerate storage at room temperature or at least at 0°C, for many days without any detectable oligomerization. Thus, the use of FVT has been restricted to the synthesis of specific unstable thioketones.

This is the case of propanethione (thioacetone) which has been obtained by cleavage of tetramethyltrithiolan **17** and by retro-ene reaction of allyl isopropyl sulfide and characterized by PE spectroscopy.[11] The same reactions were used to generate cyclohexanethione.[11,18] The retro-ene reaction has also been tested as a way to thiocamphor, which however is a stable compound.[18]

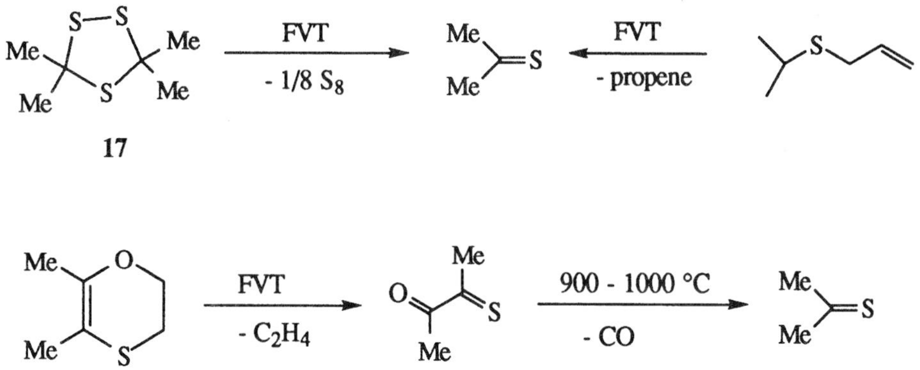

17

Scheme 11

α-Functionalized thioketones are generally unstable. This is the case of methyl vinyl thioketone. Beslin *et al.* used the retro-Diels-Alder reaction of norbornene derivatives to generate this ethylenic thioketone and some substituted homologues and have investigated their dimerization into dithiins.[41,42]

α-Oxo-thioketones have been obtained by cleavage of substituted dihydrooxathiins and by retro-ene reaction.[18,34] 3-Thioxobutan-2-one has been trapped in the gas phase by cyclopentadiene. When the thermolysis was conducted at higher temperature (Scheme 11), the transient α-oxo-thioketone was cleaved into carbon monoxide and propanethione which was trapped with cyclopentadiene. This thermolysis is thus an alternative way to thioacetone.

3. THIOKETENES

Thermolytic ways to thioketenes are numerous. When they are devoid of bulky substituents, thioketenes are very reactive species which have to be studied at low temperature or in the gas phase. Probably the most useful physical method to identify them is IR spectroscopy. A very characteristic strong $v_{C=C=S}$ band appears in their spectra between 1740 and 1795 cm^{-1}.[43]

Thioketene itself (ethenethione) has been obtained by FVT of ethanedithioic acid. The formed thioketene (and H$_2$S) was characterized by PE spectroscopy.[44] Thiocarboxylic acid derivatives, such as thionoesters and dithioesters, have also been tested as precursors for thioketenes. The thermolysis of O-trimethylsilyl thionocarboxylic esters **18** however, was found to give mainly ketenes and not thioketenes.[45] The thermolysis of methyl dithioacetate **19** (Scheme 12) was studied by the Curie point pyrolysis technique.[46] Thioketene was detected by mass spectrometry. In contrast, when heated at *ca.* 700 K, butyl dithioacetate **20** gave 1-butene elimination.[47] Elimination of carbon disulfide has also been observed in the high temperature thermolysis of various dithioesters.[48] FVT at 750°C of methyl cyclopropanedithiocarboxylate **21** led to the simplest α-unsaturated thioketene by metanethiol extrusion and ring opening. This thioketene was characterized by IR spectroscopy at 77 K (3090 (v_{C-H}), 1745 ($v_{C=C=S}$), 1607 cm^{-1} ($v_{C=C}$)).[48] Carbon disulfide (1520 cm^{-1} ($v_{C=S}$)) was also detected in the thermolysis products.

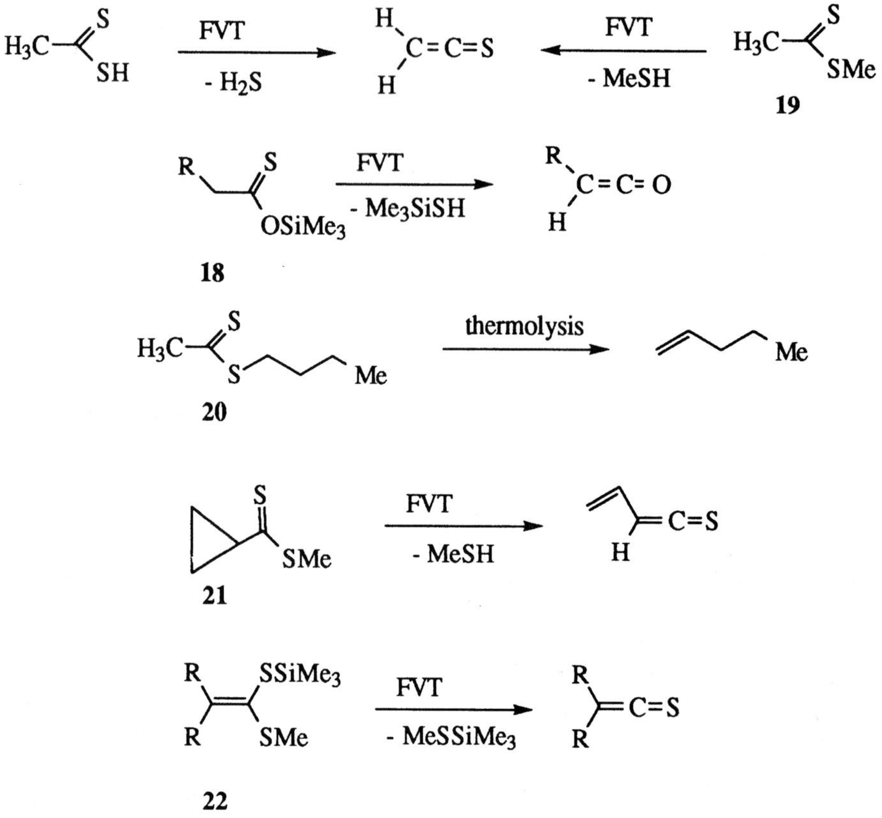

Scheme 12

In contrast with dithioesters, which generally lead to mixtures, and to mono-thioketene dithioacetals, which give ketenes upon thermolysis,[49] silylated ketene dithioacetals **22** are clean precursors for thioketenes.[50,51] When heated in the gas phase at *ca.* 930 K, they were cleaved into methylthiotrimethylsilane and the desired thioketenes. A series of thioketenes (including an α-unsaturated example) were synthesized this way, characterized by low temperature IR spectroscopy and trapped with dimethyl amine to give thioamides.

Various four membered rings have been used to give thioketenes. This is the case of tetramethyl cyclobutanedithione which upon thermolysis at 940°C gave dimethylthioketene (Scheme 13).[52] Dithietanes, such as **23** and **24**, are other useful precursors for various thioketenes.[53] Dichlorothioketene can be obtained by this method.[54] Recently, Kappe *et al.* also used this kind of precursor to obtain α-thioxo thioketenes.[55] The same authors have thermolyzed various other hetero-cycles to synthesize thioketenes.

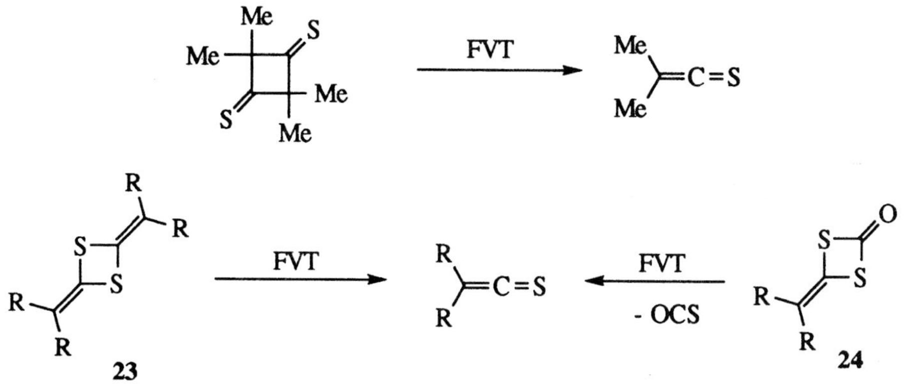

Scheme 13

1,2,3-Thiadiazoles have probably been the most studied precursors for thio-ketenes.[56–58] For instance, the thermolysis of 4-*t*-butyl-1,2,3-thiadiazole (Scheme 14) in a stream of hydrogen chloride gave a thioacyl chloride via the transient *t*-butylthioketene. R. Schulz and A. Schweig[59] used this type of cleavage to synthe-size tetramethylenethioketene from the bicyclic precursor **25**. Elimination of nitrogen is, in this case, accompanied by a ring contraction. Tetramethylenethio-ketene was characterized in the gas-phase by PE spectroscopy and by IR spectroscopy. This thioketene was also obtained by FVT of the corresponding monosilylated ketene dithioacetal.[50]

Another route to thioketenes is the retro-Diels-Alder reaction. The thermolysis of norbornenethione and of an ethanoanthracenic thione has been used to gene-rate thioketene which was detected by IR spectroscopy and trapped by an amine.[51]

Compounds bearing three cumulated double bonds have attracted much work. In sulfur chemistry, the more representative cumulenes of this type are propadienethione (H_2CCS) and ethenedithione ($SCCS$).

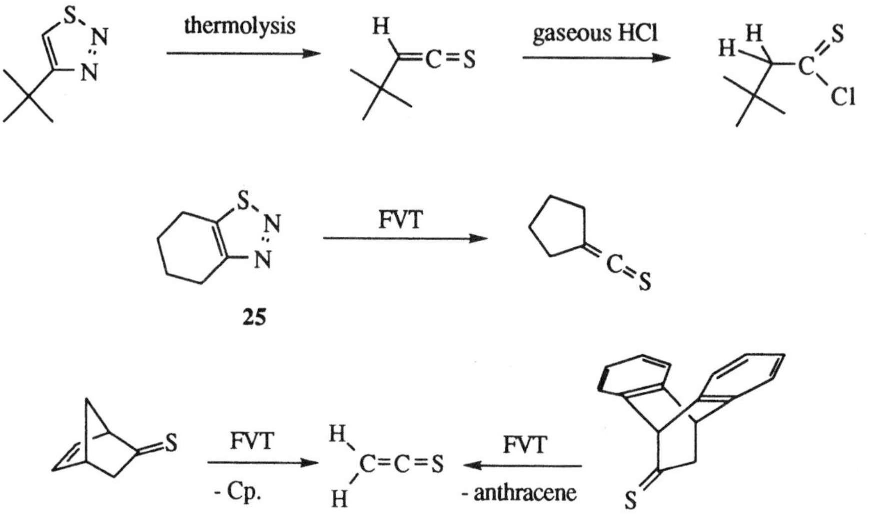

Scheme 14

Propadienethione was obtained first by a coupled rDA-MeSSiMe$_3$ elimination reaction.[51] The used precursor was the ethanoanthracenic compound **26** (Scheme 15). H$_2$CCS was detected by low temperature IR spectroscopy (2105, 2170 cm^{-1}) and later trapped with dimethyl amine. In this case the formed acrylic thioamide reacted further with the nucleophilic methanethiol formed from MeSSiMe$_3$ and HNMe$_2$. The isolated product was the methylthio-substituted thioamide **27**. Another way to propadienethione has been investigated by Brown *et al.* These authors thermolyzed the bicyclic thiadiazole **28**. Loss of nitrogen gave trimethylenethioketene which in turn was cleaved to ethene and H$_2$CCS characterized by microwave spectroscopy.[60,61]

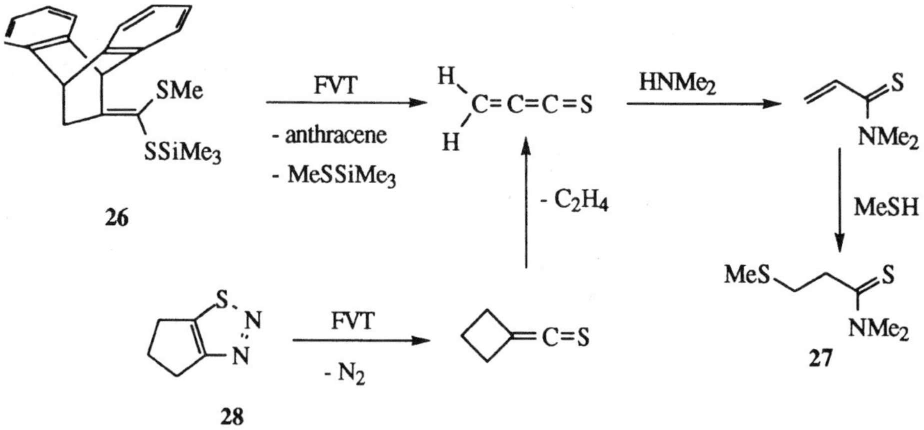

Scheme 15

In 1983, Raine *et al.* reported that, in contrast with OCCO, SCCS should be a relatively stable molecule.[62] This was experimentally demonstrated by neutralization–reionization mass spectrometry.[63] Later, ethenedithione was obtained by photolysis of various bicyclic precursors and identified by IR spectroscopy.[64] Finally, FVT was used to obtain this cumulenethione from a Meldrum's acid derivative (Scheme 16) and the isoxazolones **29**.[65] The IR spectrum of ethenedithione shows a strong band at 1180 cm^{-1} attributed to the asymmetric C=S stretch.

OC$_3$S has also been obtained by thermolysis.[66]

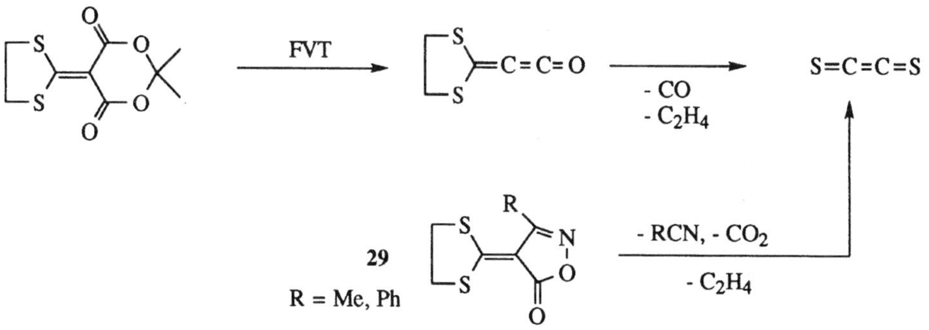

Scheme 16

4.   UNSATURATED THIONOCARBOXYLIC ACID DERIVATIVES

The derivatives of propene thionocarboxylic acid (H$_2$C=CH—C(S)OH) and propene dithiocarboxylic acid (H$_2$C=CH—CS$_2$H) are poorly stable, in contrast with their substituted homologues (R$_2$C=CH—C(S)OR, R$_2$C=CH—CS$_2$R...). For instance, the usual ways to dithioesters are not suitable for the synthesis of methyl propenedithioate. This compound has been obtained by Gosselin *et al.* by thermolysis of the ethanoanthracenic dithiocarboxylate **30** (Scheme 17).[67] The acrylic dithioester was found to dimerize at −80°C. Interestingly, not only a [4+2] dimer was isolated, but also a [4+4] dimer.

The same method has been used to obtain the corresponding acrylic thionoester.[68] This compound is relatively stable and its NMR spectra could be recorded at room temperature. When the thermolysis was effected at higher temperature, a partial thiono–thiolo rearrangement was observed and mixtures of the O-methyl and S-methyl thioesters were obtained.

Acrylic thioamides were synthezised by FVT of their ethanoanthracenic precursors (Scheme 17).[69] They were characterized by their ^{1}H and ^{13}C NMR spectra recorded and found to be stable for weeks at −20°C. Upon heating to 50–70°C they slowly polymerize. No dimers were detected. The photoelectronic spectrum

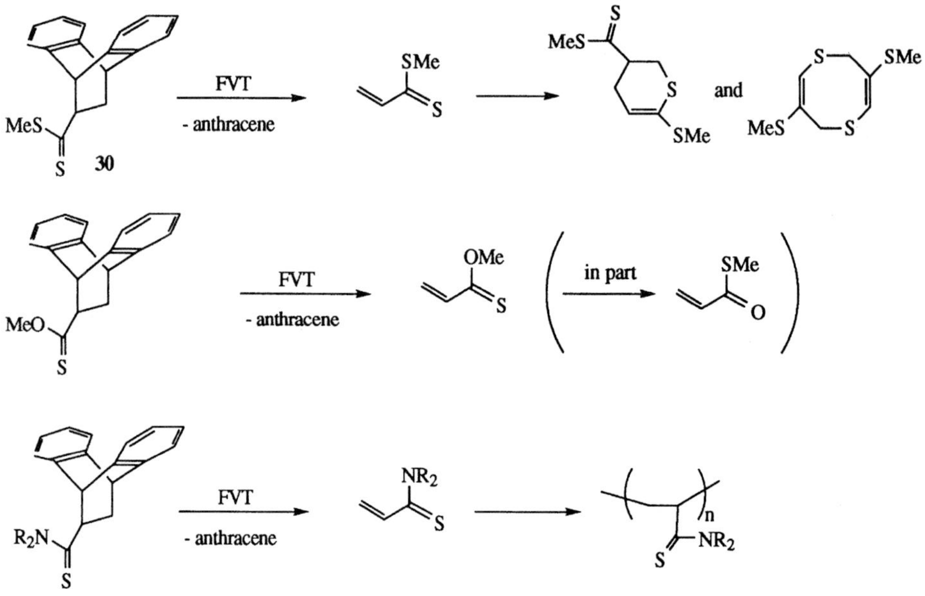

Scheme 17

of these thioamides, as well as of the acrylic dithioester and thionoester have been recorded.[67]

The behaviour of methyl penta-2,4-dienedithioate has been investigated (Scheme 18).[70] This compound is the simplest dienic dithioester. It has been generated by two ways. The first one is a simple retro-Diels-Alder reaction. The second one is a rDA reaction coupled with a sulfur dioxide cheletropic extrusion. The obtained dithiocarboxylate is not stable and undergoes a pericyclic cyclization, giving the (2H)-thiopyran **31**.

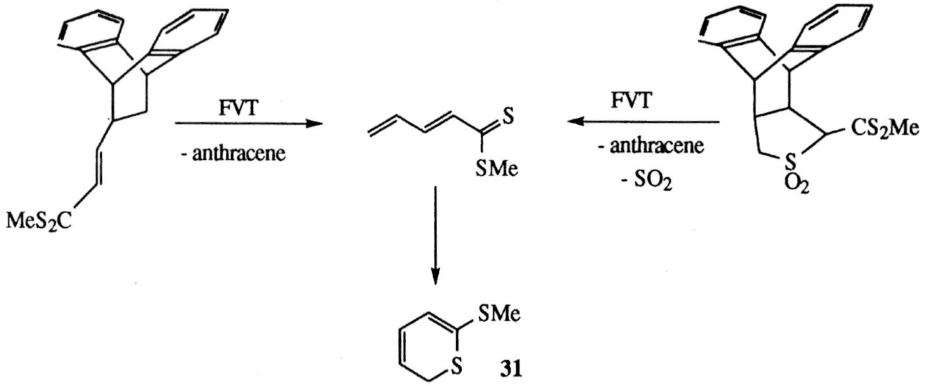

Scheme 18

5. SULFINES

The *S*-oxides of thioaldehydes and thioketones are generally more stable than their non-oxidized parent compound. This is the case for the simplest sulfine, methanethial *S*-oxide, which can be observed at −100°C by NMR spectroscopy. As methanethial, it has been the subject of lot of work.

It was first obtained by [2+2] cycloreversions from two four membered rings, thietane *S*-oxide **32** and 1,3-dithietane 1-oxide **1** (Scheme 19).[71] It was characterized by mass spectrometry and, later, by microwave and IR spectroscopy.[72,73] Its IR spectrum, recorded in an argon matrix, showed bands at 1357, 1165 and 767 cm⁻¹. 2-Deutero-1,3-dithietane 1-oxide was used to produce monodeuteromethanethial *S*-oxide. This compound, which is the simplest non symmetric sulfine, was obtained as a mixture of stereoisomers.[13]

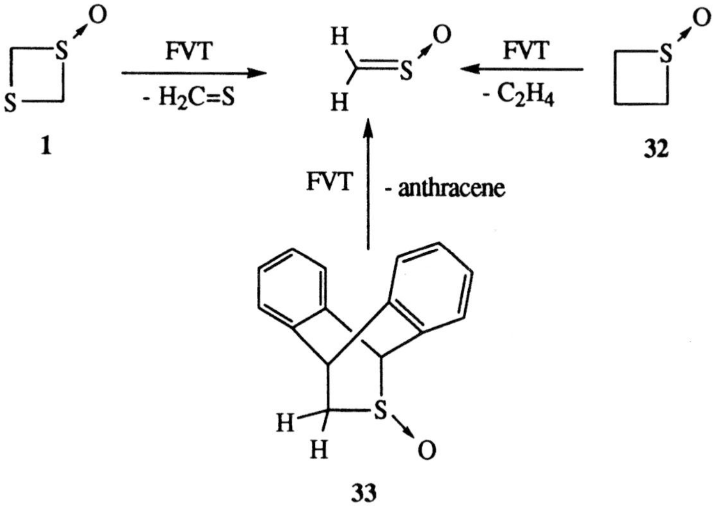

Scheme 19

Not only the [2+2] cycloreversion, but also the [4+2] cycloreversion has been applied to the synthesis of this simple sulfine. The thiaethanoanthracene which was used for the synthesis of methanethial was oxidized and the obtained sulfoxide **33** was thermolized to give methanethial *S*-oxide.[14] The NMR spectra of this sulfine were recorded at −100°C. The two protons appeared as an AB system at 7.73 and 7.84 ppm. The observation of a peak at 158.8 ppm in the ¹³C spectrum was also in agreement with the spectra previously reported for stable aldehydic sulfines.

Other heterocycles have been used as precursors for methanethial *S*-oxide. Upon thermolysis the mono-oxide of 1,3,5-trithiane **34** gave two molecules of methanethial and one molecule of the *S*-oxide (Scheme 20).[71] 1,2,4-trithiolan-4-oxide was cleaved into methanethial, its *S*-oxide and atomic sulfur.[12] Alternatively, methanethial S-oxide was synthesized by methane elimination from dimethylsulf-oxide[72] and by HCl extrusion from methanesulfenyl chloride.[12] Unexpectedly, this last method could not be extended to the synthesis of other sulfines. FVT of alkanesulfenyl chlorides does not give alkanethial oxides but alkenes, sulfur monoxide and hydrogen chloride.[71]

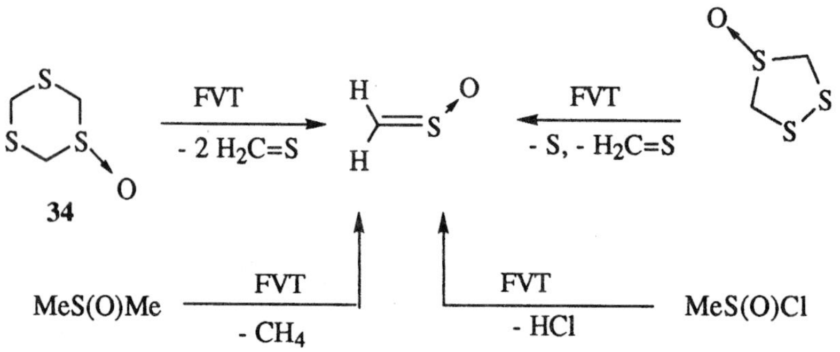

Scheme 20

The retro-ene reaction has been used to synthesize alkanethial S-oxides. Ethanethial S-oxide is of particular interest as it is the lachrimatory factor of onion.[75] It was obtained by thermolysis of t-butyl vinyl sulfoxide as a 98/2 mixture of cis and trans isomers (Scheme 21).[76] In contrast an attempt to obtain prop-2-enethial S-oxide (the sulfine of thioacrolein) by thermolysis of diallyl sulfoxide remained unsuccessful. This thermolysis was, indeed, a good way to thioacrolein.[77]

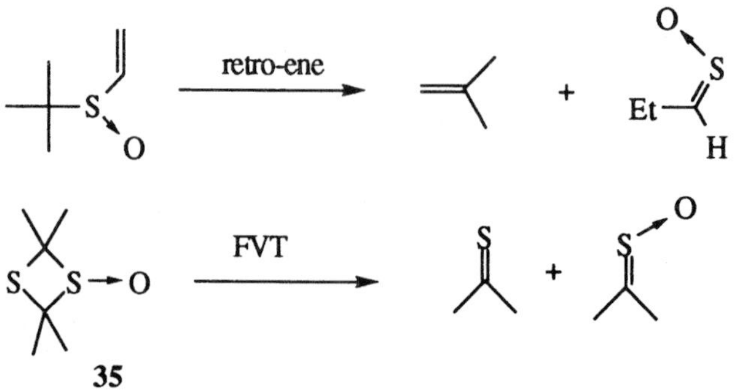

Scheme 21

The thermolysis of substituted dithietan oxide has been used to synthesize various sulfines. Propanethione S-oxide (the sulfine of thioacetone) was obtained from the tetramethyl derivative **35**.[71] Halogenosulfines were obtained from halo-substituted dithietanes. Another way to these compounds is the retro-Diels-Alder reaction.[78-81]

The rDA reaction can also be used to generate α-oxosulfines. The S-oxide of thi-oxoethanal was obtained by FVT of 2,3-dihydro-1,4-oxathiin-4-oxide (Scheme 22) and was identified by its IR and PE spectra.[82] This method was extended to the synthesis of substituted α-oxosulfines. Interestingly, the sulfine of monothio-biacetyl was found to be stable in CDCl$_3$ solution at room temperature.[82] Upon thermolysis at higher temperature (*ca.* 1000°C) α-oxosulfines were cleaved into atomic sulfur and the corresponding α-dicarbonyl derivatives.

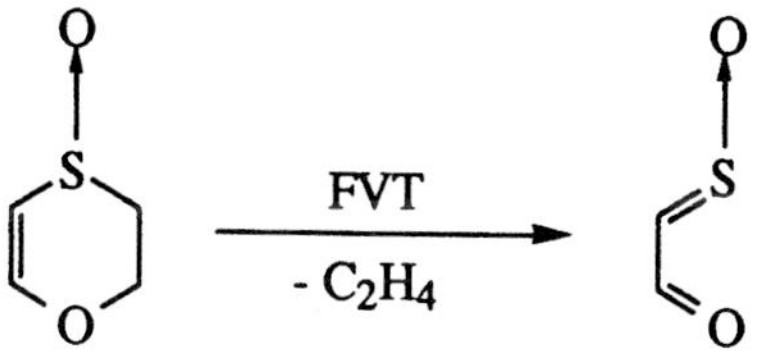

Scheme 22

6. SULFENES

Sulfenes (thioaldehyde and thioketone S,S-dioxides)[83] are thermally unstable products. They may be cleaved into sulfur monoxide and a carbonyl compound or into sulfur dioxide and a carbene. This is probably why they have been only slightly studied by FVT.

Methanethial S,S-dioxide has been obtained by FVT of chlorosulfonylacetic acid **36** (Scheme 23).[84,85] Upon thermolysis, this acid is cleaved into carbon dioxide, hydrogen chloride and the desired sulfene. Methanethial S,S-dioxide was characterized by infra-red spectroscopy (1330, 1230 cm^{-1}) and trapped with methanol to give the sulfonate **37**.

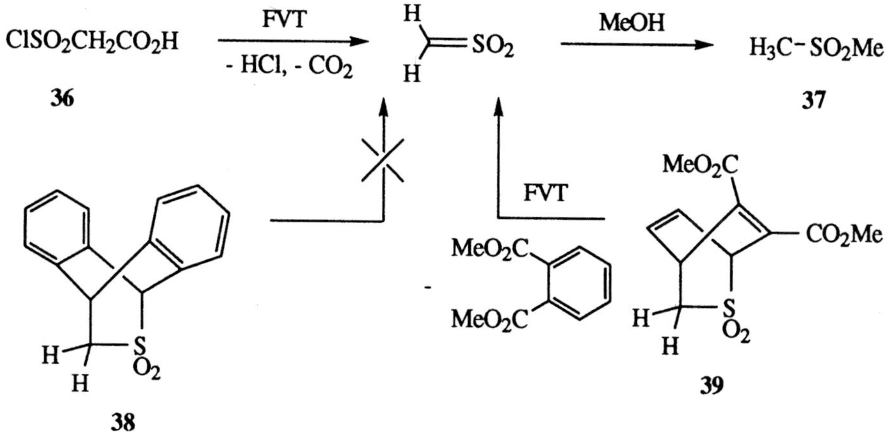

Scheme 23

As in the case of thioaldehydes and sulfines, the retro-Diels-Alder reaction has been used to generate methanethial S,S-dioxide. However, the dioxide of the thiaethanoanthracene **38** is not a suitable starting material. This compound does not give rise to rDA products (anthracene and the sulfene) but to products derived from a radical cleavage into sulfur dioxide and a diradical. Alternatively, methanethial S,S-dioxide was cleanly obtained by thermolysis of the thia-bicyclo(2,2,2)octadiene derivative **39**.[86,87]

Methanesulfonic anhydride,[85] 3-hydroxythietane dioxide[88] and N-methylsulf-onyl phtalimide[89] are useful precursors for methanethial S,S-dioxide (Scheme 24).

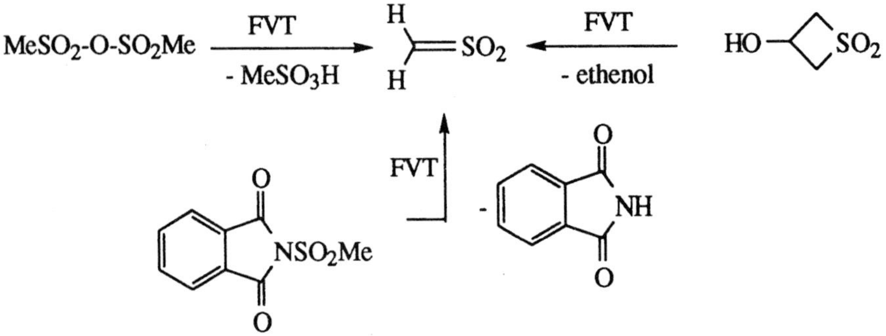

Scheme 24

The thio-Claisen rearrangement of allyl vinyl sulfone has been studied by King and Harding (Scheme 25).[90] Thus, allyl vinyl sulfone was transformed into pent-4-enethial S,S-dioxide upon heating to 800°C under vacuum. The obtained sulfene was found to be unstable and lost sulfur monoxide. Indeed, the isolated product was pent-4-enal. This is a clear example of sulfur monoxide extrusion from a sulfene. In certain cases this reaction can compete with a sulfur dioxide extrusion.

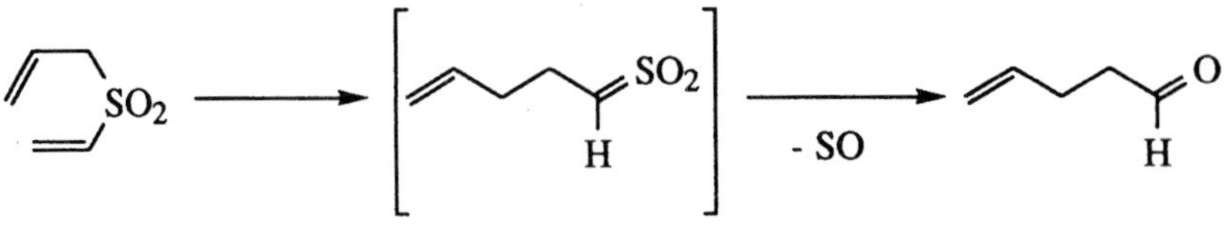

Scheme 25

Such concurrence was observed during the thermolysis of the α-diazo-sulfone **40**.[91] In this case, dinitrogen extrusion, followed by an isomerization of the intermediate carbene gave a diaromatic sulfene (Scheme 26). The isolated products, however, were the corresponding ketone and 3-methylfluorene. The ketone was formed by sulfur monoxide extrusion and the formation of the deoxygenated products was tentatively explained by the rearrangement of a carbene resulting from the loss of SO_2.

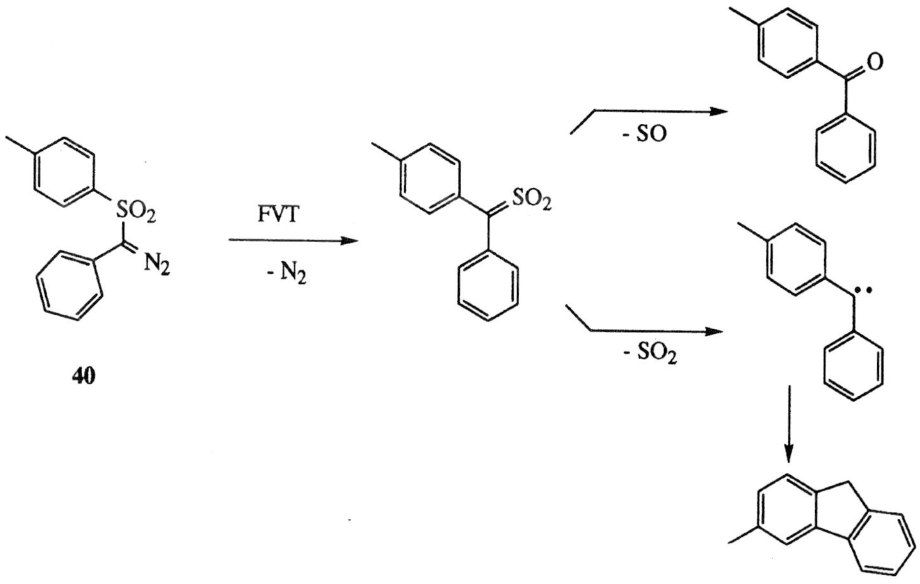

Scheme 26

7. SELENOCARBONYL COMPOUNDS

The thermolytic behaviour of selenides and diselenides is quite well documented.[92–95] However in contrast with the large amount of experiments run with thio-compounds, FVT has been only scarcely applied to the synthesis of double bonded C=Se compounds. In fact, the low stability of the simple C—Se bond renders its homolytic cleavage often unavoidable at high temperature. An illustration of this difference of reactivity between thio- and seleno-derivatives is presented in a study published by Bock and colleagues. These authors have compared the thermal decomposition of dialkyldisulfide and dialkyldiselenide.[96]

As shown in Scheme 27, under comparable conditions, pyrolysis of methyl-diselenide did not yield the corresponding selenoaldehyde whereas the sulfur analogue gave thioformaldehyde even when the temperature was raised over 1000 K. This easy C—Se bond cleavage, often resulting in Se loss from potential precursors of selenocarbonyl compounds, has probably discouraged numerous experimenters from introducing seleno-compounds into their oven. As a consequence, the literature concerning the synthesis of selenoaldehydes and sele-noketones by FVT is rather scarce.[97]

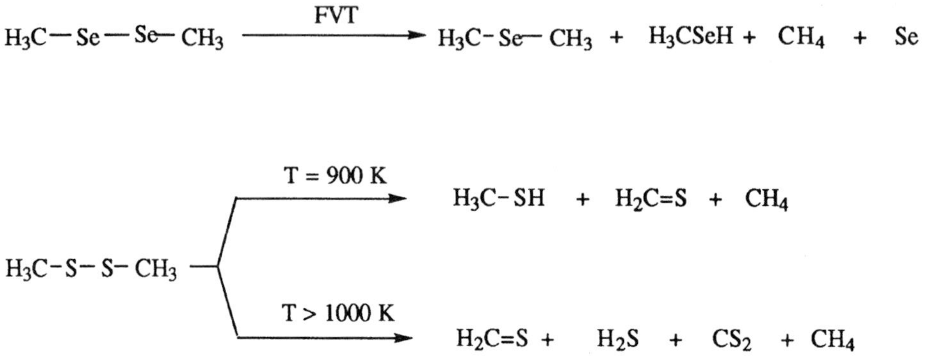

$$H_3C-Se-Se-CH_3 \xrightarrow{\text{FVT}} H_3C-Se-CH_3 + H_3CSeH + CH_4 + Se$$

$$H_3C-S-S-CH_3 \begin{cases} \xrightarrow{T = 900 \text{ K}} H_3C-SH + H_2C=S + CH_4 \\ \\ \xrightarrow{T > 1000 \text{ K}} H_2C=S + H_2S + CS_2 + CH_4 \end{cases}$$

Scheme 27

Selenoaldehydes

The preparation of the simplest selenoaldehyde by flash thermolysis was described almost simultaneously by two teams in 1984.[98,99] Since that year optical, microwave, and photoelectron spectra of methaneselenal have been described.[97,98,101,102,104]

Judge et al.[98,101] used dimethylselenide as starting material. The reaction was carried out in a pyrolysis chamber at 700°C under an optimum pressure of 1.5 Torr. The resulting products were trapped at −180°C.

$$H_3C-Se-CH_3 \xrightarrow{\text{FVT}} H_2C=Se + CH_4$$

Scheme 28

The near infrared absorption spectrum of methaneselenal was recorded within the region 400–825 nm.[98] The main band was located at 12169 cm^{-1} (822 nm) and intense absorptions were also observed at 12876, 13573 and 14266 cm^{-1}. Three years later the excitation spectrum of the same compound was obtained over the 695–750 nm region by using the combined techniques of flash pyrolysis and laser induced fluorescence (LIF).[101] Most of the obtained bands were in agreement with the results obtained from the absorption spectrum however new bands at 13635, 13950 and 14335 cm^{-1} were detected. In order to obtain a more precise spectrum of the 695–700 nm region, the trimer of methaneselenal and its deutariated analogue were used as starting materials.[101] For H_2CSe the origin of an intense singlet–triplet system was observed at 12171.0 cm^{-1} whereas for D_2CSe it was observed at 12262.7 cm^{-1}. More recently, the same authors have performed a detailed study of the high-resolution laser fluorescence excitation spectra of vibronic bands of $H_2C^{78}Se$, $H_2C^{80}Se$ and $D_2C^{80}Se$.[103]

Glinski and coworkers have demonstrated that chemoluminescence was also a useful technique in observing spectra of "hard-to-study species" like methaneselenal.[104,105] In a first publication[104] the emission spectra were recorded during the reaction of a 10% F_2/Helium mixture with dimethylselenide. Two different conditions of reactant pressure were used. A first trace was obtained from the reaction of 60 mTorr of dimethylselenide with 270 mTorr of 10% F_2/He. A characteristic band at 12146 cm^{-1} (823.3±1.2 nm) was detected, in good agreement with previous studies.[98,101] The same precursor was used to obtain H_2C=Se for vibration frequencies measurements.[105] A band observed at 1460 cm^{-1} was assigned to the CH_2 bend vibration. Other bands were observed at 860 ($v_{C=Se}$), 906 and 914 cm^{-1}.

Brown et al. have also chosen dimethylselenide to generate H_2C=Se. Under their conditions the optimum temperature was 1000°C.[100] Analysis of the microwave spectrum indicated that C=Se and C—H bond lenghts are 1.759 Å and 1.09 Å respectively. The angle HCH has also been determined (120.4°). In order to improve their results they studied the microwave spectra of isotopic species of selenoformaldehyde like $H_2{}^{13}C^{78}Se$, $D_2C^{78,80}Se$ and $DHC^{78,80}Se$.[106]

Photoelectronic spectroscopy was used by Bock et al.[99] to detect methaneselenal.

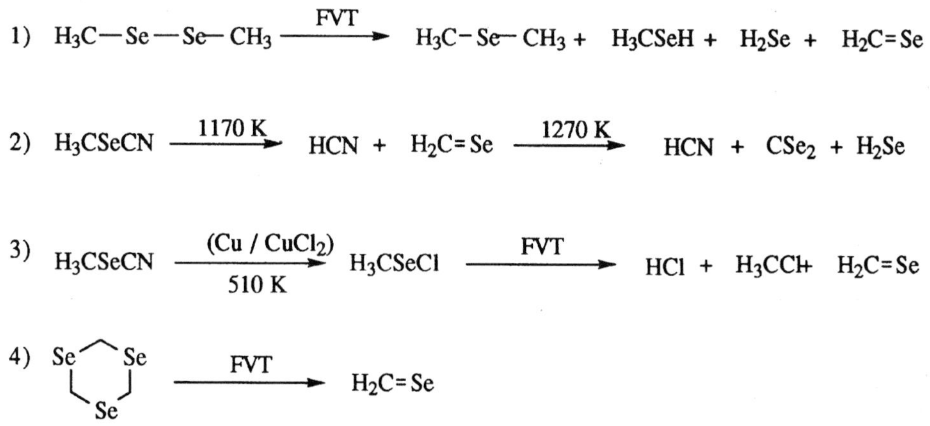

Scheme 29

As shown in Scheme 29, dimethylselenide was not a convenient precursor and yielded a mixture of compounds comprising of the selenide and methyl selenol. Methyl selenocyanate by direct pyrolysis gave methaneselenal in small amounts. When it was converted into methylselenylchlorid the reaction was not convincing. Finally, the more efficient starting material was the trimer of H_2C=Se, 1,3,5-triselenane, which yielded mainly methaneselenal.

By a similar method, 2,4,6-trimethyl-1,3,5-triselenal was used to synthezise selenoacetaldehyde at 873 K or 1100 K (Scheme 30).[99,107] The microwave spectrum (between 26.5 and 40 GHz) of this selenoaldehyde was recorded. Measurements gave indications concerning the C=Se bond length $(1.758 \pm 0.01$ Å) and CCSe angle $(125.7 \pm 0.3°)$. The spectrum showed that the internal rotation is hindered and that barrier heights for the ground and first torsionally excited states are 6.7×10^3 and 6.9×10^3 J mol^{-1}, respectively.

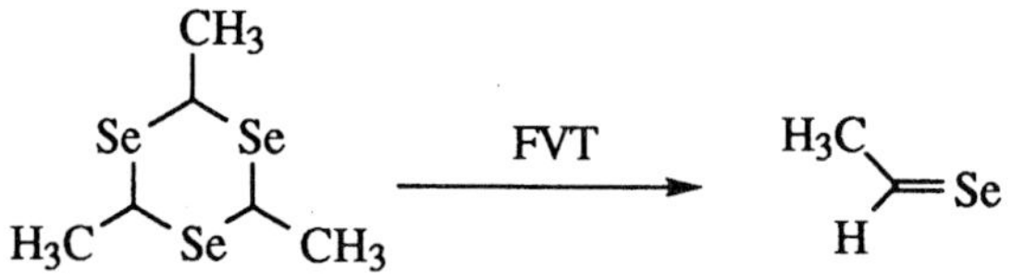

Scheme 30

As seen before, numerous work has been done on the synthesis and the possible isomerisations of α-oxothiocarbonyl compounds and α-dithiones. Much less studies have been published on their seleno analogues. Even though to the best of our knowledge, HCOCHSe has not been subjected to spectrochemical studies, it is deductible from chemical trapping experiments that α-oxoselenoal-dehydes exist in their open form.[108] α-Thioxoselenoaldehydes and α-diselenocar-bonyl compounds are poorly known. However as nearly all α-dithiones exist in their isomeric cyclic form, it is expected that their mono- and di-selenoanalogs would also cyclize (Scheme 31).

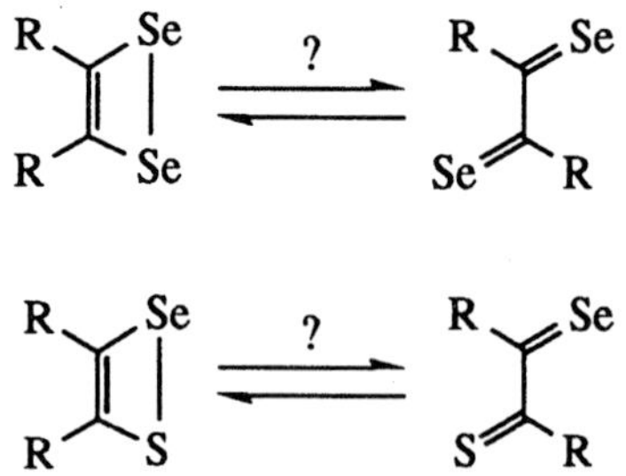

Scheme 31

This was confirmed by Diehl and Schweig's results.[109] Characterisation of the iso-meric forms of selenoxoethaneselenal and thioxoethaneselenal, 1,2-diselenete and 1,2-thiaselenete, has been performed. These compounds were obtained by che-lotropic elimination of carbon monoxide from 1,3-diselenol-2-one and 1,3-thia-selenol-2-one (Scheme 32). Various conditions of pressures and temperatures were used. For example at P=10^{-2} mbar the generation of 1,2-thiaselenete required a temperature of 600°C whereas for 1,2-diselenete 410°C was sufficient.

Photoelectronic spectroscopy was used for detection and it appeared that the obtained spectra of the two compounds were very similar to the spectrum of 1,2-dithiete previously recorded.[35]

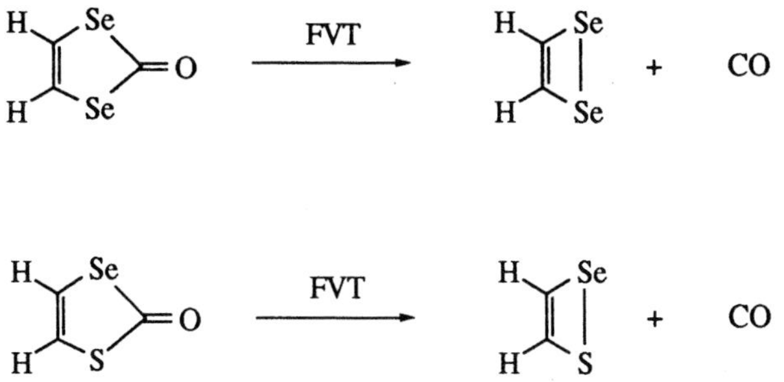

Scheme 32

IR spectra displayed for both compounds a strong absorption at $734\,\text{cm}^{-1}$, characteristic of a δ_{CH} vibration. This value was very close to the absorption obtained in the case of 1,2-dithiete ($\delta_{CH}=735\text{ cm}^{-1}$) and was in agreement with the fact that the replacement of a sulfur by a selenium generally does not change the position of the IR bands corresponding to an absorption in which the heteroatoms are not involved.[110]

Finally, as there are often many similarities in the behaviour of molecules under FVT and mass spectroscopic conditions, it is of interest to note that the mass spectrum of 1,3-diselenol-2-one and 1,3-thiaselenol-2-one, obtained by electron impact, gave fragmentations that are attributable to (1,2-diselenete)$^+$ and (1,2-thiaselenete)$^+$ or to their open form at $m/z=186$ and 138 respectively.

The synthesis of propeneselenal (selenoacrolein) was also attempted.[111] Various methods including the thermolysis of diallyl selenide, allyl selenocyanate and propaneselenal-dimethylanthracene adduct were tried. Unfortunately, characterisation of this α,β-insaturated selenoaldehyde could not be performed probably because of its too high instability.

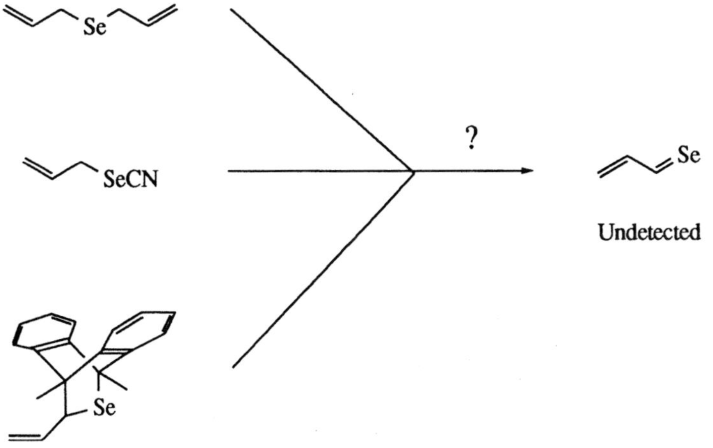

Scheme 33

Halogenated Derivatives

Selenocarbonyldifluoride (F_2CSe) has been obtained from several compounds (Scheme 34),[112–116] for example by thermolysis of its dimer in spiral tubes at 360°C[113] or 200°C.[114] Also, above 150°C, polymers of F_2CSe gave a mixture of the selenocarbonyl derivative and its dimer.[115]

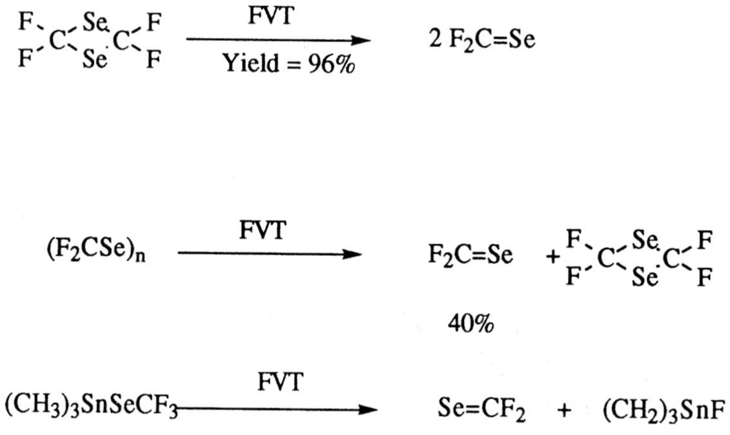

$$\underset{F}{\overset{F}{>}}C\underset{Se}{\overset{Se}{<}}C\underset{F}{\overset{F}{<}} \quad \xrightarrow[\text{Yield} = 96\%]{\text{FVT}} \quad 2\ F_2C{=}Se$$

$$(F_2CSe)_n \quad \xrightarrow{\text{FVT}} \quad F_2C{=}Se \quad + \quad \underset{F}{\overset{F}{>}}C\underset{Se}{\overset{Se}{<}}C\underset{F}{\overset{F}{<}}$$

$$40\%$$

$$(CH_3)_3SnSeCF_3 \quad \xrightarrow{\text{FVT}} \quad Se{=}CF_2 \quad + \quad (CH_2)_3SnF$$

41

Scheme 34

Another synthesis of F_2CSe has been mentioned:[116] thermolysis of the stannane derivative **41** under 10^{-3} Torr at 300–340°C provided the selenocarbonyl derivative in quantitative yield:

Spectroscopic data of selenocarbonyldifluoride are summarized below:

^{19}F-NMR: $\delta = -65$ ppm; ^{77}Se-NMR: $\delta = 810.6$ ppm ($J_{^{77}Se-^{19}F} = 256$ Hz); IR (Gas, υ cm^{-1}): 2490-2479-2410-2400-2358-2343-2005-1286-1214-1200-1125-1118-1068-66-430.[117]

Mass spectrum obtained by electron impact ionization revealed peaks at m/z = 130 (F_2CSe^+); 111 ($FCSe^+$); 80 (Se^+); 69 (CF_3^+); 50 (CF_2^+); 31 (CF^+) and 20 (HF^+).[118]

Visible and UV spectra have been recorded between 200 and 700 nm.[114] In the visible region a series of vibronic bands (between 350 and 500 nm) occured. The near ultraviolet spectrum displayed an important absorption at 240 nm and another one at 200 nm.

The behaviour of $\upsilon_{1(C=Se)}$ stretching band was studied in the ground and in the first excited states. The bond length extention was 0.077 Å. Molecular parameters of $F_2C{=}Se$ are also known.[119]

$$d_{(C=Se)} = 1.743\ \text{Å} \qquad <_{(F-C-F)} = 107.5°$$

$$d_{(C-F)} = 1.314\ \text{Å} \qquad <_{(Se-C-F)} = 126.3°$$

Selenocarbonyl dichloride (Scheme 35), another example of tetraatomic selenocarbonyl derivatives, was also generated by thermolysis. Whereas, at 360°C, the
pyrolysis of the dimer $(Cl_2CSe)_2$, in a spiral tube, gave a mixture of CSe_2 and
CCl_4,[113] the same starting material at lower temperature provided a blue compound which was characterised as $SeCCl_2$.[119,120] Mass, UV, IR and Raman spectra
of this product were recorded.

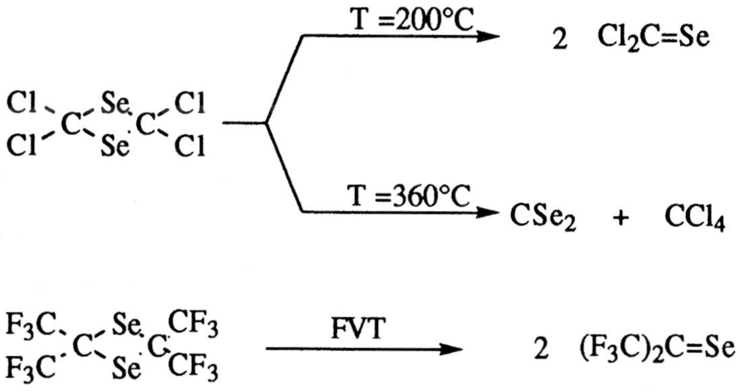

Scheme 35

More recently,[121] 2,2,4,4-tetrakis(trifluoromethyl)-1,3-diselenate **42** has been
thermolysed at 750°C and yielded a very small quantity of a purple compound
which was assumed to be hexafluoroselenoacetone. Unfortunately no spectral
data were given.

On the other hand, $(F_3C)_2C{=}Se$ was generated by reaction of hexafluoropro-
pene with Se in presence of CsF in dimethylformamide and trapped with
anthracene to give a seleno-dihydro-ethanoanthracene adduct (Scheme 36). This
compound would probably be a good precursor for gas phase studies of
$(CF_3)_2C{=}Se$.

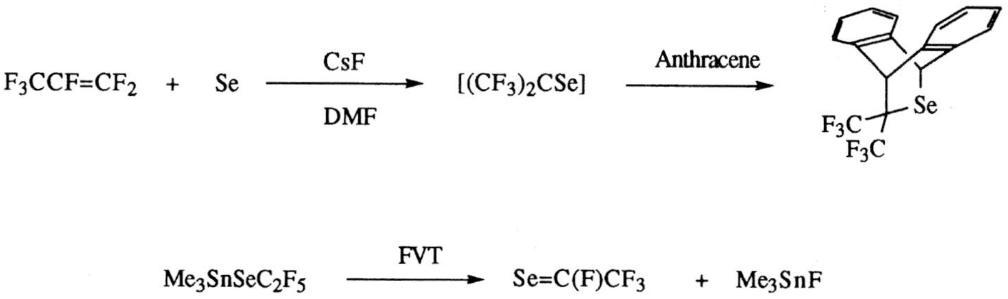

Scheme 36

Finally, trifluoromethylselenocarbonylfluoride was generated from trimethyl-stannylpentafluoroethylselane at 300°C, and trapped with dienes.[122]

Selenoketenes

Selenoketenes are generally obtained by thermolysis of selenodiazoles (Scheme 37). At 600°C, under a pressure of 30.10^{-3} Torr, 1,2,3-selenodiazole gave the simplest selenoketene *via* N_2 extrusion:

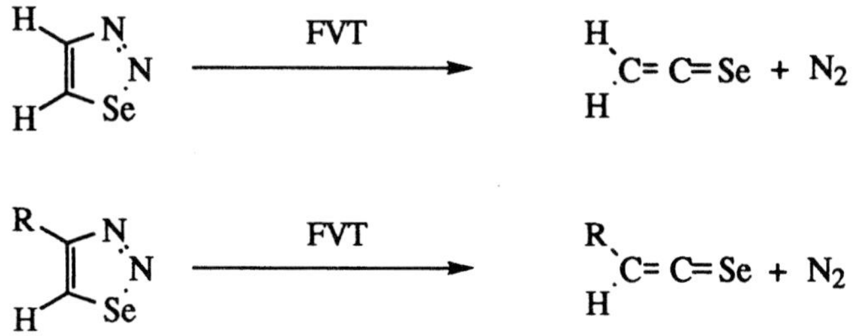

Scheme 37

Microwave studies of this molecule were carried out. C=C, C=Se, C—H bond lengths and the HCH angle were determined.[123,124] Obtained values were in good agreement with recent *ab initio* calculations.[125] Experimental and theoretical data are presented in Table 1.

Photoelectron and mass spectra of this purple species were recorded.[126,127] After thermolysis at 820 K, mass spectrum displayed peaks at m/z=108, 106, 104, 103 and 102, characteristic of a compound containing a selenium atom (isotopes ^{82}Se, ^{80}Se, ^{78}Se, ^{77}Se and ^{76}Se) that were assigned to the radical cation $H_2C=C=Se^+$.

Thermolysis of selenodiazoles can be extended to 4-alkyl substituted analogues which, at 500–600°C gave corresponding selenoketenes (Scheme 37). The obtained blue compounds were characterized by IR spectroscopy. $\upsilon_{(C=C\ str.\ vib.)}$ and $\upsilon_{(C=Se\ str.)}$ were detected at 1695 and 1700 cm^{-1}. However when 4-phenyl-selenodiazole was thermolysed, phenyl-acetylene was detected instead of phenyl-selenoketene.[128]

Table 1 Geometrical parameters of $H_2C=C=Se$. Bond lengths are given in Å and angles in degrees.

	Experimental	Theoretical
C=C	1.303	1.317
C=Se	1.706	1.699
C—H	1.0908	1.0850
<HCH	119.7	119.2

Cyclopentylidenselenoketene and cycloheptylidenselenoketene were also obtained by gas phase pyrolysis of cyclohexeno- and cycloocteno-1,2,3-selenadiazole (Scheme 38).[129]

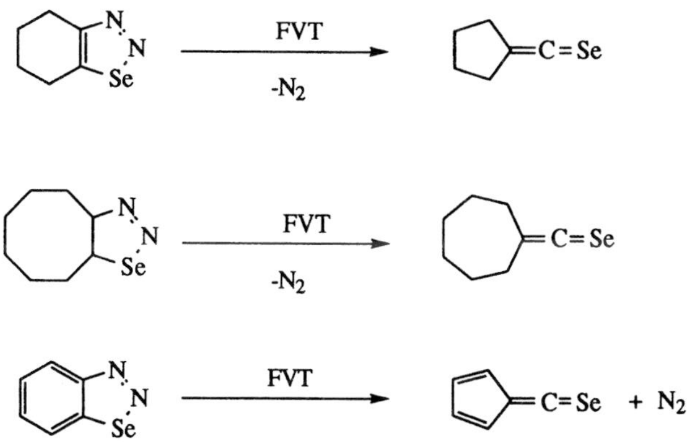

Scheme 38

1,2,3-benzoselenadiazole was used as starting material to synthesize 6-fulveneselone (Scheme 38). This red product was characterised by photoelectronic spectroscopy. Bands appeared at 8.34 and 10.76 eV.[130] Its IR spectrum has also been recorded and displayed bands at 1710 ($v_{C=C=Se}$), 1685, 1650, 1450, 1360, 1105, 1033, 843, 733 and 590 cm^{-1}.

Propadieneselone (Scheme 39) was prepared from cyclopentene-1,2,3-selenadiazole at T=700°C and P=10^{-6} Torr. The reaction involved an intermediate selenoketene which was characterized by IR spectroscopy after thermolysis at 500°C. A strong absorption at 1761 cm^{-1} characteristic to $v_{(C=C=Se\ str.)}$ was located. Other bands at 2967, 2863, 1757, 1746, 1432 and 1232 cm^{-1} appeared. At 700°C, the IR spectrum indicated that propadieneselone was formed. Bands at 1993, 1985 and 1981 cm^{-1} were assigned to the antisymmetric vibration of this cumulene ca. 240 cm^{-1} above usual selenoketene frequencies. Propadieneselone, by irradiation led to propyneselenal.[131]

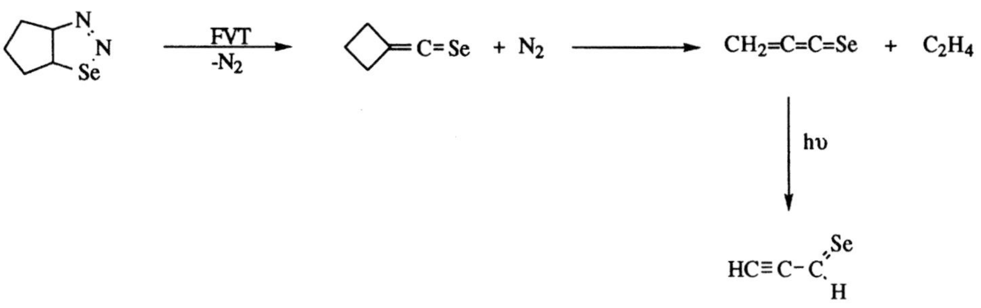

Scheme 39

Recently, a 2,2-dihydro-1,3,4-selenodiazole (Scheme 40) was pyrolysed at 450°C. Mass spectroscopy revealed intense peaks at m/z=218 and 206, characteristic of di-tert-butyl-selenoketene and di-*t*butyl selenoketone:[132]

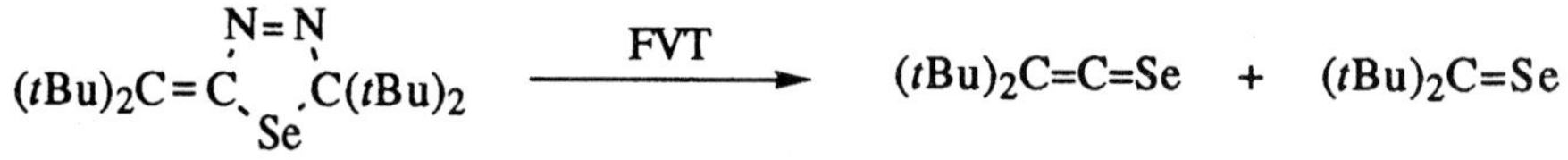

$$(tBu)_2C=C\underset{Se}{\overset{N=N}{\underset{|}{|}}}C(tBu)_2 \quad \xrightarrow{\text{FVT}} \quad (tBu)_2C=C=Se \quad + \quad (tBu)_2C=Se$$

Scheme 40

8. TELLUROCARBONYL DERIVATIVES

According to the literature, the instability of the simple C—Te bond generally prevents the generation of tellurocarbonyl derivatives by thermolysis of telluro compounds. To the best of our knowledge the only compound containing a C=Te double bond that has been generated under thermal conditions is tellurocarbonyldifluoride (F_2CTe) which was obtained in 60% yield by thermolysis of ($CH_3SnTeCF_3$) at 280°C under 10^{-3} Torr (Scheme 41).[133,134]

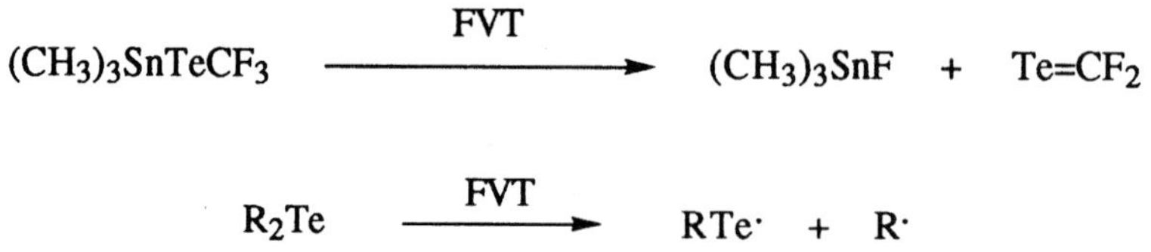

$$(CH_3)_3SnTeCF_3 \quad \xrightarrow{\text{FVT}} \quad (CH_3)_3SnF \quad + \quad Te=CF_2$$

$$R_2Te \quad \xrightarrow{\text{FVT}} \quad RTe^{\cdot} \quad + \quad R^{\cdot}$$

Scheme 41

IR and mass spectra of this species were recorded. In the region 400–4000 cm^{-1}, the IR spectrum, obtained in a solid argon matrix at 13 K, displayed vibrations at 1226.0 cm^{-1} ($v_{sym\ (C=Te)}$) and 1185.5 cm^{-1} ($v_{asym\ (CF_2)}$).

Other experiments concerning the thermolysis of symmetrical or unsymmetrical dialkyltellurium compounds did not give rise to tellurocarbonyl compounds.[135-137] To explain these results, the authors proposed an homolytic cleavage of the C—Te bond (Scheme 41).[135]

Thus, by using the chemoluminescence technique, Glinski *et al.*[105] have shown that thermolysis of dimethyldisulfide and dimethyldiselenide yielded the corresponding thioformaldehyde and selenoformaldehyde whereas CH_3TeCH_3 and CH_3TeCH_3 did not give the corresponding methanetellural.

Kirss *et al.* have studied the behaviour of various dialkyl and diallylditelluride at 300 and 400°C. They have showed that an homolytic cleavage of the C—Te bond occured and led to alkanetelluride and dialkyl ditelluride when precursors possessed reactive hydrogen in the β-position.[136] Sang *et al.* have also thermolysed $(C_3H_7)_2Te$ and have evidence by mass spectroscopy of the formation of $(C_3H_7)TeH$.[137]

All these results point to the fact that FVT is not an appropriate method to generate tellurocarbonyl compounds.

REFERENCES

1. A. Thuillier, *Phosphorus Sulfur Silicon* **23**, 253 (1985).
2. G.W. Kirby, *Phosphorus Sulfur Silicon* **74**, 17 (1993).
3. R. Okazaki, In *"Organosulfur Chemistry"*, Academic Press, pp. 225–255 (1995).
4. A.P.C. Mann and D.A. Williams, *Nature* **283**, 721 (1980).
5. A.B. Callear, J. Connor and D.R. Dickson, *Nature* **221**, 1238 (1969).
6. A. Jones and F.P. Lossing, *J. Phys. Chem.* **71**, 4111 (1967).
7. D.R. Johnson and F.X. Powell, *Science* **169**, 679 (1970); D.R. Johnson, F.X. Powell and W.H. Kirchoff, *J. Mol. Spectrosc.* **39**, 136 (1971).
8. J.W.C. Johns and W.B. Olsen, *J. Mol. Spectrosc.* **39**, 479 (1971).
9. H.W. Kroto and R. J. Suffolk, *Chem. Phys. Lett.* **15**, 545 (1972).
10. B. Solouki, P. Rosmus and H. Bock, *J. Am. Chem. Soc.* **98**, 6054 (1976).
11. H. Bock, T. Hirabayashi and S. Mohmand, *Chem. Ber.* **115**, 492 (1982).
12. H. Bock, B. Solouki, S. Mohmand, E. Block and L.K. Revelle, *J. Chem. Soc., Chem. Commun.* 287 (1977).
13. E. Block, E.R. Corey, R.E. Penn, T.L. Renken, P.F. Sherwin, H. Bock, T. Hirabayashi, S. Mohmand and S. Mohmand, *J. Am. Chem. Soc.* **104**, 3119 (1982).
14. Y. Vallée, J.L. Ripoll, C. Lafon and G. Pfister-Guillouzo, *Can. J. Chem.* **65**, 290 (1986).
15. H. Allgeier and T. Winkler, *Tetrahedron Lett.* 215 (1976).
16. J.W. Lown and R.R. Koganty, *J. Am. Chem. Soc.* **108**, 3811 (1986).
17. H.G. Giles, R.A. Marty and P. de Mayo, *J. Chem. Soc., Chem. Commun.* 409 (1974).
18. H.G. Giles, R.A. Marty and P. de Mayo, *Can. J. Chem.* **54**, 537 (1976).
19. L. Wazneh, J.C. Guillemin, P. Guenot, Y. Vallée and J.M. Denis, *Tetrahedron Lett.* **29**, 5899 (1988); G. Maier, U. Flögel, A.P. Reisenauer, B.A. Hess Jr and L.J. Schaad, *Chem. Ber.* **124**, 2609, (1991).
20. Y. Vallée, M. Khalid, J.L. Ripoll and A. Hakiki, *Synth. Commun.* **23**, 1267 (1993).
21. B. Schuler and W. Sundermeyer, *Tetrahedron Lett.* **30**, 4111 (1989).
22. Y. Vallée, unpublished results.
23. J.E. Baldwin and R.C. Gerald Lopez, *Tetrahedron* **39**, 1487 (1983).
24. H. Bock, S. Mohmand, T. Hirabayashi and A. Semkow, *J. Am. Chem. Soc.* **104**, 312 (1982).
25. H. Bock, S. Mohmand, T. Hirabayashi and A. Semkow, *Chem. Ber.* **115**, 1339 (1982).
26. K. Georgiou and H.W. Kroto, *J. Mol. Spectrosc.* **83**, 94 (1980).
27. R.D. Brown, P.D. Godfrey, R. Champion and M. Woodruff, *Aust. J. Chem.* **35**, 1747 (1982).
28. M. Bogey, C. Demuynck, J.L. Destombes, A.C. Gaumont, J.M. Denis, Yo. Vallée and J.L. Ripoll, *J. Am. Chem. Soc.* **111**, 7399 (1989).
29. G. Pfister-Guillouzo, F. Gracian, A. Senio, F. Bourdon, Y. Vallée and J.L. Ripoll, *J. Am. Chem. Soc.* **115**, 324 (1993).
30. C.O. Kappe, M.W. Wong and C. Wentrup, *Tetrahedron Lett.* **34**, 6623 (1993).
31. M. Torres, A. Clément and O.P. Strausz, *Nouv. J. Chim.* **7**, 269 (1983).
32. D. Mc Naughton and D.N. Bruget, *J. Mol. Spectrosc.* **134**, 129 (1989).
33. F. Bourdon, J.L. Ripoll, Y. Vallée, S. Lacombe and G. Pfister-Guillouzo, *J. Org. Chem.* **55**, 2596 (1990).
34. F. Bourdon, J.L. Ripoll and Y. Vallée, *Tetrahedron Lett.* **31**, 6183 (1990).
35. R. Schulz, A. Schweig, K. Hartke and J. Köster, *J. Am. Chem. Soc.* **105**, 4519, (1983); M. Rodler and A. Bauder, *Chem. Phys. Lett.*, **114**, 575, (1985); F. Diehl, H. Meyer, A. Schweig, B. Andes Hess Jr and J. Fabian, *J. Am. Chem. Soc.*, **111**, 7651.
36. Yo. Vallée, J.L. Ripoll and D. Maume, *J. Anal. Appl. Pyrolysis* **14**, 171 (1988).
37. Y. Vallée and J.L. Ripoll, *Phosphorus, Sulfur, Silicon* **59**, 121 (1991).
38. R. Arnaud, P.Y. Chavant, F. Hagoug, N. Pelloux-Léon, J.L. Ripoll and Y. Vallée, *Phosphorus, Sulfur, Silicon* **95–96**, 319 (1994).
39. R. Arnaud, N. Pelloux-Léon, J.L. Ripoll and Y. Vallée, *Tetrahedron Lett.* **35**, 8389 (1994).

40. R. Arnaud, N. Pelloux-Léon, J.L. Ripoll and Y. Vallée, *J. Chem. Soc., Perkin Trans 2*, 1077 (1995).

41. P. Beslin, D. Lagain and J. Vialle, *Tetrahedron Lett.* 2677 (1979).

42. P. Beslin, D. Lagain, J. Vialle and C. Minot, *Tetrahedron* **37**, 3839 (1981).

43. E. Schaumann, *Tetrahedron* **44**, 1827 (1988).

44. H. Bock, B. Solouki, G. Bert and P. Rosmus, *J. Am. Chem. Soc.* **99**, 1663 (1977).

45. L. Carlsen and H. Egsgaard, *J. Chem. Soc., Perkin Trans 2*, 1081 (1982).

46. L. Carlsen and H. Egsgaard, *J. Chem. Research (S)* 340 (1984).

47. N. Al-Awadi, D.B. Bigley and R.E. Gabbott, *J. Chem. Soc., Perkin Trans 2*, 1223 (1978).

48. Y. Vallée, unpublished results.

49. L. Carlsen, H. Egsgaard, E. Schaumann, H. Mrotzek and W.R. Klein, *J. Chem. Soc., Perkin Trans 2*, 1557 (1980).

50. Y. Vallée, S. Masson and J.L. Ripoll, *Tetrahedron Lett.* **27**, 4313 (1986).

51. Y. Vallée, S. Masson and J.L. Ripoll, *Tetrahedron* **31**, 3921 (1990).

52. G. Seybold, *Tetrahedron Lett.* 555 (1974).

53. G. Seybold and C. Heibl, *Chem. Ber.* **110**, 1225 (1977).

54. G. Adiwidjaja, C. Kirsch, F. Pedersen, E. Schaumann and A. Senning, *Chem. Ber.* **124**, 1485 (1991).

55. C.O. Kappe, C.Th. Pedersen, J.M. Catel and Y. Mollier, *J. Chem. Soc., Perkin Trans 2*, 351 (1994).

56. G. Seybold and C. Heibl, *Angew. Chem., Int. Ed. Engl.* **14**, 248 (1975).

57. E. Schaumann, J. Ehlers and H. Mrotzek, *Liebigs Ann. Chem.* 1734 (1979).

58. B. Bak, H. Svanholt and A. Holm, *Acta Chim. Scand.* **A34**, 625 (1980).

59. R. Schulz and A. Schweig, *Z. Naturforsch.* **39b**, 1536 (1984).

60. R.D. Brown, P.D. Godfrey, P.S. Elmes and D. Mc Naughton, *J. Chem. Soc., Chem. Commun.* 573 (1987).

61 R.D. Brown, K.G. Dyall, P.D. Godfrey, P.S. Elmes and D. Mc Naughton, *J. Am. Chem. Soc.* **110**, 789 (1988).

62. G.P. Raine, H.F. Schaefer and R.C. Haddon, *J. Am. Chem. Soc.* **105**, 194 (1983).

63. D. Sülzle and H. Schwarz, *Ang. Chem., Int. Ed. Engl.* **27**, 1337 (1988).

64. G. Maier, H.P. Reisenauer, J. Schrot and R. Janoschek, *Ang. Chem., Int. Ed. Engl.* **29**, 1464 (1990).

65. C. Wentrup, P. Kambouris, R.A. Evans, D. Owen, G. Macfarlane, J. Chuche, J.C. Pommelet, A. Ben Cheikh, M. Plisnier and R. Flammang, *J. Am. Chem. Soc.* **113**, 3130 (1991).

66. H. Bock, R. Dammel and D. Jaculi, *J. Am. Chem. Soc.* **108**, 7844 (1986).

67. P. Gosselin, S. Masson and A. Thuillier, *Tetrahedron Lett.* **21**, 2421 (1980).

68. G. Pfister-Guillouzo, A. Senio, F. Gracian, M. Khalid, J.L. Ripoll and Y. Vallée, *New J. Chem.* **19**, 1071 (1995).

69. M. Khalid, Y. Vallée and J.L. Ripoll, *Chem. Ind.* 123 (1988).

70. Y. Vallée, M. Khalid and J.L. Ripoll, *Tetrahedron Lett.* **34**, 2605 (1993).

71. E. Block, R.E. Penn, R.J. Olsen and P.F. Sherwin, *J. Am. Chem. Soc.* **98**, 1264 (1976).

72. R.E. Penn and R.J. Olsen, *J. Mol. Spectrosc.* **61**, 21 (1976).

73. D.E. Powers, C.A. Arrington, W.C. Harris, E. Block and V.F. Kalasinsky, *J. Phys. Chem.* **83**, 1890 (1979).

74. F.C. Thyrion and G. Debecker, *Int. J. Kinet.* **5**, 583 (1973).

75. E. Block, *Ang. Chem., Int. Ed. Engl.* **31**, 1135 (1992).

76. E. Block, L.K. Revelle and A.A. Bazzi, *Tetrahedron Lett.* **21**, 1277 (1980).

77a. E. Block and S.-H. Zhao, *Tetrahedron Lett.* **31**, 5003 (1990).

77b. For a recent synthesis of thioacrolein S-oxide see: N. Pelloux-Léon, R. Arnaud, J.L. Ripoll, P. Beslin and Y. Vallée, *Tetrahedron Lett.* **38**, 1385 (1997).

78. C. Maletzo and W. Sundermeyer, *Chem. Ber.* **124**, 1311 (1991).

79. W. Sundermeyer, *Synthesis*, 349 (1988).

80. B. Schuler and W. Sundermeyer, *Chem. Ber.* **123**, 177 (1990).

81. H. Fritz and W. Sundermeyer, *Chem. Ber.* **122**, 1757 (1989).

82. F. Bourdon, J.L. Ripoll, Y. Vallée, S. Lacombe and G. Pfister-Guillouzo, *New J. Chem.* **15**, 533 (1991).

83. J.F. King, *Acc. Chem. Res.* **8**, 10 (1975).

84. J.F. King, P. de Mayo and D.L. Verdun, *Can. J. Chem.* **47**, 4509 (1969).

85. J.F. King, R.A. Marty, P. de Mayo and D.L. Verdun, *J. Am. Chem. Soc.* **93**, 6304 (1977).
86. J.F. King and E.G. Lewars, *J. Chem. Soc., Chem. Commun.* 700 (1972).
87. J.F. King and E.G. Lewars, *Can. J. Chem.* **51**, 3044 (1973).
88. J.F. King, P. de Mayo, C.L. Mc Inttosh, K. Piers and D.J.H. Smith, *Can. J. Chem.* **48**, 3704 (1970).
89. W.J. Mijs, J.B. Reesink and U.E. Wiersum, *J. Chem. Soc., Chem. Commun.* 412 (1972).
90. J.F. King and D.R.K. Harding, *J. Am. Chem. Soc.* **98**, 3312 (1976).
91. B.E. Sarver, M. Jones, Jr. and A.M. Van Leusen, *J. Am. Chem. Soc.* **97**, 4771 (1975).
92. D.S. Margolis and R.W. Pittman, *J. Chem. Soc.* 799 (1957).
93. T. Otsubo, F. Ogura and H. Yamaguchi, *Synt. Commun.* **10**, 595 (1980).
94. H. Higuchi, T. Otsubo, T. Oguro, H. Yamaguchi, Y. Sabata and S. Misumi, *Bull. Chem. Soc. Jpn.* **55**, 182 (1982).
95. N.A. Korchevin, E.N. Sukhomazova, N.V. Russavkaya, E.N. Deryugia and M.G. Voronkov, *Metalloorg. Khim.* **3**, 943 (1990).
96. T. Hirabayashi, S. Mohmand and H. Bock, *Chem. Ber.* **115**, 483 (1982).
97. D. J. Clouthier and D.C. Moule, *Topics in current chemistry* **150**, 167 (1989).
98. R.H. Judge and D.C. Moule, *J. Am. Chem. Soc.* **106**, 5406 (1984).
99. H. Bock, S. Aygen, P. Rosmus, B. Soluki and E. Weissflog, *Chem. Ber.* **117**, 187 (1984).
100. R.D. Brown, P.D. Godfrey and D. McNaughton, *Chem. Phys. Lett.* **118**, 29 (1985).
101. D.J. Clouthier, R.H. Judge and D.C. Moule, *Chem. Phys.* **114**, 417 (1987).
102. R.H. Judge, D.J. Clouthier and D.C. Moule, *J. Chem. Phys.* **82**, 1807 (1988).
103. D.J. Clouthier, R.H. Judge and D.C. Moule, *J. Mol. Spectrosc.* **141**, 175 (1990).
104. R.J. Glinsky, E. Mishalanie and J.W. Birks, *J. Am. Chem. Soc.* **108**, 531 (1986).
105. R.J. Glinski, C.D. Taylor and H.R. Martin, *J. Phys. Chem.* **95**, 6159 (1991).
106. R.D. Brown, P.D. Godfrey and D. Mc Naughton, *J. Mol. Spectrosc.* **120**, 292 (1986).
107. M. Hutchinson and H.W. Krote, *J. Mol. Spectrosc.* **70**, 347 (1978).
108. G.A. Krafft and P.T. Meike, *Tetrahedron Lett.* **26**, 1947 (1985).
109. F. Diehl and A. Schweig, *Angew. Chem., Int. Ed. Engl.* **26**, 343 (1987).
110. D.L. Klayman and W.H.H. Günther, *Organic Selenium compounds: their chemistry and biology*, Wiley Interscience, New York, p. 835 (1973).
111. F. Bourdon, Thesis, University of Caen (France), 1991.
112. D. Christen, H. Oberhammer, W. Zeil, A. Haas and A. Darmadi, *J. Mol. Spectrosc.* **66**, 203 (1980).
113. A. Darmadi, A. Haas and B. Koch, *Z. Naturforsch* **35b**, 526 (1980).
114. M.Y. Bolück, D.C. Moule and D.J. Clouthier, *J. Can. J. Chem.* **61**, 1743 (1983).
115. H. Haas, B. Koch and N. Welcman, *Z. Anorg. Allg. Chem.* **427**, 114 (1976).
116. J. Grobe and D.L. Van, *Angew. Chem., Int. Ed. Engl.* **23**, 710 (1984).
117. A. Haas, H. Willner, H. Bürger and G. Pawelke, *Spectrochimica Acta* **33A**, 937 (1977).
118. B. Binnewies, J. Grobe and D.L. Van, *Phosphorus and Sulfur* **21**, 349 (1985).
119. A. Haas, *J. Fluorin. Chem.* **32**, 415 (1986).
120. A. Darmadi, A. Haas and H. Willner, *Z. Naturforsh* **36b**, 1261 (1981).
121. A. Haas, C. Limberg and M. Spehr, *Chem. Ber.* **124**, 423 (1991).
122. J. Grobe, D.L. Van and J. Welzel, *J. Organomet. Chem.* **386**, 321 (1990).
123. B. Back, O.J. Nielsen, H. Svanholt and A. Holm, *Chem. Phys. Lett.* **53**, 374 (1978).
124. B. Back, O.J. Nielsen, H. Svanholt and A. Holm, *Chem. Phys. Lett.* **55**, 36 (1978).
125. J. Leszcynski and J.S. Kwiatkowski, *Chem. Phys. Lett.* **201**, 79 (1993).
126. A. Bock, S. Aygen, P. Rosmus and B. Slouki, *Chem. Ber.* **113**, 3187 (1980).
127. A. Holm, C. Berg, C. Bjerre, B. Bak and H. Svanholt, *J. Chem. Soc. Chem. Commun.* 99 (1979).
128. R. Schulz and A. Schweig, *Angew. Chem., Int. Ed. Engl.* **19**, 69 (1980).
129. R. Schulz and A. Schweig, *Z. Naturforsh* **39b**, 1536 (1984).
130. R. Schulz and A. Schweig, *Tetrahedron Lett.* **25**, 2337 (1984)
131. W.W. Sander and O.C. Chapman, *J. Org. Chem.* **50**, 543, (1985).
132. R.H. Berg, N. Harrit, E. Larsen and A. Holm, *Acta Chem. Scan.* **43**, 885 (1989).
133. A. Hass and C. Limberg, *Chimia* **46**, 78 (1992).
134. R. Boese, A. Haas and C. Limberg, *J. Chem. Soc., Dalton trans.* **17**, 2547 (1993).
135. W.E. Hoke, P.J. Lemoniasand and R.J. Korenstein, *J. Mater. Res.* **3**, 329 (1988).
136. R.U. Kirss, D.W. Brown, K.T. Higa and R.W. Gedridge Jr, *Organometallics* **10**, 3589 (1991).
137. W.B. Sang, K. Durose, A.W. Brinkman and J. Woods, *Chemtronics* **5**, 179 (1991).

INDEX